13th International UNIMAS Engineering Conference (ENCON 2020)

IOP Conference Series: Materials Science and Engineering
Volume 1101

Kuching, Malaysia
27 - 28 October 2020

Editors:

Lee Yee Yong
Ahmad Kueh Beng Hong
Ivy Tan Ai Wei
Kho Lee Chin
D.N.S.D. Awang Salleh
Lidyana Roslan

ISBN: 978-1-7138-3395-6
ISSN: 1757-8981

Printed from e-media with permission by:

Curran Associates, Inc.
57 Morehouse Lane
Red Hook, NY 12571

Some format issues inherent in the e-media version may also appear in this print version.

Institute of Physics
Dirac House, Temple Back
Bristol BS1 6BE UK

Phone: 44 1 17 929 7481
Fax: 44 1 17 920 0979

techtracking@iop.org

Additional copies of this publication are available from:

Curran Associates, Inc.
57 Morehouse Lane
Red Hook, NY 12571 USA
Phone: 845-758-0400
Fax: 845-758-2633
Email: curran@proceedings.com
Web: www.proceedings.com

TABLE OF CONTENTS

CIVIL ENGINEERING

BUILDING AND CONSTRUCTION MANAGEMENT

MECHANICAL AND PROCESS ENGINEERING

IOP Conf. Series: Materials Science and Engineering 1101 (2021) 011001 doi:10.1088/1757-899X/1101/1/011001

Preface

A very good day and greetings to all

On behalf of the organising committee and the Faculty of Engineering, UNIMAS, I would like to extend our very warm welcome to everyone to the 13th International UNIMAS Engineering Conference 2020 (EnCON2020) on 27 – 28 October 2020 at Faculty of Engineering, Universiti Malaysia Sarawak. The year 2020 is a very interesting year where we are required to social distance, but I am delighted to say that distance had not restricted engineering academics, the industry and society to come together to share our knowledge, ideas and concepts through EnCon2020. For the first time ever in the history of UNIMAS and the Faculty of Engineering, we are hosting our flagship International UNIMAS Conference (EnCon) fully ONLINE. EnCon 2020 is the 13th Engineering Conference for UNIMAS but the very first conference to be held through an online platform. Indeed, it is through advancement in technology, engineering and science that we manage to breakthrough of all traditional constraints, unknowns and hardships of holding conferences in this tough time to gather everyone here today. EnCon2020 ONLINE with the theme "Empowering Industry and Transforming Society through Engineering, Technology and Management" has indeed taken up a whole new meaning and perspective for everyone who is joining us these two days. Together, we overcome!

This year's theme was set on a conception that efforts of academicians in universities should be shine upon and meaningfully contribute to the industry and society. Our research and works in the areas of engineering, technology and management must be shared with the industry and the society. We believe that academicians can contribute to the advancement of the engineering industry through creative problem solving and our determined nature to advance and continuously breaking through. Most importantly, we believe that our works in engineering and science will always uphold the spirit of building the nation and advancing the lives of the people. Therefore, this year we have ten tracks of research papers and five special industry tracks put together to discuss and find solutions to real-time engineering and society issues. We sincerely hope that EnCon2020 will be the new bridge between universities and the engineering industry and society. Faculty of Engineering, UNIMAS is also committed to support the education of engineering and science to young minds. Hence, our program Science, Technology, Engineering Expo (i-STEEX 2020) had also proceeded creatively in this challenging time. We received over 50 participations from young children and teachers in science to showcase their ideas/products via an online competition mode. School children from all over the state have sent us their works online to be vetted by our expert judges. It is a huge achievement indeed!

Last but not least, all these would not have materialised without the hard work, determination and commitment of our organising committee. The support of the industry sponsors, co-organisers and the management of UNIMAS are also very much appreciated. My deepest gratitude to everyone who makes EnCon 2020 ONLINE happen. I hope the participants will find this conference engaging, fruitful and beneficial for your future venture.
Stay Safe!

Ir. Dr Ting Sim Nee
Chairperson, EnCon 2020
Universiti Malaysia Sarawak (UNIMAS)

IOP Conf. Series: Materials Science and Engineering 1101 (2021) 011001 doi:10.1088/1757-899X/1101/1/011001

Disclaimer for EnCon 2020

In light of the current global health hazard, COVID-19, the committee has decided, in support of the worldwide initiative in combating the spreading of the pandemic, to conduct the conference on a virtual platform. As many countries including Malaysia has taken a stringent measure to impose not only cross-countries but also cross-states or even cross-cities travel restrictions, it is first and foremost for the welfare and health of all participants, and equally crucial for the survivability and harmonious continuity of the collaborative spirit of the scientific communities with the common theme the world all over, that the move to bring the conference online is resolved.

Amid all uncertainties in facing the current pandemic, it is in our concerted voice that the progressive nature of the scientific development, with the conference being one of the core events, should not be further deterred. Therefore, the online platform is at present the most optimal approach to ensure the continuity of such a longstanding tradition, differing only in the way the event is carried out.

The conference model follows very much the approach identical to traditionally conducted face-to-face format, the primary distinction being that all presentations and discussions are performed on the virtual atmosphere. We present in the followings some useful details about the virtual conference:

Date: 27-28 October 2020
Location of organiser: Universiti Malaysia Sarawak, Kota Samarahan, Sarawak, Malaysia

Our confirmed list of keynote speakers along with their proposed titles and affiliations are summarized below.

1. **Dato' Ir. Dr. Gue See Sew**
   Chief Executive Officer of G&P Professionals Sdn Bhd
   Title: Challenges on Soft Ground and Solutions

2. **Prof Ts Dr. Mohd Rosli Bin Hainin**
   Deputy Vice-Chancellor (Academics and International), Universiti Malaysia Pahang
   Title: Roles of Pavement Technologies in Sustaining Humanity

3. **Prof Dr. Chee Khoon Ng**
   Professor, Department of Civil Engineering, Faculty of Engineering, Universiti Malaysia Sarawak
   Title: Empowering the Readymix Concrete Industry with Advances in Concrete Technology

4. **Prof Dr. Irtishad Uddin Ahmad**
   Head of the Department of Civil Engineering, American University of Sharjah (AUS), United Arab Emirates
   Title: Driving Forces in Global Construction: Infrastructure, Technology and Sustainability

Authors/Presenters are given 20 minutes presentation slot with 5-10 minutes for Q&A. Each session is chaired by one chairperson with one personnel on standby mode for any issues on

IOP Conf. Series: Materials Science and Engineering 1101 (2021) 011001 doi:10.1088/1757-899X/1101/1/011001

technical, such as the qualities of sound and video live-streaming as well as connection, etc. Each session is conducted via a virtual meeting platform by invitation following the conference attendees list. The conference is expected to accommodate 70 paper presenters plus 40 non-paper participants from Malaysia, Indonesia, India, Australia, Middle East, as well as other southeast Asian countries. In delivering the conference, the latest virtual meeting platform, for instance Webex meeting online are employed.

It is anticipated that a post-mortem session will be conducted within one month of the formal completion of the conference for quality assurance and reference for future improvement.

IOP Conf. Series: Materials Science and Engineering 1101 (2021) 011001 doi:10.1088/1757-899X/1101/1/011001

EnCon 2020 Committee Members

Advisors	Assoc Prof Ir Dr Siti Noor Linda Binti Hj. Taib Dr Norhuzaimin Bin Julai Professor Ir Dr Andrew Ragai Henry Rigit Dr Nazeri Abdul Rahman Dr Nordiana Binti Rajaee
Chairperson	Ir Dr Ting Sim Nee
Assistant Chairperson	Assoc Prof Dr Mohammad Ibrahim Safawi Bin Mohammad Zain
Secretary	Ms Hasmida Binti Hamza
Assistant Secretary Small Committee (Program)	Dr Charles Bong Hin Joo Ir Rudiyanto Bin Philman Jong Dr May Raksmey Dr Norazlina Binti Bateni Madam Nor Azalina binti Rosli Madam Dayangku Salma Binti Awang Ismail Madam Nur Shafini Binti Hamdan Madam Rokilah Binti Bohari Khan Madam Safaraliwati Binti Ghazali
Treasurer	Ir Dr Abdul Razak Bin Abdul Karim
Assistant Treasurer	Ir Dr Norazzlina Binti M. Sa'don Mr Abdul Azim Bin Abdullah
Committee	
Publicity Small Committee (MoA and MoU)	Dr Rosmina Binti Ahmad Bustami Ir Dr David Bong Boon Liang Assoc Prof Ir Dr Lim Soh Fong Assoc Prof Dr Norsuzailina Binti Mohamed Sutan Dr Yonis M. Yonis Buswig Dr Kasumawati Binti Lias Mr Saiful bin Edi Madam Rose Sima Ak Ikau
Sponsorship	Assoc Prof Dr Mohammad Ibrahim Safawi Bin Mohammad Zain Prof Ir Dr Law Puong Ling Dr Raudhah Binti Ahmadi Dr Azida Binti Hj Rashidi Ir Dr Mohd Danial Bin Ibrahim
Logistics	Dr Alsidqi Hasan Dr Nicholas Kuan Hoo Tien Dr Abang Mohammad Nizam Bin Abang Kamaruddin Mr Mohammad Ismail Hairul Bin Abdul Latif
Publication (**Editors**)	Dr Lee Yee Yong Assoc Prof Dr Ahmad Kueh Beng Hong Ir Dr Ivy Tan Ai Wei Dr Kho Lee Chin Dr Dayang Nur Salmi Dharmiza Awang Salleh

IOP Conf. Series: Materials Science and Engineering 1101 (2021) 011001 doi:10.1088/1757-899X/1101/1/011001

Dr. Ma Chau Khun (Universiti Teknologi Malaysia)
Dr. Yong Ee Ling (Universiti Teknologi Malaysia)
Dr. Hidayah Abd. Hassan (Universiti Teknologi Malaysia)
Dr. Leng Pau Chung (Universiti Teknologi Malaysia)
Dr. Libriati Binti Zardasti (Universiti Teknologi Malaysia)
Dr. Mohd Shahrizal B Rusli (Universiti Teknologi Malaysia)
Dr. Mastura Shafinaz binti Zainal Abidin (Universiti Teknologi Malaysia)
Dr Nur Fadilah Binti Darmansah (Universiti Teknologi Malaysia)
Dr. Zuraimy Adzis (Universiti Teknologi Malaysia)
Dr. Tan Lian See (Malaysia-Japan International Institute Of Technology UTM)
Prof. Dr. Saad Mekhilef (Universiti Malaya)
Dr. Hazrina binti Haja Bava Mohidin (Universiti Malaya)
Dr. Lai Chin Wei (Universiti Malaya)
Dr. Lim Ming Han (Universiti Tunku Abdul Rahman)
Dr. Lau See Hung (Universiti Tunku Abdul Rahman)
Dr. Lee Yee Ling (Universiti Tunku Abdul Rahman)
Dr. Chew Kuew Wai (Universiti Tunku Abdul Rahman)
Dr. Chin Ren Jie (Universiti Tunku Abdul Rahman)
Dr. Chong Lee Wai (Universiti Tunku Abdul Rahman)
Dr. Chong Siaw Yah (Universiti Tunku Abdul Rahman)
Dr. Lee Ying Loong (Universiti Tunku Abdul Rahman)
Dr. Lim Poh Im (Universiti Tunku Abdul Rahman)
Dr. Loo Shi Han (Universiti Tunku Abdul Rahman)
Dr. Tan Ooi Kuan (Universiti Tunku Abdul Rahman)
Dr. Wong Mee Chu (Universiti Tunku Abdul Rahman)
Dr. Juniza Binti Md Saad (Universiti Putra Malaysia)
Dr. Mohd. Shahrizal Ab Razak (Universiti Putra Malaysia)
Assoc. Prof. Dr Sam Sung Ting (Universiti Malaysia Perlis UniMAP)
Dr. Mohamad Nur Khairul Hafizi Bin Rohani (Universiti Malaysia Perlis UniMAP)
Dr. Januar Parlaungan Siregar (Universiti Malaysia Pahang)
Dr. Mohd Adnin Bin Hamidi (Universiti Malaysia Pahang)
Dr. Nurul Hazlina Noordin (Universiti Malaysia Pahang)
Assoc Prof. Ir. Dr. Shahiron Bin Shahidan (Universiti Tun Hussein Onn Malaysia)
Assoc. Prof. Dr. Felix Ling Ngee Leh (Universiti Tun Hussein Onn Malaysia)
Assoc. Prof. Dr. Maisara Binti Othman (Universiti Tun Hussein Onn Malaysia)
Dr. Tong Yean Ghing (Universiti Tun Hussein Onn Malaysia)
Dr. Mohd Azuan Zakaria (Universiti Tun Hussein Onn Malaysia)
Dr. Junaidah Binti Jailani (Universiti Tun Hussein Onn Malaysia)
Dr Nor Haslinda Binti Abas (Universiti Tun Hussein Onn Malaysia)
Dr. Norashidah Binti Abd Rahman (Universiti Tun Hussein Onn Malaysia)
Dr. Alvin John Lim Meng Siang (Universiti Tun Hussein Onn Malaysia)
Dr. Hanita Binti Yusof (Universiti Tun Hussein Onn Malaysia)
Dr. Farahiyah Binti Mustafa (Universiti Tun Hussein Onn Malaysia)
Dr. Huda Bin Majid (Universiti Tun Hussein Onn Malaysia)
Dr. Yee See Khee (Universiti Tun Hussein Onn Malaysia)
Mr Muhamad Hanafi Bin Rahmat (Universiti Tun Hussein Onn Malaysia)

Ms Nur Nasuha Binti Abd. Salam (Universiti Tun Hussein Onn Malaysia)
Dr. Faridah Abdul Rahman (International Islamic University Malaysia)
Dr. Nur Shahida binti Midi (International Islamic University Malaysia)
Dr. Nurul Aini Amran (Universiti Teknologi PETRONAS)
Dr. Teh Hee Min (Universiti Teknologi PETRONAS)
Assoc. Prof. Dr Kelly Yong Tau Len (Universiti Kuala Lumpur)
Dr. Amelia binti Md Som (Universiti Kuala Lumpur)
Dr. Chin Lip Han (Universiti Kuala Lumpur)
Dr. Nor Zalina Kasim (Universiti Kuala Lumpur)
Dr. Sucharita Srirangam (Taylors University)
Dr Low Wen Pei (INTI International University)
Dr. Chang Fung Lung (INTI International University)
Assoc Prof Dr. Tony Hadibarata (Curtin University Malaysia)
Dr. Low Wai Wah (Curtin University Malaysia)
Dr. Lee Yeong Huei (Curtin University Malaysia)
Dr. Tina Ting Chui Huon (Curtin University Malaysia)
Dr. Yam Ke Sen (Curtin University Malaysia)
Ir. Mr Meheron Selowara Joo (Curtin University Malaysia)
Assoc Prof. Dr. Basil T. Wong (Swinburne University of Technology)
Assoc Prof Dr. Vincent Lee Chieng Chen (Swinburne University of Technology)
Dr. Choo Chung Siung (Swinburne University of Technology)
Dr. Daniel Looi Ting Wee (Swinburne University of Technology)
Dr. Elammaran Jayamani (Swinburne University of Technology)
Dr. Chai Chang Saar (Swinburne University of Technology)
Dr. Soon Kok Heng (Swinburne University of Technology)
Dr. Muhammad Rafiq Mirza (Swinburne University of Technology)
Dr. Almon Chai Wei-Yen (Swinburne University of Technology)
Dr. Hudyjaya Siswoyo Jo (Swinburne University of Technology)
Dr. Riady Siswoyo Jo (Swinburne University of Technology)
Dr. Kelvin Kuok King Kuok (Swinburne University of Technology)
Prof. Dr. Ng Chee Khoon (Universiti Malaysia Sarawak)
Prof. Dr. Amir Azam Khan (Universiti Malaysia Sarawak)
Prof Dr. Andrew Ragai Henry Rigit (Universiti Malaysia Sarawak)
Assoc OProf. Dr. Thelaha bin Hj Masri (Universiti Malaysia Sarawak)
Dr. Idawati binti Ismail (Universiti Malaysia Sarawak)
Dr. Alsidqi Hasan (Universiti Malaysia Sarawak)
Dr. Rosmina binti Ahmad Bustami (Universiti Malaysia Sarawak)
Ir Dr. Abdul Razak bin Abdul Karim (Universiti Malaysia Sarawak)
Ir Dr. Norazzlina M. Sa'don (Universiti Malaysia Sarawak)
Ir. Dr. Leonard Lim Lik Pueh (Universiti Malaysia Sarawak)
Dr. Charles Bong Hin Joo (Universiti Malaysia Sarawak)
Dr. Raudhah binti Ahmadi (Universiti Malaysia Sarawak)
Mr. Ron Aldrino Chan@Ron Buking (Universiti Malaysia Sarawak)
Ms Hasmida Binti Hamza (Universiti Malaysia Sarawak)
Dr. Jethro Anak Henry Adam (Universiti Malaysia Sarawak)
Dr. Mohamad Raduan b. Hj Kabit (Universiti Malaysia Sarawak)
Dr. Dayang Azra binti Awang Mat (Universiti Malaysia Sarawak)
Dr. Annie Joseph (Universiti Malaysia Sarawak)
Dr Ngu Sze Song (Universiti Malaysia Sarawak)
Dr. Annisa binti Jamali (Universiti Malaysia Sarawak)
Dr. Nicholas Kuan Hoo Tien (Universiti Malaysia Sarawak)

IOP Conf. Series: Materials Science and Engineering 1101 (2021) 011001 doi:10.1088/1757-899X/1101/1/011001

Dr Noor Hisyam bin Noor Mohamed (Universiti Malaysia Sarawak)
Dr Nur Tahirah binti Razali (Universiti Malaysia Sarawak)
Dr. Khairul Fikri bin Tamrin (Universiti Malaysia Sarawak)
Dr. Abang Mohammad Nizam bin Abang Kamaruddin (Universiti Malaysia Sarawak)
Dr. Marini Binti Sawawi (Universiti Malaysia Sarawak)
Dr. Mahsuri binti Yusof (Universiti Malaysia Sarawak)
Assoc Prof Dr. Afzan binti Ahmad Zaini (Universiti Malaysia Sarawak)
Dr. Nor Hazwani Md Zain (UiTM Shah Alam)
Dr. Basharudin Bin Abdul Hadi (UiTM pulau penang)
Dr. Hj. Anas Ibrahim (UiTM pulau penang)
Dr Ng Kok Shien (UiTM pulau penang)
Mr. Muhammad Hafeez Bin Osman (UiTM pulau penang)
Assoc. Prof. Dr. Juferi Bin Idris (UiTM Sarawak)
Ms Caroline anak Peter Diman (UiTM Sarawak)
Dr. Liew Fui Kiew (Universiti Teknologi MARA, UiTM)
Dr. Nur Diana binti Wakimin (Universiti Teknologi MARA, UiTM)
Ms Siti Kartina binti Abdul Karim (Universiti Teknologi MARA, UiTM)
Dr. Lee Man Djun (University College of Technology Sarawak)
Dr. Yong Ching Yee (University College of Technology Sarawak)
Dr. Siti Hasyyati Drahman (International College of Advanced Technology Sarawak, i-CATS)

IOP Conf. Series: Materials Science and Engineering 1101 (2021) 011001 doi:10.1088/1757-899X/1101/1/011001

Acknowledgement of Co-organizers and Sponsors

IOP Conf. Series: Materials Science and Engineering 1101 (2021) 011001 doi:10.1088/1757-899X/1101/1/011001

Sponsors

PLATINUM

sarawak energy

IOP Conf. Series: Materials Science and Engineering 1101 (2021) 011001 doi:10.1088/1757-899X/1101/1/011001

Sponsors

PLATINUM

CAHYA MATA SARAWAK

IOP Conf. Series: Materials Science and Engineering 1101 (2021) 011001 doi:10.1088/1757-899X/1101/1/011001

Sponsors

PLATINUM

WENHONG
PLASTIC
INDUSTRIES
SDH BHD
(337061-T)

Blue & Green Infrastructure

IOP Conf. Series: Materials Science and Engineering 1101 (2021) 011001 doi:10.1088/1757-899X/1101/1/011001

Sponsors

GOLD

PCSS® 4D The Advanced Digital Construction

The Project Management Specialist

MALTIMUR RESOURCES

Connecting for Growth

IOP Conf. Series: Materials Science and Engineering 1101 (2021) 011001 doi:10.1088/1757-899X/1101/1/011001

Sponsors

GOLD

MESB MAINLINE Enterprise Sdn Bhd

P. L. Consultants

IOP Conf. Series: Materials Science and Engineering 1101 (2021) 011001 doi:10.1088/1757-899X/1101/1/011001

Sponsors

SILVER

POLYBUILDING

Sponsors

BRONZE

HOIFU GROUP

UGI

IOP Conf. Series: Materials Science and Engineering 1101 (2021) 011001 doi:10.1088/1757-899X/1101/1/011001

Sponsors

BRONZE

JANA TERAJU SDN BHD

Pertama Ferroalloys Sdn. Bhd.

THANK YOU ENCON 2020

IOP Conf. Series: Materials Science and Engineering 1101 (2021) 011002 doi:10.1088/1757-899X/1101/1/011002

Peer review declaration

All papers published in this volume of *IOP Conference Series: Materials Science and Engineering* have been peer reviewed through processes administered by the Editors. Reviews were conducted by expert referees to the professional and scientific standards expected of a proceedings journal published by IOP Publishing.

- **Type of peer review: Double-blind Review. Reviewers do not get to see the author identities by removing author names from the paper.**
- **The papers are manage using Conference submission management system: The full-length papers were submitted through online conference management system. The assignment of at least two reviewers with the expertise in the area of each paper submitted, and no more than 3 papers should be reviewed by the same reviewer. The assigned reviewers do not come from the same university affiliated from each paper submitted. Based on the comments received from the reviewers, the papers were either (a) accepted as original submission, (b) accepted with minor corrections, (c) accepted with major corrections or (d) rejected.**
- **Conference submission management system: Conference management System, Conf Bay**
- **Number of submissions received: 59**
- **Number of submissions sent for review: 59**
- **Number of submissions accepted:47**
- **Acceptance Rate (Number of Submissions Accepted / Number of Submissions Received X 100): 80%**
- **Average number of reviews per paper: 2 reviewers**
- **Total number of reviewers involved: 100**
- **Any additional info on review process (ie plagiarism check system): The review process was carried out in accordance with the instructions given to the reviewer. All the papers are check with their plagiarism using Turnitin.**
- **Contact person for queries: Dr Lee Yee Yong, Universiti Malaysia Sarawak, yylee@unimas.my**

IOP Conf. Series: Materials Science and Engineering 1101 (2021) 012001 doi:10.1088/1757-899X/1101/1/012001

Flatwise and edgewise compression strengths of sandwich panel with silica aerogel mat

L T Tay[1], Y Y Lee[1], A B H Kueh[1] and Y H Lee[2]

[1] Department of Civil Engineering, Faculty of Engineering, Universiti Malaysia Sarawak, Sarawak, 94300 Kota Samarahan, Sarawak Malaysia
[2] Department of Civil and Construction Engineering, Faculty of Engineering and Science, Curtin University, CDT250, 98009 Miri, Sarawak, Malaysia
Corresponding author: yylee@unimas.my

Abstract. Facade of building are mainly made up from red clay brick and concrete block. However, both materials are having high thermal mass and promote high indoor thermal discomfort level. Therefore, it is necessary to invent new building material that have low thermal mass and able achieve strength required. Silica aerogel possesses properties of lightweight and low thermal conductivity as compared to other construction materials. In this study, sandwich panel with silica aerogel mat was studied where the properties of sandwich panel silica aerogel mat are rarely found in previous investigations. Before checking its reliability as thermal insulation panel, the mechanical properties of this panel was investigate. The panel was made-up by concrete wythes with type N mortar and the silica aerogel mats with different thickness. Both concrete wythes were casted and then attached together with silica aerogel mat as the cover. 3 types of panel with different insulation thickness were then tested for flatwise and edgewise compression test. From the results, it was found that core thickness of silica aerogel mat has less influence in flatwise and edgewise compression strengths of the sandwich panel. All specimens achieved minimum strength of type N mortar. Therefore, it is recommended to be used in construction that has equivalent application of type N mortar.

1. Introduction

Scholars, architect, contractor, government and developers are paying attention to the issues related with indoor thermal comfort. Building envelope is the physical divider between the indoor and outdoor of a building. There is a close relationship between building envelope and indoor thermal comfort. In order to maintain thermal comfort, absorptivity of external surfaces, thermal capacity, and thermal conductivity of the building envelope greatly influence the internal environment and consequently the energy consumption inside the building [1]. Construction materials such as brick and concrete are having criteria such as high thermal mass and thermal conductivity. Those materials tend to store heat during daytime and release the heat at night with high rate of heat transfer across the material itself. Therefore, application of thermal insulation material to control heat transfer is an easier and practical way to maintain indoor thermal comfort level. However, conventional thermal insulation materials are either non-environment friendly or causing health issues.

Composite sandwiched structures are recently being widely applied by the world especially the marine and aviation industries [2]. Sandwiched materials are knowns to be able to provide high

IOP Conf. Series: Materials Science and Engineering 1101 (2021) 012001 doi:10.1088/1757-899X/1101/1/012001

stiffness coupled with lightweight [3]. Good shock resistance, flexural rigidity, low heat and sound transfer and easy to install [4] make sandwiched structure an ideal and promising structures in the future. Precast concrete sandwiched panel is one of the types of composite structures with 2 layers of concrete wythes separated by layer of insulation. It is gaining attention of researchers recently due to its potential to become non-load bearing wall and provide thermal insulation. Concrete sandwiched panel able to act as multi purposes construction material compare to the normal brick and concrete. At the same time, Ern et al. (2017) [5] and Huang et al. (2020) [6] suggest more researches are needed on concrete sandwiched panel as concrete sandwiched panel crucial to replace conventional construction material.

The study of silica aerogel mat as core insulation materials are limited [4]; [5]; [7]. Studies by Parra and Moehle (2014) [7] and Amran et al. (2019) [8] mainly discussed about structural properties and application for load bearing panel or wall. Sandwich panels are normally applied in wall construction where they resist most of the compression actions from the loadings generated in buildings. Therefore, the flatwise and edgewise compression strengths are the important input data for building design, even though is for non-bearing wall application. To follow the suggestion of [9] and [10], this study is about silica aerogel mat sandwiched panel. This paper provides important data and information of an innovation of future product in design stage which bridge the knowledge gap and promote sustainability. The mat/core thicknesses tested are 9, 12 and 15 mm. The composite mortars are tested for flatwise and edgewise compression strength.

2. Experimental procedure

2.1. Material

Sand that pass through 600 µm sieve and Ordinary Portland cement (CEM I 42.5N) were used for the concrete wythes mixing. In the meantime, silica aerogel mat, as shown in figure 1, with the properties given by supplier, as displayed in Table 1, is casted as the insulation material at the centre of the sandwich panel specimens.

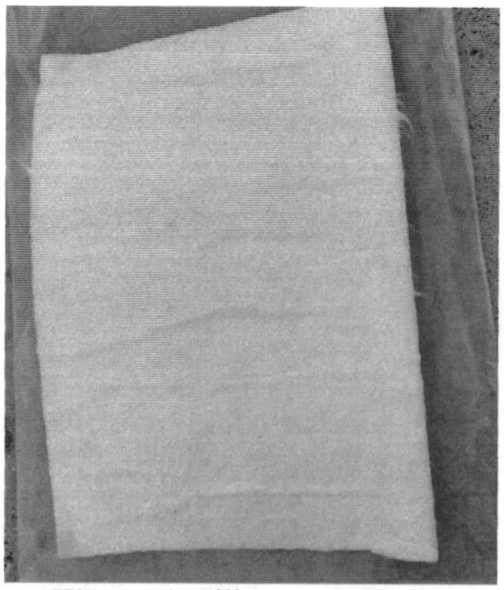

Figure 1. Silica aerogel mat.

IOP Conf. Series: Materials Science and Engineering 1101 (2021) 012001 doi:10.1088/1757-899X/1101/1/012001

Table 1. Properties of silica aerogel mat.

Properties	Value
Material	Fiberglass + Nano silica aerogel
Colour	White
Density	180-220kg/m^3
Applicable temperature	-200°C ~ + 1000°C
Thermal conductivity (25 ºC W/ m·K)	0.012-0.018
Low temperature bending	≤99%
Breaking elongation	65%
Surface chemistry	Hydrophobic
Bending strength	45kP
Compressive strength	32N

2.2. Mix design

Mix ratio for cement to sand was 1:3 to produce normal mortar for concrete wythes. 3 different core/mat with thickness of 9, 12 and 15 mm sandwiched between normal mortar were casted and tested for flatwise compression test according to ASTM C365 with the dimension of 50 mm^3. Whereas, samples testing for edgewise compression by following ASTM C364 were in the dimension of 200 mm (length) × 200 mm (width) × 100 mm (thickness) and 3 different core/mat thickness which are, 9, 12 and 15 mm. Three specimens were casted for each core thickness and testing. A total of 18 specimens were casted. All specimens were tested for 28 days of concrete ages for characteristic strength. Figure 2 shows the 15 mm thick mat sandwiched mortar for flatwise strength testing.

Figure 2. 15mm thick mat sandwiched mortar for flatwise strength testing.

2.3. Mixing procedure

Ordinary Portland cement (CEM I 42.5N), oven-dried fine aggregate (pass through 600 µm sieve) and water with the water cement ratio of 0.83 (to achieve flow rate of 110 ± 5) were weighted accordingly prior mixing. After that, OPC, fine aggregate was mixed together with addition of water. Fresh mortar was then poured into moulds prepared in advanced. After that, samples water-cured for 28 days.

IOP Conf. Series: Materials Science and Engineering 1101 (2021) 012001 doi:10.1088/1757-899X/1101/1/012001

2.4. Testing procedure

2.4.1. Flatwise compression test. Sample was placed at the center of machine (Universal Testing Machine) prior to loading. Based on standard, specimen is encouraged to be clamped at both ends to prevent movement during testing. Due to unavailable of clamps, specimen was not fixed by clamp during testing. Load was then equally distributed over the entire loading surface of specimen. A load rate of 1 kN/s was applied to specimens with pace rate of 10mm/min. However, ASTM C 365 suggest pace rate of 0.5mm/min. Load was applied until failure happened. Set up of sample for flatwise compression testing is shown in figure 3.

Figure 3. Set up of sample for flatwise compression testing.

2.4.2. Edgewise compression test. Sample was placed at the center of machine (Universal Testing Machine) prior to loading. 2kN/s of load was applied with pace rate of 10mm/min. However, ASTM C 364 suggest pace rate of 0.5mm/min. Load was equally distributed over the entire loading surface of specimen until failure happened. Despite the suggestion to clamp both ends of specimen to prevent movement, it was unable to carried out due to unavailable of clamps. Set up of sample for edgewise compression testing is shown in figure 4.

Figure 4. Set up of sample for edgewise compression testing.

3. Result and discussion

3.1. Flatwise compression test

Equation (1) shows the formula for flatwise compressive strength and results are shown in figure 5.

$$Flatwise\ compressive\ strength = P / A \tag{1}$$

where P is the maximum load in N and A is the area of loaded surface in mm^2.

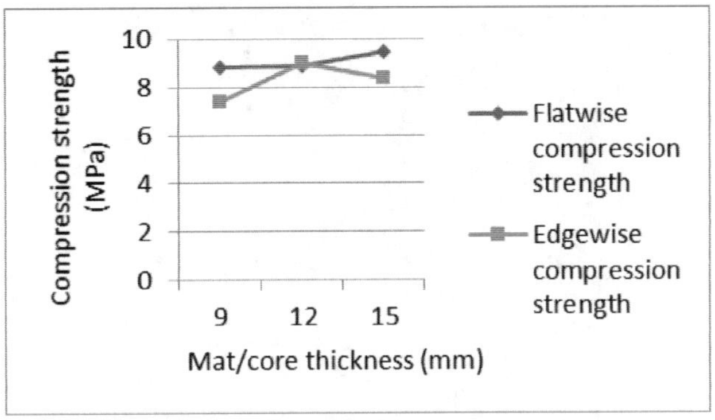

Figure 5. Flatwise and edgewise compression strength.

From the results as in figure 5, sandwich mortar with 15 mm thickness silica aerogel mat achieved the highest flatwise compression strength compared to 12 mm and 9 mm thickness mats. In the study of [11], it is observed that height or thickness of core is positively contributing to flatwise compressive strength of composite product. It is interesting to note that the higher the core thickness, the higher the load carrying capacity. This is because insulation layer is well known to play a role in thermal conductivity and not contribute to overall strength. It is no doubt that there is a lack of study about the relationship of core thickness with flatwise compression strength. Hence, relationship of core thickness with compression strength is worth to be further investigated.

3.2. Edgewise compression test

Equation (2) shows the formula for edgewise compressive strength. The results are shown in figure 5.

$$Edgewise\ compressive\ strength = P_{max} / [\, w\,(2t_{fs})\,] \tag{2}$$

where P_{max} is the maximum load in N, w is the width of sample in mm and t_{fs} is the thickness of a single facesheet in mm.

In term of edgewise compression strength, core thickness of 12 mm achieved the highest strength among the others. Meanwhile, 9 mm core thickness had the lowest edgewise compression strength. Despite of 12 mm core thickness had larger area of mortar in contact with load during testing, results showed reduction of mortar able to increase the strength. It is undeniable that, the higher the core thickness, the higher the strength to withstand load.

The trend of data obtained was increased by 1.64 N/mm^2 and then dropped from 12 mm core thickness to 15 mm core thickness. The drop of strength by 0.64 N/mm^2 for 15 mm core thickness may due to the misplacement of samples during testing. Samples may not place in the center for the testing setup, causing uneven distribution of load. The weakest point is the bonding area between the facesheet and core. When load was applied parallel to the weakest point, it further weakens the bonding. Lack of clamp at end of samples during testing also assumed to be one of the factors for concrete failure. Samples may slip without clamp to fix during application of loading, causing inaccurate of the reading.

Based on to ASTM C364 / C364M, there are 5 types of failure mode and 1 unacceptable mode can be observed. Type of failure modes for 9 mm, 12 mm and 15 mm thick mat sandwiched mortar were shown in figure 6 and it was facesheet compression failure. It is because both the facesheets have the largest contact area to load being applied and tend to compress due to the load. Due to the smaller contact area of silica aerogel mat, it was protected by mortar facesheet. Based on observation, silica aerogel mat for all 3 types of specimens still in good contact.

Figure 6. Facesheet compression failure mode.

3.3. Comparison of both compression strength test

When compared both compression strength for 9 mm core thickness, it is found that the different between both tests is only 1.42 MPa. In the meantime, the differences for 12 mm and 15 mm core thickness are 0.15 MPa and 1.1 MPa respectively. Generally, the differences between both strength test is not more than 1.5 MPa. However, compression strength for 9 mm and 15 mm samples for flatwise compression test is higher than edgewise compression test might due to the slenderness. The higher the slenderness ratio, the higher the instability and lead to buckling as observed [7] [8]. Apart from that, 12mm core thickness sample for edgewise compression observed to have higher strength by 0.15 MPa than flatwise compression strength.

3.4. Comparison with previous research

Composite sandwiched panel can be made up from various material for the cover and core. Previous researches have done by using different cover and core material such as concrete, foam concrete, polyethylene and etc based on the purpose of the final product. The aims to produce product with characteristic of lightweight, fire resistant, thermal insulation, structural strength or combination of the listed properties. Concrete cover with sandwiched insulation has the potential to act as non-load bearing wall or cladding. Even the composite product studied by Oh et al. (2013) [12] and Jennise et al. (2014) [13] are potentially attached to existing wall to provide additional functions. Moreover, shear connectors were widely used for composite panel to increase strength of the samples. Despite of that, shear connector is not used for this experiment as results came out able to reach the minimum requirement of Type N mortar. Previous researches are shown in table 2.

Table 2. Previous research of flatwise and edgewise compression tests

Reference	Material		Compression test (MPa)	
	Wythes	Insulation	Flatwise	Edgewise
Current study	Type N mortar	9 mm silica aerogel mat	8.81	7.39
	Type N mortar	12 mm silica aerogel mat	8.88	9.03
	Type N mortar	15 mm silica aerogel mat	9.49	8.39
[5]	Foamed concrete	gypsum	-	2.35
	Foamed concrete with POFA	gypsum	3.3	-
[4]	3D high density polyethylene	polyurethane foam	3.41	7.53
[7]	Glass fiber	polyurethane coconut foam	1.74 – 2.04	-

4. Conclusion

In conclusion, 3 types of core thickness had not much different in flatwise and edgewise compression strength. For flatwise compression strength, it was less than 1N/mm^2 and only 1.64N/mm^2 in term of different among the 3 type of thickness. Besides that, when compared with type N mortar, all 3 types of core thickness achieved minimum strength of 6.2N/mm^2 at 28 days of concrete curing age. Furthermore, flatwise and edgewise strength obtained are higher than previous researches, even though edgewise compression from study of [4] was higher by 0.14 MPa compared to 9 mm core thickness mortar. It is concluded that mortar panel is potentially to replace type N mortar in construction industry. It is recommended to further study the flexural strength of silica aerogel sandwich mortar as well as its thermal properties.

5. References

[1] Hwaish A N A 2015 Impact of heat exchange on building envelope in the hot climates, *International Journal of Emerging Technology and Advanced Engineering* **5**(2) 47-57

[2] Scudamore R J and Cantwell W J 2002 The effect of moisture and loading rate on the interfacial fracture properties of sandwich structures, *Polym. Comp.* **23** 406-417

[3] Kumar M V and Soragaona B 2014 Fabrication and evaluation of multilayered polyurethane foam core sandwich panels for static flexural stiffness, *Procedia Engineering* **97** 1227-1236

[4] Sharafi P, Nemati S, Samali B, Mousavi A, Khakpour S, and Aliabadizadeh Y 2018 Edgewise and flatwise compressive behaviour of foam-filled sandwich panels with 3-D high density polyethylene skins, *Engineering Solid Mechanics* **6**(3) 285-298

[5] Ern P A S, Ling L M, Kasim N, Abd Hamid Z and Masrom M A N B 2017 Thermal Performance of Precast Concrete Sandwich Panel (PCSP) Design for Sustainable Built Environment, *MS&E* **245**(3) 032073

[6] Huang J, Jiang Q, Chong X, Ye X, and Wang D 2020 Experimental study on precast concrete sandwich panel with cross-shaped GFRP connectors, *Magazine of Concrete Research* **72**(3) 149-162

[7] Parra P F and Moehle J P 2014 Lateral buckling in reinforced concrete walls, *In Tenth US Natl Conf Earthq Eng*, (*Anchorage, Alaska)*
[8] Amran Y M, Alyousef R, Alabduljabbar H, Alrshoudi F and Rashid R S 2019 Influence of slenderness ratio on the structural performance of lightweight foam concrete composite panel, *Case Studies in Construction Materials* **10** e00226
[9] Huang J, Jiang Q, Chong X, Ye X, and Wang D 2020 Experimental study on precast concrete sandwich panel with cross-shaped GFRP connectors, *Magazine of Concrete Research* **72**(3) 149-162
[10] Sanpo N, Jitputti J, Prakulpawong P, Srikamut C, Yodyingyong S and Triampo D 2020 Silica Aerogel Thermal Insulation Coating as Commodity Usage, *IOP Conference Series: Materials Science and Engineering (Incheon)* vol 811 (South Korea: IOP Publishing Ltd)
[11] Li Z and Ma J 2020 Experimental Study on Mechanical Properties of the sandwich composite structure reinforced by basalt fiber and nomex honeycomb, *Materials* **13**(8) 1870
[12] Oh T S, Jang S J, Lee K M and Yun H D 2013 Insulation type effect on the direct shear behavior of concrete sandwich panel (csp) with non-shear connectors, *In Advanced Materials Research* **663** 154-158
[13] Jennise T, Sihombing H, and Yuhazri M 2014 FLATWISE COMPRESSION ANALYSIS OF SANDWICH STRUCTURES MADE FROM CHOPPED STRAND MAT COCONUT FIBER FILLED POLYURETHANE FOAM AS A CORE MATERIAL, *Global Engineers & Technologists Review* **4**(3) 6-12

Acknowledgments
Authors would like to acknowledge the financial support from research grant of F02/SGS/1797/2019 for publication. Besides that, the authors would like to extend their appreciation to the technicians in department of civil engineering for their support in running the experiment.

Hydrodynamic forces on tubular cylinders fitted with sacrificial anodes – numerical analysis

A M Al -Yacouby[1*], M H Talhah[1], M S Liew[1] and K U Danyaro[2]

[1] Department of Civil and Environmental Engineering, Universiti Teknologi PETRONAS, Seri Iskandar, 32610, Perak, Malaysia
[2] Offshore Engineering Centre (OEC), Universiti Teknologi PETRONAS, Seri Iskandar, 32610, Perak, Malaysia
[*]Corresponding author email: ahmad.alyacouby@utp.edu.my

Abstract. In offshore engineering, the design of effective marine platforms compatible with the offshore environmental condition is a challenging task, especially accurate estimation of hydrodynamic coefficients that affect the viscous forces. Practically, the well-known Morison's equation is often adopted by designers to calculate wave loads on slender structures, where the values of drag and inertia coefficients are selected as constant values based on the design code of practice's recommendations. However, the mass and drag coefficients must be determined empirically based on specific met-ocean data for the operation location. Thus, the objective of this study is to evaluate the Hydrodynamic Forces on Circular Cylinders fitted with sacrificial anodes using empirical methods, and validate the results using Computational Fluid Dynamic (CFD). It is found that the major parameters that affect the drag and inertia coefficients are the water depth, wave heights, wave frequencies, the pipe diameters, and the presence of anode fittings.

Keywords. Hydrodynamic Forces, Tubular Cylinders, Sacrificial Anodes, Hydrodynamic coefficients.

1. Introduction

Fixed offshore structures are made of tubular steel members, welded together to form a three-dimensional rigid frame. These steel platforms are subjected to corrosion, and their successful use in engineering and commercial applications depends on the corrosion protective mechanisms. Thus, impressed current cathodic protection systems, corrosion allowance and galvanic anodes techniques are commonly adopted by designers for corrosion protection of offshore structures, besides organic surface coatings [1]. Cathodic protection is a common method used to protect immersed parts of steel surfaces from corrosion. However, added hydrodynamic loading due to sacrificial anodes is not well understood [2]. Accurate estimation of wave forces on cylindrical elements are of considerable interest in the design of offshore platforms. The simplest and most efficient method used by the engineers to predict the wave forces on submerged vertical tubular cylinders with a diameter less than half of the wavelength was proposed by Morison [3]. This equation consists of two components: drag force due to the water particle velocity and inertia force due to the water particle acceleration [3]. However, the application of this equations is limited to linear wave theory. In addition, the total hydrodynamic force on circular cylinders is a function of drag and inertia coefficients. Read et al. [4] examined the effect of the drag and inertia coefficients used in Morison's equation as predicted by some calibration approaches on the short-term wave-induced force acting on a fixed vertical cylinder. Sarpkaya [5] has presented a comprehensive review of Morison's equation and its dependence on various factors. Boccotti et al. [6] conducted a field experimental investigation to determine the accuracy of Morison equation for estimating wave forces on circular cylinders. Accurate estimation of drag (CD) and inertia coefficient (CM) plays an important role in the numerical prediction of the wave forces on circular cylinders, thus, a series of experimental investigations have been carried out by Sarpkaya [7], Troesch and Kim [8] to estimate the values of CD and CM. The design codes of practices such as API [9], DNV [10] normally recommend specific values for CD and CM to be used by engineers during estimation of hydrodynamic forces of tubular cylinders. However, accurate values for CD and CM are not well established when the cylinders are fitted with sacrificial anodes. Fixed offshore structures are typically protected against corrosion by installing sacrificial anodes. Anodes are highly active metals that are used to prevent a less active material surface from corroding. When sacrificial anodes are used, the anodes are physically connected to the structures, and an electrochemical cell is formed between the anode and the cathode through the surrounding seawater. The review of the existing literature shows that, very limited studies have been conducted to determine the effects of sacrificial anodes on wave forces using CFD tools.

Thus, the objective of this paper is to determine the hydrodynamic forces on circular cylinders fitted with sacrificial anodes using Morison equations and the findings are validated using CFD simulations. The results provide insights into better understanding of the effects of sacrificial anodes on the hydrodynamic forces of tubulars cylinders fitted with anodes for offshore applications

2. Morison's equation

2.1. Morison's equation

To estimate wave forces on slender structures, designers normally use the well-known Morison Equation. For a stationary cylinder in a plane flow field, with flow stream, with a free-stream velocity, U=U(t) the total hydrodynamic force per unit length on the cylinder can be determined by superimposing the drag and inertia forces using the equation (1).

$$\bar{q} = C_D\rho\frac{D}{2}|u|u + C_M\rho\pi\frac{D^2}{4}\dot{u} \tag{1}$$

This equation assumes that the total in-line hydrodynamic force on a structure consists of semi-empirical values of inertia and drag forces supplemental linearly. The inertia force is developed because the water taking possession a wave carries a momentum with it, whereas the drag force part is caused because of the presence of wake region on the stream side of the cylinder. The total hydrodynamic force on the jacket member will be calculable from equation (2).

$$F = \int_0^d f ds = \int_0^d [\frac{C_m\rho\pi D^2}{4}\frac{dy}{dt} + C_d\frac{\rho D}{2}|U + U_c|(U + U_c)]ds \tag{2}$$

where, d = water depth, Cm = hydrodynamic coefficient of mass, Cd = hydrodynamic coefficient of drag, ds = integration is done over the full wetted length of the cylinder, ρ = density of water (1000kg/m3), D = pipe diameter, U = horizontal water particle velocity, dy/dt = water particle acceleration, and U_c = current velocity.

As reported by Wilson [11], the Morison's equation is deduced by applying the principle of conservation of linear fluid momentum. This principle is based on Newton's second law applied to fluid of fixed control volume V at any time t. Here, V is a rectangular box surrounding a disk of a submerged circular cylinders, where V can be determined as D2Δz, where D is the disk diameter, and Δz is the height of the disk. Therefore, for water flow along the x-directions, with horizontal velocity u, the net horizontal shear force on the disc can be determined as equation (3).

$$\sum F_x = \frac{\partial}{\partial t}\int_V \rho u dV + \int_{A0} \rho u.u dA_0 \quad (3)$$

Where dV is a small element of the control volume, and dA0 is the element of area on the surface of the control volume, perpendicular to the flow direction. General derivation of equation (3) is discussed by Munson et al. [12].

2.2. Linear wave theory

Linear wave theory is a first order, small amplitude theory developed by Airy [13]. This theory forms the basis for the probabilistic spectral description of waves. The governing equations and the solutions for the waves are essentials for the prediction of wave forces on offshore structures [11].

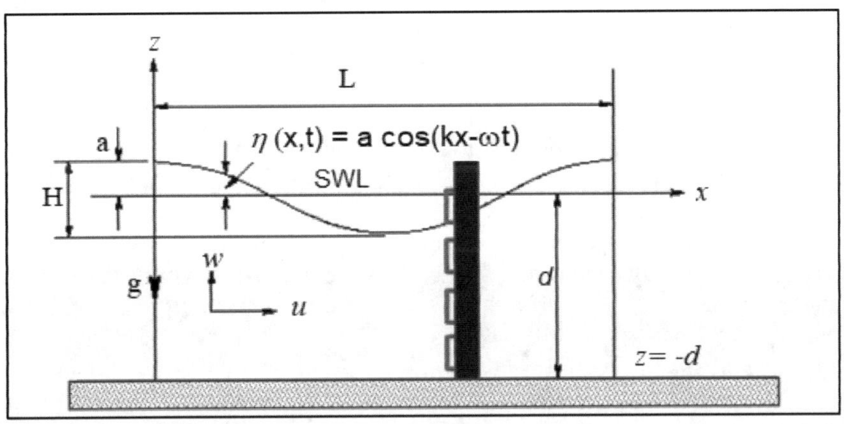

Figure 1. Definition of simple harmonic waves past a cylinder fitted with sacrificial anodes.

Assuming the simple harmonic plane wave defined in figure.1, is propagating along the positive x direction. The origin of the coordinate is located at still water level (SWL), the vertical coordinate is z, directed upward as the positive direction. The density of water ρ is taken as 1025 kgm^{-3}. The differential equations and the boundary conditions for the water particles velocity and pressure can be expressed as follows:

$$\frac{\partial u}{\partial z} - \frac{\partial \omega}{\partial x} = 0 \quad (4)$$

$$\frac{\partial u}{\partial x} + \frac{\partial \omega}{\partial z} = 0 \quad (5)$$

$$\frac{\partial u}{\partial t} = -\frac{1}{\rho}\frac{\partial p}{\partial x} \quad (6)$$

$$\frac{\partial \omega}{\partial t} = -\frac{\partial p}{\partial z} - g \quad (7)$$

$$\omega = -\frac{\partial \eta}{\partial t} \ at \ z = 0 \quad (8)$$

$$p = p_a \quad at\ z = 0 \tag{9}$$

$$\omega = 0 \quad at\ z = -d \tag{10}$$

Equation (4) represents the irrotational condition, which is also known as the zero velocity. Equation (5) is the well-known continuity equation. Equations (11) and (12) are the Eulerian equations of motions. Similarly, the boundary conditions at the water surface as presented in equations (8) and (9), while equation (10) represents the boundary condition at seabed. The solution that satisfy the linear equation is the Laplace Equations, for surface elevation $\eta = \eta(x,t)$ presented in equation (11) which describes a progressive harmonic waves moving in the positive x direction.

$$\eta(x,t) = Asin(kx - \omega t) \tag{11}$$

Using equation (11), together with equations (4) until (10), the water particles velocities, u, and ω, and the corresponding accelerations $\partial u/\partial t = \dot{u}$ and $\partial \omega/\partial t = \dot{\omega}$ are determined.

3. Methodology

3.1. Empirical methods

In this phase of the research, the total hydrodynamic forces on tubular members fitted with sacrificial anodes are evaluated using the well-known Morison's equation. The wave kinematics were estimated using Air Wave theory, then the total forces are determined considering the appropriate drag and inertia coefficients recommended by design code of practices. The met-ocean data such as water depth, wave heights, wave periods used in the theoretical analysis represents the operation criteria for selected operation locations in the Malaysian water. A MATLAB code was developed for estimating all the necessary parameters influencing the hydrodynamic coefficients of circular cylinders fitted with anodes.

3.2. Computational fluid dynamic (CFD) simulation

CFD is a numerical technique used for the solution of the complex equations governing fluid flow and heat transfer problems inside a defined flow geometry [14]. In this study, CFD was adopted for comparison of the theoretical results. Wave forces acting on the circular cylinder were determined, using ANSYS Fluent software, which is used to solve conservation equations. The incompressible equation Navier-Stokes Equations is set as follows:

$$\nabla \cdot \boldsymbol{u} = 0$$

$$\frac{\partial \boldsymbol{u}}{\partial t} + \nabla \cdot (\boldsymbol{u} \otimes \boldsymbol{u}) = -\frac{1}{\rho}\nabla p + \frac{u}{p}\Delta u + g + f, \tag{12}$$

where u is the fluid velocity, ρ is the density of the fluid, p is the pressure, μ is the molecular velocity, g is the gravitational force, and f denotes additional momentum sources. The general overview of CFD setup is shown in figure 2. The CFD simulation consists of Geometry, Engineering data, Model setup, meshing, specifications of the component names, boundary setup, solution of complex equations describing the interaction of fluid with the cylinder, and results presentation.

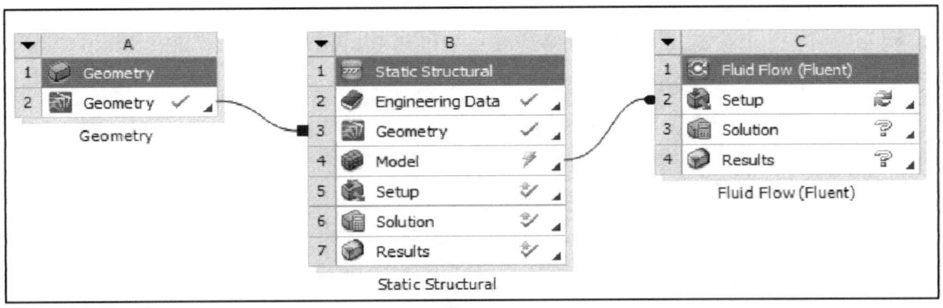

Figure 2. CFD simulation setup in ANSYS Fluent.

Although the theoretical analysis addressed several values for the outer diameters of the cylinders, effects of numerical simulation was limited to one cylinder only. The numerical simulation was limited to a circular cylinder with outer diameter, $D = 0.8$ m, and the wave height was varied from 5.0 to 14.0m while the wave frequency was varied from 3 to 21.6 seconds. The model geometry, meshing and the model setup implemented for the CFD simulation is depicted in figure. 3.

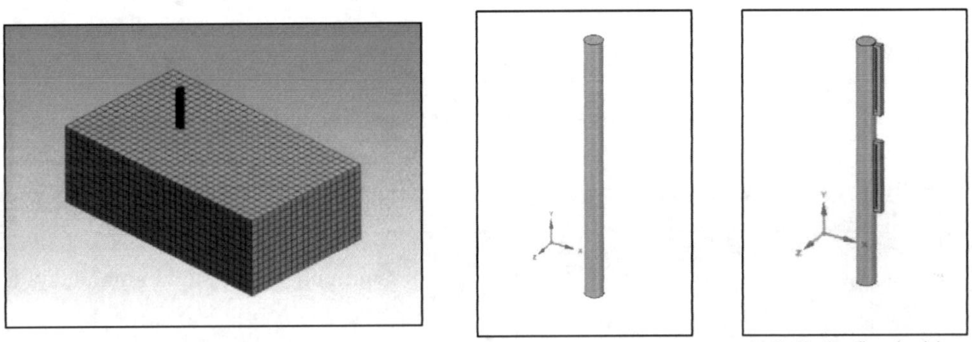

a) Geometry b) Smooth cylinder c) Cylinder fitted with sacrificial anodes

Figure 3. Geometry and meshing details of circular cylinders used for CFD simulation.

To ensure consistency of the boundary conditions, and it is strongly reflecting the application conditions that is being simulated, the boundary conditions are specified as following. A uniform flow velocity U is specified in the streamwise direction at the inlet boundary. At the outlet, the Neumann boundary condition was adopted for the velocity, and the pressure is specified as a reference value of zero. Non-slip boundary condition was applied on the perpendicular walls, the floor bed and the vertical cylinder [15, 16].

4. Results and Discussion

4.1. The effect of wave heights and wave periods on total hydrodynamic forces

In this section, the effects of wave heights and wave periods on the hydrodynamic forces are discussed. A MATLAB code was developed to estimate the hydrodynamic forces based on the well-known Morison Equations. Figure 4 shows the variation of total hydrodynamic forces with the wave height, H_{max}. The value of H_{max} was varied from 3 to 21.6 seconds. From the graph, one can observe that increasing the wave height has significantly influenced the corresponding forces. The graph also shows that the wave forces decreased with the increasing wave periods.

IOP Conf. Series: Materials Science and Engineering 1101 (2021) 012002 doi:10.1088/1757-899X/1101/1/012002

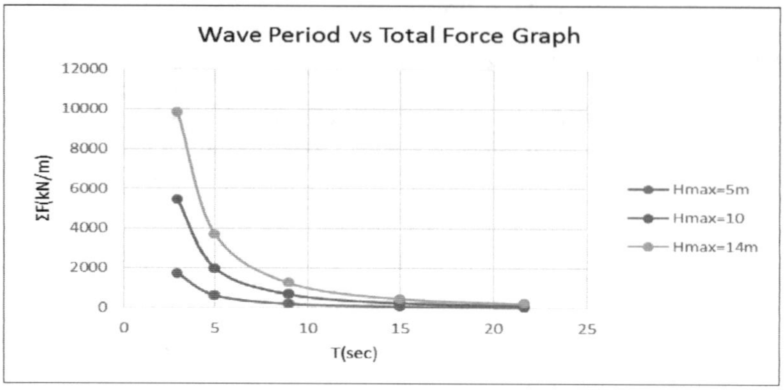

Figure 4. Variation of wave forces as a function of wave height, for D = 2 m.

4.2. The effect of water depth on hydrodynamic forces

Figure 5 shows the variation of wave forces as a function of water depths. In this study, the water depth was varied as d =30m, 50m and 75m. These water depths values represent typical offshore operation locations within the Malaysian water. The figure shows that for a given water depth, the values of hydrodynamic forces increased with the increasing wave height, H_{max}. Further, the graphs show that the maximum force was determined as 1210 kN, corresponding to water depth of 75 m, and T_{ass} =14 seconds. Generally, the trend of the graphs suggests a consistent increase in the wave forces as the wave heights increase. Similarly, one can observe that the values of wave forces increased exponentially as the water depth increases.

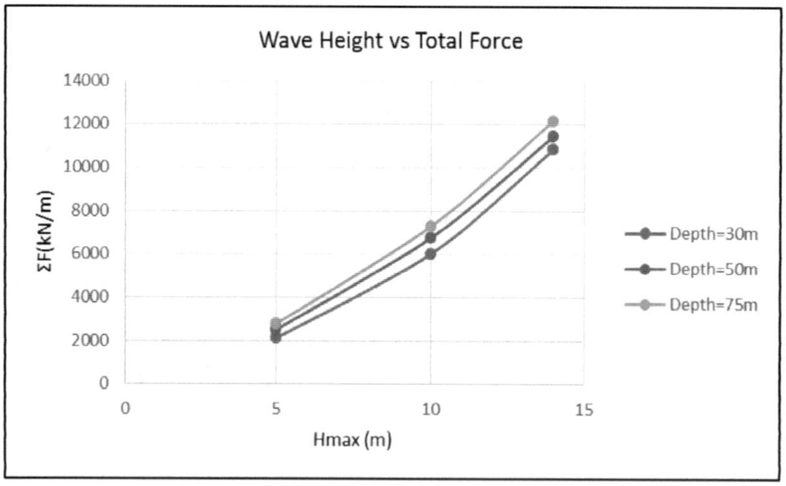

Figure 5. Variation of wave forces with water depth, d = 30 m, 50 m, 75m, at different H_{max}.

4.3. The effect of pipe diameter on hydrodynamic forces

Figure 6 shows the variation of wave forces with respect to the outer diameter of the pipe. The pipe diameter was varied as D = 0.8 m, 1.5 m and 2 m. It can be observed that as the pipe diameter increases, the total hydrodynamic forces increased as well. In addition, the graph also shows that the maximum wave forces are recorded at low wave periods, and the forces decreased gradually to reach negligible values at T = 22 seconds.

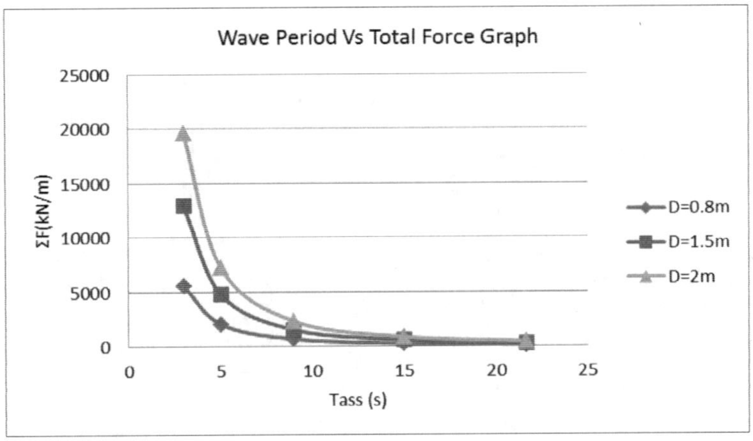

Figure 6. Variation of wave forces with outer diameter, D = 0.8 m, 1.5 m, 2 m, for different wave periods.

4.4. Comparison between clean and sacrificial anode fitted members

In this section a comparison of the total hydrodynamic forces on smooth cylinders and cylinders fitted with anodes was conducted. The waves forces are estimated using the specified values of drag and inertia coefficients in PTS [17].

Figures 7 - 9 show the time series records for drag, inertia and the total forces, for smooth and anode fitted cylinders, together with the corrected force determined using the C_D and C_M values determined using CFD simulations. The total force for the smooth cylinder and the one fitted with sacrificial anodes are determined as 1670 kN and 3199 kN respectively, while the total forces on the cylinder fitted with sacrificial anodes determined using C_D and C_M values determined from CFD was 2719 kN. The percentage different for the total force between the corrected values of hydrodynamic coefficient and the ones determined from PTS is 8.12%. This shows that the values of hydrodynamic coefficients recommended by the code of practices are more conservatives. Thus, cost effective platforms can be designed if more accurate values of hydrodynamic coefficients are used to estimated wave forces on anode fitted circular members.

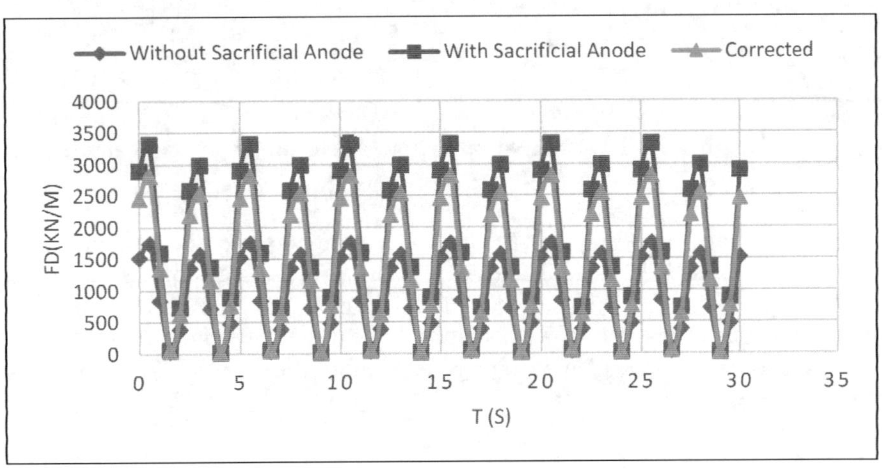

Figure 7. Time series records for drag forces on smooth and anode fitted cylinders.

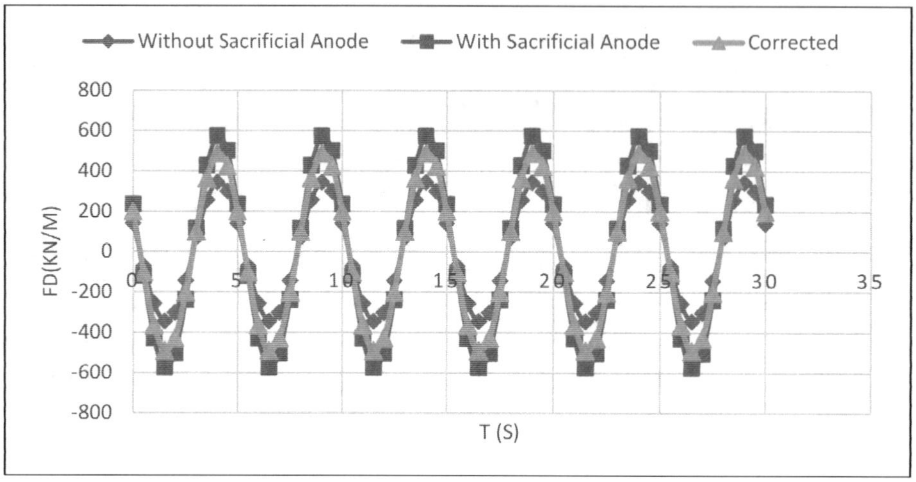

Figure 8. Time series records for inertia forces on smooth and anode fitted cylinders.

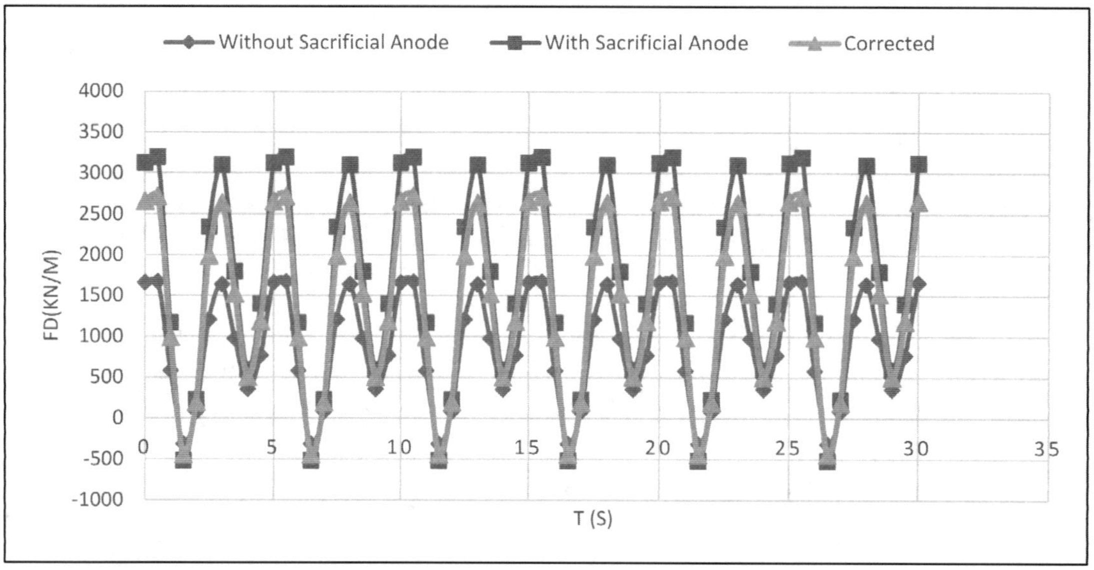

Figure 9. Time series records for the total hydrodynamic forces on smooth and anode fitted cylinders.

5. Conclusion

Wave forces acting on smooth and anode fitted circular cylinders are determined by employing the Morison equation and CFD simulations. The effects of sacrificial anodes on the hydrodynamic forces was initially determined using the well-known Morison equation and the findings are compared using CFD simulation techniques. The effect of wave height, wave period, water depth and the outer diameters on the hydrodynamic forces are investigated. The general conclusions drawn from this parametric study are as follows:

- The numerical calculations show that increasing the wave height has significantly increased the total hydrodynamic, while increasing the wave periods resulted in a comparatively smaller hydrodynamic forces.
- The study shows that the values of hydrodynamic coefficient determined using the numerical simulations are generally smaller that the values spefified in the code of practice. A decrease

of 8.15% was observed on C_D and C_M values.

- The results show that water depth, wave heights, wave frequencies, and the pipe diameters have major influences on the hydrodynamic forces of circular cylinders fitted with sacrificial anodes.

6. References

[1] Talbot, D. E. and Talbot, J. D. 2018 Corrosion Science and Technology, CRC Press.
[2] Downie, M., Murray, B., Bettess, P. and Haswell, J. 1993 Added hydrodynamic loading due to sacrificial anodes *Wave Kinematics and Environmental Forces*, 191-207.
[3] Morison, J., Johnson, J. and Schaaf, S. 1950 The force exerted by surface waves on piles *J. Petroleum Technol.* **2(05)** 149-54.
[4] Raed, K. and Soares, C. G. 2018 Variability effect of the drag and inertia coefficients on the Morison wave force acting on a fixed vertical cylinder in irregular waves *Ocean Eng.* **159** 66-75.
[5] Sarpkaya, T. 2010 Wave Forces on Offshore Structures, Cambridge University Press.
[6] Boccotti, P., Arena, F., Fiamma, V. and Barbaro, G. 2012 Field experiment on random wave forces acting on vertical cylinders *Probabilistic Eng. Mechanics* **28** 39-51.
[7] Sarpkaya, T. 1986 Force on a circular cylinder in viscous oscillatory flow at low Keulegan—Carpenter numbers *J. Fluid Mechanics* **165** 61-71.
[8] Troesch, A.W. and Kim, S. 1991 Hydrodynamic forces acting on cylinders oscillating at small amplitudes *J. Fluids Structures* **5(1)** 113-26.
[9] API RP2A-WSD, A. 2000 Recommended Practice for Planning, Designing, and Constructing Fixed Offshore Platforms—Working Stress Design, Washington, DC: American Petroleum Institute.
[10] DNV, G. 2017 Structural Design of Offshore Units—WSD Method, DNV GL: Oslo, Norway.
[11] Wilson, J. F. 2003 Dynamics of Offshore Structures, John Wiley & Sons.
[12] Munson, B.R., Okiishi, T.H., Huebsch W.W. and Rothmayer, A.P. 2013 Fluid Mechanics, Wiley Singapore.
[13] Airy, G.B. 1845 Tides and Waves, B. Fellowes.
[14] Scott, G. 1992 Computational Fluid Dynamics for the Food Industry, Campden & Chorleywood Food Research Association.
[15] Son, H.A., Lee, S. and Lee, J. 2020 Numerical analysis of drag force acting on 2D cylinder immersed in accelerated flow *Water* **12(6)** 1790.
[16] Yang. F., Zhou, Z., Lou, X. and Lu, L. 2020 Flow around a near-bed horizontal circular cylinder mounted on a vertical wall *Ocean Eng.* **216** 107586.
[17] PETRONAS Technical Standards 2014.

Acknowledgments
The authors would like to acknowledge the financial support provided by Universiti Teknologi PETRONAS (UTP) through YUTP grant (015LC0-095).

Stress analysis of freestanding risers subjected to waves and current loadings - Numerical modelling

A M Al-Yacouby[1*], Hakim Arifin[1], M S Liew[1] and Zahari Razak[2]

[1]Department of Civil and Environmental Engineering, Universiti Teknologi PETRONAS, 32610 Seri Iskandar, Perak, Malaysia
[2]Kuala Lumpur City Centre, 50088 Kuala Lumpur, Malaysia
[*]Corresponding author email: ahmad.alyacouby@utp.edu.my

Abstract. Free Standing Risers are vertical steel pipes tensioned by a near surface buoyancy can. These risers are subjected to different types of loads that lead to fatigue damages. Therefore, this study aims to predict the stresses of freestanding risers subjected to hydrodynamic loads. The methods applied consists of numerical analysis and computer simulation. In this paper, the wave and current loads acting on the risers have been determined using Morison equation, then the stress analysis was conducted using ANSYS software. The important parameters investigated are the pipe diameter D, the wall thickness, t, the wave height, H_{max}, the wave period, T_{ass}, the tensile strength, fy and the water current, v. The outcome of the study shows that wave height, wave period, pipe diameter and the wall thickness are affecting the maximum stress on the risers.

Keywords. Stress Analysis, Freestanding Risers, Waves loads, Current.

1. Introduction

Production risers have been used worldwide for transportation of oil and gas. The modern history of oil and gas started since the 18th century when oil refinery and oil-works were commercially formed for the first time. Since then, transportation of oil and gas become crucial issue to dealing with. In the U.S., 70% of crude oil and petroleum products are transported by pipeline and riser. Riser is important in subsea oil/gas production systems that acts as oil and gas conductor pipe between an offshore platform or vessel and a well at seabed [1]. Just like the pipeline, riser is a vertical pipe system connecting a floater at the water surface and the wellheads at the seabed [2]. In deep water, freestanding risers are subjected to many loads that lead to fatigue. Fatigue is associated with many factors [3]. VIV is considered as a critical among all factors that affect the stress of riser significantly [4]. This paper will discuss about the effect of pipe diameter, D, wall thickness of riser, t, wave height, H_{max} and wave period, T_{ass} on riser stress.

Production risers are fabricated variously; there are risers that can be sat on site, and pre-installed risers that can be connected to the pipe on the seabed by a subsea tie-in arrangement [5]. The famous four of them includes Top Tensioned Risers, Flexible Riser, Steel Catenary Risers and Free-Standing Hybrid Risers. Free Standing Hybrid Riser (FSHR) is comprised of vertical steel risers and Flexible Jumpers (FJ). They are jointly connected to a submerged Buoyancy Can (BC) [6]. The original philosophy for this system, developed by Cameron Iron Works in early 1983, was to design a system, for small or marginal fields, that could drain a reservoir and then be moved on to a second or even a third location [7]. Nowadays, freestanding hybrid riser, FSHR have been commonly deployed in offshore oil field such as Gulf of Mexico, offshore Brazil and offshore Indonesia [8].

A number of studies related to stress assessment of tubular steel has been conducted in the past decades with the aim to develop a simplified yet accurate empirical formulation. At present, the most simplified empirical formulation for stress assessment of tubular steel often considers axial, hoop and shear stress as these are the most significant controlling factor on the ultimate stress of tubular steel structure. In order to improvise the accuracy of existing empirical formulations, the effect of different parameters on stress such as pipe diameter, D, wall thickness, t, wave height, H_{max} and wave period, T_{ass} are to be considered. However, the developed empirical formulation tends to become more complex as more parameters are considered and consequently leads to human errors during application or calculation.

2. Theoretical background

2.1. Morison's equation

Hou [9] once conducted a stress analysis on Buoyancy Can of FHSR to predict fatigue damage. Upper area of the riser is very sensitive due to the currents and wave velocities. Thus, axial stress is significant at this portion as the moment is higher. At lower assembly, it is considered as fix. Stress at the bottom region is higher because of its connectivity. In many cases, stress analysis of tubular structure addressed hoop stress, axial stress and shear stresses. The combination of the three fundamental stresses is well known as Von Mises stress. In certain case i.e. riser structure shear stress is insignificant. K. Wang et al. [10] considered only combination of axial and bending/hoop stress given by $\sigma = \sigma_a + \sigma_M$. Based on PTS 20.214 shear stress is excluded [11]. The Maximum Von Mises stress in a riser is calculated using equation (1).

$$2{\sigma_V}^2 = 2{\sigma_h}^2 + 2{\sigma_L}^2 - \sigma_h\sigma_L + 3{\sigma_S}^2 \tag{1}$$

where σ_V = Von Mises,
σ_h = Hoop stress (due to pressure)
σ_L = Longitudinal stress (due to pressure, thermal expansion, bending)
σ_S = Shear stress

Basically, the shear stress is neglected for riser because riser is in vertical position. However, shear stress is considered in the API [12]. All stresses are calculated according to Von Mises stress failure criterion as following equation (2).

$$2{\sigma_V}^2 = (\sigma_L - \sigma_h)^2 + (\sigma_h - \sigma_S)^2 + (\sigma_S - \sigma_L)^2 \tag{2}$$

2.2. Computer simulation

Evaluation of the hydrodynamic stresses on fluid-structure interaction simulation is a great challenge, because the grid around the particle is usually not fine enough to properly resolve the velocity gradients. Software simulation is relevant for stress analysis as it solves all the complex formulation. Most well-known Navier-Stokes equation always been used for ages to model the incompressible flows of fluid motion [13].

$$\frac{\partial \boldsymbol{u}}{\partial t} + \nabla \cdot (\boldsymbol{u} \otimes \boldsymbol{u}) = -\nabla p + \frac{1}{Re}\nabla^2 u + f \tag{3}$$

where u is the non-dimensional velocity
p is the non-dimensional pressure
t is non-dimensional time

IOP Conf. Series: Materials Science and Engineering 1101 (2021) 012003 doi:10.1088/1757-899X/1101/1/012003

3. Methodology

3.1. Hydrodynamic forces
Table 1 presents the parameters used to estimate the hydrodynamic forces using Morison Equation. Riser diameter was as 0.244 m with 0 m corrosion allowance. Thus, total wall thickness is equal to thickness of the main pipe. Mass and drag coefficient and current velocities are in accordance with PTS [11].

Table 1. Environment parameters.

Parameters	Values
Inertia force per unit length, f_I (kNm^{-1})	
Drag force per unit length, f_D (kNm^{-1})	
Seawater density, ρ (kNm-^{3})	1025
Mass coefficient, C_m	1.6
Drag coefficient, C_d	0.65
Wave velocity, (ms^{-1}), U	
Current velocity, (ms^{-1}), u_c	
Wave acceleration, (ms^{-2}), $\dot{u}$	
Riser diameter, D (m)	0.244

3.2. Software simulation
Software simulation was conducted to assess the stresses of riser subjected to hydrodynamic forces. There are few steps to complete the simulation:

1. Input the Engineering Data.
2. Riser Modelling.
3. Quality of Meshing.
4. Input the Parameter.
5. Solve the Simulation.

A riser was designed 75 m long standing vertically, with the tension on the top provided by the buoyancy can at the top and fixed by the lower connector at the bottom. The coordinate system is selected as: x direction coincides with riser axis, y is the current direction, and z is the crossflow direction. For actual case, the maximum deflection occurs in the middle of the riser. However, in this simulation, the maximum riser deflection occurs at top portion of the riser as the buoyancy can was not included. Physical properties of the riser are shown in the table 2.

The riser was set straight and vertically standing in the still water. The simulation was performed under current velocities of 1.8 m/s and 1.43 m/s at 1.0 L m and 0.5 L m from the seabed respectively based on PTS [11]. The riser was fixed at the bottom of the riser and total forces of hydrodynamic forces, gravitational acceleration, and ocean current were applied along the riser body. The stress and deflection were observed for different case and riser.

Figure 1 shows the texture of meshing on riser model that had been applied hydrodynamic loads. Sensitivity of meshing is important for assurance of measurement quality. The finest size of meshing, the more accurate the result will be. The smallest size of meshing that computer capable was applied until it reached the limit. Hydrodynamic loads were applied in many directions. Figure 2 shows the direction of applied loads along the riser.

Table 2. Physical properties of riser.

Parameter	Values
Riser outer diameter, D (m)	0.244
Riser inner diameter, Di (m)	0.178
Internal Corrosion Allowance (mm)	3
Material Grade (API 5L)	X-70
Material Yield Stress (MPa)	483
Ultımate Tensile Strength (MPa)	565
Young Modulus (MPa)	206843
Poission Ratio	0
Steel Density (kg/m3)	7850

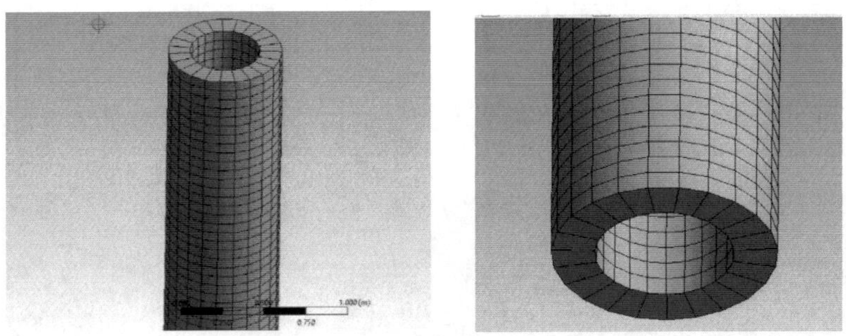

Figure 1. Meshing of 3D model.

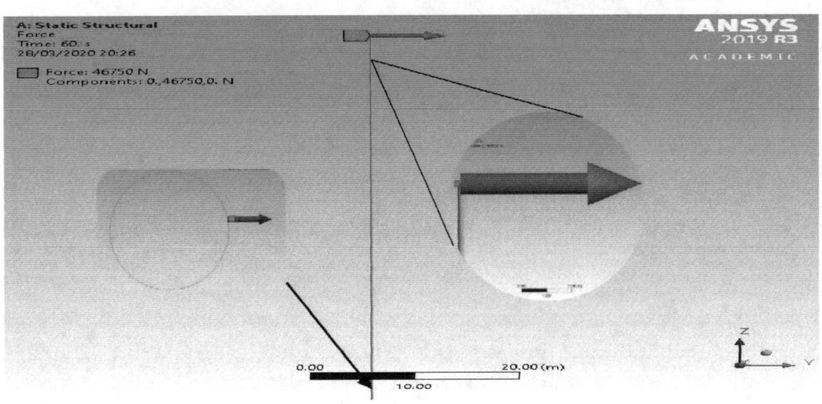

Figure 2. Applied forces direction.

Simulations were performed for several times until ultimate precision achieved. The steps are repeated for a few times using different parameter as shown in table 3. Each parameter is denoted as case 1, case 2 case 3 and case 4.

Table 3. Manipulated parameter.

	Case 1	Case 2	Case 3	Case 4
	Do (m)	t (m)	H_{max} (m)	T_{ass} (m)
Riser 1	0.244	0.033	11.65	9.64
Riser 2	0.2684	0.0363	12.815	10.604
Riser 3	0.2806	0.03795	13.3975	11.086
Riser 4	0.2928	0.0396	13.98	11.568
Riser 5	0.3172	0.0429	15.145	12.532
Riser 6	0.366	0.0495	17.475	14.46

4. Results and Discussion

4.1. The effect of wave heights on total hydrodynamic forces
Figure 3 shows the effects of diameter on total hydrodynamics forces respectively. The value of the initial H_{max} was increased by 10%, 15%, 20%, 30% and 50%. The graph shows that increasing the wave height has resulted in higher drag and inertia forces. For instance, the wave height with 50% increment has resulted the highest hydrodynamic force with maximum increment of 53.32% as the value of the force increased from its initial value of 46750 kN to 71677 kN. We can also observe that the effect of wave heights on the drag forces is higher than the effects on the inertia forces.

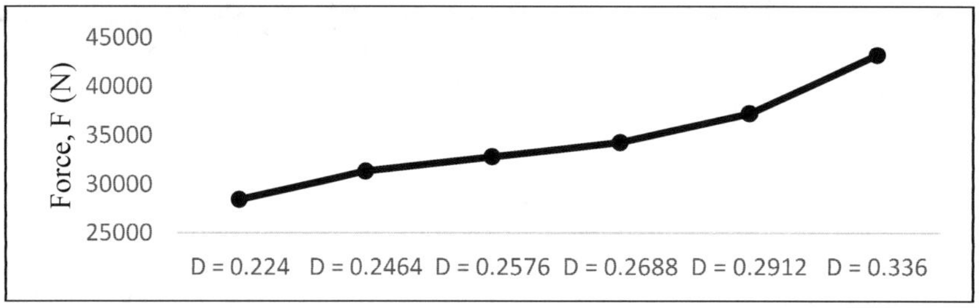

Figure 3. Drag forces of different diameter for t = 0.033 m, H = 11.65 m, T = 9.64 s, & fy = 483 MPa.

4.2. The effect of wave periods on hydrodynamic forces
Figure 4 shows the effects of wave periods on drag, inertia and the total hydrodynamics forces respectively. Based on the graphs above the effect of drag forces is greater than inertia forces. The value of the initial wave period was increased by 10%, 15%, 20%, 30% and 50%. The graphs show that increasing the wave period has resulted in higher drag and inertia forces. For instance, the wave height with 50% increment has resulted the highest hydrodynamic force with maximum increment of 32.8% as the value of the force increased from its initial value of 46750 kN to 62082.4 kN. We can also observe that the effect of wave periods on the drag forces is higher than inertia forces.

4.3. The effect of wave heights on hydrodynamic forces
Figure 5 shows the effects of wave heights on drag, inertia and the total hydrodynamics forces respectively. The value of the initial H_{max} was increased by 10%, 15%, 20%, 30% and 50%. The graph shows that increasing the wave height has resulted in higher drag and inertia forces. For instance, the wave height with 50% increment has resulted the highest hydrodynamic force with maximum increment of 74.83% as the value of the force increased from its initial value of 46750 kN to 81732 kN. We can

IOP Conf. Series: Materials Science and Engineering 1101 (2021) 012003 doi:10.1088/1757-899X/1101/1/012003

also observe that the effect of wave heights on the drag forces is higher than the effects on the inertia forces.

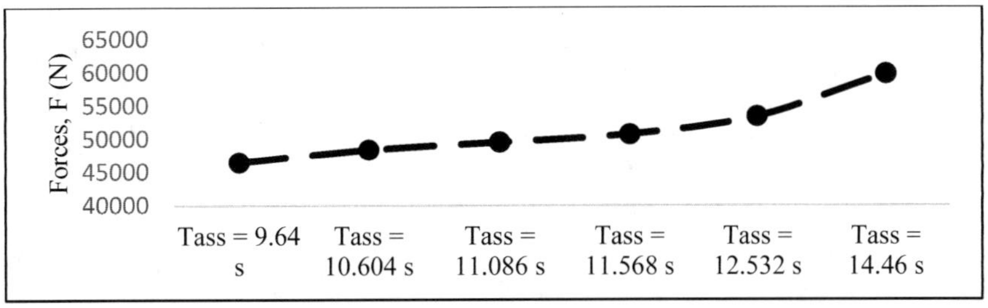

Figure 4. Inertia forces of different wave period for D = 0.224 m, t = 0.033 m, H = 11.65 m & f_y = 483 MPa.

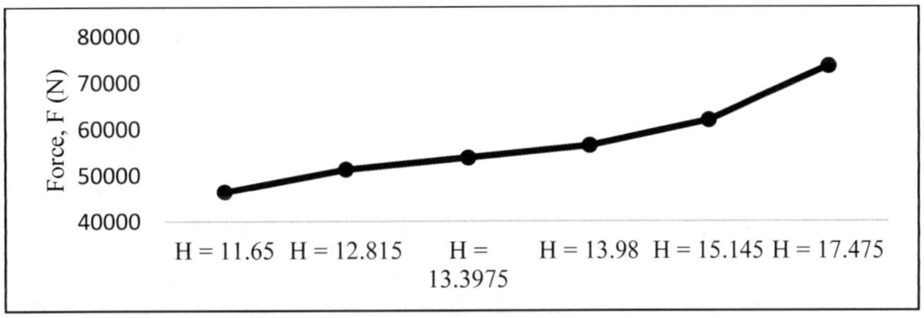

Figure 5. Total forces of different wave height for D = 0.224 m, t = 0.033 m, T = 9.64 s & f_y = 483 MPa.

4.4. The effect of pipe diameter, D on stress

Figure 6 shows the effects of pipe diameter on riser stresses. The value of the initial diameter was increased by 10%, 15%, 20%, 30% and 50%. The graph shows that increasing the diameter has resulted in stresses. For instance, the diameter with 50% increment has resulted the lowest stress with maximum increment of 33.83% as the value of the stress increased from its initial value of 194498 MPa to 128695 MPa. We can also observe that the riser stresses inversely change with the diameter size

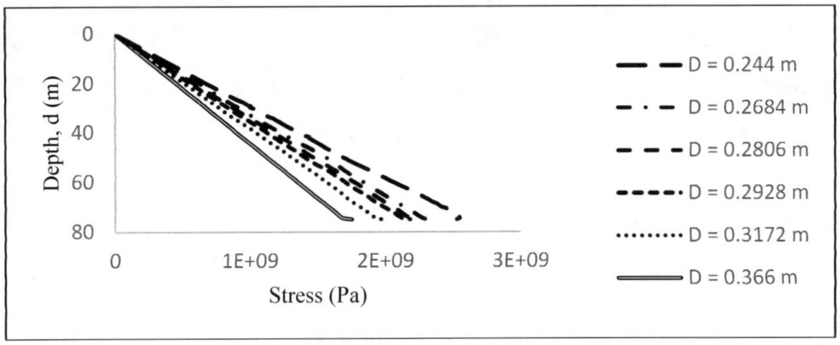

Figure 6. Stresses of riser of different pipe diameter for t = 0.033 m, H = 11.65 m, T = 9.64 s & f_y = 483 MPa.

4.5. The effect of wall thickness, t on stress
Figure 7 shows the effects of wall thickness on riser stresses. The value of the initial diameter was increased by 10%, 15%, 20%, 30% and 50%. The graph shows that increasing the diameter has resulted in stresses. For instance, the diameter with 50% increment has resulted the lowest stress with maximum increment of 88.50% as the value of the stress increased from its initial value of 194498 MPa to 366637 MPa. We can observe that the riser stresses decrease when thickness increase. Thus, the change in thickness affect the stress significantly among the others.

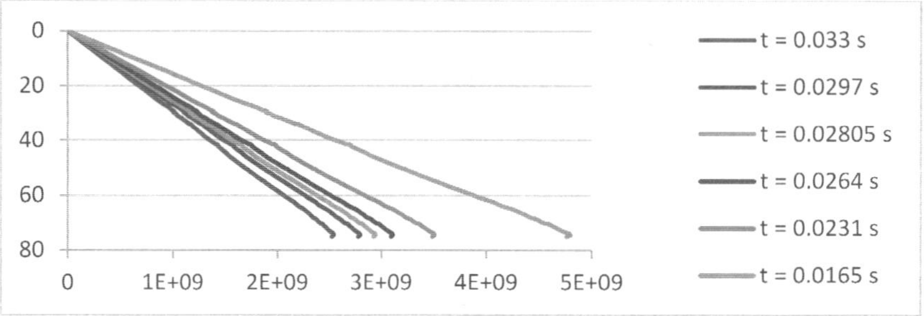

Figure 7. Stresses of riser of different wall thickness for D = 0.244 m, H = 11.65 m, T = 9.64 s & f_y = 483 MPa.

4.6. The effect of wave height, H_{max} on stresses
Figure 8 shows the effects of wave height on riser stresses. The value of the initial diameter was increased by 10%, 15%, 20%, 30% and 50%. The graph shows that increasing the diameter has resulted in stresses. For instance, the diameter with 50% increment has resulted the lowest stress with maximum increment of 74.66% as the value of the stress increased from its initial value of 194498 MPa to 339709 MPa. We can also observe that the riser stresses increase when wave height increases.

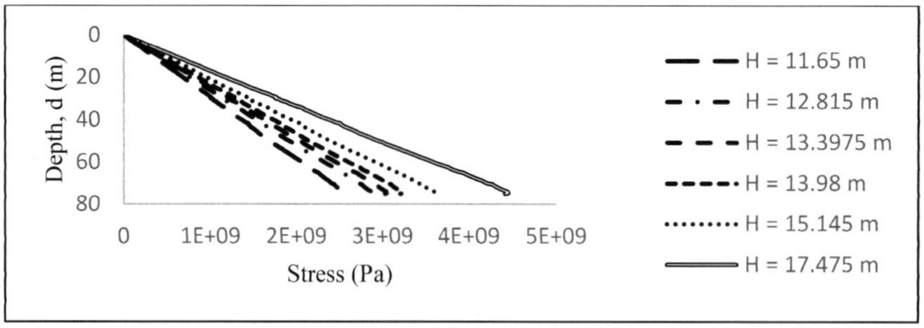

Figure 8. Stresses of riser of different wave height for D = 0.244 m, t = 0.033 m T = 9.64 s & f_y = 483 MPa.

4.7. The effect of wave period, T_{ass} on stresses
Figure 9 shows the effects of wave periods on riser stresses. The value of the initial diameter was increased by 10%, 15%, 20%, 30% and 50%. The graph shows that increasing the diameter has resulted in stresses. For instance, the diameter with 50% increment has resulted the lowest stress with maximum increment of 32.72% as the value of the stress increased from its initial value of 194498 MPa to 258146 MPa. We can also observe that the riser stresses increase when wave period increases.

IOP Conf. Series: Materials Science and Engineering 1101 (2021) 012003 doi:10.1088/1757-899X/1101/1/012003

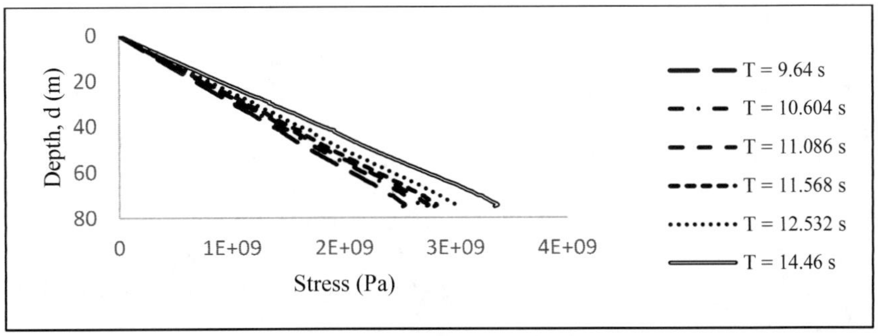

Figure 9. Stresses of riser of different wave periods for D = 0.244 m, t = 0.033 m H = 11.65 m & f_y = 483 MPa.

4.8. The effect of current on stresses

Figure 10 shows the graphs of the effect of current velocity, v on stress. The value of the initial diameter was increased by 10%, 15%, 20%, 30% and 50%. The graph shows that increasing the diameter has resulted in stresses. For instance, the diameter with 50% increment has resulted the lowest stress with maximum increment of 15.36% as the value of the stress increased from its initial value of 194498 MPa to 224381 MPa. We can also observe that the riser stresses increase when wave period increases.

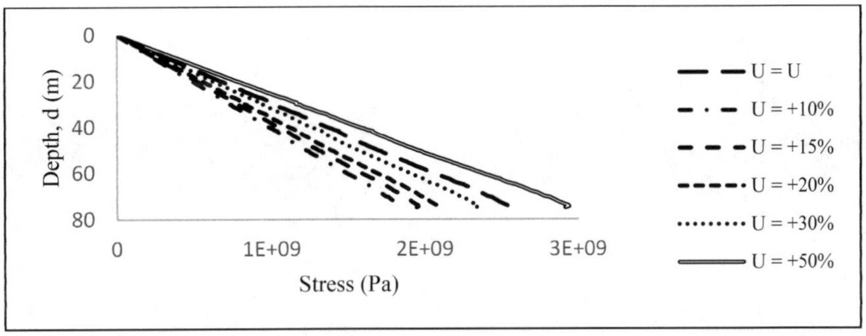

Figure 10. Stresses of riser of different velocities for D = 0.244 m, t = 0.033 m, H = 11.65 m & T = 9.64 s.

5. Conclusion

In this paper, the stress analysis of freestanding risers subjected to waves and current loadings has been investigated using numerical modelling. Upon completion of numerical analysis, stress analysis was conducted. The hydrodynamic forces are calculated and assessed. The riser stresses are assessed by simulation using ANSYS software. The important parameters involved are the pipe diameter, the wall thickness, the wave height, the wave period and the water current. The analysis results show that the stresses are linearly proportional to water depth. The wave periods have more significant effect on the stress as compared to wave height. Generally, the pipe diameter, wall thickness, wave heights and wave periods have significant impact on the riser's stresses.

6. References

[1] Chakrabarti, S.K. and R.E. Frampton 1982 Review of riser analysis techniques *Applied Ocean Research* **4(2)** 73-90.
[2] Veritas, D.J.D.N. 2010 DNV-RP-F204 Riser Fatigue.
[3] Veritas, D.N.J.D. 2010 R.P.D.-R.-C., Fatigue design of offshore steel structures.

[4] Kim, D.K., Incecik, A., Choi, H.S., Wong, E.W.C., Yu, S.Y. and Park, K.S.2018 A simplified method to predict fatigue damage of offshore riser subjected to vortex-induced vibration by adopting current index concept *Ocean Eng.* **157** 401-11.
[5] Fishe, E. and P. Hoolley 1995 *Development and deployment of a freestanding production riser in the gulf of Mexico* Offshore Technology Conference.
[6] Kim, K.-S., Choi, H.-S. and Kim, K.S. 2018 Preliminary optimal configuration on free standing hybrid riser *Int. J. Naval Architecture Ocean Eng.* **10(3)** 250-8.
[7] Tellier, E. and Thethi, R. 2009 *The evolution of free standing risers* ASME 2009 28th International Conference on Ocean, Offshore and Arctic Engineering, American Society of Mechanical Engineers.
[8] Bai, Y. and Bai, Q. 2005 Subsea Pipelines and Risers, Elsevier Ltd.
[9] Hou, T. 2014 Numerical Simulation of Free Standing Hybrid Risers, Texas A&M University.
[10] Wang, K., Xue, H., Tang, W. and Guo, J. 2013 Fatigue analysis of steel catenary riser at the touch-down point based on linear hysteretic riser-soil interaction model *Ocean Eng.* **68** 102-11.
[11] PTS 20.214 2008 Pipeline and Riser Engineering, PETRONAS CARIGALI Sdn. Bhd. (PCSB).
[12] API, R. 1993 16Q, Recommended Practice for Design, Selection, Operation and Maintenance of Marine Drilling Riser System, Washington, D.
[13] Armfield, S. and Street, R. 2002 An analysis and comparison of the time accuracy of fractional-step methods for the Navier–Stokes equations on staggered grids *Int. J. Numer. Methods Fluids* **38(3)** 255–82.

Acknowledgments
The authors would like to acknowledge the financial support provided by Universiti Teknologi PETRONAS (UTP) through YUTP grant (015LC0-095).

IOP Conf. Series: Materials Science and Engineering 1101 (2021) 012004 doi:10.1088/1757-899X/1101/1/012004

Estimation of the buried pipeline displacement using distributed fibre optic sensing: An experimental study

Hamzh Alarifi[1*] **and Hisham Mohamad**[1]

[1]Department of Civil and Environmental Engineering, Universiti Technology PETRONAS, Bandar Seri Iskandar, 32610 Tronoh, Perak, Malaysia.
[*]hamzh_hmd@yahoo.com

Abstract. Pipe displacement (bending) is one of the most prominent risks caused by the soil movements in the landslide areas. The lateral soil movement causes damages to the buried pipelines, which might explode due to leakage of flammable materials. Employing fibre optic sensors to monitor the buried pipes can maintain the integrity of piping systems and the surrounding environment. The fibre optic sensor has already demonstrated its capabilities to provide a continuous monitoring system for the structures. This paper aims to evaluate the buried pipeline displacement caused by the lateral soil movement. Brillouin Optical Time Domain Analysis (BOTDA) is used to measure the strain values from the fibre optic cable. A small physical model is used to demonstrate the bending of High-density polyethylene (HDPE) pipe under the ground movement. This paper concludes that the fibre optic sensor could be employed to evaluate the horizontal and vertical displacements of the buried pipeline.

Keywords: Buried pipeline, displacement, landslide, physical model, fibre optic sensors, BOTDA.

1. Introduction

Over the world, civil Infrastructures such as bridges, buildings, and pipelines are subject to different kinds of geohazards (earthquakes, landslides, etc.). Some of these infrastructures might completely collapse and leave environmental disasters. The pipeline is a primary entity for the civil infrastructures which is used to transport fluids (oils, water, and gas) over countries and between cities. Sometimes, the pipeline needs to cross the rugged areas that might be subject to natural hazards such as landslides. Buried pipes may damage due to external loads resulted from lateral earth pressure increment. Many countries have spent billions of dollars on maintenance and repair of the buried pipelines due to the lack of monitoring and early warning systems for hazard detection [1].

The most important factors that must be emphasised during the buried pipeline design stage are soil displacement effect on the pipes and estimation of the maximum strains in the pipeline [2, 3]. A continuous monitoring method to pipeline systems should be applied to avoid the risk of pipeline failure, especially those buried underground. However, using separate conventional sensors is an expensive solution for monitoring the pipeline systems because it requires placing thousands of sensors and cables on the pipes. Optical Fiber Sensor (OFS) is one of the current and standard sensors that provide low cost, continuous long-distance monitoring [4].

The optical fibre sensor is a technology that uses a light wave to obtain the deformation of the measurement change. This sensor is already applied in many fields of civil engineering. Distributed optical fibre strain sensor (DOFSS) is one of the latest technologies being used in Geotechnical engineering to provide real-time monitoring. The basis of this technique is a continuous strain measurement using an optical fibre distributor sensor known as Brillouin Optical Time-Domain Reflectometer (BOTDR) or Brillouin Optical Time-Domain Analysis (BOTDA) [5].

Both BOTDA and BOTDR can record the strain measurements by using various cables of fibre optic such as telecommunication fibre optic cable. The fibre optic sensors can provide strain measurement along the cable (up to 50 km long) by achieving a high-resolution which reaches up 10 macrostrains. Moreover, the strains of various pipes materials can be measured by the fibre optic sensor with a similar accuracy of strain gauge measurements [6]. Fibre optic monitoring system based on distributed measurements can provide continuous strain measurement when it is applied along the entire length of a pipeline [7].

Several studies and researches performed on the use of optical fibre sensors to monitor the buried pipelines are subjected to different loads. Moore and Take carried out experiments on various pipes to monitor them under the effect of a normal fault [8]. The strain values along the pipes were obtained by both fibre optic strain sensors and strain gauges sensors at the same moment. Their findings indicated that fibre optic strain sensors have an advantage of providing more data provision than the conventional strain gauge. In addition, Ravet used a distributed Brillouin sensor system to measure the tensile and compressive strains and monitor the buckling of a steel pipe [9].

Glisic and Yao completed a large-scale experiment to monitor the health of buried concrete pipes based on Brillouin light scattering (fibre optic sensor). They discovered that the fibre optic sensor could be used as an early warning system for pipes bends. Additionally, it can be utilised to determine the location of pipe damage caused by soil movement [10]. When the pipeline crosses active landslide areas, the soil movement can lead to developing strains and displacements in the pipes. The challenge is how to employ the fibre optic sensors to estimate the displacement of the buried pipelines. In this study, a new method to detect and estimate pipe displacement is presented based on fibre optic sensors.

2. Materials and Equipment

In this study, a small physical model is used to monitor pipe behaviour using fibre optic sensors. This model was developed by Alarifi, to study the behaviour of a buried pipeline under the landslide effect using strain gauge sensors [11]. Plastic acrylic sheets are used to fabricate the model of a box of 500 mm long , 250 mm wide, and 100 mm deep. Movable plates are fixed in the middle of the box by four linear bearing blocks which help the plates move smoothly. A High-Density polyethylene pipe (480 mm long and 20 mm diameter) is installed by simple support at both ends. A manual pulley and rope used to pull the movable plates out of the box, as shown in figure 1.

A fibre optic telecommunication cable of 8 m long is used to measure the strain along the pipe at four positions (bottom, top, front and backside). A linear variable differential transformer (LVDT) with the capacity of 100 mm is used to measure the displacement of the movable plate. Fully dry sand is utilised as the backfill material for this experiment. The materials properties are listed in table 1.

3. Method Description

The fibre optic cable was installed along the HDPE pipe at four locations, as shown in figure 2. The cable is well-attached to the pipe using an adhesive tape and instant adhesive which is made of cyanoacrylate sold as "Super Glue". The fibre optic cable was tested by a laser light inspector to check if there was a breakpoint before starting the experiment.

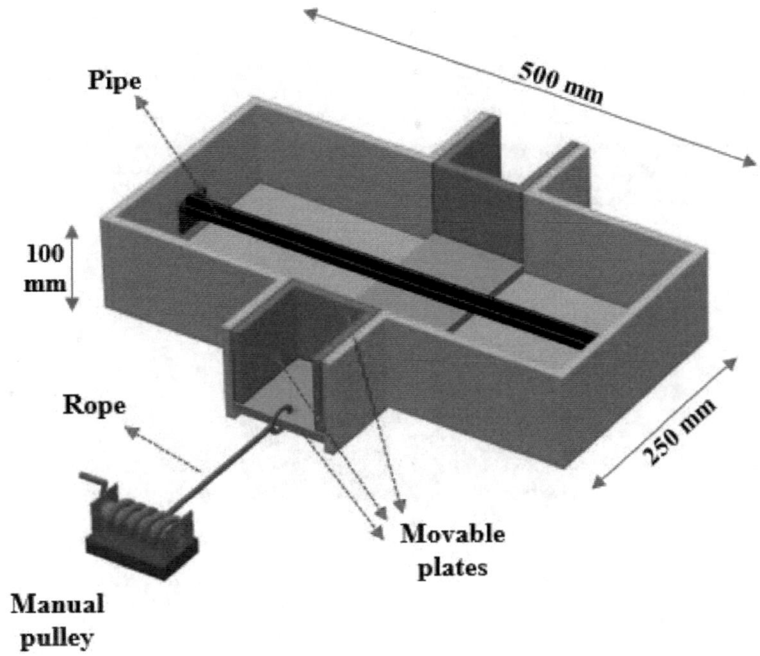

Figure 1. The physical model.

Table 1. The materials properties.

Soil		HDPE pipe	
class	poorly graded sand	class	(HDPE) PN 16
D_{10} D_{30} D_{60}	0.37 mm 0.72 mm 1.04 mm	Tensile strain (break)	600%
Unit weight	15.49 kN/m^3	Density	0.951 g/cm^3
Friction angle	30.8°	-	-

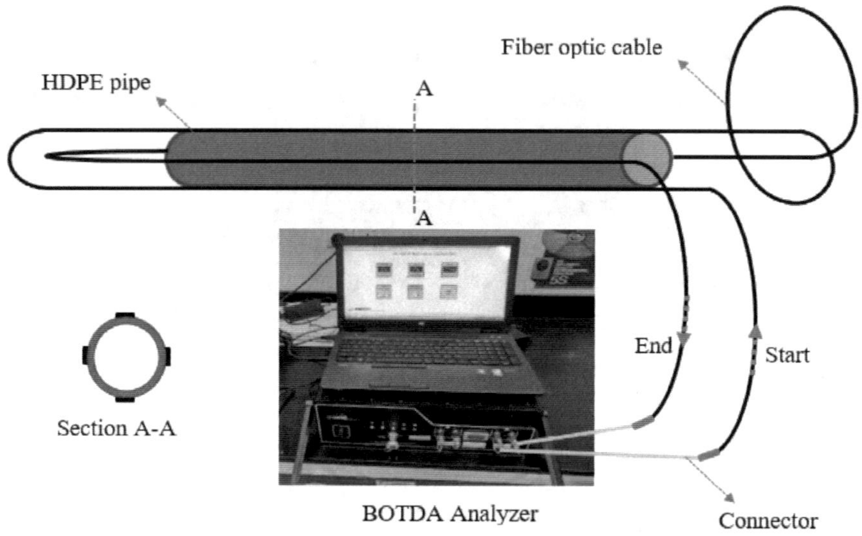

Figure 2. The fiber optic cable attached to HDPE pipe.

The HDPE pipe is conveniently placed in the middle of the model after filling the box with sandy soil at the bottom level of the pipe (invert level), as shown in figure 3 (a). A second layer of the soil (250 mm in depth) is added above the pipe, and the soil layers are compacted by using a small manual rammer. The LVDT was installed in front of the movable plate and connected to a data logger to record the plate displacement immediately. Moreover, the optical fibre cable connected to the BOTDA analyser channel and the baseline of the strain measurements is recorded.

The movable plate was pulled out (5 mm) by a manual pulley and the first measurement of the strain was recorded. Moving plate displacement increased gradually by 10 mm to 75 mm, and strain measurements were recorded at each stage of movement to monitor strain changes. figure 3 (b) & (c) illustrate the initial stage and lateral soil movement induced by pulling the movable plate at the fifth stage (45 mm), respectively. In figure 3 (c), the blue dashed line shows the active soil zone while the red dashed line represents the boundary of the passive soil zone.

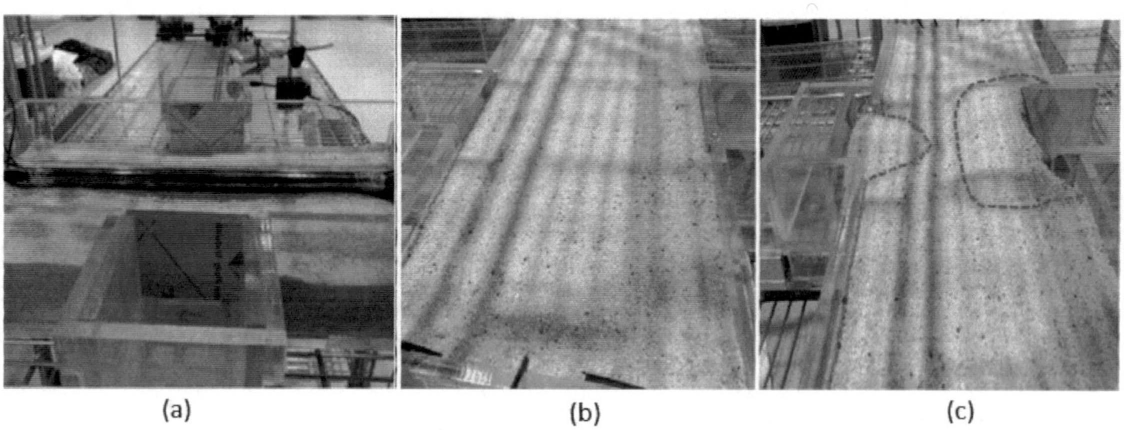

Figure 3. (a) Placing the pipe in the model (b) Initial stage (c) Lateral soil movement (45 mm).

4. Results & Data Analysis

The fibre optic sensors are used to monitor the buried (HDPE) pipe which is subjected to lateral ground movement in the small-scale model. The strains values were obtained by using BOTDA analyser via fibre optic cable. Figure 4 demonstrates the recorded strain values along the cable which include parts that are placed at various paths on the pipe at different stages of soil movement resulting from pulling out the movable plate. The movable plate displacement was controlled and it started with 5 mm and increased by 10 mm at each step until it reached 75 mm.

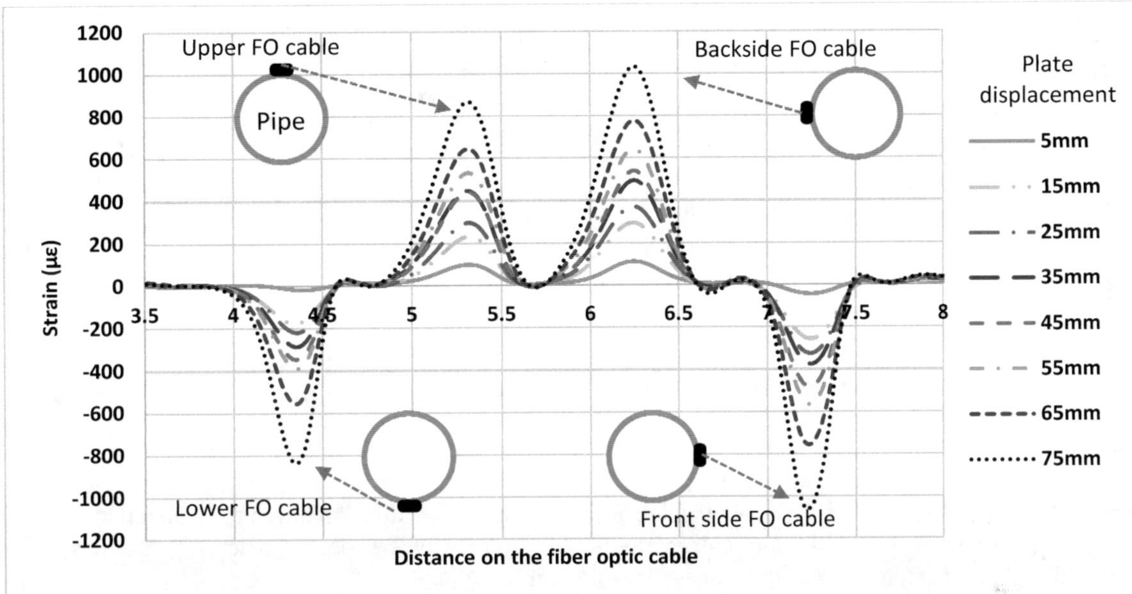

Figure 4. The strain values along the fibre optic cable at different plate displacement.

*The front side is located in active soil while the backside located in passive zone

In general, the strain values grew-up steadily along the buried pipe as the soil displacement increased. The highest strain values for all sides (top, bottom, backside, and front side) were recorded in the middle of the pipe. However, the strain values of the top and bottom sides were very close with different signals of positive and negative, respectively. It also happened to the strain values obtained from the front and backside sensors, as shown in figure 5.

Both top and backside fibre optic sensors were in the positive zone, and they had almost the same behaviour, while the bottom and front sides were in the negative area, and they had similar behaviour. Positive strains values indicate the deformation of tensile force in the upper and backward parts of the tube as a result of the movement of soil towards the front side of the pipe. In contrast, negative values indicate the compressive force at the bottom and front side of the pipe.

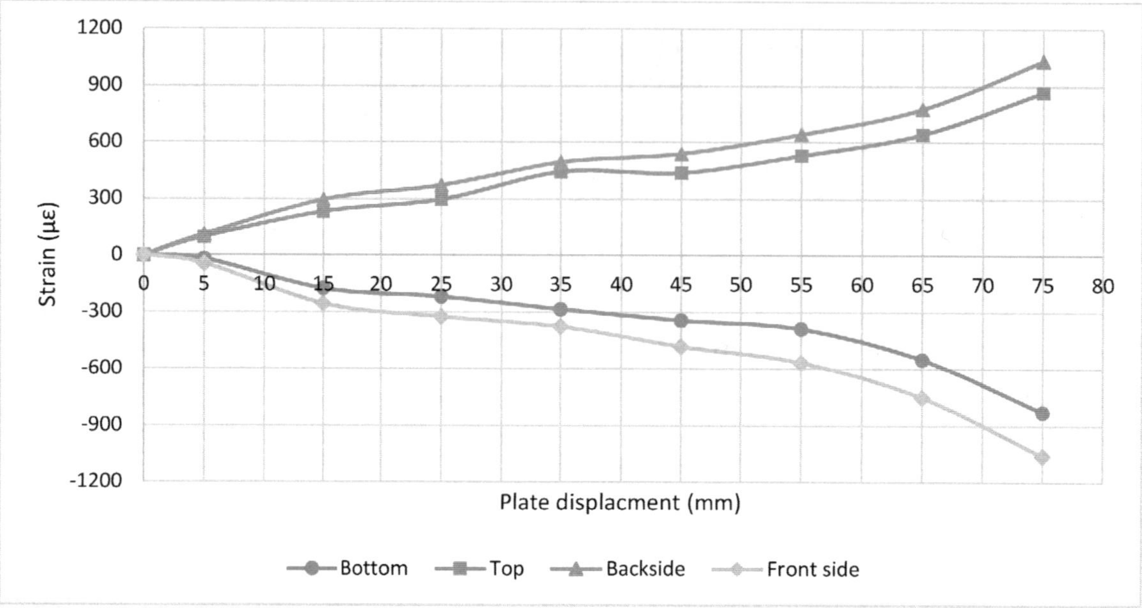

Figure 5. Strain values in the middle of the pipe.

The strain values increased as a result of increased pressure on the pipe, and pipe displacement is developed in both horizontal and vertical directions. Figure 6 illustrates the pipe position before and after running the experiment and the analysis is based on the strain values obtained from attached sensors along the pipe. After pulling the movable plate, the buried pipe was exposed to active lateral pressure due to the soil movement toward the front side of the pipe.

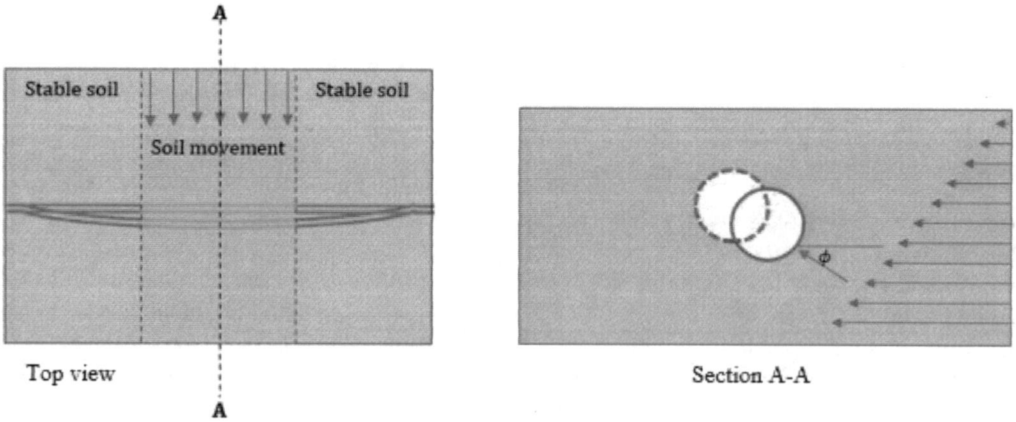

Figure 6. The Pipe position before and after running the test.

Due to the lateral soil movement towards the pipe, a curved surface was developed similar to what happened to the simple support beam when it was exposed to the external loads (figure 7). This curvature is equal to the total strain values of the two opposite directions (e.g. top and bottom) and is divided by the pipe diameter. The horizontal displacement was determined by analysing the front and backside strains while the bottom and top strains calculated the vertical displacement. Both horizontal and vertical pipe displacements (in the middle of the pipe) are calculated by integrating the curvature twice. The derivation of pipe deformation from distributed fibre optic sensors was done by analysing the strain on two fibres that placed symmetrically concerning the axis. The analysis was done by comparing the strain data between two opposing fibres (strain ε_1 and strain ε_2) (figure 7), which resulted in bending curvature. The horizontal and vertical displacement (y) of the pipe was determined based on equation (1).

$$y = \iint \frac{\varepsilon_1 - \varepsilon_2}{D} \, d^2x \qquad (1)$$

y= pipe displacement (Horizontal or vertical)

ε_1 = strain value (Bottom or front side)

ε_2 = strain value (Top or back side)

D= diameter of the pipe

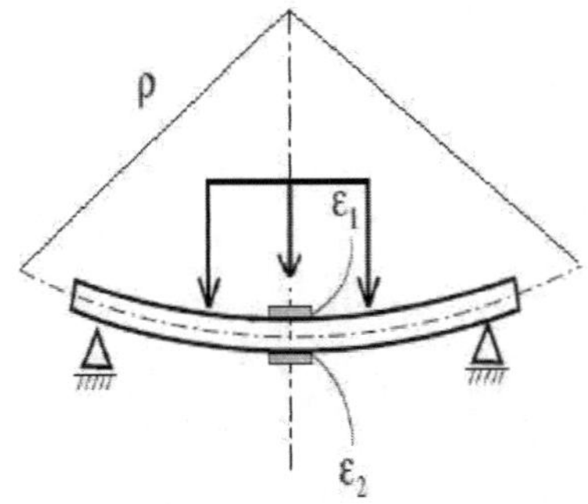

Figure 7. The plane view of the pipe.

Figure 8 illustrates the maximum horizontal and vertical pipe displacement at each stage of the plate motion. The maximum displacement values of the pipe at each stage of plate displacement were calculated and listed in table 2.

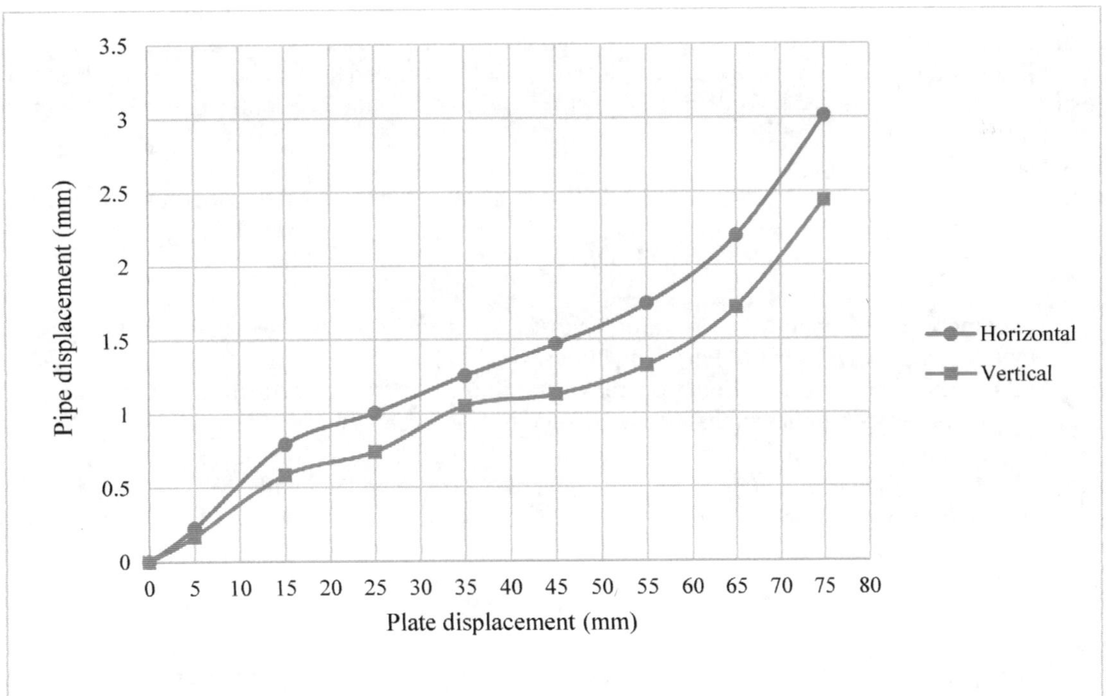

Figure 8. The horizontal and vertical displacement at middle of the pipe.

Table 2. Values of horizontal and vertical displacement.

Plate Displacement (mm)	Bottom Strain (με)	Top strain (με)	Backside strain (με)	Front side strain (με)	Pipe Vertical Displacement (mm)	Pipe Horizontal Displacement (mm)	pipe motion angle ϕ
5	-18.77	97.10	110.50	-44.50	0.167	0.223	36.83
15	-176.03	231.53	294.33	-256.33	0.587	0.793	35.5
25	-218.67	296.36	371.77	-325.18	0.742	1.004	36.46
35	-284.84	443.89	493.89	-376.65	1.049	1.254	39.93
45	-343.29	438.51	538.97	-481.47	1.126	1.469	37.45
55	-387.86	529.59	641.62	-566.66	1.321	1.740	37.2
65	-550.13	642.41	776.03	-751.55	1.717	2.200	38
75	-828.49	865.47	1032.24	-1059.31	2.439	3.012	39.01

The movable plate was pulled-out starting at 5 mm and gradually grew 10 mm up to 75 mm each. The maximum horizontal displacement of 3.01 mm was reached at 75 mm while the maximum vertical displacement of 2.44 mm achieved at the same stage of plate movement (75 mm). In a nutshell, horizontal displacements of the pipe were higher than vertical displacements at all the stages of the plate motion. In addition, noted that the vertical displacement did not exceed 84% of horizontal displacement. The displacement angle of the buried pipe ranges from 35° to 40° to the horizontal axis.

5. Conclusion

In this study, a lab experiment uses the fibre optic sensor to estimate the horizontal and vertical displacements of the buried pipeline. The findings indicate the possibility of employing the fibre optic sensor as an early warning system. The conclusions of this study are as follow:

- A comprehensive monitoring solution can be implemented by using the fibre-optic sensor to monitor the displacement of buried pipelines.
- Fibre optic sensors can help the operator to minimise the time spent in site inspection, as it pinpoints the specific locations where the pipe is moving.
- Fibre optic sensors provide continuous monitoring of pipelines deformation in real-time.
- The computation of pipeline displacement could be measured by the fibre optic sensor based on the analysis of strain measurements.
- The sensitivity of this sensor to detect the pipe displacement at a very early stage when the soil movement cannot be visually assessed.

6. References

[1] Sinha, S.K. and Knight, M.A. 2004 Intelligent system for condition monitoring of underground pipelines. *Computer-Aided Civil Infrastructure Eng.* **19(1)** 42–53.
[2] American Lifelines Alliance-ASCE 2005 *Guidelines for the design of buried steel pipe*. American lifelines alliance USA.
[3] EN 2006 *Eurocode 8: design of structures for earthquake resistance—part 4: silos, tanks and pipelines.* European Committee for Standard UK.
[4] Inaudi, D. and Glisic, B. 2010 Long-range pipeline monitoring by distributed fiber optic sensing. *J. Press. Vessel Technol.* **132(1)** 1–9.

[5] Mohamad, H. 2008 *Distributed optical fibre strain sensing of geotechnical structures*, University of Cambridge.
[6] Simpson, B., Hoult, N.A. and Moore, I.D. 2015 Distributed sensing of circumferential strain using fiber optics during full-scale buried pipe experiments *J. Pipeline Syst. Eng.* **6**(**4**) 1–10.
[7] Frings, J. 2011 Enhanced pipeline monitoring with fiber optic sensors. *In Proc. of the 6th Pipeline Technology Conf.* **45** 1-12.
[8] Ni, P., Moore, I.D. and Take, W.A. 2018 Distributed fibre optic sensing of strains on buried full-scale PVC pipelines crossing a normal fault *Géotechnique* **68(1)** 1–17.
[9] Ravet, F., Zou, L., Bao, X., Chen, L., Huang, R.F. and Khoo, H.A. 2006 Detection of buckling in steel pipeline and column by the distributed brillouin sensor *Optical Fiber Technol.* **12**(**4**) 305–11.
[10] Glisic, B. and Yao, Y. 2012 Fiber optic method for health assessment of pipelines subjected to earthquake-induced ground movement *Structural Health Monitoring* **11(6)** 696–711.
[11] Alarifi, H., Mohamad, H., Isah, B.W. and Wahab, M.A. 2019 Physical modelling of lateral soil displacement and pipeline movement due to landslide. *Proc. of Int. Conf. on Landslide and Slope Stability (SLOPE 2019)*, Bali, 1–9.

Acknowledgement
The first author is thankful to Universiti Teknologi PETRONAS (UTP) for the financial support given through the graduate assistantship scheme.

Shear failure of the Meraka hardwood in bolted connections loaded parallel to the timber grain

Abdul Razak Abdul Karim[1*], Pierre Quenneville[2] and Norazzlina M. Sa'don[1]

[1] Faculty of Engineering, Universiti Malaysia Sarawak, 94300 Kota Samarahan, Sarawak, Malaysia.
[2] Department of Civil and Environmental Engineering, The University of Auckland, Auckland, New Zealand.
*Corresponding author email: akarazak@unimas.my

Abstract. This current study was conducted to establish a bolted connection experimental database on Malaysia woods. The effort of improving the retrofit guideline in designing the wall-diaphragm connections of masonry buildings can be continuously done. Brittle failure verifications on the Meraka hardwood to verify the occurrence of row shear failure are presented in this paper. The hardwood species was selected due to its common use as structural rafters and joists in the construction of roof and floor diaphragms of masonry buildings, respectively. Ten groups of a single row of steel-wood-steel (SWS) connections loaded parallel to the timber grain were tested. Each group was prepared with a total of ten replicates of specimens. Comparisons of bolted connection strength between the experimental results and the predictions of the existing design equations are discussed. It was found that the timber standard of Malaysia (MS544) too conservative in estimating the strength, whilst the Row Shear Model (RSM) is more comparative.

1. Introduction

Quenneville [1] stated that the community of the international timber engineering has agreed in principle that the bolted connection capacity determination in design standards should be based on the models that are capable to identify possible failure modes of the bolted connections. Both modes of bearing and fracture wood failures that exhibit ductile and brittle, respectively, are to be considered. Any of these failure modes provide the lowest strength should be the one that govern the connection performance. To review the existing design equations in estimating the timber bolted connection strength loaded parallel to the grain, the descriptions of the timber code of Malaysia and the Row Shear Model (RSM) are given in the following paragraphs.

In accordance to the Malaysian timber standard [2], the permissible load (F_{adm}) of a laterally loaded bolted connection system can be referred to the Section 11.2.3 of the document and as per given below.

$$F_{adm} = k_1 k_2 k_{16} k_{17} F \quad (1)$$

where,

k_1 = duration of load factor, see Table 4 of [2];
k_2 = i) dry timber is equal to 1.0 or ii) wet timber is equal to 0.7;
k_{16} = i) bolts that transfer load through metal side plates of adequate strength and the bolts are a close fit to the holes in these plates provided that b/d > 5 (where b

denotes the effective timber thickness and d is the bolt diameter) is equal to 1.25 or ii) otherwise is equal to 1.0;

k_{17} = factor of multiple bolted joints, refer Table 15 of [1]; and

F = basic working load, given in Section 11.2.2 of [1].

Referring to table 12 of [2], the basic working load values, F, are estimated based on the selection of the size of bolt diameter, d, and the effective thickness of timber, b. This clearly shows that the capacity of the timber bolted connection is predicted based on the ductile behaviour. The brittle behaviour is only considered by taking into account the value of the factor of multiple bolted joints less than 1, $k_{17} < 1$. The bolted connection is assumed to be brittle if only consisting more or equal to five fasteners.

From many published research works [1]-[6], it can be identified that the timber bolted connections can be failed in brittle behaviour. Figure 1 shows the illustrations of the brittle modes of failures observed in bolted timber connections loaded parallel to the timber grain. [6] stated that a set of equations was developed by Quenneville [1] to predict the timber bolted ultimate strength based on the observed failure modes during tests such as group tear out, row shear, splitting and net tension. [1] describes each of the equations representing the failure mode in details. Due to the study test specimens were designed as a single row bolted connection type, this gives the row shear failure to be the critical brittle mode to occur on the bolted joints compared to the aforementioned failure modes. Thus, this paper only describes in details the Row Shear Model.

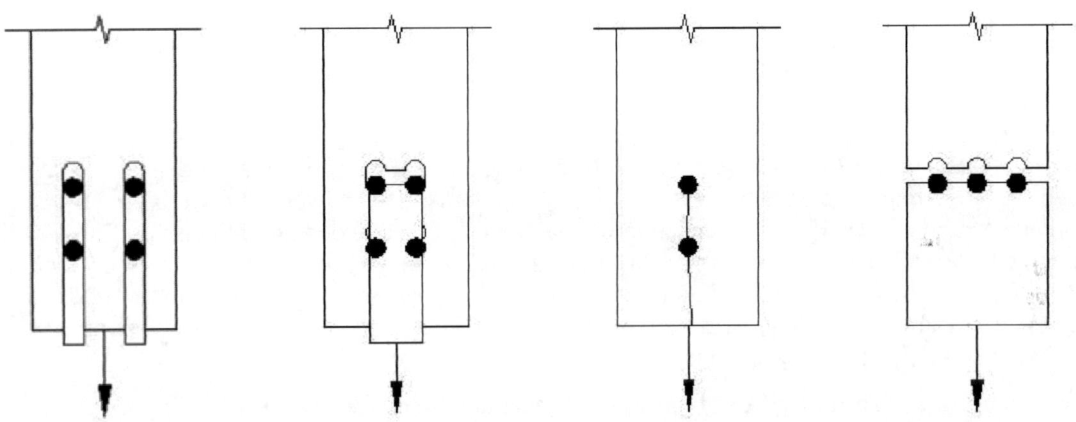

(a) row shear (b) group tear-out (c) splitting (d) net tension

Figure 1. Brittle failure modes for timber bolted connections loaded parallel-to-grain [1].

The capacity of a group of dowel fasteners that fail in row shear is calculated using the equation given by [1] and the details are provided as follows:

$$R_{r\,rs} = RS_{i\,min} n_r \tag{2}$$

where,

n_r = number of rows in the joint as per load component;

$RS_{i\,min}$ = minimum (RS_1, RS_2, …, RS_{nr}), in N;

RS_i = shear capacity along two shear planes of fastener row "i", in N;

$$= \frac{2(f_v)K_{ls}\, t\, n_{fi}\, a_{cr\,i}}{CF};$$

f_v = member shear strength, in MPa, equal to $17.8G^{1.24}$;

G = 5th percentile relative density of timber in the oven dry condition;

K_{ls} = factor for member loaded surfaces (0.65 for side member, 1 for internal member);

t = member thickness, in mm;

n_{fi} = number of fasteners in row "i";

IOP Conf. Series: Materials Science and Engineering 1101 (2021) 012005 doi:10.1088/1757-899X/1101/1/012005

$a_{cr\,i}$ = minimum of e_t and s_b for row "i" (see Figure 2), in mm; and
CF = calibration factor.

The expression relating shear strength to specific gravity of hardwoods (i.e. $f_v = 17.8G^{1.24}$), which was adopted from [7], was used to calculate the shear strength value for the Meraka hardwoods. In the results and discussion part of this paper, the standard structural value of 1.61 MPa of the shear strength parallel to the timber grain given in table 4 of [8] was also taken into consideration and adopted to the RSM equation for the purpose of comparisons. Thus, the use of the specific gravity to shear strength relationship provided by [7] can be examined.

Due to the initiation of the improvement of the retrofit guideline in designing the wall-diaphragm connections of masonry buildings, an establishment of experimental database of bolted connections on Malaysia woods is essential to be performed. This database is very useful for the evaluation of the existing design equations in predicting the strength of timber bolted connections. The effectiveness verification of the design equations discussed above in designing the connections can be made. Because of the main objective is for retrofitting the connection between the masonry wall and timber diaphragm of masonry buildings, the common species of woods that are used in the construction of the structural members of the roof and floor of diaphragms is selected. By reviewing the published document of [9], it can be found that the Meraka species is highly potential to be used as beams or joists as per explained in the next section of this paper.

2. Experimental Programme

2.1. Materials used

The Shorea albida is referring to the botanical name of the Meraka hardwood, which is known in Peninsular Malaysia as Red Balau and in Brunei as Alan Batu [10]. The common name in Sarawak is known as Alan Batu, whilst other names known as Meraka, Sengawan, and Seringawan [11]. In accordance with table 3 of [2] the Meraka hardwood is categorized under strength group of SG3, which is suitable for the application of beams or joists as per given by table 1 that extracted from the document of [9]. Table 1 describes the specific strength groups of timber and their applications. Based on the findings stated above, the Meraka hardwood can be typically used to construct the floor and roof diaphragms in the existing masonry buildings.

Table 1. Strength groups of timber and their applications [9].

Strength Group	Applications
	Structural Components
• SG1, SG2, SG3 AND SG4	• Columns, beams, bearer, studs, joists, ties and struts
• SG1, SG2, SG3, SG4 AND SG5	• Form work
	Roofing
• SG1, SG2, SG3 AND SG4	• Rafters, ties, struts, purlins and bracing
• SG1, SG2, SG3, SG4 AND SG5	• Battens

The cross section of 50 mm × 100 mm, representing the typical size of roof rafters, was used for the preparation of wood specimens for the bolted connection tests. To avoid the threaded part of the bolt to be in contact with the timber specimens, a total of 127 mm bolt length (with 85 mm of shank length) was chosen. The bolts used were a mild steel of 4.6 grade with a 13 mm of bolt diameter size. The number of side plates used for each extremity was two in total, whereas the thickness was 15 mm with 400 MPa of ultimate tensile strength, f_{up}. This thickness of steel plate was chosen due to avoid a bearing failure of the plate. Therefore, the strength of the wood can be evaluated by ensuring the plate is intact and does not fail before the wood element.

2.2. Specimen configuration details

The bolted connection test specimens were prepared as a steel-wood-steel configuration. In order to remove the eccentricity influence on the connection when the specimens loaded in tension during testing, this three-member connection or double shear joint was selected. A total of ten different groups of connections with at least ten replicates of each group was tested. An identical bolted connection at both ends of each specimen was tested to ensure the desired connection is the one that failed. A minimum distance of 400 mm between connection at both extremities was complying with [12] that is 30 times the diameter of the fastener shank, d. This is to ensure both connections at both ends of specimens are two independent joints during testing. The number of fasteners, n_f, for all groups of connections was varied from one two three bolts, but with only one fastener row, n_r. Thus, the results presented in this present work are for double shear joints with a single fastener row that consisted of up to three bolts in numbers. All groups of connections were also designed to have a minimum of 50 mm of either their bolt spacing (s_b) or size of end distance (e_t). This is purposely done to induce the row shear failure on all specimens, so that the number of observations of this type of failure mechanisms can be maximized. Comparisons between the experimental results and the connection strength values predicted by both MS544 and RSM can be done. The details of variables used to prepare the specimens of the bolted connection test are illustrated in figure 2.

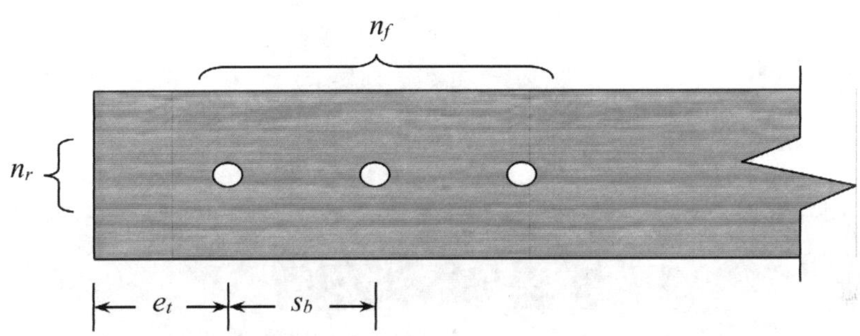

Figure 2. Variables of timber bolted connections.

2.3. Testing methods

A Shimadzu universal testing machine with a 300 kN maximum capacity was utilized in performing the timber bolted connection tests. Tensile loading parallel to the timber grain was applied to all bolted connection specimens up to the ultimate strength of any of connection ends. All bolts were fastened with a finger-tight force for ensuring a self-alignment of the test specimens was permitted. This is also to avoid the occurrences of friction forces between the head or nut of bolts and the steel plate. A monotonic tension load at a displacement-control rate of 0.015 mm per second was applied to each connection specimen through the steel side plates until failure of the wood specimen occurred [13]. The break point of each specimen can be clearly seen from the plotted graph shown in the computer monitor during testing when the load dropped with no recovery. The load-displacement data collected from the data logger was used to plot the load versus displacement graph and the maximum load recorded was determined as the ultimate capacity of the tested connection. Each specimen type of failure observed after testing was also recorded. From figure 3, one can see that the typical setting up of the bolted timber connection tests.

The basic parameter of the Meraka hardwood, which is the dry density, was determined by conducting the wood density tests because it is essential for the prediction of the strength values of the bolted connection using the Row Shear Model equation. Due to the effect of the wood moisture content on the ultimate resistance of the bolted connection, a monitoring of the water content was performed by conducting the wood moisture content testing for all specimens. The procedures conducted for both density and moisture content tests are in compliance with the techniques outlined in the Australian/New Zealand Standard of [14] and [15], respectively. As per recommendation of the testing standards, all samples for the density and moisture content tests were obtained by extracting the tested timber specimens of the bolted connection experiments.

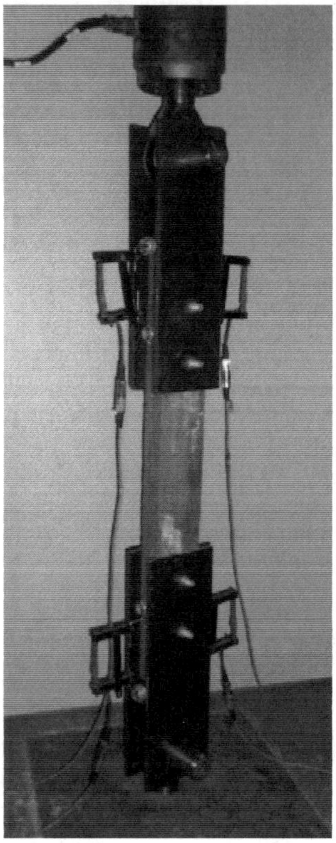

Figure 3. Bolted connection test typical setup.

3. Results and Discussion

3.1. Basic timber properties

From the results of the wood density tests conducted, both the average and the 5th percentile values of the dry density of the Meraka hardwood were calculated. The average dry density is designated as ρ_{avg}, meanwhile the 5th percentile dry density is labelled as $\rho_{5th\%}$. Refer to table 2 for the values of the dry density. One should note that the 5th percentile value given in the table was computed by assuming a normal data distribution. The density of Meraka obtained from the test is found to be satisfactorily comparable with the data given by [11], which is 732 kg/m^3 at 18% moisture content. The moisture content results gathered give an average value of 18%, which is less than 19%, considers as dry condition as per stated in the timber standard of [2].

Table 2. Density of Meraka hardwood.

Species	ρ_{avg} [a] (kg/m^3)	CoV [b]	$\rho_{5th\%}$ [c] (kg/m^3)
Meraka	696	15.0	524

[a] Average dry density.
[b] Coefficient of variations.
[c] 5th percentile dry density.

IOP Conf. Series: Materials Science and Engineering 1101 (2021) 012005 doi:10.1088/1757-899X/1101/1/012005

3.2. Observed failure mode

Referring to the mode of failure observed on the wood specimens of the bolted connection testing, the row shear failure parallel to the timber grain was found to be dominant as shown in figure 4. This mode of failure obviously can be seen due to the minimum values between the end distance and bolt spacing. Thus, all specimens of connections designed with the end distance or bolt spacing of 50 mm to fail predominantly in this row shear failure mode. Many published experimental works [5], [6], and [16] have reported a similar finding to the present observation. From both figures 4(a) and 4(b), an enlargement of the hole diameter is not seen on the tested specimens that are failing in the row shear. Also, a bending failure is not visible on the steel bolts or fasteners. Thus, any wood bolted joints with 50 mm and less, either its end distance or bolt spacing, can be expected to fail in brittle mode.

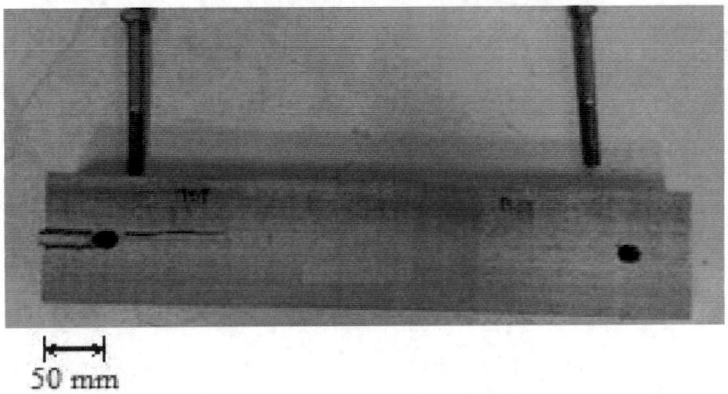

(a) end distance equal to 50 mm

(b) spacing between bolts equal to 50 mm

Figure 4. Tested specimen failed in row shear of wood.

From figure 5(a), it shows the typical load versus displacement curves of bolted timber connections that are illustrating the brittle failure mode. The similar figure also shows that a clear sudden drop of failure load at a low displacement value was observed for the brittle failure of wood specimens. In comparison to the ductile failure of wood specimens shown in figure 5(b), it can be noticeably seen that the development of strain hardening of the bolted timber connection system is established. The occurrence of the plateau of the load-displacement curves with high displacement values are obviously visible in the graph of bolted connections failed in ductile. These describe the sequences of the ductile failure of the wood specimens that are initiated by the first mode of wood bearing failure due to in contact with the fasteners that causing the enlargement of the size of the hole diameter in wood. The second mode of the bending failure of the steel bolts is then followed as the embedding strength of the timber exceeded the bending resistance of the steel fasteners. When the bending capacity of the steel bolts exceeding the embedding strength of the wood, the third mode of the secondary brittle failure of the wood is expected to be seen. Figure 6 shows the tested specimen with 100 mm end distance that failed in ductile behaviour, which all sequences of failures can be clearly seen in the specimen.

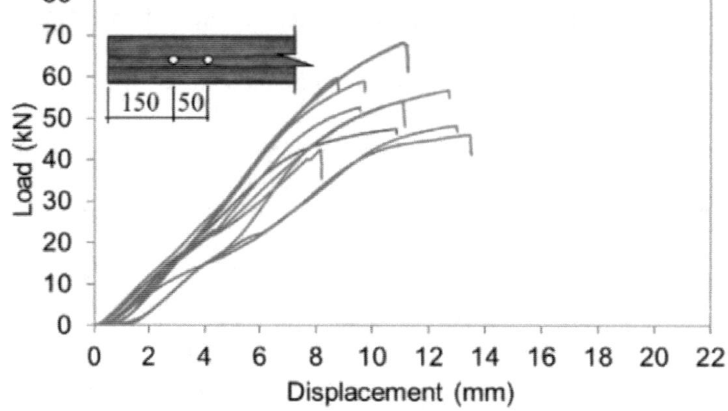

(a) brittle behaviour

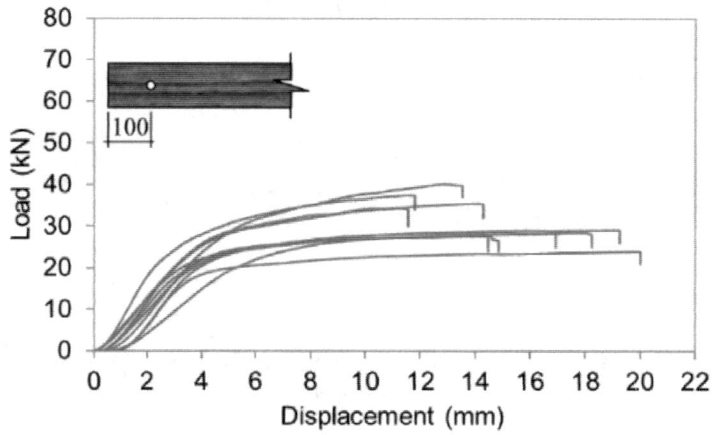

b) ductile behaviour

Figure 5. Typical load versus displacement profiles.

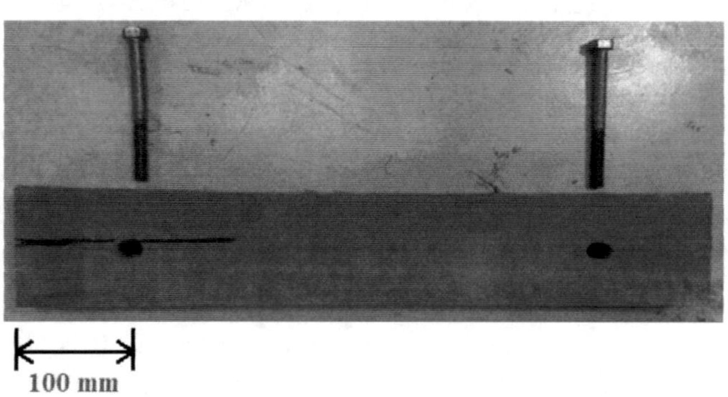

Figure 6. Tested specimen failed in ductile behaviour.

3.3. *Ultimate strength of connections*

From the timber bolted joint test results obtained (refer table 3), it can be seen that the ultimate strength of the connections was influenced by the increment of the end distance. Generally, the ultimate strength increases when the end distance increases. This can be seen by comparing the 5th percentile experimental values of Groups 8 and 9 with Group 10. An increment of the end distance from 50% to 100% may resulting the ultimate strength to increase from about 4% to 17%. In addition, by comparing the 5th percentile experimental values of Groups 3, 4, 5 and 6 with Group 7, the ultimate strength values were increased from about 11% to 28%, except a reduction of 20% of ultimate strength was occurred in Group 4. From this, it can be observed that even if the end distance was increased the ultimate strength may also be reduced due to the bolt spacing of 50 mm. By comparing Group 2 and Group 7, in order to see the effect of bolt spacing, the 5th percentile experimental value of Group 2 is found to increase about 18% with the spacing between bolts was doubled from 50 mm to 100 mm. From the discussion made above, the geometrical configuration parameters of both the end distance and the bolt spacing are significantly affecting the ultimate strength of the bolted connection strength. Thus, the consideration of the minimum measurement between the end distance and the bolt spacing may cause a decrement of the ultimate strength as per recommended by the Row Shear Model, RSM. In other word, the minimum value between the end distance and the bolt spacing is the interdependent parameter to control the ultimate strength capacity of the connection with more than one bolt.

Table 3. Strength predictions versus experimental results.

						Predictions			Experimental			Ratio		
Group	d (mm)	e_t (mm)	s_b (mm)	n_r	N[a]	*RSM*[b] (kN)	*MS544* (kN)	*RSMa*[c] (kN)	R_{avg} (kN)	*COV* (%)	$R_{5th\%}$ (kN)	*RSM* / $R_{5th\%}$	*MS544* / $R_{5th\%}$	*RSMa* / $R_{5th\%}$
(1)	(2)	(3)	(4)	(5)	(6)	(7)	(8)	(9)	(10)	(11)	(12)	(13)	(14)	(15)
1	13	50	-	1	1	16	8.2	3	26	11	21	0.77	0.40	0.16
2	13	50	100	1	2	32	16.4	6	53	11	43	0.74	0.38	0.15
3	13	150	50	1	2	32	16.4	6	53	14	41	0.78	0.40	0.16
4	13	125	50	1	2	32	16.4	6	47	23	29	1.10	0.56	0.22
5	13	100	50	1	2	32	16.4	6	50	9	43	0.75	0.38	0.15
6	13	75	50	1	2	32	16.4	6	55	9	47	0.68	0.35	0.14
7	13	50	50	1	2	32	16.4	6	48	14	37	0.87	0.45	0.18
8	13	100	50	1	3	48	24.6	10	79	6	71	0.67	0.35	0.14
9	13	75	50	1	3	48	24.6	10	71	7	63	0.75	0.39	0.15
10	13	50	50	1	3	48	24.6	10	65	4	61	0.79	0.40	0.16

[a] Number of fasteners
[b] Calculated using $f_v = 17.8G^{1.24}$ [7]
[c] Estimated using f_v = shear strength parallel to grain value given in Table 4 of [8]

For the effect of number of bolts on the ultimate strength or capacity of connections, the 5th percentile of experimental strength values was determined to be increased with an increment of the number of bolts. These can be seen when comparing the followings: (i) Group 7 and Group 1 (+76%); (ii) Group 8 and Group 5 (+65%); (iii) Group 9 and Group 6 (+34%); and (iv) Group 10 and Group 7 (+65%). For the multiple bolted joints, it should be noted that the bolt spacing configuration can be as small as 50 mm. This is not only causing a decrement of the ultimate strength, but it can induce a brittle failure mode of the connection as stated in Section 3.2 of this paper (i.e. All tested bolted connection specimens with 50 mm bolt spacing in the present experimental study are primarily failed in row shear failure). As mentioned previously in the third paragraph of Section 1 of this article, the brittle behaviour is only considered in the design code for the multiple bolted joints consisting of more or equal to five fasteners. This is clearly an inaccurate assumption as the brittle failure occurrence can be caused by the minimum value between the end distance and bolt spacing configurations of the bolted connections as observed in this present work. This is in agreement with many published works [3]-[6], and [16].

3.4. Comparison between experimental results and prediction values

Referring to columns 7 to 9 and columns 10 to 12 of table 3, a comparison of the bolted connection strength between the prediction from the existing design equations and the experimental data can be seen, respectively. Column 7 shows the predicted 5th percentile strength values estimated by the Row Shear Model, designated as RSM, using equation 2. The readers must note that the predicted RSM values were calculated by adopting the expression relating shear strength to specific gravity of hardwoods given in the sixth paragraph of Section 1 of this article. The 5th percentile strength values predicted using [2], labelled as MS544, are presented in column 8. By applying the standard structural value of 1.61 MPa of the shear strength parallel to the timber grain taken from Table 4 of [8] into the Row Shear Model equation, designated as RSMa, the estimated 5th percentile strength values are given in column 9. The average strength values of experimental results, R_{avg}, and the 5th percentile strength values of the test results, $R_{5th\%}$, are shown in columns 10 and 12, respectively. Column 11 gives the coefficient of variation values of the test data. Note that the $R_{5th\%}$ values were calculated by assuming a normal distribution of statistical data analysis. To measure the difference of the strength prediction values given by the RSM, MS544, and RSMa in comparison to the experimental results, columns 13, 14 and 15 show the calculated ratio of different, respectively.

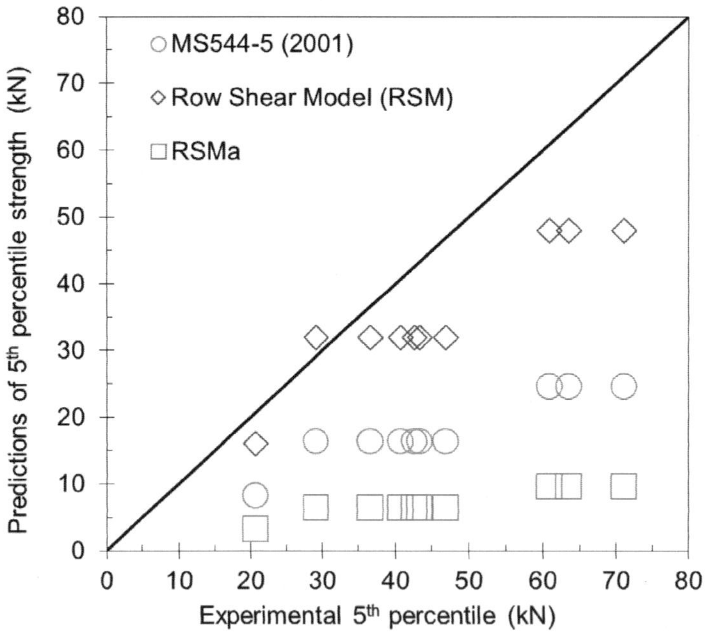

Figure 7. Effectiveness of MS544 and Row Shear Model in predicting the strength of bolted connections.

In order to see clearly the effectiveness of the existing design equations compared to the test results, a graph of predictions versus experimental data is plotted (see figure 7). The strength predictions values of the existing design equations are considered to be safe if plotted below the 45° line. However, any predicted value that found too far below the 45° line is identified to be too conservative. From figure 7, it can be verified that the current standard of [2] provides the most conservative prediction of design strength values with ratio varies from 0.35 to 0.56. By adopting the standard structural value of 1.61 MPa of the shear strength parallel to the timber grain taken from table 4 of [8] to RSM equations, the strength prediction values, RSMa, become an unacceptably under-designed the connection capacity with a range of ratio from 0.14 to 0.22. This can cause the design output to be over-designed and not economical as more fabrications of steel materials are required due to either the number of fasteners or its diameter is needed to be increased. Therefore, this can make the strengthening design for wall-diaphragm connections in masonry buildings to be uneconomical as the design output produces bigger diameter of steel bolts to withstand the demand force. By using the Row Shear Model equation, RSM, a better prediction of connection strength can be achieved. The ratio given by the RSM is a range from 0.67 to 1.10. From a comparison of prediction values between RSM and RSMa, it can be seen that the use of the correlation relating shear strength to specific gravity of hardwoods (i.e. $f_v = 17.8G^{1.24}$) adopted from the [7] was found to be more reliable in estimating the shear strength parallel to the timber grain of the Meraka hardwood compared to the standard structural value of the shear strength taken from table 4 of [8]. In overall, from the evaluations made above, the use of the existing Malaysian timber standard and Row Shear Model to estimate the bolted connection strength in Meraka hardwood was verified. The use of the Row Shear Model equation in estimating the design strength of the bolted connection is found to be more comparative, meanwhile the predictions given by [2] and [8] are too conservative.

4. Conclusions

From the present findings, the brittle failure occurrence was observed for all connection specimens with either the end distance or bolt spacing of 50 mm and less. Besides the number of bolts, the minimum values of both end distance and bolt spacing was found to be dominantly affecting the ultimate strength of timber bolted connections. The efficacy of the strength predictions by both the existing Malaysian wood standard and the Row Shear Model was also assessed. The code provides too conservative values of design strength with an average of 41% of effectiveness in comparison to the experimental 5th percentile results. By adopting the standard structural value taken from table 4 of [8] to the Row Shear Model equation, it even reduces the design efficacy to an average of 16%. This makes the design outputs to be uneconomical due to more cost is required for steel fabrication as an increment of size of the steel bolt diameter or an addition of bolt numbers to be used in the connection design. Better predictions of design strength values of the bolted connection can be achieved using the Row Shear Model equation. It provides an average effectiveness of 79% in comparison to the experimental 5th percentile strength results. Therefore, the Row Shear Model equation is highly recommended to complement the existing timber design code in the bolted connection design.

5. References

[1] Quenneville, P. 2009 Design of bolted connections: A comparison of a proposal and various existing standards *J. Struct. Eng.* **22(2)** 57-62.

[2] Malaysian Standard 2001 MS 544: Part 2: 2001 Code of Practice for Structural Use of Timber: Part 2: Permissible Stress Design of Solid Timber (First Revision), Department of Standards Malaysia.

[3] Quenneville, J.H.P., Smith, I., Aziz, A., Snow, M. and Ing, H.C. 2006 Generalised Canadian approach for design of connections with dowel fasteners, *CIB-W18 meeting Proceedings*, paper CIB-W18/39-7-6, Florence, Italy.

[4] Quenneville, P. and Bickerdike, M. 2006 Effective in-row capacity of multiple-fastener connections, *CIB-W18 meeting Proceedings*, paper CIB-W18/39-7-1, Florence, Italy.

[5] Mohammad, M. and Quenneville, J.H.P. 2001 Bolted wood-steel and wood-steel-wood connections: Verification of a new design approach *Can. J. Civ. Eng.* **28** 254-63.

[6] Quenneville, J.H.P. and Mohammad, M. 2000 On the failure modes and strength of steel-wood-steel bolted timber connections loaded parallel-to-grain *Can. J. Civ. Eng.* **27** 761-73.

[7] Forest Products Laboratory 1987 Wood Handbook: Wood as an Engineering Material, Forest Service U.S. Department of Agricultural, Washington, D.C.

IOP Conf. Series: Materials Science and Engineering 1101 (2021) 012005 doi:10.1088/1757-899X/1101/1/012005

[8] Malaysian Standard 2001 MS 544: Part 2: 2001 Code of Practice for Structural Use of Timber: Part 2: Permissible Stress Design of Solid Timber (First Revision), Department of Standards Malaysia.
[9] Malaysian Timber Industry Board 2013 Wood Wise: Strength Groups of Timber and Their Applications. http://www.mtib.gov.my/index.php [Retrieved on 29/04/13].
[10] Malaysia Timber Council 2013 MTC Wood Wizard: Malaysian Timbers at Your Fingertips. http://www.mtc.com.my/resources-WoodWizard.php [Retrieved on 31/01/13].
[11] Forest Department Sarawak 1999 Handbook of Some Sarawak Timbers. Forest Department Sarawak, Kuching, Sarawak, Malaysia.
[12] International Organization for Standardization 2008 Draft International Standard ISO/DIS 10984-2 Timber structures - Dowel-type fasteners - Part 2: Determination of embedding strength and foundation values. International Organization for Standardization, Geneva, Switzerland.
[13] International Organization for Standardization 2003 International Standard ISO 16670 Timber structures – Joints made with mechanical fasteners – Quasi-static reversed-cyclic test method. International Organization for Standardization, Geneva, Switzerland.
[14] Australian/New Zealand Standard 2000 AS/NZS 1080.3: 2000 Timber - Methods of Test - Method 3: Density. Standards New Zealand (electronic copy).
[15] Australian/New Zealand Standard 1997 AS/NZS 1080.1: 1997 Timber - Methods of Test - Method 1: Moisture content. Standards New Zealand (electronic copy).
[16] Abdul Karim, A.R., Quenneville, P. M.Sa'don, N. and Ingham, J. 2013 Assessment Guidelines of Wall-Diaphragm Connections for Masonry Buildings. *New Developments in Structural Engineering & Construction Volume II*, Eds. S. Yazdani and A. Singh, Research Publishing, Singapore.

Acknowledgments
The special thanks from the authors are conveyed to the Research Innovation and Enterprise Centre of Universiti Malaysia Sarawak (UNIMAS) for providing a financial support to this project under the Special Short Term Grant of F02/SpSTG/1372/16/14. Many thanks to all team members, especially to Nazmi Ainuddin Johan, Kiu Pey Ing, Ng Sin Rui, Goh Yan Man, Hii Kai Ling and Chai Chin Wen, for their sincerity and perseverance in ensuring the completion of this research project.

IOP Conf. Series: Materials Science and Engineering 1101 (2021) 012006 doi:10.1088/1757-899X/1101/1/012006

Sri Aman peat: Settlement observation and geotechnical properties investigation

Norazzlina M. Sa'don[1*], Abdul Razak Abdul Karim[1], Siti Noor Linda Taib[1], Edward Anak Muol[2], Benson Ling Jin Yaw[2] and Angela Ee[2]

[1]Faculty of Engineering, Universiti Malaysia Sarawak (UNIMAS), 94300 Kota Samarahan, Sarawak, Malaysia.
[2]Department of Minerals & Geoscience (Sarawak), Jalan Wan Abdul Rahman, Kenyalang Park, 93712 Kuching, Sarawak, Malaysia.
[*]Corresponding author email: msazzlin@unimas.my

Abstract. This paper presents the findings from the field observation and site exploration for construction on peat in Sri Aman, Sarawak, Malaysia. A visit to Balai Bomba and Pejabat Metrologi in Sri Aman has been done in August 2019. From the observation and measurement conducted on the two locations of the constructed area on peat shows the settlement, δ recorded ranges from 100 mm to 150 mm. A field sampling for the determination of geotechnical properties of peat has been done in Balai Bomba Sri Aman. The depth of the peat in the area is about 2.88 m and the results show that the peat has high natural moisture content ranges from 900 % to 1400 %. The organic content (OC) for the site is in the range of 70 %-90 % for a depth of 0.5 m to 2.88 m, and categorized as H3, fibrous peat except for the first 0.5 m is 63 % which fall under H5, hemic peat group according to Von Post classification.

1. Introduction

Peat is extremely soft and defined as a non-homogeneous deposit of partially decomposed vegetative matter saturated with water and encountered in low-lying areas where the water table is near or above the ground surface [1]. High moisture content in peat is the main factor contributing to the characterization of peat with low bearing capacity and high compressibility behaviour. In Sarawak, about 90 percent of the peat is classified as deep peat with a depth of more than 1.5 meters and the depth of peat layers increases from the coast towards the inland [2]. The content of peat differs from location to location due to the factor such as origin fiber, temperature and humidity [3]. Peat can be classified in accordance to Von Post Classification and fiber content [4]. The Von Post 10 degree of humidification is based on the appearance of peat water upon squeezing by hand. The Von Post scale ranging from H1 (least decomposed-very fibrous) to H10 (most decomposed-very few fibers). Table 1 and figure 1 summarized the classification of peat type based on [5] degree of humification and fiber content to ease the understanding of the reader.

In general, peat exhibits an immediate 'elastic' settlement as soon as it is loaded, and a 'consolidation' settlement thereafter. The compression behavior of peat varies from the compression behavior of other types of soils in two ways. First, the compression of peat is much larger than that of other soils. Secondly, the creep portion of settlement plays a more significant role in determining the

Table 1. Classification of peat [4], [6], [7]

Von Post Scale	Fibre Content (ASTM D1997)	Type	Peat Microstructure	Description
$H_1 - H_3$	Peat with greater than 67% fibers	Fibric (Fibrous)	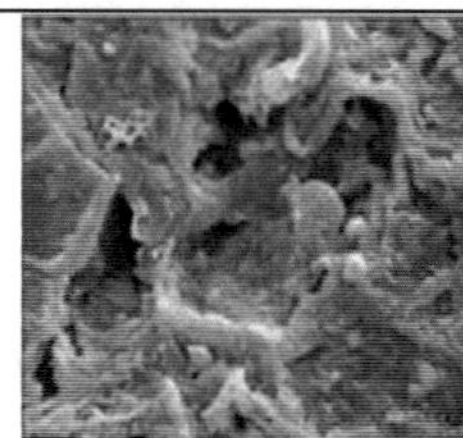	Least decomposed; recognizable plant structure and extremely acidic
$H_4 - H_6$	Peat with between 33% - 67% fibers	Hemic (Semi-fibrous)		Intermediate decomposed between fibrous and sapric peat
$H_7 - H_{10}$	Peat with Less than 33% fibers	Sapric/ (Amorphous)		Highly decomposed with original plant fibers have mostly disappeared. Very dark grey to black in colour

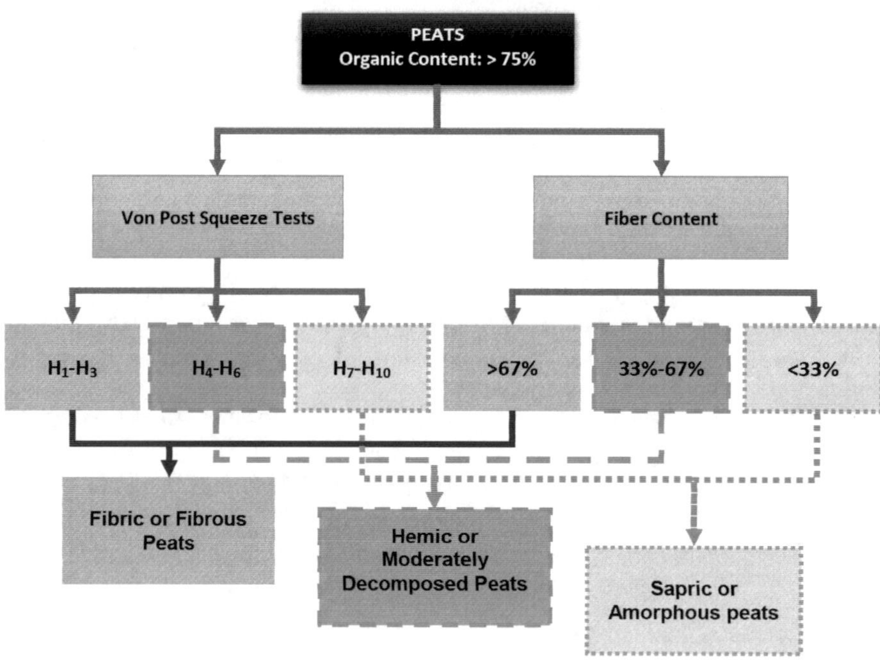

Figure 1. Flowchart for determination of peat type.

total settlement of peat than of other soil types [8]. The compression and settlement processes may take a considerably longer time and it increases when the ground water level decreases, thus this high compressibility of peat and their potential for long creep have posed difficulties for engineers, which results in a large deformation. A study by [9] showed that the secondary compression is prominent in peat deposits because it exists at a high void ratio, which exhibit high values of Cc and displays a high ratio of Cα/Cc. Table 2 presents the summary of the natural moisture content with the relevant Cα/Cc values for different peat deposits [9]-[10].

Table 2. Summary of natural moisture content and Cα/Cc for peat deposits [10].

Peat	Natural Moisture Content	Cα/Cc	Reference
Fibrous peat	850	0.060-0.100	Hanrahan (1954)
Peat	520	0.061-0.078	Lewis (1956)
Armorphous & fibrous peat	500 -1500	0.035-0.083	Lea and Brawner (1963)
Canadian muskeg	200-600	0.090-0.100	Adam (1965)
Amorphous to fibrous peat	705	0.073-0.091	Keene and Zawodniak (1968)
Peat	400-700	0.075-0.085	Weber (1969)
Fibrous peat	605-1290	0.052-0.072	Samson and LaRochell (1972)
Fibrous peat	613-886	0.060-0.085	Berry and Vickers (1975)
Amorphous to fibrous peat	600	0.042-0.083	Edil and Dhowian (1981)
Fibrous peat	660-1590	0.060	Lefebvre et al. (1984)
Dutch peat	370	0.060	Den Haan (1994)
Fibrous peat	610-850	0.052	Karunawardena (2007)

This paper briefly presents the observation of the settlement for structures/infrastructures constructed on peat in Sri Aman, Sarawak. A field sampling has been done in Balai Bomba, Sri Aman for basic properties determination of the site. The tests have been performed in Geotechnical Laboratory, UNIMAS and all the experimental investigations were performed in accordance with [11] unless stated otherwise.

2. Site Settlement Observation in Sri Aman, Sarawak

2.1. Site 1: Balai Bomba Sri Aman

Settlement of the infrastructures and perimeter slab are observed during the site visit and most of the measured settlement, δ, is in the range of 100 mm to 150 mm (refer to figure 2 and figure 3). However, there is no obvious crack sign is witnessed on the structures as shown in Figure 4; double-storey terrace, semi-detached quarters and multipurpose hall.

Figure 2. Settlement measured in the range of 100 mm to 150 mm.

Figure 3. Settlement observed at the car porch and the perimeter slab.

(a) Double Storey Semi-D

(b) Double Storey Terrace

IOP Conf. Series: Materials Science and Engineering 1101 (2021) 012006 doi:10.1088/1757-899X/1101/1/012006

(c) Administrative Building (d) Training Tower

Figure 4. No obvious crack were observed in the main structures.

2.2. Site 2: Pejabat Meteorologi Sri Aman

Settlement of the perimeter slab is observed during the site visit and most of the measured settlement, δ, is in the range of 50 mm to 100 mm (refer to figure 5). No peat sampling has been taken except the field vane shear test conducted with recorded undrained shear strength of 4 kPa to 6 kPa for a depth of 1.0 m and 1.3 m, respectively.

3. Sri Aman Peat Augering and Sampling

Figure 6 shows the preparation of peat augering and sampling using peat sampler. The peat depth is about 2.88 m and the sample were collected at every 0.5 m interval behind the terrace double storey building (table 3). All collected peat samples were brought to the UNIMAS Geotechnical Laboratory for geotechnical properties determination.

IOP Conf. Series: Materials Science and Engineering 1101 (2021) 012006 doi:10.1088/1757-899X/1101/1/012006

Figure 5. Settlement observed in Meteorology Department Building, Sri Aman.

Figure 6. Preparation of peat augering and sampling.

IOP Conf. Series: Materials Science and Engineering 1101 (2021) 012006 doi:10.1088/1757-899X/1101/1/012006

Table 3. Collected peat samples at the designated depth.

Depth (m)	Collected Samples	Depth (m)	Collected Samples
0.00-0.50		1.50-2.00	
0.50-1.00		2.00-2.50	
1.00-1.50		2.50-2.88	

4. Geotechnical Properties of Sri Aman Peat

Table 4 summarized the laboratory testing results for basic properties of peat collected at Balai Bomba Sri Aman. The organic content for the site is in the range of 70 %-90 % for a depth of 0.5 m up to 2.88 m with high natural moisture content ranges from 900 % to 1400 %. The peat is categorized as fibrous peat, H_3 except for the first 0.5 m, which fall under the hemic peat, H_5 based on the Von Post classification (squeeze) test.

Table 4. Basic properties of peat at various depths in Balai Bomba Sri Aman.

Depth (m)	Field Vane Shear Test $c_{u, field}$ (kPa)	Natural Moisture Content (%)	pH Value	Specific Gravity Gs	Loss on Ignition LOI (%)	Organic Content OC (%)	Von Post Classification
0.00-0.50	10	920	4.23	1.540	64.9	63.5	H_5
0.50-1.00	11	1086	4.92	1.478	78.8	77.9	H_3
1.00-1.50	14	1338	4.00	1.666	77.2	76.3	H_3
1.50-2.00	-	1463	4.35	1.611	81.1	80.3	H_3
2.00-2.50	-	1044	5.15	1.789	95.1	94.9	H_3
2.50-2.88	-	1151	4.97	1.940	90.6	90.2	H_3

IOP Conf. Series: Materials Science and Engineering 1101 (2021) 012006 doi:10.1088/1757-899X/1101/1/012006

4.1. Moisture content of peat

[12] stated that the moisture content of peat is determined at various temperature ranges between 60 °C and 105 °C. However, from the study conducted, it is proven that at 105 °C during 24 h, loss of organic matter is limited, while at lower temperature, small amounts of free water are retained. Thus, in this study the 105 °C temperature is use for 24 h oven-dried for determination of moisture content of peat. The natural moisture content of the collected Sri Aman peat ranges from 900% to 1400% with depth is determined through performing the oven-dried techniques at 105°C for a period of 24 hours. Peat collected is classified as very high in moisture content, and the findings is an agreement with [13] and [14] which stated that the typical water content of the peat range from 200% to 2000% (refer table 5).

4.2. Specific gravity of peat

The specific gravity of peat is greatly affected by its composition and percentage of inorganic component, which related to the degree of decomposition and mineral content of peat. The high specific gravity indicates a higher degree of decomposition and higher mineral content. According to [15] and [16], for peat with an organic content of 75 % and greater, the specific gravity is in the range from 1.3 to 1.8 with an average of 1.5. For the Sri Aman peat specific gravity, Gs was measured in accordance with the procedure described in BS1377 using kerosene and about 10 g of oven-dried peat gave a value ranges from 1.5 to 1.9 for a depth up to 2.8 m.

4.3. Organic Content of Peat

The Organic Content (OC) is determined from *Loss on Ignition Test*, in accordance to ASTM D2974 (440°C) and BS1377: Part-3 (550°C) by [17]. In order to achieve destruction of organic matter, ignition of the soil at high temperature, 550°C with 5 h duration of firing is used to determine the organic content directly [1]. The correlations are most simply directly in terms of ignition loss and calculated in accordance to an equation proposed by [12]:

$$\textit{Organic Content}, \text{OC} (\%) = 100 - \text{C} (100-(\text{LOI})) \quad (1)$$

where: Correction factor C = 1 for 440°C; & C=1.04 for 550°C [1]; and
Loss on Ignition, LOI (%) = 100.0 – Ash Content

In this study, the organic content (OC) of peat is determined through Loss on Ignition (LoI) test by igniting specimens of the powdered materials in a muffle furnace at a temperature of 550°C. The organic content (OC) obtained from the site is in the range of 70 %-90 % for a depth of 0.5 m to 2.88 m, which tabulated in Table 4.

4.4. pH value of peat

The pH values obtained are in the range of 4 to 5 and classified as highly to moderate acidic peat in accordance to [4]. Table 5 presents the criteria of peat acidity level extract from [4].

Table 5. Classification of peat acidity [4].

Level of Acidity	Criteria
Highly Acidic	Peat with a pH less than 4.5
Moderate acidic	Peat with a pH between 4.5 and 5.5
Slightly acidic	Peat with a pH greater than 5.5 and less than 7
Basic	Peat with a pH equal or greater than 7

4.5. *Field Vane Shear Test*
From the field vane shear test conducted the undrained shear strength, cu, ranges from 10 kPa to 14 kPa with increasing of depth at 0.5 m interval up to 1.5 m. According to Huat [13], the average undrained shear strength for tropical peat in Malaysia obtained through field vane shear test is between 3 kPa and 15 kPa which is much lower than that of the mineral soil.

5. Conclusions
The observation of the settlement for structures/infrastructures constructed on peat in Sri Aman, Sarawak is in the range of 50 mm to 150 mm. From the laboratory test conducted for the collected samples, the site has high natural moisture content and organic content ranges from 900 % to 1400 % and 66 % to 95 %, respectively. The Sri Aman peat is categorized as H_3, fibrous peat except for the first 0.5 m, which fall under H_5, hemic peat group according to Von Post classification.

6. References
[1] Edil, T.B. 2003 Recent advances in geotechnical characterization and construction over peats and organic soils. *2nd International Conference on Advances in Soft Soil Engineering and Technology*, Putrajaya, Malaysia, pp. 1-16.

[2] Sa'don, N.M., Abdul Karim, A.R., Taib S.N.L. and Yusof, M. 2018 Strength properties of reinforced peat using fiber-polyester and shredded rubber-crumb as reinforcement material *Int. J. Eng. Technol.* **7** 26-30.

[3] Huat, B.K., Asadi, A. and Kazemian, S. 2009 Experimental investigation on geomechanical properties of tropical organic soils and peat. *Am. J. Eng. Appl. Sci.* **2(1)** 184-8.

[4] ASTM 2007 D2974: Standard test methods for moisture, ash, and organic matter of peat and other organic soils. *ASTM International*, West Conshohocken.

[5] von Post L 1922 Sveriges geologiska undersoknings torvinventering och nagre av dess hittills vunna resultat *Sr. Mosskulturfor* **1** 1–27.

[6] Jarrett, P.M. 1995 Site investigation for organic soils and peat, Geoguide 6, JKR Document 20709-0341-95, Public Works Department Malaysia, Malaysia.

[7] Huat, B.B., Kazemian, S., Prasad, A. and Barghchi, M. 2011 State of an art review of peat: General perspective *Int. J. Phys. Sci.* **6(8)** 1988-96.

[8] Kazemian, S. and Huat, B.B.K. 2009 Compressibility characteristics of fibrous tropical peat reinforced with cement column *Electronic J. Geotechnical Eng.* 14C.

[9] Mesri, G., Stark, T.D., Ajlouni, M.A. and Chen, C.S. 1997 Secondary compression of peat with or without surcharging *J. Geotechnical Eng.* **123(5)** 411-21.

[10] Sa'don, N.M., Abdul Karim, A.R., Ahamad, Z. and Marriappan, A. 2016 Sarawak hemic peat consolidation settlement and shear strength behaviour *15th International Peat Congress 2016*, Kuching Sarawak, Malaysia, pp. 630-634.

[11] BSI 1990 BS 1377-3: Methods of test for soils for civil engineering purposes. Chemical and electro-chemical tests.

[12] Skempton, A.W. and Petley, D.J. 1970 Ignition loss and other properties of peats and clays from Avonmouth, King's Lynn and Cranberry Moss *Geotechnique* **20(4)** 343-56.

[13] Huat, B.B.K. 2004 Organic and Peat Soils Engineering, Universiti Putra Malaysia Press, Serdang, pp 146–150.

[14] Mesri, G. and Ajlouni, M.A. 2007 Engineering properties of fibrous peat *J. Geotechnical Geoenvironmental Eng.* **133(7)** 850-66.

[15] Davis, J.H. 1997 The peat deposits of Florida their occurrence, development and uses, Florida Geological Survey. *Geol. Bull.* **3**.

[16] Ajlouni, M.A. 2000 Geotechnical Properties of Peat and Related Engineering Problems, *University of Illinois at Urbana-Champaign*. 15-189.

[17] Jarrett, P.M. 1995 Geoguide 6. Site investigation for organic soils and peat, *JKR Document 20709-0341-95*, Public Works Department Malaysia, Malaysia.

Acknowledgments
The authors appreciate and thankful to Jabatan Mineral & Geosains (JMG), Kuching and Universiti Malaysia Sarawak (UNIMAS) for the research workplace and facilities provided. To all team members, thank you for the commitment and dedication given towards completion of the current study. Finally, to Tn Hj Affandi Othman, Cik Nur Fatin Amiera Razali and those who are contributing directly and indirectly during the site visit and peat exploration in Sri Aman, Sarawak.

IOP Conf. Series: Materials Science and Engineering 1101 (2021) 012007 doi:10.1088/1757-899X/1101/1/012007

Numerical modelling of seepage analysis using SEEP/W: A case study for the Kerian River Flood Mitigation Project (Phase 3) in Bandar Baharu, Kedah

N F Mohd Nordin[1,2*], H Mohamad[1] and H Alarifi[1]

[1] Department of Civil and Environmental Engineering, Universiti Teknologi PETRONAS, Bandar Seri Iskandar, 32610 Tronoh, Perak, Malaysia
[2] Department of Irrigation and Drainage Malaysia, Jalan Sultan Salahuddin, 50626 Kuala Lumpur, Malaysia
*Corresponding author email: nor_16000645@utp.edu.my

Abstract. The levee stretch near Dataran Bandar Baharu was a part of the Kerian River Flood Mitigation Project (Phase 3) in Kedah to address frequent flooding problems at the area. It has been facing recurring levee failure despite several repair attempts in the past. As the research area was subjected to water level fluctuations, a numerical model using SEEP/W was analysed to study the effect of drawdown and seepage on the failed levee. Variables considered in this study include soil permeability, groundwater table and time dependent river water levels on the upstream side of the levee. The transient analysis carried out indicated that the levee was susceptible to failure due to the relatively high phreatic line within the levee during drawdown and a seepage face occurrence on the upstream side of the levee.

1. Introduction

With a total area of 1,621.4 km^2, the Kerian River basin is located predominantly in the state of Perak, with the northern portion of the Kerian River basin encroaching into the state of Kedah and mainland of Pulau Pinang. The 104 km Kerian River originates from the Bintang Range and flows south-westward towards the Malacca Straits near the Nibong Tebal town. Tributaries of the Kerian River include Selama River, Ijok River, Samagagah River and Semang River. After the 2014 major flood, a series of projects divided into 30 phases between the three states, namely Perak, Kedah and Pulau Pinang had been initiated by the Federal Government of Malaysia to prevent catastrophic flood reoccurrences in the Kerian River basin.

The Kerian River Flood Mitigation Project (Phase 3) located in Bandar Baharu, Kedah (figure 1) was one of the above-mentioned projects with its construction area starting from the Sungai Tepus vicinity until Ampang Jajar (Kerian River barrage). Despite project completion in early 2019, the 350 metres levee stretch between CH17,200 and CH17,550 faced persisting levee failure where the first indication of levee failure had occurred in August 2015. Since then, the levee failed separately in December 2015, February 2017, October 2017 and June 2018. Though several repair attempts had been carried out amounting to approximately RM400,000.00, the problem was not addressed justifiably. Slides and horizontal cracks along both the upstream and downstream slope of the levee was observed along the referred stretch. Figure 2 presents an aerial view of the failure and figure 3 depicts a close-up view of the failure.

Figure 1. The location of the Kerian River Flood Mitigation Project (Phase 3) in Bandar Baharu, Kedah.

Figure 2. Aerial view of the levee stretch between CH17,200 and CH17,550 for the Kerian River Flood Mitigation Project (Phase 3) in Bandar Baharu, Kedah.

Figure 3. Close up view of the failed levee stretch between CH17,200 and CH17,550 for the Kerian River Flood Mitigation Project (Phase 3) in Bandar Baharu, Kedah.

The Kerian River at this area was subjected to water level fluctuations due to the varying river water levels caused by changes in rainfall intensity, the river's role in the agricultural sector for paddy fields irrigation; and its location nearby to the river estuary which causes tide-induced fluctuations. As the water level fluctuation at the area was moderate, it was possible that the initial design of the constructed levee was primarily based on steady state seepage analyses and precluded transient analyses. Hence, numerical modelling was carried out using SEEP/W to study the effect of drawdown and seepage on the failed levee.

2. Literature Review

Transient flow is defined as the condition during water flow where pore water pressure, and consequently total head, varies with time [1]. The change of the hydraulic boundary conditions and total stresses that occurs whilst in transient state causes: (i) saturated seepage through relatively pervious foundation strata [2], [3]; (ii) unsaturated seepage through earth embankments [4], [5]; and (iii) shear induced pore water pressures [6], [7]. The rate of pore water pressure response to these changes are mainly controlled by the hydraulic conductivity and compressibility of the soil [8].

Past literature regarding analyses on rapid drawdown include [6], [9]–[12] with many variations. Determination of the pore water pressure or the phreatic line through the levee is essential to further analyse its stability [13]. An increase of the pore water pressure will result in a decrease of the effective stress in the soil thus leading to a reduced factor of safety for the slope [14]. Most geotechnical engineering practice adapt a simplified approach to determine the phreatic line. However, complex numerical approach such as the finite element method provides a more thorough analysis [15], [16].

Numerical approach using SEEP/W have been utilised by past researchers to simulate both saturated and unsaturated flow in soils [17]–[19]. SEEP/W is a numerical model software that can mathematically simulate the real physical process of water flowing through a particulate medium using the finite element method; with results that are consistent with results from physical modelling in the laboratory [20]. Numerical modelling has multiple advantages over physical modelling whereby some of the evident advantages include a quick set up, readily available to investigate a wide range of scenarios, provides ease in accounting for gravity, ensure safety and prevents danger of physical harm to personnel as well as able to provide information and results at any location within the configured cross section [20].

3. Geometry and Material Properties

Survey works that was carried out previously before the commencement of the Kerian River Flood Mitigation Project (Phase 3) served as a basis for the model geometry. The survey cross section at CH17,400 in figure 4 was chosen as it was located approximately mid-length of the levee stretch experiencing failure, thus providing appropriate representative of the study area.

Adopting a scale of 1:250, polylines were drawn starting at the middle of the Kerian River riverbed up until the levee cross section on the right side of the riverbank. In accordance with the constructed levee on site, a 1V:4/3H slope was drawn for both upstream and downstream side of the levee, with a height of 1500 mm and crown width of 3000 mm. The subsurface foundation of the levee was divided into four layers based on laboratory test results of borehole samplings that had been carried out prior to this study. A sketch of the utilised SEEP/W model geometry is presented in figure 5 with the soil material properties for each layer defined in table 1.

The soils were classified using the British Soil Classification System where laboratory tests had identified the levee fill (Layer 1) as SILT with low plasticity; Layer 2, 3 and 4 as CLAY of extremely high plasticity; and Layer 5 as SAND with presence of fines. Laboratory tests had also identified the soil hydraulic conductivity and compressibility parameters for Layer 1, 2, 3 and 4 respectively. However, the soil sample retrieved for Layer 5 was inadequate to carry out advanced laboratory tests. Thus, correlated Standard Penetration Test (SPT) N-values and published literature values were used [21]–[24].

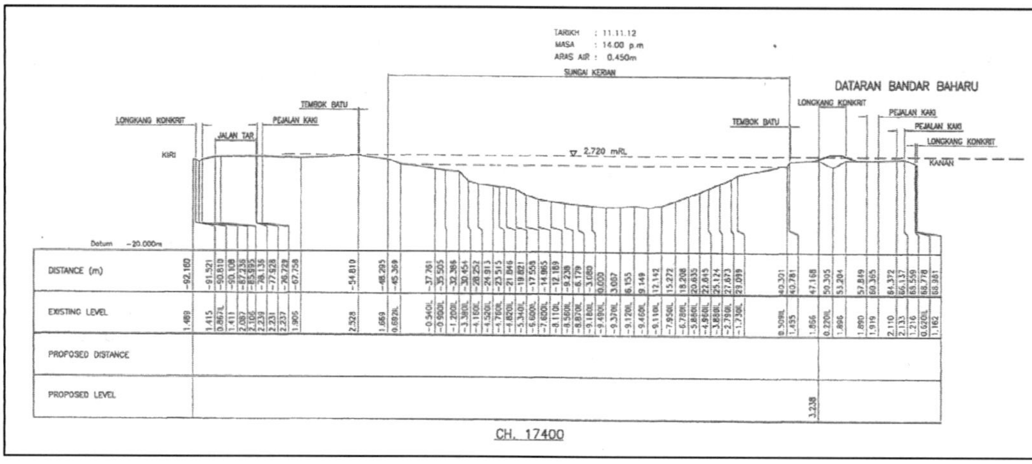

Figure 4. Survey cross section at CH17,400 which was located approximately mid-length of the levee stretch experiencing failure.

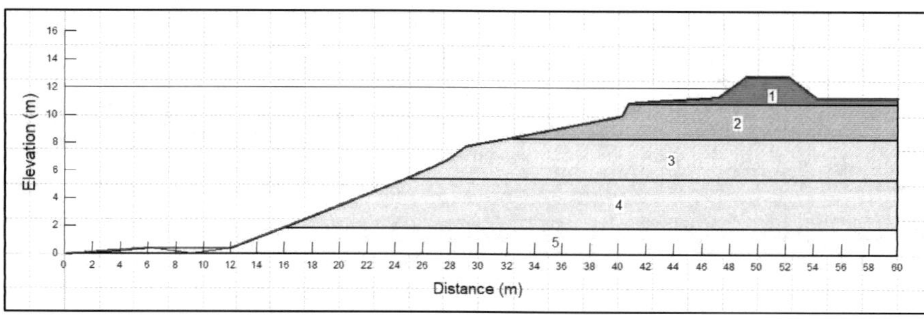

Figure 5. Model geometry of the study area utilised in SEEP/W.

Table 1. Soil material properties adopted for the SEEP/W model.

Parameter \ Material	Layer 1 Levee fill material	Layer 2 BHF 3.00-3.50	Layer 3 BH G 6.00-6.50	Layer 4 BH G 9.00-9.50	Layer 5 Correlated, literature values
BS 5930	FML	CE	CE	CE	S-F
Saturated conductivity (K), m/s	1.42×10^{-8}	7.77×10^{-10}	7.91×10^{-10}	7.37×10^{-10}	1.68×10^{-4}
Residual water content, m^3/m^3	0.125	0.150	0.225	0.140	0.030
Saturated water content, m^3/m^3	0.2528	0.4290	0.4047	0.4272	0.4000
Coefficient of compressibility (Mv), /kPa	0.0003000	0.0001833	0.0001767	0.0001767	0.0001000

IOP Conf. Series: Materials Science and Engineering 1101 (2021) 012007 doi:10.1088/1757-899X/1101/1/012007

4. Geometry and Material Properties

Seepage analysis was carried out using finite element in 2-dimensional with an element size of 0.5 m as its mesh configuration. An unstructured pattern mix of quads and triangles finite element mesh were created for the models. As this levee was modelled for transient analysis, time steps were created using river water levels that were recorded at the Kerian River barrage located downstream of the study area.

Based on the chronology failure, the levee had shown its first signs of failure in early August 2015. Assuming that there were no drastic changes of the river water levels in the same months of the subsequent few years, river water level data from the 1st of August 2018 until the 7th of August 2018 were used. The maximum head of the river water level selected for modelling was limited to 2.51 metres above sea level (masl)* based on observation on site, taking into account the upstream location of the research area and the closure of the Kerian Barrage Gate during high water tide to prevent flooding. A boundary condition of zero pressure was also set at 500 mm below ground level nearby the downstream levee toe appertaining to visual site observation during boring and sampling.

Numerical modelling for the levee began with a steady state analysis at 2.28 masl followed by a transient drawdown to 0.87 masl to a transient increment to 2.51 masl, and so forth using the recorded river water levels in table 2. A total of 27 models were carried out where the river water level fluctuations were modelled for a duration of 5 to 7 hours each and seepage analysis were carried out at approximately 30 minutes interval with the river water level increment set linearly. Based on each recorded river water level, the level variation (Δ) was calculated and the phreatic line (represented by the water total head) development throughout the levee fill structure (toe-to-toe) were modelled (figure 6 and figure 7).

Table 2. Variation of the river water levels and phreatic line development.

No.	Date	Time	River water level, H (masl)	Variation, Δ (m)
1	1/8/2018	1.15 am	2.28	–
2	1/8/2018	7.14 am	0.87	-1.41
3	1/8/2018	1.07 pm	2.66*	+1.64
4	1/8/2018	7.55 pm	0.36	-2.30
5	2/8/2018	1.58 am	2.4	+2.04
6	2/8/2018	7.59 am	0.82	-1.58
7	2/8/2018	1.49 pm	2.69*	+1.69
8	2/8/2018	8.35 pm	0.30	-2.21
9	3/8/2018	2.40 am	2.47	+2.17
10	3/8/2018	8.44 am	0.81	-1.66
11	3/8/2018	2.30 pm	2.66*	+1.70
12	3/8/2018	9.15 pm	0.31	-2.20
13	4/8/2018	3.22 am	2.49	+2.18
14	4/8/2018	9.28 am	0.85	-1.64
15	4/8/2018	3.12 pm	2.56*	+1.66
16	4/8/2018	9.55 pm	0.38	-2.13
17	5/8/2018	4.05 am	2.45	+2.07
18	5/8/2018	10.14am	0.93	-1.52
19	5/8/2018	3.54 pm	2.41	+1.48
20	5/8/2018	10.35pm	0.52	-1.89
21	6/8/2018	4.50 am	2.37	+1.85
22	6/8/2018	11.03am	1.04	-1.33
23	6/8/2018	4.38 pm	2.21	+1.17
24	6/8/2018	11.18 pm	0.70	-1.51
25	7/8/2018	5.39 am	2.26	+1.56
26	7/8/2018	12.00 pm	1.17	-1.09
27	7/8/2018	5.28 pm	1.99	+0.82

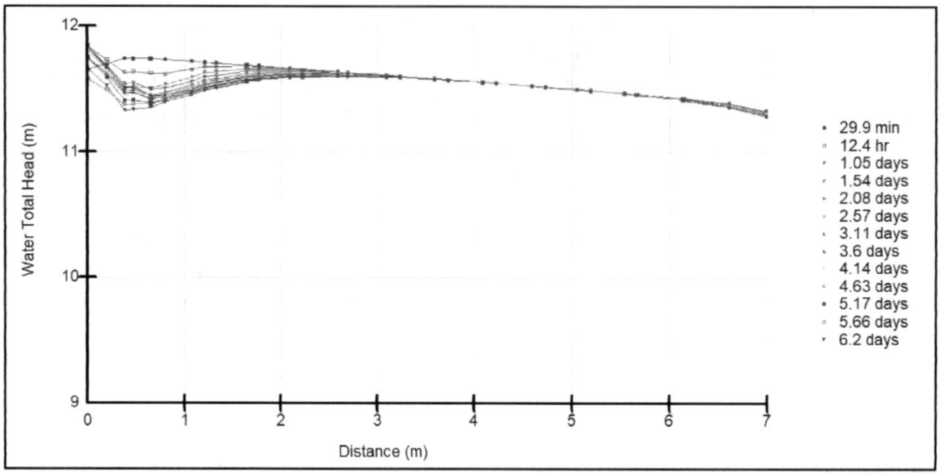

Figure 6. Simulation of water total head during water level increment throughout the levee fill structure (toe to toe).

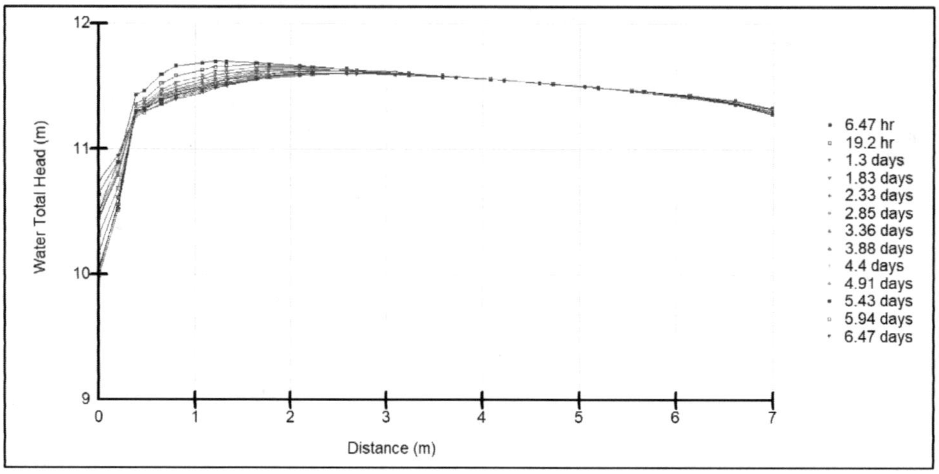

Figure 7. Simulation of water total head during drawdown throughout the levee fill structure (toe to toe).

No seepage face was observed at the downstream face of the levee for all of the models, indicating an acceptable outcome. However, it was noted that during drawdown, the lowering rate of the phreatic line was not concurrent with the decreasing river water level. This indicates presence of excess pore water pressures in soils above the river water level during drawdown that affected the soil's stresses and caused the levee failure. Figure 8 shows the plot of the pore water distribution throughout the levee fill structure (toe-to-toe) during water increment instances whilst figure 9 shows the plot of the pore water distribution throughout the levee fill structure (toe-to-toe) during drawdown instances. Positive pore pressure was observed for all the models during water level increment except for model 13 and 27 (after 3.09 days and 6.68 days respectively), which had shown a slight existence of negative pore water pressure beyond 300 mm from the levee toe. Contrarily for drawdown instances, negative pore pressure was observed in all the models within 400 mm from the levee toe with positive pore pressure observed beyond the 400 mm range. Negative pore pressure signifies the presence of partially saturated soils above the river water level, whereas positive pore pressure signifies fully saturated soils. This condition indicates a delayed pore water reduction as the water level decreased. A slow decline of the phreatic line or the water table suggests a large lambda slope coefficient which can cause instability to the levee slope [25].

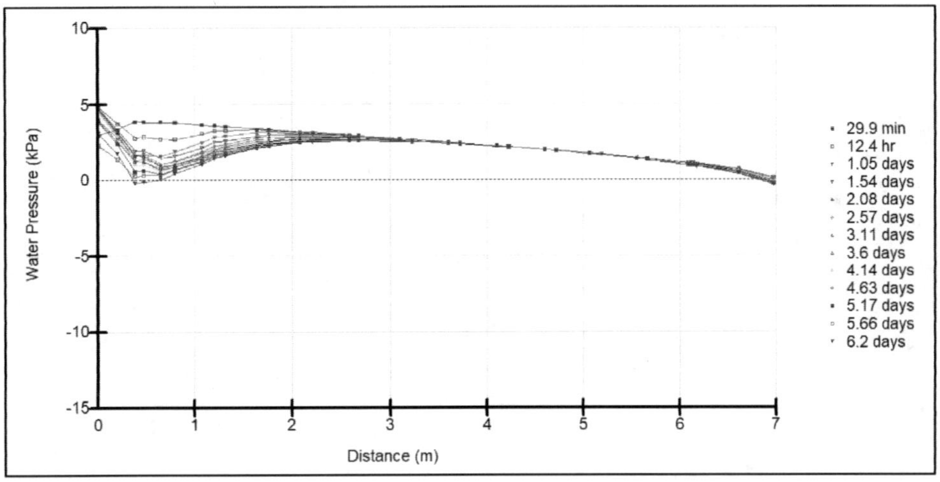

Figure 8. Simulation of pore water pressure during water level increment throughout the levee fill structure (toe to toe).

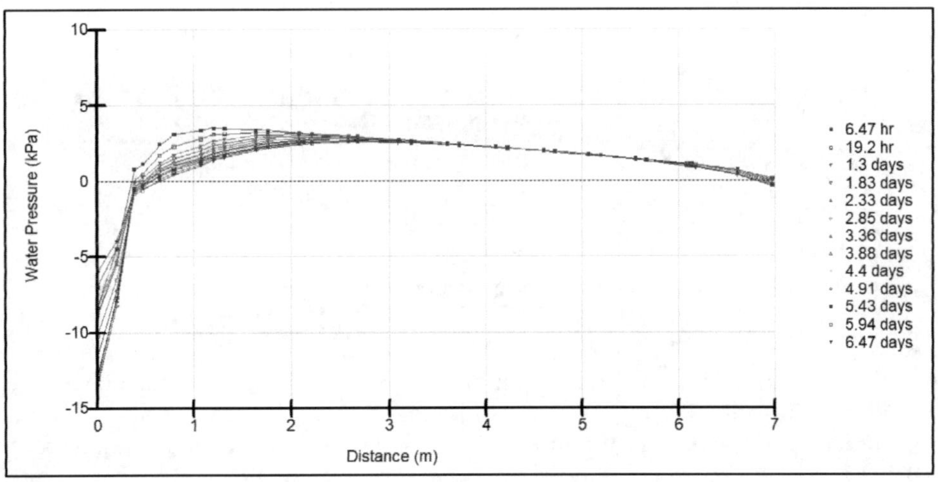

Figure 9. Simulation of pore water pressure during drawdown throughout the levee fill structure (toe to toe).

To better understand seepage and flow characteristics at the upstream side of the levee that experienced failure, a graph showing the rate at which water flux was accumulated on the upstream face area of the levee was plotted (figure 10). Water flux is defined as the volumetric flow rate per unit area per time. Based on Figure 10, the total water flux was influenced by the fluctuations of the river water level. When there is an increment of the river water level, the total water flux at the upstream face of the levee increased significantly. Subsequently, seepage flux also occurred during drawdown. However, the amount was substantially lesser, given that the river water level was near the base of the levee, reducing the unit area.

The results also showed a seepage face occurrence in all the drawdown models along the upstream side of the levee, represented by the velocity vectors as illustrated in figure 11. In SEEP/W, velocity vectors indicate where flow is occurring and its' quantity. The presence of seepage at the face of the levee upstream slope reduces its cohesive strength and alternately increases shear stress, causing failure on the levee structure. This is in conformance with the study carried out by [26] in developing an empirical relationship between the seepage forces acting on the side slope of a channel and the stability of the channel with various flow conditions.

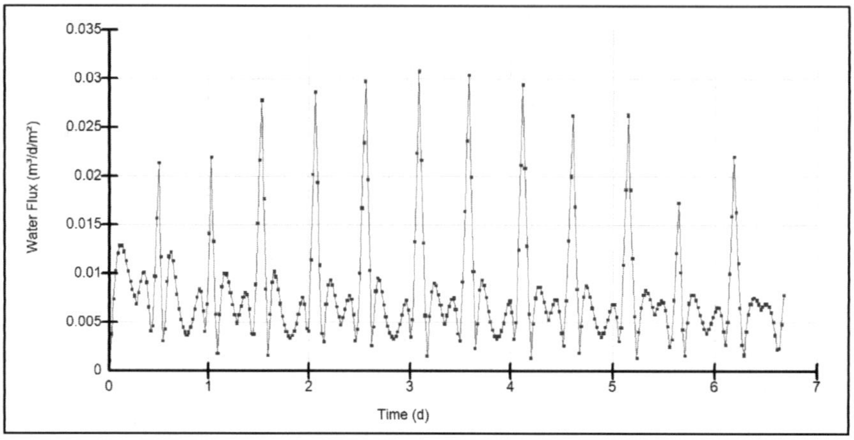

Figure 10. Total water flux at the upstream face of levee.

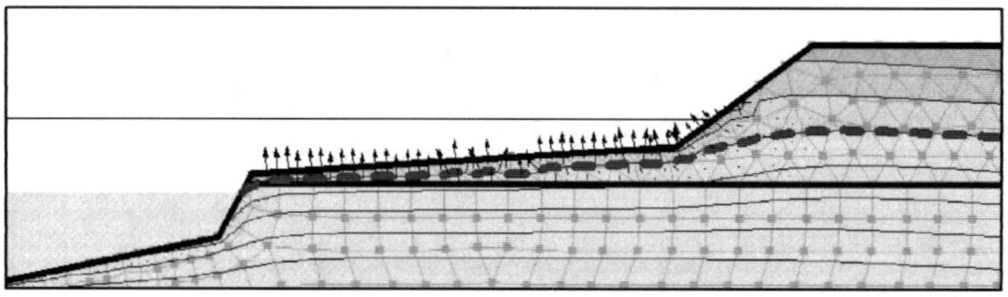

Figure 11. Illustrative example of the vectors generated in SEEP/W representing probable seepage face along the upstream side of the levee during drawdown.

5. Conclusion

Seepage analysis was carried out using numerical models in SEEP/W for the failed levee stretch in the Kerian River Flood Mitigation Project (Phase 3) in Bandar Baharu, Kedah. A total of 27 models were developed using the finite element method to simulate the levee in transient conditions for an overall period of 7 days. Results indicated that during river water level increment, the seepage rate into the levee structure was noticeably higher compared to the rate at which it was seeping out during drawdown. The slow decline of the phreatic line within the levee structure during drawdown and the occurrence of seepage face at the upstream side of the levee reflects an excess in pore water pressure thus causing the levee susceptible to instability and failure.

6. References

[1] Lambe T W and Whitman R V 1969 *Soil Mechanics* (New York: John Wiley & Sons)
[2] Casagrande A 1961 Control of seepage through foundations and abutment of dam, first Rankine lecture *Géotechnique* **11** 161–82
[3] Mansur C I, Postol G and Salley J R 2000 Performance of relief well systems along Mississippi river levees *J. Geotech. Geoenvironmental Eng.* **126** 727–38
[4] Lam L, Fredlund D G and Barbour S L 1987 Transient seepage model for saturated–unsaturated soil systems: a geotechnical engineering approach *Can. Geotech. J.* **24** 565–80
[5] Le T M H, Gallipoli D, Sanchez M and Wheeler S J 2011 Stochastic analysis of unsaturated

seepage through randomly heterogeneous earth embankments *Int. J. Numer. Anal. Methods Geomech.* **36** 1056–76

[6] Duncan J M, Wright S G and Wong K S 1990 Slope Stability during Rapid Drawdown *Proc. of the H. Bolton Seed: Memorial symposium* pp 253–72

[7] Alonso E E and Pinyol N M 2011 Landslides in reservoirs and dam operation *Dam maintenance and rehabilitation II* pp 3–27

[8] Pauls G J, Sauer E K, Christiansen E A and Widger R A 1999 A transient analysis of slope stability following drawdown after flooding of a highly plastic clay *Can. Geotech. J.* **36** 1151–71

[9] Bishop A W 1954 The use of pore-pressure coefficients in practice *Géotechnique* **4** 148–52

[10] Moregenstern N 1963 Stability charts for earth slopes during rapid drawdown *Géotechnique* **13** 121–31

[11] Lowe J and Karafiath L 1960 Effect of anisotropic consolidation on the undrained shear strength of compacted clays *Research Conf. on Shear Strength of Cohesive Soils* (ASCE) pp 837-58

[12] Baker R, Frydman S and Talesnick M 1993 Slope stability analysis for undrained loading conditions *Int. J. Numer. Anal. Methods Geomech* **17** 15–43

[13] Yan Z L, Wang J J and Chai H J 2010 Influence of water level fluctuation on phreatic line in silty soil model slope *Eng. Geol.* **113** 90–98

[14] Perri J F, Shewbridge S E, Cobos-Roa D A and Green R K 2012 Steady state seepage pore water pressures' influence in the slope stability analysis of levees *GeoCongress 2012: State of the Art and Practice in Geotechnical Engineering* pp 604–13

[15] Huang M and Jia C Q 2009 Strength reduction FEM in stability analysis of soil slopes subjected to transient unsaturated seepage *Computers and Geotechnics* **36** 93–101

[16] Sun G, Yang Y, Cheng S and Zheng H 2017 Phreatic line calculation and stability analysis of slopes under the combined effect of reservoir water level fluctuations and rainfall *Can. Geotech. J.* **54** 631–45

[17] Arshad I, Babar M M, and Vallejera C A E 2019 Computation of seepage and exit gradient through a non-homogeneous earth dam without cut-off walls by using Geo-slope (SEEP/W) software *PSM Biological Research* **4** 40–50

[18] Aribudiman I N, Redana I W, Harmayani K D and Ciawi Y 2019 Seepage in soil from the difference of water viscosity using Geo-studio SEEP/W program *Int. Research Journal of Engineering, IT & Scientific Research* **5** 15–26

[19] Zhang L L, Zhang L M and Tang W H 2005 Rainfall-induced slope failure considering variability of soil properties *Géotechnique* **55** 183–8

[20] GEO-SLOPE International Ltd. 2012 *Seepage modelling with SEEP/W* (Alberta: GEO-SLOPE International Ltd.) pp 4-5

[21] Jorgensen D G 1980 *Relationships between basic soils-engineering equations and basic ground-water flow equations* (US Department of the Interior, Geological Survey)

[22] Ng C W W and Shi Q 1998 A numerical investigation of the stability of unsaturated soil slopes subjected to transient seepage *Computers and Geotechnics* **22** 1–28

[23] Wilson G V, Periketi R K, Fox G A, Dabney S M, Shields F D and Cullum R F 2007 Soil properties controlling seepage erosion contributions to streambank failure *Earth Surface Processes and Landforms: The Journal of the British Geomorphological Research Group* **32** 447–59

[24] Chu-Agor M L, Wilson G V and Fox G A 2008 Numerical modeling of bank instability by seepage erosion undercutting of layered streambanks *J. Hydrol. Eng.* **13** 1133–45

[25] Zheng Y R, Shi W M, Kong W X and Lei W J 2005 Determination of the phreatic-line under reservoir drawdown condition *Slopes and Retaining Structures Under Seismic and Static Conditions* pp 1-10

[26] Burgi P H and Karaki S 1971 Seepage effect on channel bank stability *J. Irrig. Drain. Eng.* **97** 59–72

IOP Conf. Series: Materials Science and Engineering 1101 (2021) 012007 doi:10.1088/1757-899X/1101/1/012007

IOP Conf. Series: Materials Science and Engineering 1101 (2021) 012008 doi:10.1088/1757-899X/1101/1/012008

Trend analysis of potential evapotranspiration in peninsular Malaysia

S L S Yong[1], J L Ng[1], Y F Huang[2], C K Ang[3]

[1] Department of Civil Engineering, Faculty of Engineering, Technology, and Built Environment, UCSI University Kuala Lumpur, Malaysia
[2] Department of Civil Engineering, Lee Kong Chian Faculty of Engineering and Science, Universiti Tunku Abdul Rahman Kuala Lumpur, Malaysia
[3] Department of Mechanical Engineering, Faculty of Engineering, Technology and Built Environment UCSI University Kuala Lumpur, Malaysia
Corresponding authors: ngjl@ucsiuniversity.edu.my

Abstract. Potential evapotranspiration (PET) plays a prominent part in hydrologic water cycle which is responsible for the water from the Earth's surface travel back to the atmosphere. Climate change brought contradictory changes in PET which led to the inconsistency of PET trend globally. Hence, the present study carried out the trend of potential evapotranspiration using meteorological data from 7 different meteorological stations located in Peninsular Malaysia. Mann-Kendall and Sen's slope estimator tests were employed in this article for PET trend analysis. Bayan Lepas experienced significant increasing trend for both FAO 56-Penman Monteith and Mahringer models during Southwest monsoon. For Mahringer model, significant increasing trends were observed at Ipoh and Pulau Langkawi stations during Northeast and inter monsoon 2, respectively. There were only a few meteorological stations experienced decreasing trend throughout the period. Although the combinations of increasing and decreasing PET trends were observed, it can be highlighted that majority of the PET time series exhibited increasing trend in Peninsular Malaysia.

1. Introduction

Climate change has proved to be a global concern over the years due to its catastrophic impacts. From the agricultural perspective, widespread changes in climate pattern threaten the fresh water supply and increase the risk of floods. Intergovernmental Panel on Climate Change (IPCC) [1] pointed out that the uncommon changes in climate has been observed globally. There is a surge in frequency of warm temperature extremes and heavy precipitation globally [2]. There is a surge in global average temperature across land and ocean surface areas (0.85°C warming) for the period 1880-2012. Consequently, the contradictory changes in potential evapotranspiration (PET) components (aerodynamic, radiative and thermal) has led to the inconsistency of PET trend globally. A better comprehension of PET trend is of paramount importance to the management of water resources in tropical regions which are rich in biodiversity. The study of PET trend could contribute to a better understanding for meteorologists, design engineers and other related professions to develop better project planning in dealing with climate change phenomenon. Any changes in PET could substantially affect biodiversity in tropical forest, particularly Peninsular Malaysia. Approximately 60% of Peninsular Malaysia land is covered by forests [3]. Evapotranspiration (ET) has been acknowledged as a core process in hydrological cycle. The term ET refers to the combination process of evaporation and transpiration that happen concurrently. [4] expressed the term PET as the water quantity that can potentially evaporate and transpire from vegetation without any constraint. PET can be measured using empirical model which integrated solar radiation, temperature, relative humidity and wind speed.

Trend detections in time series data such as inconsistency and non-homogeneity can be performed through the application of parametric and non-parametric statistical methods. The expected distribution of the data set is the decisive factor when it comes to the selection of statistical method. Parametric methods require data that follow the normal distribution, free from outliers and independent. Non-parametric methods, as opposed to parametric methods require data sets that are independent [5]. A number of PET trend studies had been carried out globally and locally. [6] analyzed PET to detect the trends in Karnal district of Haryana, India. Statistically significant decreasing trends of PET were observed over the period of 1981-2011. In Malaysia, [7] conducted the trend analysis for rainfall, PET and rainfall deficit at Alor Setar station. Although there was no statistically significant trend detected for PET, an increasing PET was observed from Jun to November at 5% significant level. This indicated that the high ET rate in these months led to the occurrence of rainfall deficit. [8] carried out a study on trend analysis of PET and climatic water balance for the entire Peninsular Malaysia. The results demonstrated that an increasing trend of PET at majority of the stations especially during Southwest monsoon. This is due to the rising monthly temperature and drier climate condition experienced during Southwest monsoon. In contrast, a downward trend of climatic water balance was revealed. This could lead to water deficiency which triggers the drought episodes especially for the central regions (i.e. Selangor and Negeri Sembilan) and the regions that have never encountered drought event. [9] studied the trend analysis of PET and its driving factor for the change in PET in Peninsular Malaysia. The authors claimed that there was a surge in annual and seasonal PET trend. Among all the meteorological parameters, minimum temperature was found increasing rapidly which contribute to the increase in PET. This could affect the atmospheric water balance in Peninsular Malaysia.

Accordingly, the study goal is to determine the trend of potential evapotranspiration on seasonal scale in Peninsular Malaysia. The historical meteorological data of 7 stations for the period of 2009 to 2018 were acquired. Two empirical PET models, namely FAO-56 Penman Monteith and Mahringer models were selected to estimate the PET.

2. Materials and methods

2.1 Study Area

Peninsular Malaysia is situated in Southeast Asia (latitude 1° N to 7 ° N and longitude 100 ° E to 104 ° E) and it covers a total area of 130,590 km2. Peninsular Malaysia experiences two tropical weather phenomenon, which is Northeast monsoon (November-February) and Southwest monsoon (May-August). In between March-April and September-October, two inter-monsoon seasons happen. During the Northeast monsoon period, major rainy season will be taken place especially on the east side of Peninsular Malaysia (e.g. Kelantan) and western part of Sarawak will be exposed to large amount of daily rainfall during the Northeast monsoon. Drier and warmer climate condition will be experienced during the Southwest monsoon [10,11,12]. Peninsular Malaysia receives sufficient solar radiation, which is about 12 hours of daylight hours throughout the year. Malaysia is exposed to rainy, hot and humid in the course of the year with 23-32°C of average temperature and and 2.5-5 mm/day of relative humidity. Generally, the wind speed in Peninsular Malaysia varies between 0.9 and 2.3 m/s [3]. The selected meteorological stations are shown in Table 1. For this study, data set of 10 years consisting of 7 meteorological stations were collected from Malaysia Meteorological Department (MMD) to estimate the PET time series. Four different meteorological parameters, namely the relative humidity (%), wind speed (m/s), solar radiation (MJ/m2), air temperature (°C) were used to compute PET.

Table 1. List of meteorological stations.

Station Name	Record Period	Latitude	Longitude
Bayan Lepas	2009-2018	05° 18' N	100° 16' E
Ipoh	2009-2018	04° 34' N	101° 06' E
Kota Bahru	2009-2018	06° 10' N	102° 18'
Kuantan	2009-2018	03° 46' N	103° 13' E
Muadzam Shah	2009-2018	03° 03' N	103° 05' E
Pulau Langkawi	2009-2018	06° 20' N	99° 44' E
Subang	2009-2018	03° 08' N	101° 33' E

2.2 Calculation of potential evapotranspiration (PET)

In this study, two empirical PET models were chosen in accordance with input data. The empirical models were categorized into two different groups, namely combination based and mass-transfer based. Combination-based models are the models that involve both aerodynamic theory and energy balance. Mass-transfer based models utilize the concept of eddy motion transfer of water vapor from an evaporating surface to the atmosphere. Table 2 lists the equation for the empirical models. The globally accepted PET model, FAO 56-PM model which recommended by the Food and Agricultural Organization in 1988 was selected in this study [13]. The input parameters were comprised of minimum and maximum temperature, sunshine hours, wind speed, and relative humidity. FAO 56 -PM model has proved to have outstanding performance in estimating PET under different climate conditions owning to the consideration of both meteorological and physiological crop variables in PET estimation. Mahringer model is one of the mass-transfer based models that is fundamentally based on Dalton's gas law. Unlike FAO 56-PM model, Mahringer model requires less input parameters such as wind speed, relative humidity and temperature.

Table 2. Methods for potential evapotranspiration (PET) estimation.

Classification	Method (abbreviated)	Equation	
Combination	FAO 56-PM [13]	$PET = \dfrac{0.408\Delta(R_n - G) + \gamma \dfrac{900}{T_{mean} + 273} u_2 (e_s - e_a)}{\Delta + \gamma(1 + 0.34\, u_2)}$	(1)
Mass-transfer	Mahringer [14]	$PET = 2.86u^{0.5}(e_s - e_a)$	(2)

Note: R_n indicates the net radiation of the crop surface ($MJm^{-2}day^{-1}$); Δ represents the slope vapor curve ($kPa°C^{-1}$); T_{mean} represents the daily mean air temperature at 2 m height (°C); U_2 represents the wind speed at 2m height (m/s); G represents the soil heat flux density ($MJm^{-2}day^{-1}$); e_a represents the actual vapor pressure (kPa); e_s is the saturation vapor (kPa); γ is the psychrometric constant ($kPa°C^{-1}$); u is the wind speed (m/s).

IOP Conf. Series: Materials Science and Engineering 1101 (2021) 012008 doi:10.1088/1757-899X/1101/1/012008

2.3 Mann-Kendall (M-K) test

To explore the trend analysis of PET time series, Mann-Kendall (M-K) test is employed to acquire the significance of the PET change. [15]. World Meteorological Organization (WHO) has suggested M-K test to determine the trend in hydrology time series owning to its simplicity and wide application [16]. The Mann-Kendall statistic S is expressed as below:

$$S = \sum_{k=1}^{n-1} \sum_{j=k+1}^{n} sgn(x_{j\_}x_k) \tag{3}$$

where n is the length of the time series; xj and xk are the PET values at years j and k, given that j>k, accordingly and sgn represents signum function where

$$sgn(x_j - x_k) = \begin{cases} 1 \; if \; x_j - x_k > 0 \\ 0 \; if \; x_j - x_k = 0 \\ -1 if \; x_j - x_k < 0 \end{cases} \tag{4}$$

The variance of PET data statistic is expressed as:

$$VAR(S) = \frac{1}{18}\left[n(n-1)(2n+5) - \sum_{p=1}^{q} t_p(t_p - 1)(2t_p + 5)\right] \tag{5}$$

where, t_i represents the number of ties up to PET data sample i. The Z_c test statistics is expressed as:

$$z = \begin{cases} \frac{S-1}{\sqrt{VAR(S)}} & if \; S > 0 \\ 0 & if \; S = 0 \\ \frac{S+1}{\sqrt{VAR(S)}} & if \; S < 0 \end{cases} \tag{6}$$

2.4 Sen's slope estimator test

Sen's slope estimator test provides an indication to deduce the PET trend's magnitude [17]. It estimates the rate of change per unit time within the PET time series. The slope can be calculated as:

$$Q_i = \frac{x_j - x_k}{j - k} \tag{7}$$

where j>k. The median of Sen's slope is expressed as follow:

$$Q = Q_{[(N+1)/2]}, if \; N \; is \; odd \tag{8}$$

or

$$Q = \frac{1}{2}\left(Q_{[N/2]} + Q_{[(N+1)/2]}\right), if \; N \; is \; even. \tag{9}$$

IOP Conf. Series: Materials Science and Engineering 1101 (2021) 012008 doi:10.1088/1757-899X/1101/1/012008

3. Result & Discussion

Table 3 and 4 demonstrated the overall results of M-K and Sen's slope estimator tests for PET computed using both FAO 56-PM and MA models. The Z values that were positive depicted an increasing PET trend for the period 2009-2018. As can be seen from Table 3 and 4, majority of all seasonal PET values exhibited positive Z value which indicated increasing trend. If the p value is less than 0.05, the trend is significant. Bayan Lepas experienced significant increasing trend for both FAO 56-PM and MA models during Southwest monsoon. For MA model, significant increasing trends were observed at Ipoh and Pulau Langkawi stations during Northeast and inter monsoon 2. These trends are consistent with the Sen's slope values which showed reasonably higher rate of change in comparison with other meteorological stations. Bayan Lepas station exhibited highest value of Sen's slope value which were 4.575 and 9.757 at Southwest monsoon for both FAO 56-PM and MA models.

Table 3. Summary of MK test and Sen's estimator for PET calculated by FAO-56 PM model.

Station	Northeast Monsoon series			Inter Monsoon 1 series		
	p-value	Z	Sen's slope	p-value	Z	Sen's slope
Bayan Lepas	0.727	0.111	0.955	1.000	0.022	0.151
Ipoh	0.601	0.156	1.857	0.862	-0.067	-0.595
Kota Bahru	1.000	0.022	1.357	0.862	0.067	0.42
Kuantan	0.862	-0.067	-0.394	0.484	-0.200	-0.851
Muadzam Shah	0.601	0.156	0.572	1.000	0.022	0.050
Pulau Langkawi	0.484	0.200	5.730	0.727	0.111	0.407
Subang	0.073	0.467	3.715	1.000	0.022	0.070

Station	Inter Monsoon 2 series			Southwest Monsoon series		
	p-value	Z	Sen's slope	p-value	Z	Sen's slope
Bayan Lepas	0.727	0.111	1.959	**0.041**[a]	**0.289**[a]	**4.575**[a]
Ipoh	0.601	0.156	1.744	0.156	0.378	2.463
Kota Bahru	0.108	0.422	3.866	1.000	0.022	0.171
Kuantan	0.601	0.156	1.225	0.484	0.200	1.242
Muadzam Shah	0.108	0.422	1.610	0.727	0.111	0.369
Pulau Langkawi	0.108	0.422	5.219	0.727	0.111	0.707
Subang	1.000	0.022	0.043	0.601	0.156	0.879

[a] Bolded value indicates significant increasing of PET.

Table 4. Summary of MK test and Sen's estimator for PET calculated by MA model.

Station	Northeast Monsoon series			Inter Monsoon 1 series		
	p-value	Z	Sen's slope	p-value	Z	Sen's slope
Bayan Lepas	0.484	0.200	5.682	0.216	0.333	2.325
Ipoh	**0.029** [a]	**0.556** [a]	**7.97** [a]	0.484	0.200	1.713
Kota Bahru	0.381	-0.244	-2.614	0.381	0.244	1.892
Kuantan	0.727	-0.111	-2.507	0.601	-0.156	-1.190
Muadzam Shah	0.216	-0.333	-2.956	0.381	-0.244	-1.152
Pulau Langkawi	0.291	0.289	6.219	0.381	0.244	1.495
Subang	0.047	0.511	9.006	0.727	-0.111	-1.172

Station	Inter Monsoon 2 series			Southwest Monsoon series		
	p-value	Z	Sen's slope	p-value	Z	Sen's slope
Bayan Lepas	0.156	0.378	6.434	**0.009** [a]	**0.644** [a]	**9.757**
Ipoh	0.156	0.378	5.564	0.156	0.378	9.714
Kota Bahru	0.484	0.200	3.797	1.000	-0.022	-0.445
Kuantan	0.484	0.200	2.216	1.000	0.022	1.157
Muadzam Shah	0.073	0.467	5.013	0.727	-0.111	-1.058
Pulau Langkawi	**0.029** [a]	**0.556** [a]	**7.365** [a]	0.601	0.156	1.452
Subang	0.484	0.200	3.417	0.216	0.333	3.582

[a] Bolded value indicates significant increasing of PET.

On the other hand, negative Z value represents decreasing trend in PET time series. Kuantan station experienced decreasing PET trend (Z values of -0.067 and -0.20 for FAO 56-PM model, -0.111 and -0.156 for MA model) for Northeast monsoon and inter monsoon 1. It can be highlighted that decreasing trends were insignificant for all meteorological stations. Although the combinations of increasing and decreasing PET trends were observed, increasing PET trends were being dominant and it can be concluded that majority of PET time series exhibited increasing trend in Peninsular Malaysia.

The graphical representation of the Sen's linear estimates for the PET time series are shown in Fig. 1. Sen's slope test with positive value depicts increasing PET trend while the negative value depicts decreasing PET trend. From Figure 1, Kuantan experienced decreasing trend during Northeast monsoon and inter monsoon 1 and positive trend during Southwest monsoon and inter monsoon 2.

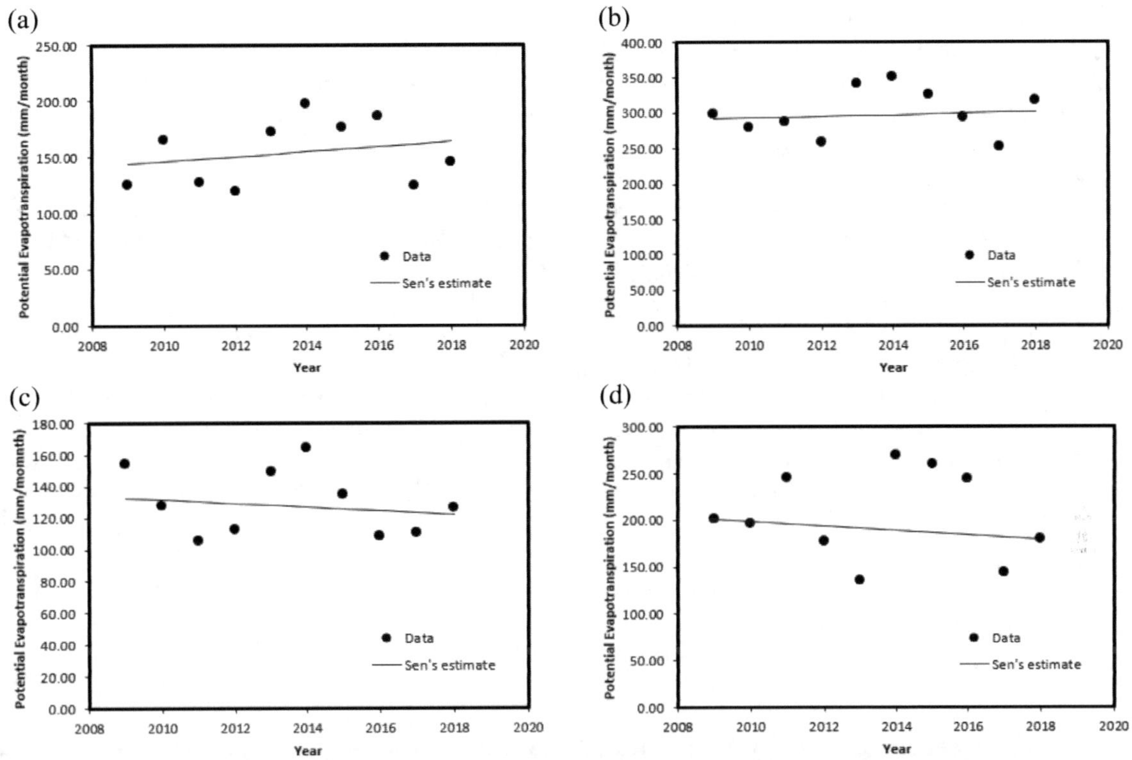

Figure 1. Sen's linear estimate of seasonal PET time series at Kuantan Station (Mahringer model) (a) Inter monsoon 2 (b) Southwest monsoon (c) Inter monsoon 1 (d) Northwest monsoon.

As mentioned previously, significant increasing trends were experienced for both PET models during Southwest monsoon. This is due to the rising temperature during Southwest Monsoon. Malaysian Meteorological Department (MMD) stated that the rising in temperature was by virtue of Southwest Monsoon which hot and drier climate will be experienced. [18]. The results are consistent with findings of [7] who discovered increasing PET trend from June to September (Southwest monsoon) in Alor Setar. Furthermore, the results are consistent with the study conducted by [8] where majority of the Peninsular Malaysia regions showed more significant increasing PET trend than climatic water balance (CWB). This might cause water deficiency which consequently contribute to the occurrence of drought events for Peninsular Malaysia.

4. Conclusion

In short, the detection of PET trend has been successfully performed in Peninsular Malaysia. Mann-Kendall test was utilized to access the PET trend at seasonal scale for the period of 2009-2018. Sen's slope estimator test was utilized to discover the PET trend's magnitude. The results will convey relevant information for the management of water resources in tropical region, particularly in Peninsular Malaysia. Results indicated that majority of the stations experienced increasing trends throughout the period. FAO-56 Penman Monteith and Mahringer models demonstrated significant increasing trends especially during Southwest monsoon. This is due to the surge in temperature which contributes to the increasing PET trend. Moreover, there was no significant decreasing PET trend for the entire period. The number of stations with decreasing trend are far less than those with increasing trend. Kuantan station experienced decreasing PET trend during Northeast monsoon and inter monsoon 1.

This study provides an overview of trends in PET in Peninsular Malaysia. Further studies should be carried out by including more regions especially the eastern part of Malaysia in order to provide more valuable information. Also, more trend analysis methods such as Spearman's Rho test should be included for better comparison.

5. References

[1] IPCC, Climate change 2014: synthesis report. *Contribution of Working Groups I, II and III to the fifth assessment report of the intergovernmental panel on climate change* [Core Writing Team, Pachauri RK, Meyer LA (eds)] IPCC, Geneva, Switzerland 151

[2] Ng J L, Abd Aziz S, Huang Y F, Wayayok A, and Rowshon M K 2016 Analysis of annual maximum rainfall in Kelantan, Malaysia *In III International Conference on Agricultural and Food Engineering* **1152** pp 11-18

[3] Muhammad M K I, Nashwan M S, Shahid S, Ismail T B, Song Y H and Chung E S 2019 Evaluation of empirical reference evapotranspiration models using compromise programming: a case study of Peninsular Malaysia *Sustainability* **11** pp 4267

[4] Lu J, Sun G, McNulty S G, and Amatya D M 2005 A Comparison of Six Potential Evapotranspiration Methods for Regional Use in the Southeastern United States *J. Am. Water Resour. Assoc.* **41** pp 621-633

[5] Jha M K, and Singh A K 2013 Trend analysis of extreme runoff events in major river basins of Peninsular Malaysia *Int J Water* **7** pp 142-158

[6] Patle G T, Singh D K, Sarangi A, Rai A, Khanna M, and Sahoo R N 2013 Temporal variability of climatic parameters and potential evapotranspiration *Indian J. Agric. Sci.* **83** pp. 518-524

[7] Ahmad A A, Yusof F, Mispan M R, and Kamaruddin H 2017 Rainfall, evapotranspiration and rainfall deficit trend in Alor Setar, Malaysia. *Mal. J. of Fund. Appl. Sci.* **13** pp 400-404

[8] Hui-Mean F, Yusop Z, and Yusof F 2018 Drought analysis and water resource availability using standardised precipitation evapotranspiration index *Atmos. Res.* **201** pp 102-115

[9] Pour S H, Abd Wahab A K, Shahid S, and Ismail Z B 2020 Changes in reference evapotranspiration and its driving factors in peninsular Malaysia. *Atmos. Res.* 105096

[10] Lin N J, Abd Aziz S, Feng H Y, Wayayok A, and Kamal M R 2015 Homogeneity analysis of rainfall in Kelantan, Malaysia *J. Teknol.* **76**

[11] Ng J L, Abd Aziz S, Huang Y F, Mirzaei M, Wayayok A, and Rowshon M K 2019 Uncertainty analysis of rainfall depth duration frequency curves using the bootstrap resampling technique. *J. Earth Syst. Sci.* **128** 113

[12] Ng J L, S. Y. Yap S Y, Huang Y F, Noh N M, Al-Mansob R A, and Razman R 2020 Investigation of the best fit probability distribution for annual maximum rainfall in Kelantan River Basin. *IOP Conf. Ser: Earth Environ. Sci.* **47** p 012118

[13] Allen R G, Pereira L S, Raes D, and Smith M 1998 *Crop evapotranspiration: Guidelines for computing crop water requirements. Irrigation and Drainage Paper No 56.* Food and Agriculture Organization of the United Nations (FAO), Rome, Italy.

[14] Mahringer W 1970 Verdunstungs Studien am Neusiedler See *Theor. Appl. Climatol.* **18** pp. 1–20

[15] Kendall M G 1975 *Rank correlation methods* 4th edition Charles Griffin& Co. Ltd, London

[16] Tian Q, Wang Q, Zhan C, Li X, and Liu X 2012 Analysis of climate change in the coastal zone of eastern China, against the background of global climate change over the last fifty years: case study of Shandong peninsula, China *Int J. Geosci.* **3** pp 379

[17] Sen P K 1968 Estimates of the regression coefficient based on Kendall's tau *J. Am. Stat. Assoc.* **63** pp 1379–1389

[18] Hamzah F M, Saimi F M, and Jaafar O 2017 Identifying the Monotonic Trend in Climate Change Parameter in Kluang and Senai, Johor, Malaysia *Sains Malays.* **46** 1735-1741

Acknowledgement
The authors are grateful to the Malaysian Meteorological Department (MMD) for the provision of meteorological data. The authors would like to acknowledge the sincere appreciation towards the financial support from the UCSI University through Pioneer Scientist Incentive Fund (PSIF) under project code Proj-2019-In-FETBE-065.

An Experimental Study on The Correlation of Natural Rainfall Intensities and Raindrop Size Distribution Characteristics

S L G Law[1], K K Kuok[1] and S G Trinidad[2]

[1] Faculty of Engineering, Computing and Science (FECS), Swinburne University of Technology, Sarawak, Malaysia.
[2] Department of Civil and Construction Engineering, Swinburne University of Technology, Melbourne, Australia.
Corresponding author: SLaw@swinburne.edu.my

Abstract. This paper aims to study the natural raindrop size distribution characteristics based on the experimental works for three (3) different rainfall intensities, which are 32.41 mm/h, 56.84 mm/h and 85.73 mm/h. A tipping bucket rain gauge was used to record rainfall data during the field study and a professional DSLR camera (Sony α6000) was used to the capture raindrop distribution during rainfall events. A sufficient number of photographs were taken with necessary adjustments on camera setting to produce the required sharpness of the images and to significantly reduce noise disturbances. Sharp images were selected for image segmentation using a customized coding script to further process the images on MATLAB. The segmented or processed images presented in the form of numerical data would provide the details on the number of drop counts in addition to the diameters of the individual drops captured. The raindrop sizes or diameters ranging from 0.5 mm to 4.5 mm were divided into group of 0.5 mm intervals. It is found that lower rainfall intensity at 32.41 mm/h has the highest volumetric drop distribution of 1.0–1.5 mm interval drop group 78.64%. The relatively higher rainfall intensity at 85.73 mm/h tends to produce higher counts in larger droplet sizes with increments of 6.90% for 2.5–3.0 mm and increment of for 3.45% 4.0–4.5 mm. The study also recorded a significant increment of median droplet sizes (D_{50}) ranging from 1.18 mm to 1.33 mm as the rainfall intensity increased from 32.41 mm/h to 85.73 mm/h, respectively.

Keywords. *Rainfall Intensity, Raindrop Size Distribution, Volumetric Drop Distribution, Median Drop Size.*

1. Introduction

Raining (precipitation) is a natural phenomenon in the water cycle process that composes parameters such as humidity, atmospheric pressure, temperature, etc. [1]. Formation of precipitation consists of two actions; the action of raindrop breaking apart (fragmentation) and combining of raindrop particles (coalescence) [2]. It is noted that the occurrence and frequency of coalescence and fragmentation of raindrops would deviate significantly in every rainfall event. Hypothetically, higher rainfall intensities would increase the chances of formation of larger raindrops and thus would increase the number of raindrops. Thus, the average raindrop size would be comparatively higher as rainfall intensity increases. Villermaux & Bossa [3] suggested that the formation of larger droplets are highly possible in the larger and deeper convective clouds.

The median drop size (D_{50}) is a common practice to be taken as a benchmark of the raindrop size distribution of a specific rainfall event. Several authoritative researchers in the field such as Mason & Andrews [4], Kelkar [5] and Laws & Parsons [6] published their research works on the relationship of D_{50} and rainfall intensities that were presented in the form of empirical equations. Although numerous studies had been conducted to determine D_{50}, the studies were confined to investigate the different ranges of natural rainfall intensities at localised and regionalized areas. The studies were mainly carried out in the temperate region of the United States of America and negligible number of studies on the equatorial region.

Rainfall characteristics in the equatorial region are distinguished by extremely high rainfall volume in the range of 2,000–6,000 mm/yr [7], [8], which can be translated to rainfall intensities of approximately 50–200 mm/h [9], [10]. To date, there is a negligibly low number of studies conducted on the correlation of natural rainfall intensities and raindrop size distribution characteristics.

Table 1 shows the classification of rainfall intensities stated by several authoritative researchers, including Llasat [11], Yakubu, Yusop & Fulazzaky [12], Varikoden, Samah & Babu [13] and Yang [14]. As shown in Table 1, it is noted nearly all studies were conducted on the relatively low rainfall intensities that highly resembled the characteristics of rainfall in the temperate regions. This research looks into the rainfall intensities and drops size distribution for rainfalls experienced in the equatorial region. Literature reviews showed that there are very limited information or citations on high equatorial rainfall intensities. This paper aims to determine the raindrop distribution characteristics with respect to high rainfall intensities encountered in the equatorial zone.

Table 1. Rainfall Intensity Classification.

Source	Llasat [9]	Yakubu, Yusop & Fulazzaky [10]	Varikoden, Samah & Babu [11]	Yang [12]
Location	UK	Malaysia	Malaysia	Malaysia
Low	$I < 2$	$I < 4$	$I < 4$	$0 < I \leq 10$
Medium	$2 < I \leq 15$	$4 < I \leq 15$	$4 < I \leq 8$	$10 < I \leq 30$
High	$15 < I \leq 30$	$15 < I \leq 30$	$8 < I \leq 12$	$30 < I \leq 60$
Very High	$30 < I \leq 60$	$30 < I \leq 60$	$I > 12$	$I > 60$
Extreme High	$I > 60$	$I > 60$	-	-

Note: I = Rainfall Intensity (mm/h)

2. Materials and Methods

In this study, natural rainfall characteristics with respect to intensity and size distribution patterns were investigated experimentally. The surroundings of the experimental site would require sufficient clearance and without high structures nearby. The selected site was located in Kuching (Sarawak), a city with tropical rainforest climate [15], which was approximately 2.46 km from Swinburne University of Technology Sarawak Campus (Figure 1).

This site is accessible by motor vehicles via Jalan Three Hills Park and Jalan Pending. The coordinates of the site are 1° 32' 47.99" N and 110° 22' 26.37" E. A standard tipping bucket rain gauge has a standard 350mm high with 200 mm diameter opening as shown in Figure 2 (a). It was installed onsite to collect rainfall data; rainfall duration and daily rainfall amount. The rain gauge was manufactured by Hydrological Services P/L at Liverpool, NSW, Australia. The requirements on the orientation of the tipping bucket rain gauge include its mounting on approximately 2 meters from the ground. For this study, three (3) different natural rainfall intensities of 32.41 mm/h, 56.84 mm/h and 85.73 mm/h were investigated for determination of the properties and characteristics of raindrop size

IOP Conf. Series: Materials Science and Engineering 1101 (2021) 012009 doi:10.1088/1757-899X/1101/1/012009

distribution with respect to the three different rainfall intensity groups. The results were compared with the raindrop size classification put forward by Yakubu, Yusop & Fulazzaky [12] and Llasat [11].

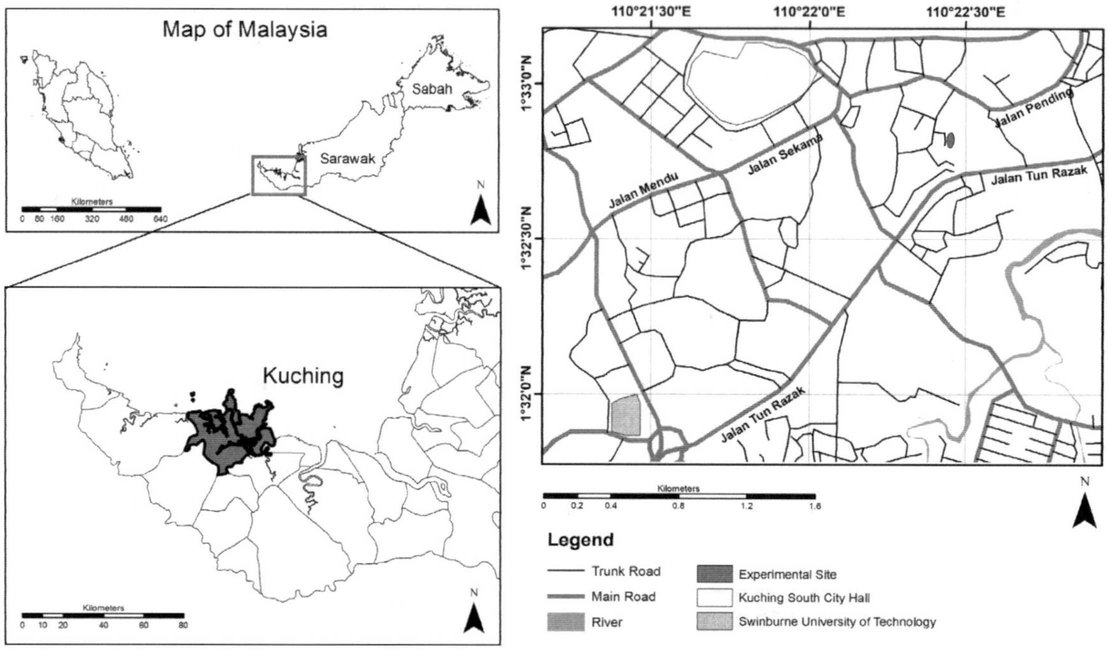

Figure 1. Locality of Study Site.

A professional DSLR camera (Sony α6000) in Figure 2 (b) was used to capture the natural rainfall droplet distribution. High shutter speed with precise camera settings was chosen for the photo shooting on fast raindrop movements. Adjustments on camera settings (ISO, aperture, shutter speed, focal length, etc.) were performed during the photo shooting process in order to produce sharp images of raindrop sizes and to significantly reduce noise disturbances in the pictures. A larger number of photographs or pictures were taken for the individual rainfall events in order to capture representative raindrop distribution pattern. A customized rainfall photobooth (Figure 2 (c)) with a dark colour background was fabricated to capture the individual droplet sizes during a rainfall event.

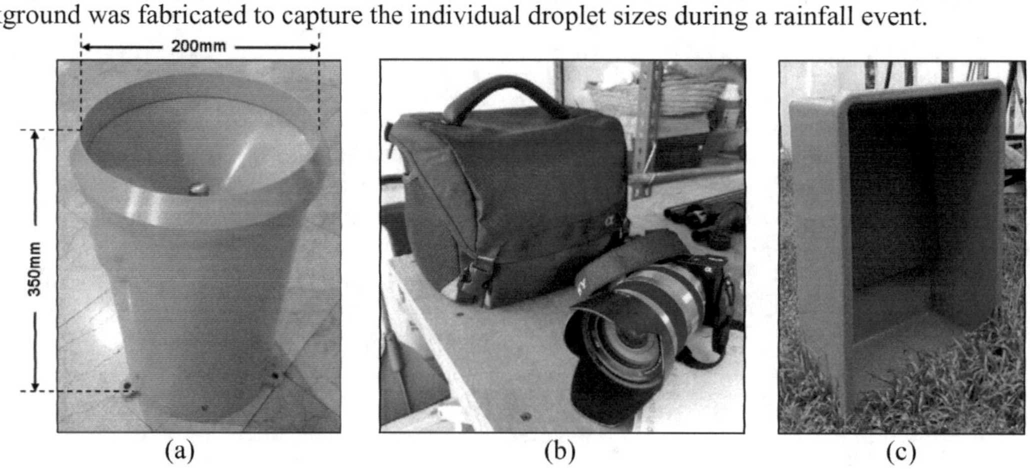

(a) (b) (c)

Figure 2. (a) Tipping Bucket Rain Gauge, (b) DSLR Camera (Sony α6000), and (c) Customized Photobooth.

IOP Conf. Series: Materials Science and Engineering 1101 (2021) 012009 doi:10.1088/1757-899X/1101/1/012009

To determine raindrop size distribution properties, the selected images include the image of the droplet were processed by a customized coding script program and to be run in MATLAB for image segmentation. Based on the pixel measurements, the processed or segmented images provided the information on raindrop count, drop diameter, drop area, drop perimeter and coordinates of the raindrops. The numeral data were then extracted from MATLAB in Excel Format and the droplet sizes were subsequently categorised into several intervals for raindrop size distribution analysis. During the data sorting process, drop sizes below 0.3 mm were deemed as a cut-off point or considered negligible because of the misting effect for drop size <0.3 mm where are detectable during the image segmentation process. Hurley et al. [16] stated that misting particle sizes could range from 10 μm to 100 μm (0.01 mm to 0.10 mm), and even up to 300 μm (0.3 mm). The drop ranges are grouped at 0.5 mm intervals from 0.0 mm to 4.5 mm. The drop distributions and volumetric drop median in each of the groups with respect to different rainfall intensities were compared and analysed.

3. Results and Discussions

The three (3) rainfall intensities are plotted in bar chart format as shown in Figure 3, and the details on volumetric drop distribution, mean and standard deviations with respect to intensities are illustrated in Table 2. The results show similar trends for all intensities with the highest volumetric distribution mean (VDM) of 55.33% for 1.0–1.5 mm drop size and the lowest VDM of 1.66% for 4.0–4.5 mm drop size. Based on the results, the rainfall intensity of 32.41 mm/h was found to produce highest volumetric drop distribution of 78.64% for drop size of 1.0–1.5 mm, as compared to 56.84 mm/h of 47.69% and 85.73 mm/h of 39.66%. For the higher intensity at 85.73 mm/h, it is shown that there is a significant increase in the larger drop sizes of the 3.0–3.5 mm and 4.0–4.5 mm groups that constitute about 5.17% and 3.45%, respectively. From this study, it is also found that incremental of rainfall intensity from 56.84 mm/h to 85.73 mm/h would result in significant increase (15.52% increase) in larger drop sizes of 3.0–4.5 mm, as well as significant reduction (13.57% decrease) in the distribution of smaller drop sizes between 1.5–2.5 mm. Table 2 shows the details of volumetric drop distribution (%), mean and standard deviation.

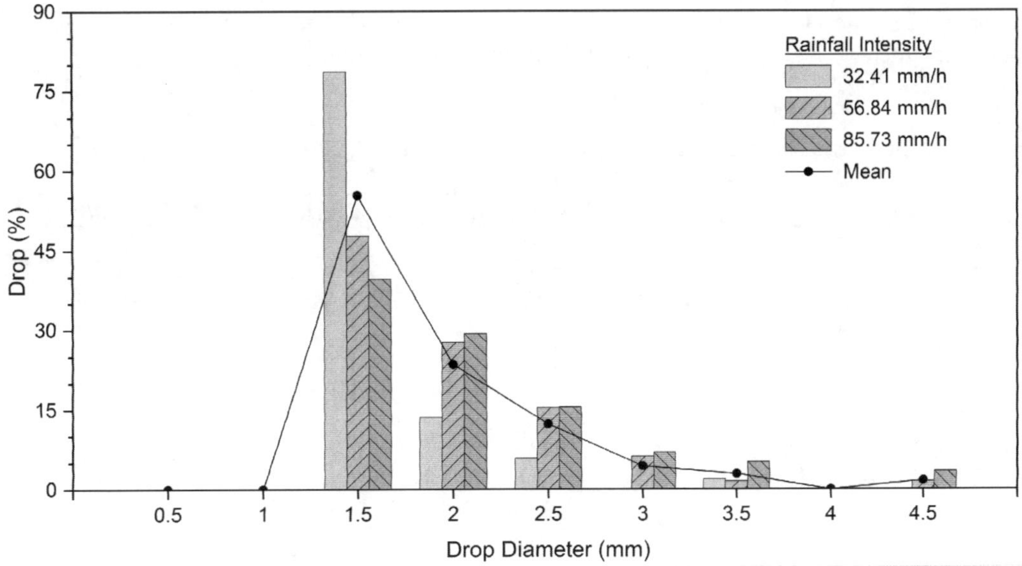

Figure 3. Natural Rainfall Intensities and Drop Size Distribution.

Table 2. Rainfall Intensity Classification.

Drop Size (mm)	32.41 mm/h	56.84 mm/h	85.73 mm/h	Mean	St. Dev.
0.5	0.00	0.00	0.00	0.00	0.00
1	0.00	0.00	0.00	0.00	0.00
1.5	78.64	47.69	39.66	55.33	20.58
2	13.59	27.69	29.31	23.53	8.65
2.5	5.83	15.38	15.52	12.24	5.56
3	0.00	6.15	6.90	4.35	3.79
3.5	1.94	1.54	5.17	2.88	1.99
4	0.00	0.00	0.00	0.00	0.00
4.5	0.00	1.54	3.45	1.66	1.73

A comparison of the current findings on intensity versus drop size distribution with Elbasit, Yasuda & Salmi [17], a similar trend is noted with respect to incremental larger drop sizes and decreases in smaller drop sizes. It is noted that there are glaring or discernible differences between the findings of the current study conducted in the equatorial area and previous researches in the temperate zone. A comparatively higher volumetric drop distribution for smaller raindrop size (1.0–1.5 mm) is recorded in this (equatorial) as compared to the findings of Elbasit, Yasuda & Salmi [17], i.e., higher margin on the larger drop sizes of >4.0 mm. Because of the potential limitations, it is noted that rainfall distribution patterns and raindrop size distribution patterns are highly dependent on the regional locality in different parts of the world and climatic characteristics. Generally, this study shows rainfall intensities correlate well with raindrop size distribution patterns.

The best fit plots of cumulative volumetric drop of the three (3) different rainfall intensities are depicted in Figure 4 and the quantile regression details are shown in Table 3. It is shown that the overall cumulative distribution pattern versus intensity is consistent with respect to drop size distribution. It is also found that the median drop size increases as rainfall intensity increases. From the results, the median drop size (D_{50}) for intensities at 32.41 mm/h, 56.84 mm/h, 85.73 mm/h are 1.18 mm, 1.57 mm and 1.70 mm, respectively. As shown in Table 3, an increase in rainfall intensities would result in a general increase in drop sizes, as indicated in the percentile diameter (D_{10} and D_{90}) and quartile diameter (D_{25} and D_{75}) for all the 3 rainfall intensities understudy.

Table 3. Raindrop Size Distribution - Percentile and Quartile Drop Diameters.

Intensity	32.41 mm/h	56.84 mm/h	85.73 mm/h
D_{10}	0.86	1.01	1.03
D_{25}	0.92	1.28	1.33
D_{50}	1.18	1.57	1.70
D_{75}	1.49	2.14	2.37
D_{90}	2.14	2.72	3.36
Span ($D_{90} - D_{10}$)	1.28	1.71	2.32
Relative Span (Span / D_{50})	1.08	1.09	1.36
Quartile Ratio (D_{75} / D_{25})	1.61	1.67	1.79

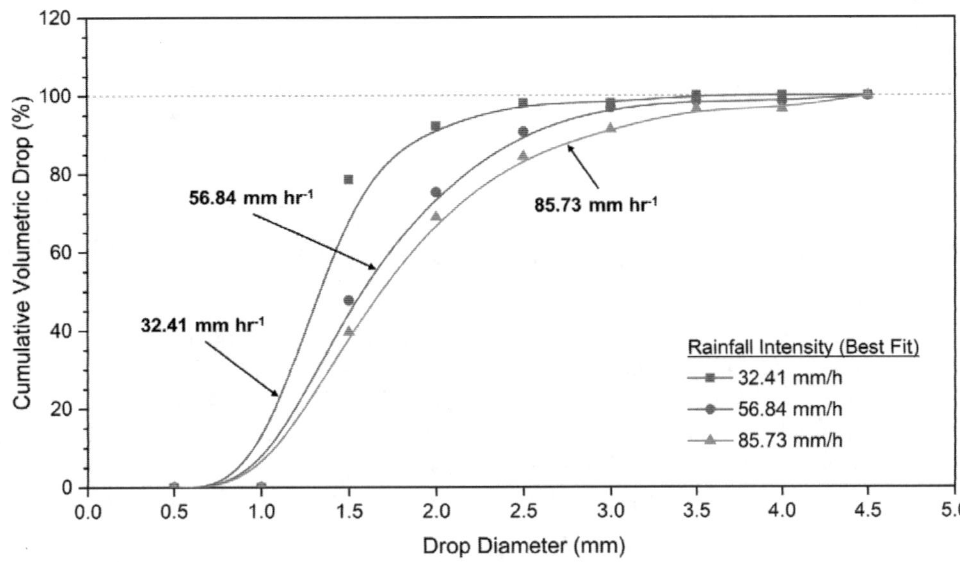

Figure 4. Best Fit Plots of Drop Diameter versus Cumulative Volumetric Drop.

Based on the findings, it can be statistically be interpreted that higher rainfall intensities are always characterized by having closer distances among raindrop particles and thus higher chances for smaller raindrops to collide and to coalesce into larger droplets before falling from the cloud. Related studies conducted by Laws & Parsons [6], Zanchi & Torri [18] and Van Dijk, Bruijnzeel & Rosewell [19] in the temperate region at low rainfall intensities and different study approaches show that an increase of raindrop sizes with the increment of rainfall intensities.

4. Conclusion

The experimental outcomes of the study are summarized in Table 4 showing intensity versus drop size and volumetric drop weightage (%). To put the field study results in perspective, the drop sizes are simplified into 6 ranges or intervals from <0.3 mm to >4.0 mm (Table 4).

Table 4. Intensity Versus Volumetric Drop Weightage (%).

Rainfall Intensity	**I = 32.41 mm/h**	**I = 56.84 mm/h**	**I = 85.73 mm/h**
Drop Size (mm)	Drop (%)	Drop (%)	Drop (%)
$d \leq 0.3$		Negligible (Misting Effect)	
$0.3 < d \leq 1.0$	0.00	0.00	0.00
$1.0 < d \leq 2.0$	92.23	75.35	68.97
$2.0 < d \leq 3.0$	5.83	21.53	22.42
$3.0 < d \leq 4.0$	1.94	1.54	5.17
$d > 4.0$	0.00	1.54	3.45

From this study, the following conclusions can be drawn with respect to rainfall intensities between 32.41 mm/h and 85.73 mm/h. Generally, it is confident to conclude that, about 95% of the cumulative volumetric drop tend to fall within the 1.0 mm and 3.0 mm, while the constituents of drop sizes of <1.5 mm and >4.0 mm are negligible. Drop sizes between 1.5 mm and 2.0 mm constitute for about 92.23%, 75.53% and 68.97% of the cumulative volumetric drop weightages for intensities at 32.41 mm/h, 56.84 mm/h and 85.73 mm/h, respectively. The 2.0 mm and 3.0 mm drop sizes account for approximately 5.83%, 21.53% and 22.42% of the total volumetric drop weightages for intensities at

32.41 mm/h, 56.84 mm/h and 85.73 mm/h, respectively. The 3.0 mm to 4.0 mm raindrop size particles recorded as low as 1.94%, 1.54% and 5.17% for intensities at 32.41 mm/h, 56.84 mm/h and 85.73 mm/h, respectively. The incremental number of drop sizes between 3.0 mm to 4.0 mm particles is not proportional to the increment of rainfall intensities.

5. References

[1] M. Sumathi, K. Kumarasamy, M. Thiyaagarajan, and J. Punithanvathi, "A study on the temperature and rainfall conditions of pudukkottai district, tamil nadu, india.," *Int. J. Dev. Res.*, vol. 1, no. 3, pp. 022–026, 2011.

[2] Y. Liao and D. Lucas, "A literature review on mechanisms and models for the coalescence process of fluid particles," *Chem. Eng. Sci.*, vol. 65, no. 10, pp. 2851–2864, 2010.

[3] E. Villermaux and B. Bossa, "Single-drop fragmentation determines size distribution of raindrops," *Nat. Phys.*, vol. 5, no. 9, pp. 697–702, 2009.

[4] B. J. Mason and J. B. Andrews, "Drop-size distributions from various types of rain," *Q. J. R. Meteorol. Soc.*, vol. 86, no. 369, pp. 346–353, 1960.

[5] V. N. Kelkar, "Size distribution of raindrops, part II," *Indian J. Meteorol. Geophys.*, vol. 11, pp. 323–330, 1960.

[6] O. Laws and D. A. Parsons, "Relationship of raindrop size and intensity," *Trans. Am. Geophys. Union*, vol. 24, no. 2, pp. 452–460, 1943.

[7] L. L. G. Samuel and K. K. Kuok, "Sensitivity Analysis of the Revised Universal Soil Loss Equation's Rainfall Erosivity Factor (R-Factor)," *Test Eng. Manag.*, vol. 83, no. May-June, pp. 6809–6815, 2020.

[8] S. M. Kueh and K. K. Kuok, "Precipitation downscaling using the artificial neural network BatNN and development of future rainfall intensity-duration-frequency curves," *Clim. Res.*, vol. 68, no. 1, pp. 73–89, 2016.

[9] C. A. Alvares, J. L. Stape, P. C. Sentelhas, J. L. De Moraes Gonçalves, and G. Sparovek, "Köppen's climate classification map for Brazil," *Meteorol. Zeitschrift*, vol. 22, no. 6, pp. 711–728, 2013.

[10] I. J. Jackson, *Climate, water and agriculture in the tropics.* London, UK: Longman, 1977.

[11] M. C. Llasat, "An objective classification of rainfall events on the basis of their convective features: Application to rainfall intensity in the Northeast of Spain," *Int. J. Climatol.*, vol. 21, no. 11, pp. 1385–1400, 2001.

[12] M. L. Yakubu, Z. Yusop, and M. A. Fulazzaky, "The influence of rain intensity on raindrop diameter and the kinetics of tropical rainfall: case study of Skudai, Malaysia," *Hydrol. Sci. J.*, vol. 61, no. 5, pp. 944–951, 2016.

[13] H. Varikoden, A. A. Samah, and C. A. Babu, "Spatial and temporal characteristics of rain intensity in the peninsular Malaysia using TRMM rain rate," *J. Hydrol.*, vol. 387, no. 3–4, pp. 312–319, 2010.

[14] H. H. Yang, "Analysis of hydrological processes of Langat River sub basins at Lui and Dengkil," *Int. J. Phys. Sci.*, vol. 6, no. 32, 2011.

[15] K. K. Kuok, Y. S. Mah, M. A. Imteaz, and S. M. Kueh, "Comparison of future intensity duration frequency curve by considering the impact of climate change: case study for Kuching city," *Int. J. River Basin Manag.*, vol. 14, no. 1, pp. 47–55, 2016.

[16] M. J. Hurley *et al.*, *SFPE handbook of fire protection engineering, fifth edition*. 2016.

[17] M. A. M. A. Elbasit, H. Yasuda, and A. Salmi, "Application of piezoelectric transducers in simulated rainfall erosivity assessment," *Hydrol. Sci. J.*, vol. 56, no. 1, pp. 187–194, 2011.

[18] C. Zanchi and D. Torri, *Evaluation of rainfall energy in central Italy.* Chichester, UK: John Wiley and Sons Ltd, 1980.

[19] A. I. J. M. Van Dijk, L. A. Bruijnzeel, and C. J. Rosewell, "Rainfall intensity-kinetic energy relationships: A critical literature appraisal," *J. Hydrol.*, vol. 261, no. 1–4, pp. 1–23, 2002.

Acknowledgement
The writers wish to express their gratitude to Professor Ir. Dr. Law Puong Ling of the University of Malaysia, Sarawak (UNIMAS) for providing technical supports and valuable comments on the experimental works. This work was supported in part by the Fundamental Research Grant Scheme (FRGS) (Reference Code: FRGS/1/2019/TK01/SWIN/02/1) by Ministry of Eduction Malaysia (MOE).

Experimental Study on Effects of Plot Length on Runoff Depth Under Natural Precipitation

S L G Law[1], K K Kuok[1] and S G Trinidad[2]

[1] Faculty of Engineering, Computing and Science (FECS), Swinburne University of Technology, Sarawak, Malaysia.
[2] Department of Civil and Construction Engineering, Swinburne University of Technology, Melbourne, Australia.
Corresponding author: SLaw@swinburne.edu.my

Abstract. This paper focuses on the experimental study to determine the relationship of slope length on runoff depth and runoff coefficient. Based on the thorough review of different plot sizes (lengths) for soil loss estimation, the slope of the current experimental plots was designed with an angle of 22.78° from horizontal. The plots were grouped into Plot A (1 m wide x 1 m slope long), Plot B (1 m wide x 2 m slope long) and Plot C (1 m wide x 3 m slope long). Homogenous soil samples were used for all the 3 experimental plots and a tipping bucket rain gauge with automatic data logger was installed at the experimental plot for rainfall intensities collection for the individual rainfall events. Runoff volumes were recorded for the individual rainfall events for determination of runoff coefficients. The rainfall intensities are grouped into ≤30 mm/h, 30-60 mm/h and ≥60 mm/h. It was found that the shorter plot (Plot A) has the highest runoff coefficients of 96.34%, 69.89% and 96.26% for 3 different precipitation groups ≤30 mm/h, 30-60 mm/h and ≥60 mm/h, respectively. Based on results, the longest plot (Plot C) shows that an increase in mean values of 33.59%, 38.30% and 49.45% as the intensity increased from ≤30 mm/h to ≥60 mm/h. From this study, it can be concluded that comparatively longer slope length tends to result in lower runoff depth.

Keywords. *Plot Length, Rainfall Intensity, Runoff Depth, Runoff Coefficient.*

1. Introduction

Some of the principal factors that would interfere with outcomes of erosion estimation include rainfall intensities, surface conditions, slope length, slope gradient and soil conditions [1]. It is mainly due to the momentum of the raindrops that would generate shear stresses on the soil surface [2], [3]. Meyer [4] demonstrated the importance of rainfall intensities on soil erosion rate including the factors of rainfall impacts and surface runoff. Surface runoff is defined as the water flow that occurs when excess sources of flow over the ground surface and this process is also called "infiltration excess overland flow" [5].

Rainfall-runoff phenomena is one of the most significant and complex relationships [6] whereby it involves a lot of factors, such as infiltration rate, evaporation rate, etc [7]. Tokar & Johnson [8] pointed out that many authoritative researchers studied for years on the transformation of rainfall to runoff in order to forecast water sources. A more suitable elaboration is that during storm events, the ground experiences infiltration process, whereby the infiltration rate would be slowed down when the moisture content of soil reaches its full saturation. Excess amount of water is not able to infiltrate the ground and thus the formation of surface runoff or surface overflow. Runoff coefficient is the ratio of the total amount of runoff to total rainfall amount. The higher the runoff coefficient, the higher the runoff amount on the surface due to the infiltration properties of the surface. For example, a concrete surface has a relatively higher runoff coefficient than a grass surface, because the grass surface has a higher infiltration rate and, thus lower runoff.

The rainfall-runoff relationship and runoff coefficient are highly attributed to size of the area understudy, whereby a different area brings different results on runoff effects under various precipitation rates. Although the application of the experimental plots is widely practised by several researchers [9], [10], there have been very limited studies on the effects or the relationships of plot length on runoff depth and runoff coefficient. To date, there is no well-defined classification on plot lengths and boundary conditions and there are, to numerous terminologies and definitions for different plot sizes or lengths.

For instance, some specific terminologies would refer to the microplot and large plot that would lead to different interpretations on the application of plot dimensions. Based on the description of previous researchers, there are no clear description on the characteristics and the definitions of soil loss plot as microplot, small plot, large plot, small microplot, large microplot and so on. There should have more precise definitions and justifications that qualitatively define plot size for various research and study purposes. Lal [11] put forward the clarification and understanding of the types of plots that are generally adopted by researchers including small plot, Universal Soil Loss Equation (USLE) plot and large plot.

Table 1 summarizes the characteristics of the plots in addition to the details and functions of the relevant plots, including its advantages and disadvantages. The most important determining elements are the relationship of experimental outcomes and different types of plot adopted, including runoff coefficient, soil loss amount outside the site boundary, average soil movement rate on the plots, rainfall intensities, slope steepness, ground conditions, and the ratio of errors with respect to other plots.

Table 1. Summary of Characteristics of Types of Plot.

Types of Plot	Descriptions	Sources
Small Plot	• Used to study fundamental erosion process that is harder to be observed on bigger plots. • Provides more detailed information on inter-rill erosion and sediment yield. • Recommended for use below 40 m^2.	Meyer & Harmon [12] Bagarello & Ferro [13] Moldenhauer [14]
USLE Plot	• Standard erosion and sediment yield plots developed from USLE with the dimension of 22.1 m long and 9% slope. • The size is suitable to study combine erosion process (rill and inter-rill).	Wischmeier & Smith [15] Renard et al. [10]
Large Plot	• It is more representative of the outcomes on the study of sediment yield at a practical scale. • Recommended to be between 40 m^2 to 100 m^2, and an area larger than 100 m^2.	Mayerhofer et al. [16] Van Es, Van Es & Cassel [17] Bagarello & Ferro [13]

2. Materials and Methods

The site is located within the residential area of Pending Area, Kuching, Malaysia, with coordinate of 1°32' 47.99" N, 110° 22' 26.37" E and the climate is humid most of the time that contributes to high rainfall intensities [18], [19]. The site provided enough space for the setup of experimental plots and the surroundings of the plots and the surroundings of the plots are low rise residential buildings (detached houses with a maximum of two-story), which provides sufficient clearance for the installation of the tipping bucket rain gauge. The site is flat terrain to ensure no topographical factors (ground slope = 0%) would affect the gradient of the soil plots.

The experimental plots were constructed with reference to the required dimensions by the plot, which are (1) 1 m x 1 m plot, (2) 1 m x 2 m plot, and (3) 1 m x 3 m plot, which represent Plot A, Plot B and Plot C, accordingly. The structures and dimensions were designed to adequate height so that runoffs are collected. The soil loss plot and runoff collection points were designed to stand at 550 mm and 300 mm from the ground, respectively to ensure that the sampling of the runoff can be easily executed and collected at the runoff collection point. An angle of 22.78° from horizontal was designed for the plot to ensure that the runoff during low rainfall intensity storm event. Surfaces of the formwork (wood structures) were painted with waterproof paint and the gaps and openings on the formworks are sealed with waterproof silicone to prevent moisture (water) from entering the formwork. All the layers and structures were installed and checked before the formworks were filled up with soil samples. Runoff collection points were designed to be located at the ends (lowest point) of the individual plots. Aluminium plates were shaped to guide the runoff from the plots and plastic containers were used to collect the runoff during a rainfall event. The containers measured 250 mm (H), 550 mm (W) and 850 mm (L) or equivalent to 117 Litres each. Transparent roofing was installed for observation purposes and at the same time preventing rainwater from getting into the containers. The final setup of the experimental plots is shown in Figure 1.

The criteria of the Three (3) plots are shown in Table 2. All the soil samples fall under the definition as homogenous topsoil with no addition of sand or organic materials. The sample soils were loaded onto the test bay could be considered as disturb condition, whereby poor compaction applied on the soil samples but only flatten the surface for the simulation of natural flat ground with disturbed soil condition. The continuous erosion process would occur to the surface of the plot and that would increase the number of granular particles protruding on the surface. Removal of granular particles and a small portion of the soil to flatten the surface and fresh soil samples were refilled to replace the volume removed.

A standard rain gauge was used for determination of rainfall depth for a certain time period or a rainfall event (Figure 1). It is possible to get the rainfall intensity with the recorded duration of rain and the collected amount of rainwater the rain gauge. For this tipping bucket rain gauge, it was designed as 0.5 mm per tick. Every tick made by the lever arm was recorded by a contact a sensor underneath the lever and the signal was transmitted to the data logger.

During the experimental stage, the total runoff volume for each plot was determined after the rainfall event. Five different positions from the containers were marked and used as water level measuring points by using measuring tape. Average readings of the measuring points were taken as records to minimise errors during the readings. The rainfall data from the data logger were plotted in intensity-duration graph for perspective data display. The rainfall intensities were classified into 3 groups; <30 mm/h, 30-60 mm/h and >60 mm/h. The data were analysed and presented in box and whisker plot (Figure 2) for comparison of different plot lengths under various ranges of rainfall intensities on the runoff coefficient by using Equation 1, where RC is runoff coefficient (%), D_{runoff} is total runoff depth (mm) and $D_{rainfall}$ is total rainfall depth (mm).

$$RC = \frac{D_{Runoff}}{D_{Rainfall}} \times 100\% \quad (1)$$

Figure 1. A View on the Setup of Experimental Plot.

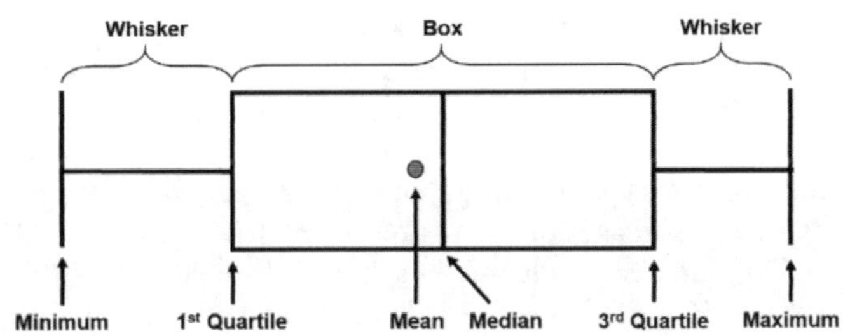

Figure 2. Interpretation of Results in Box and Whisker Plot.

Table 2. Design and Experimental Setup Details of Plot A, Plot B And Plot C.

Criteria	Plot A *(1 m x 1 m)*	Plot B *(1 m x 2 m)*	Plot C *(1 m x 3 m)*
Soil Type	Homogenous	Homogenous	Homogenous
Plot Length	1 m	2 m	3 m
Plot Width	1 m	1 m	1 m
Slope Steepness	22.78°	22.78°	22.78°
Plot Surface Condition	Flat Surface	Flat Surface	Flat Surface

3. Results and Discussions

The rainfall data were extracted from the data logger as the number of raw data from the logger in "tick-timing" and the "total ticks" as continuously recorded. Figure 3 shows the data displayed in bar chart format for rainfall depth. The data logger recorded rainfall details from 1st December 2019 to 21st January 2020. The highest daily rainfall depth recorded as high as 115 mm (230 ticks) over 24-hour period. As shown in Figure 3, from December 2019 to January 2020, the experimental plots experienced high-intensity rainfall events and low-to-medium rainfall intensities but long rainfall duration. A total of 26 storm events were recorded by the plots.

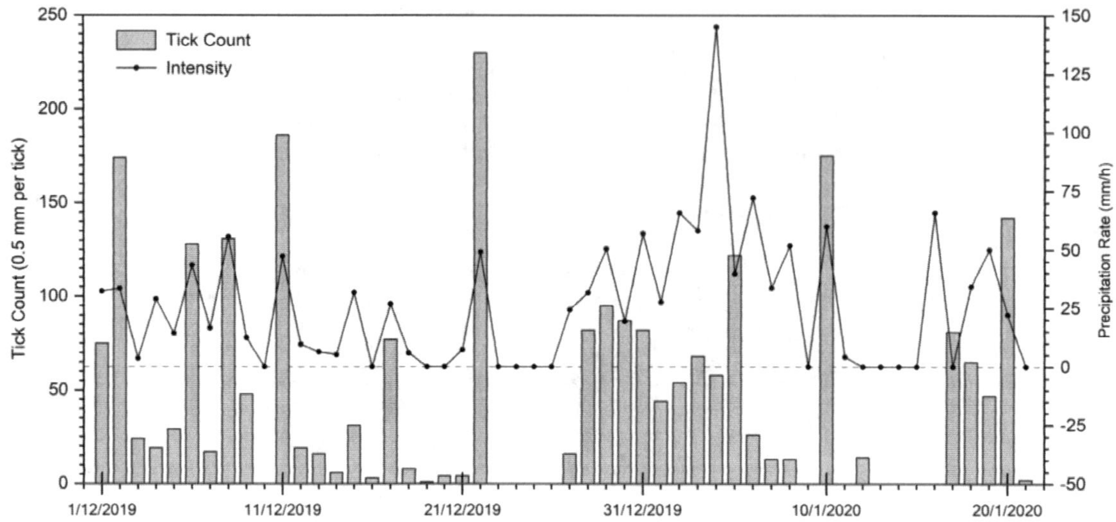

Figure 3. Measured Rainfall Intensity-Duration Data (01/12/2019 to 21/01/2020).

The rainfall-runoff relationship for Plot A (1 m x 1 m), Plot B (1 m x 2 m) and Plot C (1 m x 3 m) for this study are shown in Figure 4 and the regression results for the plots are shown in Table 3. The results shown in Figure 4 illustrate that the data distribution in similar ranges for all plots under the same precipitation. Plot A shows a comparatively higher response in the linear relationship of precipitation (P) and runoff depth (R), suggesting that intensity generates comparatively higher runoff amount as the precipitation rate increases, as compared to Plot B and Plot C. For Plot B, it tends to produce relatively higher R compared to Plot C with higher P, but Plot C has comparatively higher runoff production compared to Plot B for lower P. The relationship equation of R and P for the plots are listed in Table 3.

Table 3. Regression Results for Plot A, Plot B and Plot C.

Plot Type	Plot A	Plot B	Plot C
Dimension	1 m x 1 m	1 m x 2 m	1 m x 3 m
Pearson's r	0.916	0.935	0.911
R-Square (COD)	0.839	0.875	0.830
Adj. R-Square	0.827	0.863	0.816
Equation	$R=0.813P-9.175$	$R=0.627P-7.062$	$R=0.686P-9.686$

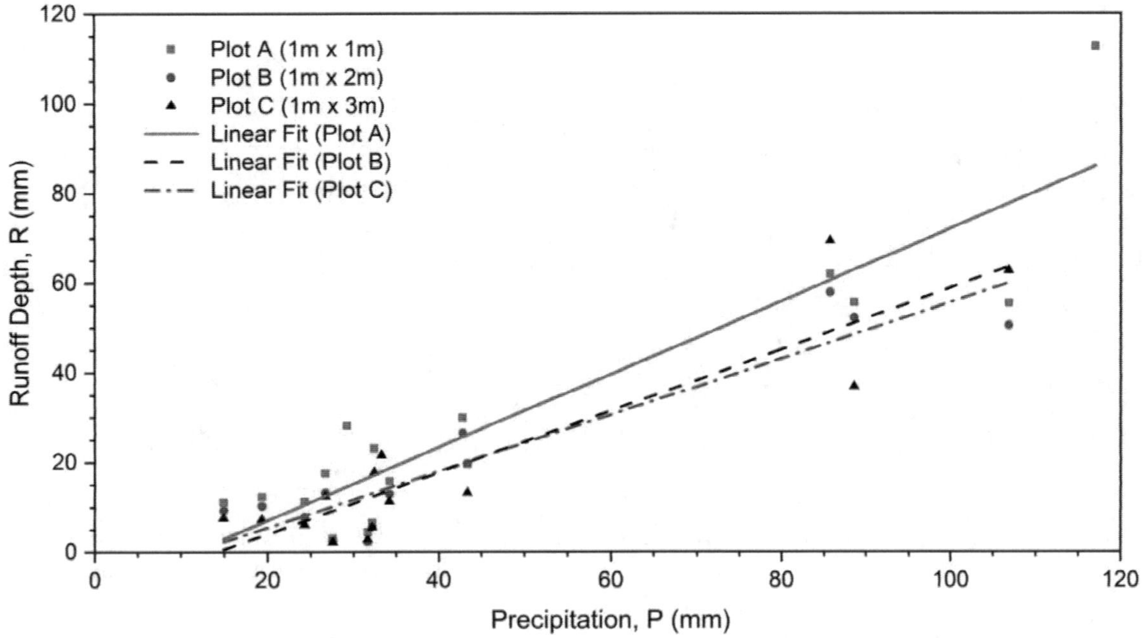

Figure 4. Scatter Plot and Linear Fit of Plot A, Plot B and Plot C.

Plot A, Plot B and Plot C showed high linear correlation between R and P with Pearson correlation coefficient, PCC (Pearson's r) of 0.916, 0.935 and 0.911, respectively. All the plots show a positive linear correlation of R and P, which indicates that the plots are appropriate for the data set. The R^2 value of 0.839, 0.875 and 0.830 for Plot A, Plot B and Plot C, respectively, suggests that all the variability responses data are around the mean value (insignificant deviation from the mean value). To date, there are only a few researchers studied on the relationship plot length and runoff volume. However, generally the results are comparable with the theory as part forward by Beven [5]. For shorter plot (Plot A), it would achieve soil moisture content saturations faster due to smaller areas as compared to longer plots (Plot B and Plot C). From this study, it is shown that the longer the plot length, the higher the area for infiltration to occur that results in lower runoff volume.

For this study, the data from natural rainfall were collected for a wide range of rainfall intensities and they are plotted in groups on runoff coefficients of the individual plots for the rainfall intensities (Figure 5) There is a common perception that larger areas of the plot could bring about a relatively higher runoff depth than a smaller plot. From this study, the results demonstrate that it is not necessarily true. From the results of precipitation groups of ≤30 mm/h, 30-60 mm/h and >60 mm/h, Plot A shows the highest runoff coefficient of 96.34%, 69.89% and 96.26%, respectively. The increments in the mean values for Plot C are 33.59%, 38.30% and 49.45% for precipitation groups of ≤30 mm/h, 30-60 mm/h and >60 mm/h, respectively.

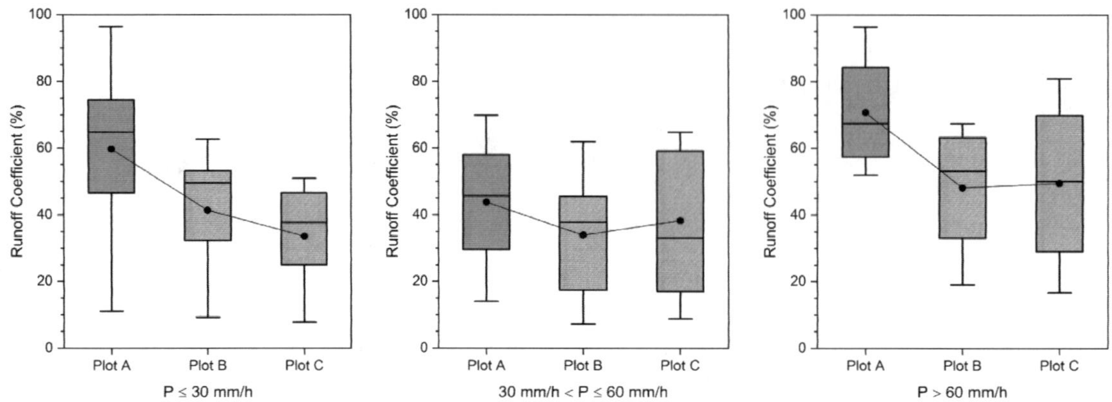

Figure 5. Box and Whisker Plots for Different Ranges of Rainfall Intensity.

Based on the results, the shorter plots would result in high runoff coefficients under various rainfall intensities as compared to longer plots. Alternatively, it could simply mean that during the runoff process, longer plot trends to have higher rates of infiltration along the contact surfaces on the plot. The shorter plot has lesser contact time with the plot surface resulting in high runoff depth and low infiltration rate during the storm events. It is notable that the data distribution characteristics depict a similar pattern with small data differences; indicating minor variations of median and minimum values. However, it is also found that there are minority outliners that could affect the distribution curves, especially in Plot A, even though insignificant on overall findings of various plots. With respect to the contact durations and infiltration processes on the plots, it can be concluded that plot size or length can be one of the most significant factors on the outcome of the runoff coefficients.

4. Conclusion

From this study, the following conclusions can be drawn: As shown in Fig. 4, shorter plot (Plot A) has relatively higher runoff volume as compared to the comparatively longer plots (Plot B and Plot C). The relationship of precipitation (P) and runoff depth (R) can be expressed as: Plot A: $R = 0.813P - 9.175$; Plot B: $R = 0.627P - 7.062$; Plot C: $R = 0.686P - 9.686$. The correlation of precipitation (P) and runoff (R) shows are showing positive linear relationships with R^2 of 0.839, 0.875, 0.830 for Plot A, Plot B and Plot C, respectively. Comparatively shorter slope length of Plot A shows the highest runoff coefficient in all the precipitation groups, with the highest runoff coefficients of 96.34%, 69.89% and 96.26% for ≤30 mm/h, 30-60 mm/h and >60 mm/h, respectively. Comparatively longer slope length of Plot C shows an increase in runoff coefficient in the precipitation groups of ≤30 mm/h, 30-60 mm/h and >60 mm/h with the mean values of 33.59%, 38.30% and 49.45%, respectively.

5. References

[1] T. J. Toy, G. R. Foster, and K. G. Renard, *Soil Erosion*. New York: John Wiley & Sons, 2002.
[2] W. S. Merritt, R. A. Letcher, and A. J. Jakeman, "A review of erosion and sediment transport models," *Environ. Model. Softw.*, vol. 18, no. 8–9, pp. 761–799, 2003.
[3] L. L. G. Samuel and K. K. Kuok, "Sensitivity Analysis of the Revised Universal Soil Loss Equation's Rainfall Erosivity Factor (R-Factor)," *Test Eng. Manag.*, vol. 83, no. May-June, pp. 6809–6815, 2020.
[4] L. D. Meyer, "How rain intensity affects interrill erosion," *Trans. ASAE*, vol. 24, no. 6, pp. 1472–1475, 1981.
[5] K. Beven, "Robert E. Horton's perceptual model of infiltration processes," *Hydrol. Process.*,

vol. 18, no. 17, pp. 3447–3460, 2004.
[6] K. K. Kuok, S. Harun, and C. P. Chan, "Hourly runoff forecast at different lead-time for a small watershed using artificial neural networks," *Int. J. Adv. Soft Comput. its Appl.*, vol. 3, no. 1, pp. 68–86, 2011.
[7] K. K. Kuok, S. Harun, S. M. Shamsuddin, and C. P. Chan, "Evaluation of Daily Rainfall-Runoff Model Using Multilayer Perceptron and Particle Swarm Optimization Feed Forward Neural Networks," *J. Environ. Hydrol.*, vol. 18, no. May, pp. 1–15, 2010.
[8] A. S. Tokar and P. A. Johnson, "Rainfall-runoff modeling using artificial neural networks," *J. Hydrol. Eng.*, vol. 4, no. 3, pp. 232–239, 1999.
[9] W. H. Wischmeier and D. D. Smith, "Predicting Rainfall Erosion Losses – A Guide to Conservation Planning," *Agric. Handb. No. 537*, no. Washington, DC: United States Department of Agriculture (USDA), 1978.
[10] K. G. Renard, G. R. Foster, G. A. Weesies, D. K. Mccool, and D. C. Yoder, *Predicting soil erosion by water: a guide to conservation planning with the revised universal soil loss equation (RUSLE)*. 1998.
[11] R. Lal, *Soil erosion research methods*. Soil and Water Conservation Society, 1994.
[12] L. D. Meyer and W. C. Harmon, "Interrill runoff and erosion: effects of row-sideslope shape, rain energy, and rain intensity," *Trans. ASAE*, vol. 35, no. 4, pp. 1199–1203, 1992.
[13] V. Bagarello and V. Ferro, "Plot-scale measurement of soil erosion at the experimental area of Sparacia (southern Italy)," *Hydrol. Process.*, vol. 18, no. 1, pp. 141–157, 2004.
[14] W. C. Moldenhauer, "Procedure for Studying Soil Characteristics Using Disturbed Samples and Simulated Rainfall," *Trans. ASAE*, vol. 8, no. 1, pp. 0074–0075, 1965.
[15] W. H. Wischmeier and D. D. Smith, *Predicting Rainfall Erosion Losses A Guide to Conservation Planning*. 1987.
[16] C. Mayerhofer, G. Meißl, K. Klebinder, B. Kohl, and G. Markart, "Comparison of the results of a small-plot and a large-plot rainfall simulator – Effects of land use and land cover on surface runoff in Alpine catchments," *Catena*, vol. 156, no. April, pp. 184–196, 2017.
[17] H. M. Van Es, C. L. Van Es, and D. K. Cassel, "Application of regionalized variable theory to large-plot field experiments," *Soil Sci. Soc. Am. J.*, vol. 53, no. 4, pp. 1178–1183, 1989.
[18] K. K. Kuok and P. C. Chiu, "Space-saving Rainwater Harvesting Tanks for Double Story Houses in Kuching, Sarawak," *Int. J. Eng. Technol.*, vol. 8, no. 1, pp. 38–43, 2019.
[19] K. K. Kuok, Z. Z. Liew, and P. C. Chiu, "Flood map development by coupling satellite maps and three-dimensional drafting software: Case study of the Sarawak River Basin," *Water SA*, vol. 39, no. 1, pp. 175–182, 2013.

Acknowledgement
The writers would like to thank Professor Ir. Dr. Law Puong Ling of the University of Malaysia, Sarawak (UNIMAS) for providing technical supports and valuable comments on the experimental works. This work was supported in part by the Fundamental Research Grant Scheme (FRGS) (Reference Code: FRGS/1/2019/TK01/SWIN/02/1) by Ministry of Education Malaysia (MOE).

Quantifying Carbon Emission from Campus Transportation: A Case Study in Universiti Kebangsaan Malaysia

Nurul Farhana Filza binti Zakaria[1], Muhamad Razuhanafi bin Mat Yazid[1] and Nur Fatma Fadilah Yaacob[1]

[1] Department of Civil and Structural Engineering, Faculty of Engineering and Built Environment, Universiti Kebangsaan Malaysia, 43600 Bangi, Selangor, Malaysia
Corresponding author: razuhanafi@ukm.edu.my

Abstract. Carbon footprint is a global concern discussed throughout the globe. The daily activities on university campuses cause emission of greenhouse gases that contribute to the increasing amount of carbon dioxide (CO_2) in the atmosphere. This research seeks to determine the CO_2 emissions by members of the campus community of Universiti Kebangsaan Malaysia (UKM), Bangi Campus. This study will also analyse the factors influencing CO_2 emission from the vehicles used on the campus. The findings of this study will serve as a basis to propose methods for reducing carbon footprint, especially from the transportation used on the campus. The study is carried out on the UKM campus to gather the data on the travel activity on the campus, where the members of the campus community serve as the study population. The DEFRA Model is used to calculate the CO_2 emission and the multivariable linear regression model is used to identify the factors influencing CO2 emission. In the car category, petrol cars with 2.0L engine have the highest mean CO2 emission of 1181.0932 kg CO2/year/capita. In the motorcycle category, petrol motorcycles with engine capacity larger than 500cc recorded the highest emission of 999.1072 kg CO2/year/capita. The CO2 emissions by both categories are below the national mean carbon dioxide emission value of 7900 kg CO2/year/capita recorded by the 2019 National Transport Policy and the mean carbon dioxide emissions of 5400 kg CO2/year/capita in high-income countries. In this study, the critical factors influencing the daily and annual CO2 emissions are the mode of transportation, type of fuel, and travel distance

1. Introduction

The recent years have seen an increasing concern regarding the harmful effect of carbon footprint on the community and environmental sustainability. According to Abdul-Azeez & Ho [1], the use of fossil fuels in the daily activities of humans have resulted in the release of an excessive amount of carbon dioxide (CO2). Since CO2 is one of the components of greenhouse gases (GHGs), the consumption of a higher amount of fossil fuels contributes significantly to global warming and climate change, which hampers the effort to achieve sustainable development. Besides contributing to global warming, other adverse effects of transportation on the environment include the production of harmful pollutants such as sulphur dioxide, nitrogen oxide, lead, hydrocarbon, carbon monoxide, and other harmful particles.

Information on the level of greenhouse gases emissions and the impact measurement of carbon footprint is critical in the effort to reduce the impact of GHG emissions [2]. Reduced carbon footprint contributes to the mitigation of global climate change. Carbon footprint is a measure of the carbon emission associated with a person, institution, or country. Knowing the carbon footprint of a person can contribute to increasing the awareness of the impact of global problems and climate change on the life of an individual [3]. The population of a university comprises the members of the staff, students, and the public that commute within the campus either using private vehicles or the transportation provided by the university, where each person has their own carbon footprint. A large university campus has a greater carbon footprint.

Abdul-Azeez & Ho [1] discovered that the highest source of CO2 emissions is from the vehicles used for commuting on the campus and delivering goods and equipment, where the carbon emission from these vehicles is equivalent to that of the national average. The key factor hindering the achievement of the sustainability goals of a university is the vehicles used by the campus community. The strategic plan to achieve campus sustainability must take into account the extended daily use of vehicles, travel distance, and the use of private and heavy vehicles. A study conducted at Universiti Kebangsaan Malaysia in 2011 discovered that about 18% of the members of campus community preferred to walk, 32% preferred to take the campus bus, and the remaining 50% preferred to use their private vehicle [4]. This research measured the CO2 emissions from the vehicles used by the members of the campus community and identified the key factors influencing CO2 emissions

2. Methodology

The research methodology is a critical component of research and is an approach adopted to achieve the key research objectives.

2.1. Research instrument

The research instrument in this study is a questionnaire comprising two sections. Section A gathers the demographic data of the respondents, including gender, age, race, occupation, and if the respondents are students, their study level, faculty, and college of residence. Section B gathers the information on the vehicles used on the campus, including the type of vehicle, type of fuel used, engine capacity, daily travel distance, and weekly travel frequency. The questionnaires were distributed to the respondents on the campus. The researcher gave a brief explanation of the survey to the respondents to ensure that they understand all items in the questionnaire.

2.2. Study area and sample size

The target population in this study is the campus community of Universiti Kebangsaan Malaysia (UKM), Bangi Campus. The 2019 population data provided by the Centre for Academic Management, UKM showed that the total number of staff and registered students at the UKM Bangi Campus is 26,601. This study adopted the method recommended by Burmeister and M. Aitken [5] for determining the sample size for studies that use regression analysis, where the sample size is obtained using the formula $N \geq 50 + 8P$ (where p is the total number of variables). This study considered 18 variables, namely method of commuting (car, motorcycle, bus, and walking), type of fuel (petrol or diesel), engine capacity (cars with 1.4L, 1.4-2.0L and 2.0L petrol engines, cars with 1.7L, 1.7-2.0L and 2.0L diesel engines, motorcycles with 125 cc, 125-500 cc, and 500 cc petrol engines, and diesel bus), and the daily and annual travel distance. Even though the formula gave a sample size of 194 respondents, 375 respondents participated in this study.

2.3. Calculation of CO_2 emissions

There are two methods for calculating carbon dioxide emissions. The straightforward and frequently used method to calculate CO_2 emissions multiplies the travel distance with the CO_2 emission factor for each vehicle [6]. The second method multiplies the amount of energy consumed during the travel with the emission factor of CO_2 of the energy. This study calculated the total CO_2 emissions by multiplying

IOP Conf. Series: Materials Science and Engineering 1101 (2021) 012011 doi:10.1088/1757-899X/1101/1/012011

the travel distance with DEFRA emission factor for each type of vehicle. Azliyana Azhari et al. [8] used the DEFRA model to analyze the carbon emission from a vehicular source in selected industrial areas in Malaysia. This study used the guidelines developed by the United Kingdom Agricultural Engineers Association for the Department of Energy and Climate Change (DECC) and the Department of Environment, Food and Rural Affairs, United Kingdom (DEFRA) [9] to calculate CO_2 emissions by using equation 1.

$$\text{Total Carbon Footprint (kg } CO_2 \text{ eq)} = \text{Travel distance (km) x Emission Factor (kg } CO_2 \text{ eq/ km)} \quad (1)$$

The data for the travel distance gathered through the questionnaire was multiplied with the emission factor to obtain the estimated daily and annual CO_2 emissions. In the case of a bus, this study assumed that the bus operated at a full capacity of 52 passengers (40 sitting and 12 standing passengers) and the emission factor was divided by 52 people to obtain the emission factor per person. Table 1 shows the emission factor for vehicles with different engine capacities and fuel types. As an example, the emission factor for a respondent commuting with a petrol car with a 1.6L engine is 0.21280 kg CO_2 eq/km.

Table 1. Emission Factor for each mode of transportation.

Mode of Transportation	Type of Fossil Fuel	Engine Size Capacity	Emission Factor (kg CO_2 eq/km)
Car	Petrol	1.4 L	0.17985
		1.4 L-2.0 L	0.21280
		2.0 L above	0.29549
	Diesel	1.7 L	0.15095
		1.7 L-2.0 L	0.18757
		2.0 L above	0.25580
Motorcycle	Petrol	125 cc	0.08499
		125-500cc	0.10316
		500 cc above	0.13724
Bus	Diesel	-	0.10351

Source: DEFRA [9]

2.4. Data analysis
This study used the Statistical Package for Social Sciences (SPSS) to analyze the data and obtain daily and annual mean emission for each method of commuting. The mean annual emission of each vehicle was compared with those of the national emissions and emissions in high-income countries. The study used the multivariate linear regression model to determine the critical factors in CO_2 emissions. The independent variables are the demographic factors, namely gender, age, occupation, mode of commuting, type of fossil fuel, engine capacity, and the travel distance. The dependent variables, namely the respondent's daily and annual CO_2 emissions, were calculated using the DEFRA model.

3. Result and Discussion

3.1. Carbon dioxide emissions

The first step in the analysis of carbon dioxide emissions is identifying the method of commuting on the campus.

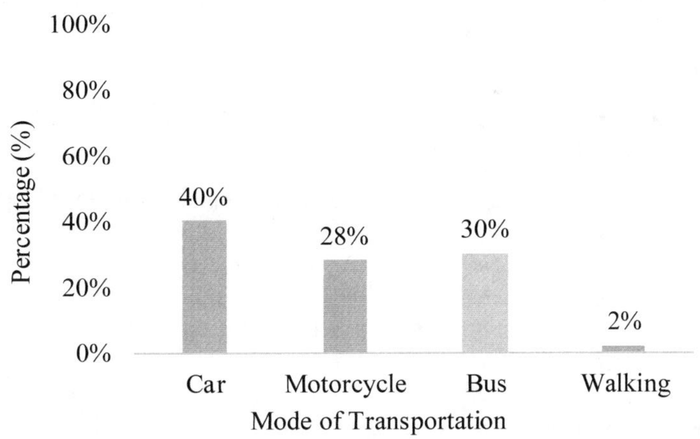

Figure 1 Graph of percentage of Mode of Transportation by respondent.

The bar chart in Figure 1 shows that 40% (152) of the respondents drove their cars, while 28% (104) preferred to commute by motorcycles. The remaining 30% (112) preferred to take the bus, and only 2% (7) chose to walk.

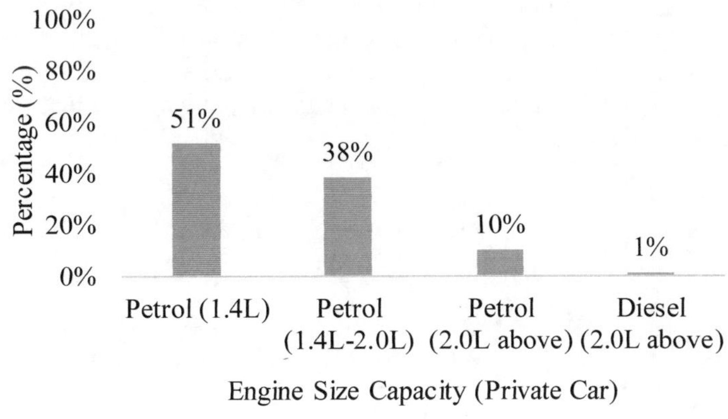

Figure 2 Graph of percentage of Engine Size Capacity (Private Car).

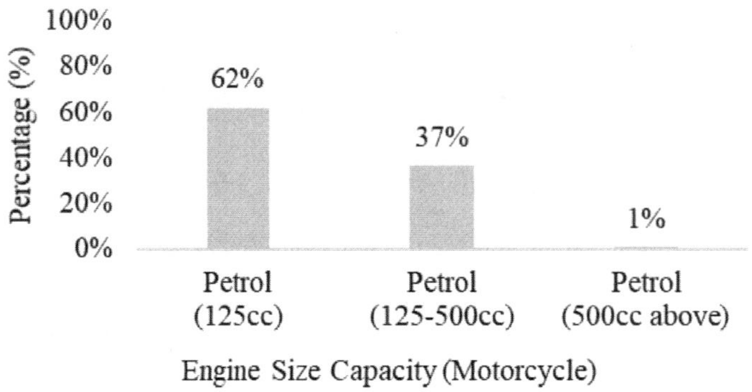

Figure 3 Graph of percentage of Engine Size Capacity (Motorcycle).

Figure 2 shows the breakdown of the cars driven by the respondents, where 51% are petrol cars with 1.4L engine, 38% are petrol cars with 1.4-2L engine, 10% are petrol cars with 2.0L engine, and 1% are diesel cars with 2.0L engine. Figure 3 shows the breakdown of the motorcycle used by 104 respondents, where 62% are petrol motorcycles with 125cc engine, 37% are petrol motorcycle 125-500cc engine, and 1% is petrol motorcycles with 500cc engine. The DEFRA model was used to calculate the daily and annual carbon dioxide (CO_2) emission from each vehicle.

Figure 4 shows that petrol cars with 1.4-2L engine released the highest amount of carbon dioxide annually (38607.8784 kg CO_2 or 33.23%), followed by petrol cars with a 1.4L engine, which released 30.33% or 35239.09 kg CO_2 annually. Cars with the highest engine capacity released the highest amount of carbon dioxide annually compared to other types of vehicles. Petrol and diesel cars with an engine capacity larger than 2L contributed 15.25% and 0.34% of the annual carbon dioxide emissions, respectively. The low percentage of CO_2 emission from cars with large engine capacity is because only a small proportion of respondents commute by this type of car. Petrol motorcycles with 125-500 cc engine released the highest amount of carbon dioxide (11694.22 kg CO_2 or 10.06%) compared to 125 cc petrol motorcycles (9.69%) and 500 cc petrol motorcycles (0.86%). Bus passengers contributed the least to CO_2 emissions, where each passenger contributed about 0.24% (276.9653 kg CO_2) of the annual carbon dioxide emissions. Walking is the most sustainable way for commuting since there is no carbon dioxide emission.

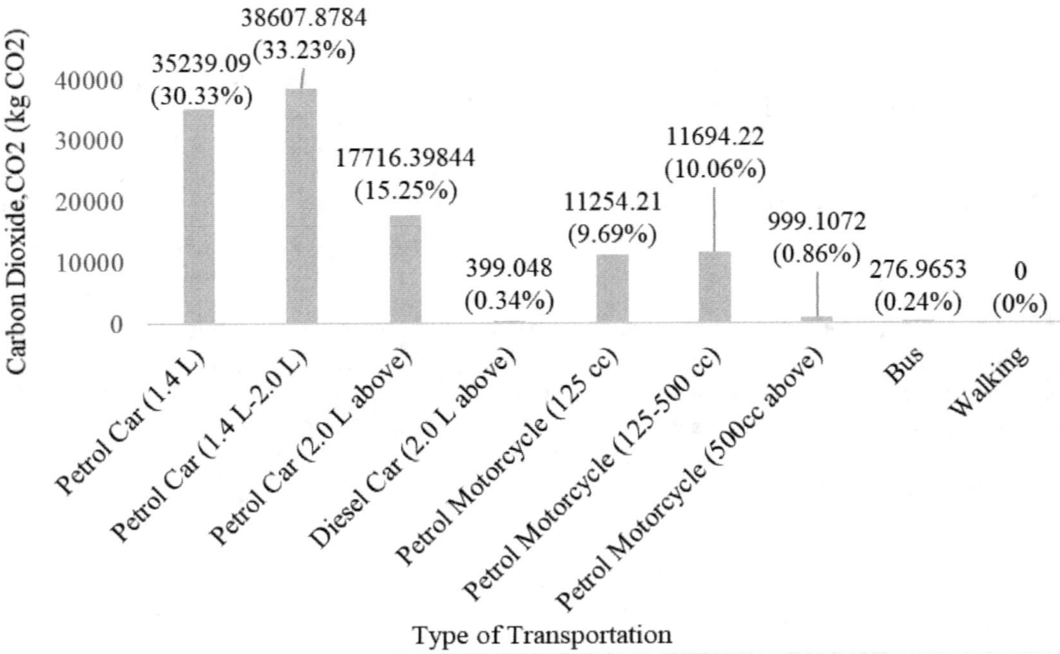

Figure 4 Graph of total annual emissions of Carbon Dioxide (kg CO_2/ year).

The descriptive analysis identified low carbon dioxide emissions per capita by each vehicle. The mean annual emission was compared with the national emission provided the National Transportation Policy, Malaysia and the mean emissions in high-income countries [10].

Table 2. Mean annual emissions of Carbon Dioxide (CO_2).

Transportation	Fuel type	Engine size capacity	Mean Carbon Dioxide (CO_2) annual (Kg CO_2 / km)
Car	Petrol	1.4 L	451.7832
		1.4 L-2.0 L	665.6531
		2.0 L above	1181.0932
	Diesel	2.0 L above	399.048
Motorcycle	Petrol	125 cc	173.1416
		125-500cc	307.7426
		500 cc above	999.1072
Bus	Diesel	-	2.4729
Walking	-	-	0
Mean CO_2 emission Malaysia =7900kg CO_2/km			
Mean CO_2 emission for high income country =5400kg CO_2/km			

Table 2 shows the mean annual carbon dioxide emissions (kg CO_2/ year/capita) for each type of vehicle relative to those of the mean national carbon dioxide emission and the mean value for high-income countries. Petrol cars with engine capacity larger than 2.0L have the highest mean annual emission (1181.0932 kg CO_2/ year/capita), followed by petrol motorcycle with engine capacity larger than 500cc (999.1072 kg CO_2/ year/capita). It is worth noting that the high mean values both transportation modes may not be accurate because of the small number of respondents using these vehicles to commute. Relative to other categories of cars, petrol cars with 1.4L-2.0L engine have the second-highest mean emissions of 665.6531 kg CO_2/year/capita. In the case of petrol motorcycles, those with 125-500 cc engine emitted the second-highest amount of carbon dioxide of 307.7426 kg CO_2/year/capita. Bus passengers emitted only 2.4729 kg CO_2/year/capita. Generally, the mean annual CO_2 emissions (kg CO_2/year/capita) for all vehicles are lower than the national mean carbon dioxide emissions (7900 kgCO_2/year/capita) and the mean carbon dioxide emission in high-income countries (5400 kg CO_2/year/capita).

3.2. Multivariate linear regression model

This study used multivariate linear regression model in SPSS software to identify the critical factors influencing carbon dioxide (CO_2) emissions from the vehicles. These factors were evaluated using p-values, which indicate the significance of the factors in CO_2 emissions. Factors with p-value <0.05 shows are significant, while those with p-value >0.05 are not significant. Table 3 presents the p-values of factors in daily and annual carbon dioxide emissions.

Table 3. Influenced Factor.

Factor	Daily			Annually		
	Unstandardized Coefficient	Standard Coefficient	Sig (p-value)	Unstandardized Coefficient	Standard Coefficient	Sig (p-value)
	B	Beta		B	Beta	
Gender	0.017	-0.005	0.869	4.417	0.005	0.883
Age	0.242	0.108	0.034	54.068	0.087	0.105
Occupation	0.089	0.022	0.671	44.665	0.039	0.464
Mode of Transportation	-1.537	-0.815	0.000	-433.949	-0.827	0.000
Fossil Fuel	0.525	0.298	0.000	133.450	0.272	0.000
Engine Size Capacity	0.131	0.225	0.073	40.353	0.249	0.059
Distance Travelled	2.679	0.672	0.000	696.784	0.652	0.000
Gender: Male	0.108	0.032	0.486	38.805	0.041	0.371
Occupation: Staff	1.954	0.478	0.000	528.685	0.465	0.000
Car	2.156	0.645	0.000	605.016	0.650	0.000
Motorcycle	0.797	0.217	0.136	230.265	0.226	0.120
Bus	0.009	0.002	0.987	2.473	0.002	0.987
Petrol Car (1.4L)	1.646	0.407	0.001	451.783	0.401	0.001
Petrol Car (1.4L-2L)	2.320	0.511	0.000	665.653	0.527	0.000
Petrol Car (2L)	4.216	0.503	0.000	1181.093	0.506	0.000
Diesel Car (2L)	1.535	0.048	0.260	399.048	0.045	0.288
Petrol Motorcycle (125cc)	0.575	0.133	0.257	173.142	0.143	0.216
Petrol Motorcycle (125-500cc)	1.124	0.207	0.033	307.743	0.203	0.034
Petrol Motorcycle (500cc above)	2.745	0.086	0.044	999.107	0.113	0.008
Bus	0.009	0.002	0.986	2.473	0.002	0.986

Table 3 shows the key factors of the model. The model was run for overall factors to determine the effects of the independent factors, namely gender, age, occupation, type of vehicle, fossil fuel, engine capacity, and travel distance, on the daily and annual carbon dioxide emissions. Generally, the type of vehicle, type of fossil fuel, and travel distance are the critical factors influencing the daily and annual emission. The age factor is only influential in the model for daily carbon emissions because the p-values of all factors are less than 0.05, which means that any change in these factors will cause higher carbon dioxide emissions. Gender, occupation, and engine capacity did not influence the daily and annual carbon dioxide emissions. In conclusion, the type of vehicle, type of fossil fuel, and travel distance are the factors having the most influence on the daily and annual CO_2 emissions. The outcomes of this study are congruent with those obtained by Triantafyllidid et al. [11], which showed that different modes of transportation and travel distance influenced CO_2 emissions.

A detailed analysis showed that the total daily and annual carbon dioxide emission is not affected by gender. However, the occupation of the staff member influenced the daily and annual carbon emissions, indicating that greater participation of the staff members in the survey resulted in higher daily and annual carbon dioxide emissions. In this study, female respondents contributed to higher CO_2 emissions relative to male respondents. On the contrary, P. Wei and H.pan [6] discovered that male respondents contributed to higher CO_2 emissions from transportation. The result of this study also showed that age and occupation are not critical factors in CO_2 emission, which is in line with the findings made by Y. Lee [12].

Further analysis of the type of vehicle showed that cars contribute to the daily and annual carbon emissions, where an increase in car usage will increase CO_2 emissions. In terms of engine capacity and type of fossil fuels, all petrol cars contribute significantly to the daily and annual carbon dioxide emissions, where both factors have a p-value of less than 0.05. The p-values for petrol motorcycles with engine capacities of 125-500cc and 500cc are less than 0.05, which means that these vehicles influenced the amount of daily and annual carbon dioxide emissions. S. Mustapa and H. Bekhet [13] have shown that petrol vehicles emitted a high amount of carbon dioxide. Given that 70% of the cars in Malaysia run on petrol, one effective way to reduce carbon emissions on the UKM campus is by providing an alternative commuting method that releases less carbon dioxide into the environment. In summary, an efficient method for reducing carbon emissions is by providing a viable travel alternative on campus.

4. Conclusion

This study used the DEFRA model to determine the CO_2 emissions of each type of vehicle used for commuting on the UKM Bangi campus. Petrol cars with engine capacity of between 1.4 to 2.0L emitted 33.33% or 38607.8784 kg CO_2 per year. Respondents who commuted by bus contributed the least to CO_2 emission (about 276.9653 kg CO_2 per year or 0.24%) and have the lowest emission rate per capita of 2.4729 kg CO_2. This low emission rate makes buses the most sustainable method for commuting. However, the mean annual CO_2 emissions of all types of vehicles used on the campus are lower than the 7900 kg CO_2 mean annual carbon emission in Malaysia and 5400 kg CO_2 mean annual emission in high-income countries.

The result of multivariate linear regression showed that, unlike the respondents who commute by bus and motorcycle, respondents who commute by car have a p-value of less than 0.05. Petrol cars with different engine capacities of 1.4L, 1.4-2.0L, and 2.0L and motorcycles with 125-500 and 500 cc engines have significant p-values. The type of vehicle, type of fuel, and distance travelled have significant p-values of less than 0.05 and are the critical factors influencing the daily and annual carbon dioxide emissions.

In conclusion, all members of the campus community, including the management, staff and students, must play a role in the effort to reduce CO_2 emissions by reducing the number of vehicles on the campus by taking the campus bus and carpooling and using vehicles with low carbon emissions. Another way to reduce carbon emissions is by providing bicycles or electric motorcycles at a low rental fee. These measures are critical in ensuring comprehensive and sustainable campus transportation that would ultimately benefit the campus community and ensure environmental sustainability.

5. References

[1] Abdul-Azeez I and Ho C 2015 Realizing Low Carbon Emission in the University Campus towards Energy Sustainability *Open Journal of energy efficiency* p 15-27
[2] Gocer O and Gocer K 2018 The Effects of Transportation Modes on Campus Use: A Case Study of a Suburban Campus *Case Studies on Transport Policy* (Elsevier vol 7) p 37-47
[3] Sippel M, Meyer D and Scholliers N 2018 What about greenhouse gas emissions from students? An analysis of lifestyle and carbon footprints at the University of Applied Science in Konstanz, Germany *Carbon Management* p 201-211
[4] Norzalwi N and Ismail A 2011 Public Approach Towards Sustainable Transportation in UKM's Campus *Australian Journal of Basic and Applied Sciences* (volume 5(5)) p 1332-1337
[5] Burmeister E and Aitken, M Leanne 2012 Sample size: How many is enough? *Australian Critical Care* 25(4) p 271-274
[6] Wei P and Pan H 2017 Research on individual carbon dioxide emissions of commuting in peri-urban area of metropolitan cities - an empirical study in Shanghai *Transportation Research Procedia* (volume 25) p 3463-3482
[7] Zhang X, Liu P, Li Z and Yu H 2013 Modeling the Effects of Low-carbon Emission Constraints on Mode and Route Choices in Transportation Networks *Procedia - Social and Behavioral Sciences* (volume 96) p 329-338
[8] Azhari A, Mohamed A Farid and Latif, M Talib 201. Carbon emission from vehicular source in selected industrial areas in Malaysia *International Journal of the Malay World and Civilisation* (Volume 4 (1)) p 89-93
[9] Defra 2010. DECC's GHG Conversion Factors for Company Reporting
[10] Kementerian Pengangkutan Malaysia 2019 Dasar Pengangkutan Negara 2019-2030
[11] Triantafyllidis S, Ries, J Robert and Kaplanidou K 2018 Carbon Dioxide Emissions of Spectators' Transportation in Collegiate Sporting Events: Comparing On-Campus and Off-Campus Stadium Locations *Sustainability Journal* (Volume 10(1)) p 12–17
[12] Lee Y 2019 Carbon Footprint and Built-up Land Footprint: Taipei Case Study E3S Web of Conferences 118 (EDP Sciences)
[13] Mustapa S. Indati and Bekhet H Ali 2018 Factors Influencing CO_2 Emissions and Strategies for Emissions Reduction in Malaysian Transportation Sector *International Journal of Engineering and Technology* p 823-829.

Acknowledgments
The author would like to thank Universiti Kebangsaan Malaysia (UKM) and the Ministry of Education for their financial support under research grant FRGS/1/2019/TK08/UKM/02/1.

IOP Conf. Series: Materials Science and Engineering 1101 (2021) 012012 doi:10.1088/1757-899X/1101/1/012012

The potential of cenospheres production from Malaysian coal power plants

Salmia Beddu[1*], Mahyun Zainoodin[1], Amalina Basri[1], Zarina Itam[1], Raudhah Ahmadi[2] and Teh Sabariah Abd Manan[3]

[1] Civil Engineering, Universiti Tenaga Nasional (UNITEN), Kajang, Malaysia
[2] Civil Engineering, Universiti Malaysia Sarawak (UNIMAS), Kota Samarahan, Malaysia
[3] Institute of Tropical Biodiversity and Sustainable Development, Universiti Malaysia Terengganu (UMT), Kuala Terengganu, Malaysia
*Corresponding author: drsalmiabeddu@gmail.com

Abstract. Coal Combustion Waste (CCW) fly ash, main component of coal ash waste generated at the Coal-Fired Power Plant throughout the year that causes the presents of toxic metals in the ash. Hence, pose significant health threats towards humans and environment. Cenosphere is a coal combustion by-product obtained from fly ash in boilers of coal fired power plants. The properties of cenosphere including spherical, lightweight, good thermal stability, chemical proof, heat resistance making it to have improved insulation. The aim of this research was to investigate the potential of cenospheres production from three different coal fired power plants in Malaysia namely Jimah, Kapar and Manjung. The wet method was used for cenospheres separation from the fly ash. The percentages of cenospheres ranged from 1 to 1.2% of every 1000 gram of fly ash. The mean sizes were 95.86μm, 88.64μm, 49.04μm for Jimah, Kapar and Manjung respectively. Cenosphere evacuated by utilizing water as a medium, it will expend a lot of water and lead to water contamination cause of leaching of toxic material while extracting the cenosphere from fly ash. Nevertheless, any health and environmental impacts from the value-added physical application of cenospheres in civil engineering materials must be screened via strict monitoring and legislated guidelines.

1. Introduction

Recently, there have been several studies and research conducted to reuse the waste produced by municipal solid waste [1]. Incineration is a growing form of waste management which is to reuse waste energy to produce electricity. To provide energy recovery, it can reduce the excess mass by 70 percent and the waste volume by 90 percent. Coal dependency in the energy sector is also observed in the Asia Pacific region, where coal has been used to cover more than 45 percent of energy consumption [1].

Electricity power is generated by combustion in thermal power plants. Coal combustion by products (CCPs) are commonly known as solid wastes produced during the combustion process. For instance, as fly ash, bottom ash, boiler slag, and flue gas desulphurization materials [2-5].

Figure 1 shows the by-product produced from burning pulverized coal in an electrical generating layout in a station, commonly known as fly ash. It's also gathered by mechanical or electrostatic separators, an incomplete combustion residue, and carried away by flue gases from the burning zone in the boiler [6]. Fly ash was reported to have various value-added applications by previous researchers as proposed in figure 2.

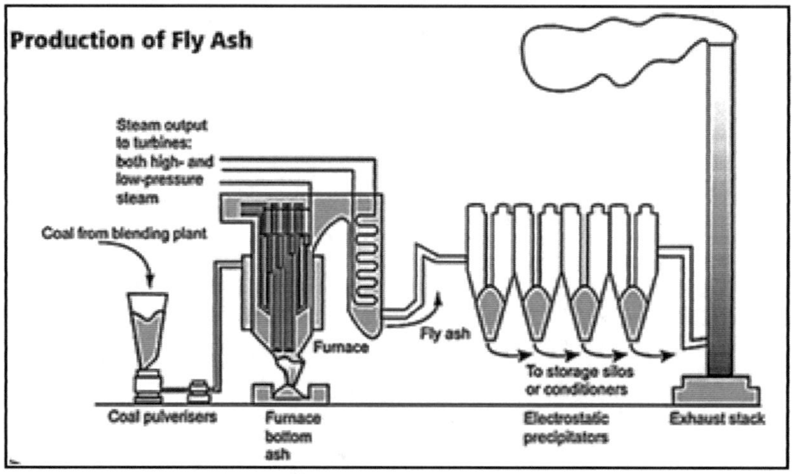

Figure 1. Coal-fired electrical generating layout in station [6].

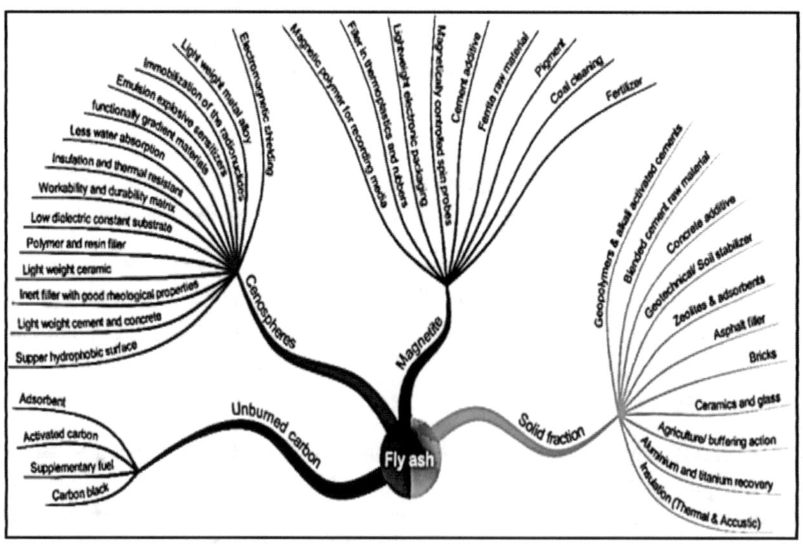

Figure 2. Components of fly ash and their potential applications [7].

One part of fly ash is the cenosphere. Along with its unique properties it has immense potential that it would be compatible with a wide variety of industrial applications.

1.1 Cenosphere

Cenosphere is one of the components that can be extracted from fly ash and has immense potential because of its unique properties including spherical, lightweight, heat resistant, have enhanced insulation, good thermal stability, and chemical inertness. In addition, the cenosphere will be conformable to a wide range of industrial applications such as, engineering industry, highway engineering, oil, and gas industry, foundry engineering as well as in the chemical industry [7].

There are some factors that lead to an increase in demand for the cenosphere such as, low density, excellent mechanical strength and reasonable cost. A limited study of the cenosphere from the Malaysian sources was found from previous studies [8]. The characteristic of the cenospheres from each coal power plant is slightly different. The formation of the cenosphere is occurred starting from thermochemical results, and phase of the transformations due to mineral coal formed in the combustion. Granulometric, chemical, and phase-mineral compositions of the cenospheres are depending on the composition of the original coal, type of furnaces employed, conditions used for cooling melt droplets, and etcetera [12]. As mentioned earlier, cenosphere means hollow and spherical. These hollows are formed during coal combustion at high temperatures. The larger size of hollow formed by a thin wall size (wall thickness is less than 10 percent of the diameter), and represented about 1-2 percent (weight) of total fly ash, the higher the cenosphere 's ability to float in water [13].

As the sizes of hollow varies, density varies. The density of the cenosphere differs from silica which can be as high as (~2600 kg/m^3) down to 200 kg/m^3 and even lower, depending on the size and number of the gas inclusions in each particle [14]. According to Hirajima, the remaining 10% of the cenosphere have diameters lower than 100 μm, while 90 % of the cenosphere have diameters between 100 – 420 μm [15-16]. Several sizes of cenospheres were reported in the previous studies were in the ranges of 10 μm to 100 μm [8], 1 μm to 300 μm [9], and 1 μm to 600 μm [10]. These ranges of sizes can be compared with the size for cenospheres collected from the Malaysia power plants. The particle size of the cenospheres were large compared to the particle size of the fly ash due to the entangled air inside the rigid hollow sphere of the cenospheres [11].

Figure 3 shows spherical shapes of the cenosphere which lead to superior properties include: increasing flow characteristics; chemical inertness; less water absorption; lightweight; good packing factor; enhanced insulation; reduced shrinkage; excellent mechanical strength; good thermal resistance; and good electrical properties. Cenosphere can be used in a wide range of industrial processes to offer product enhancements, significant manufacturing advantages and cost reductions [17]. Due to its significant characteristics like shape, size and density become highly demanded in industrial applications. Unique properties of cenosphere make it suitable for wide range industrial applications [18].

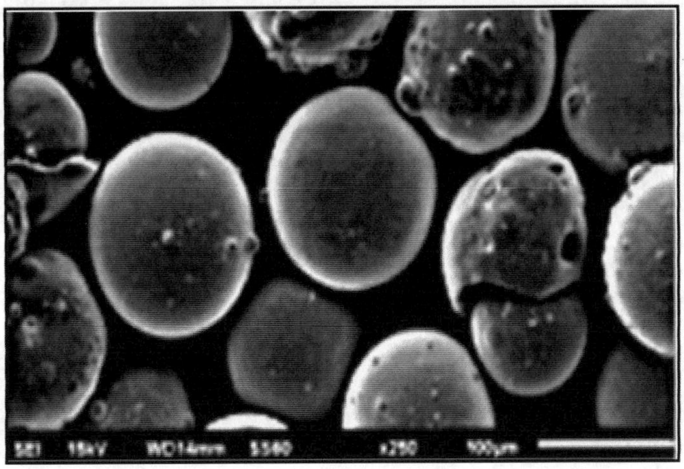

Figure 3. Spherical particles shape formation of cenosphere in fly ash [7].

1.2 Cenosphere collection

Fly ash is a waste product, produced in large quantities. Cenosphere, one of the materials that has value-added fractions in fly ash, whereas it is already well-known that it is produced as a product of coal combustion at thermal power plants. The cenosphere that is used in this research is extracted from fly ash using the wet method (Floatation / Float and Sink Method).

Briefly, a method involving dispersing the samples of dry fly ash collected from the power plants in water (distilled water or tap water) at room temperature. By the gravity separation processes, the cenosphere is extracted and removed by using water as a medium in the form of a hydraulic separation illustrated in figure 4(b). The full drawing for the wet separation apparatus method can be referred in figure 4(a). by pond skimming. Commonly, several methods are used to distinguish cenospheres from fly ash, including sink-float, centrifugal separation, and hydro- and aero-dynamic separation combination.

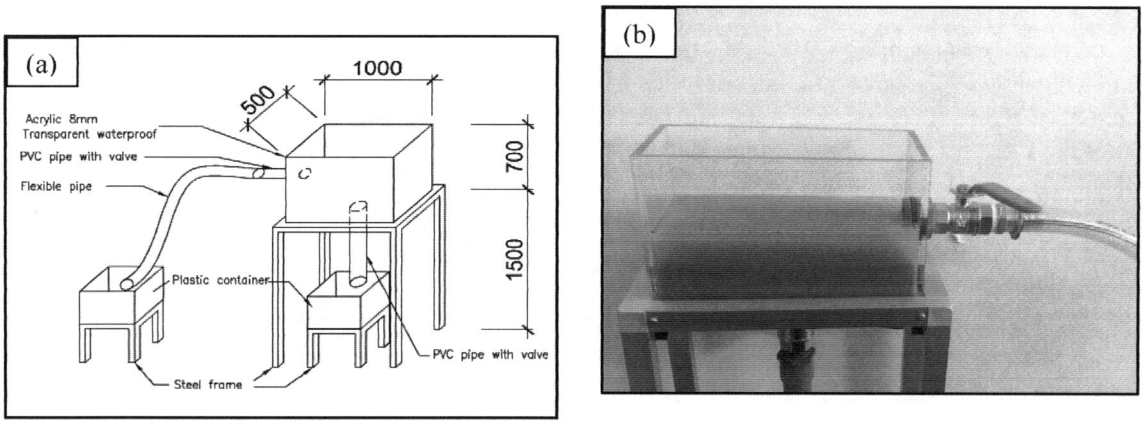

Figure 4. Full drawing for the wet separation apparatus method in (a) and (b).

The aim of this research is to investigate the potential of production of cenospheres from existing coal fired power plants in Malaysia. The percentages of cenospheres per fly ash were reported. The particle sizes of cenospheres and fly ash were compared. However, the mobilization of heavy metals fluxes and rich compounds like fly ash are harmful to the environment and health [1]-[4]. As stated by Thomas [5], the leaching properties of Jimah power plant fly ash and bottom ash were in compliance with the allowable limits of metal metal (Lead (Pb) and Arsenic (As)) and industrial effluent (Zinc(Zn) and Copper (Cu)) hazardous waste guidelines.

As cenospheres are part of components from fly ash, the heavy metals fluxes mobilization might probably be similar. However, further studies on leaching properties of cenospheres are highly recommended. Any health and environmental implications from the value-added physical application of cenospheres in civil engineering materials should be strictly monitored according to the legislated guidelines.

2. Methodology

In this research, cenospheres were extracted from fly ash using the wet method (floatation / float and sink). This method involved dispersing of dry fly ash samples in distilled water at room temperature. The cenospheres were extracted and removed by using water as a medium in the form of a hydraulic separation method using the gravitational separation processes. Fly ash was sedimented upon continuous stirring leaving floated cenospheres. The cenospheres were collected via decantation followed by the filtration process at figure 5(a) and (b).

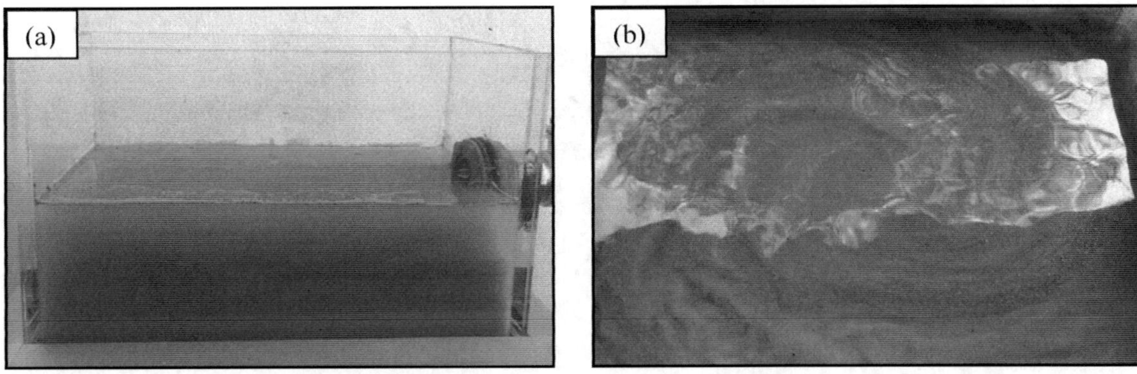

Figure 5. Cenospheres extraction methods: (a) wet method and (b) filtration method used in this study.

In Peninsular Malaysia, there are four well-known Power Plants, state-owned power companies that are assigned to supply coals shown in figure 6. Figure 7 shows cenosphere extracted from fly ash at Jimah, Kapar and Manjung power plants.

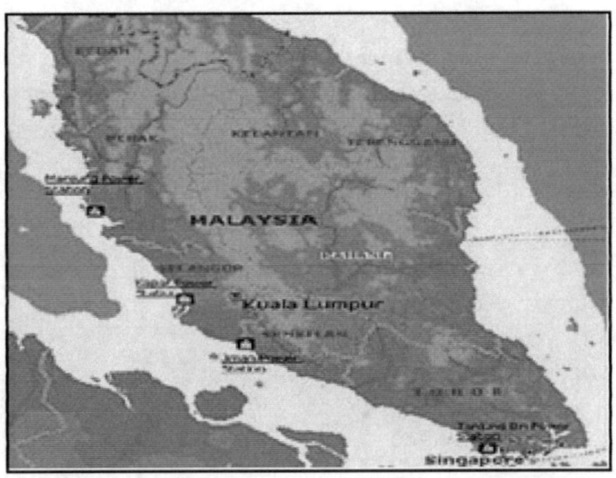

Figure 6. Power Plant Locations in Malaysia.

3. Result and discussion

Table 1 shows the percentage of cenospheres extracted using wet methods. The percentages of cenoshphere were 1.2% (Jimah), 1.1% (Kapar) and 0.9% (Manjung). The percentage of the extracted cenospheres were comparable to previous literatures [14].

Table 1 Percentage of Cenosphere extracted using the wet method.

Power Plants	Mass Fly Ash (g)	Mass Cenosphere (g)	Percentage Cenosphere (%)
Jimah	1000	12	1.2
Kapar	1000	11	1.1
Manjung	1000	9	0.9

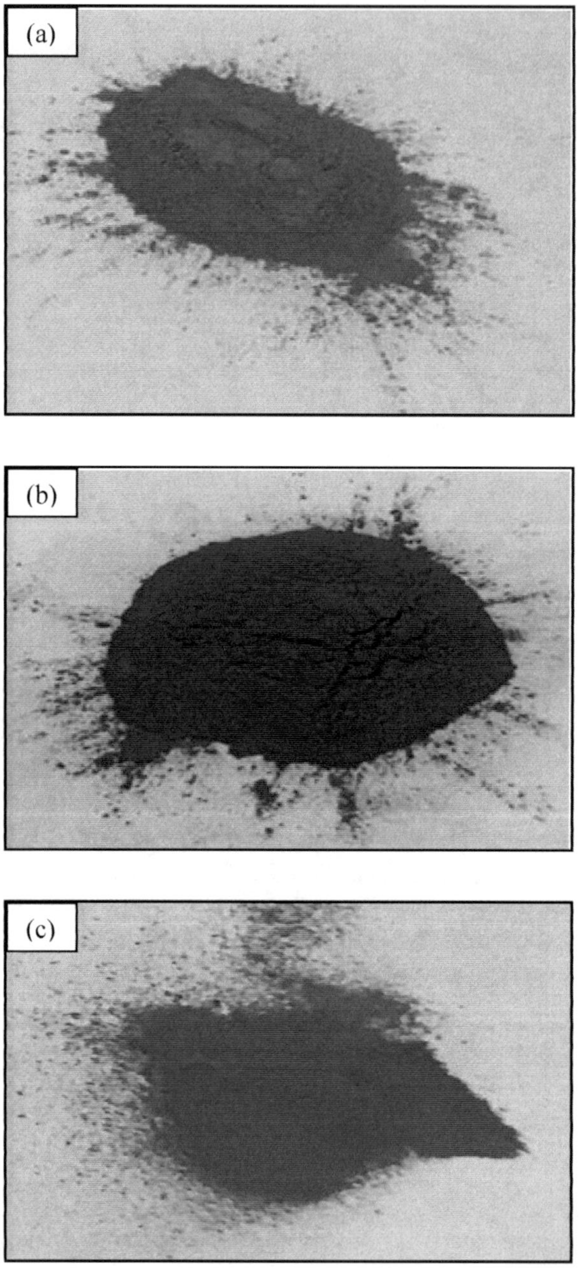

Figure 7 Cenospheres from (a) Jimah, (b) Kapar and (c) Manjung Power Plant.

Fly ash indicates that the median diameter (d50 is in the normal range from different sources between 2.5 μm and 40 μm in figure 8 (a). Similar to the extracted cenospheres in figure 8(b), it shows that the d50 ranged between 40 μm and 100 μm.

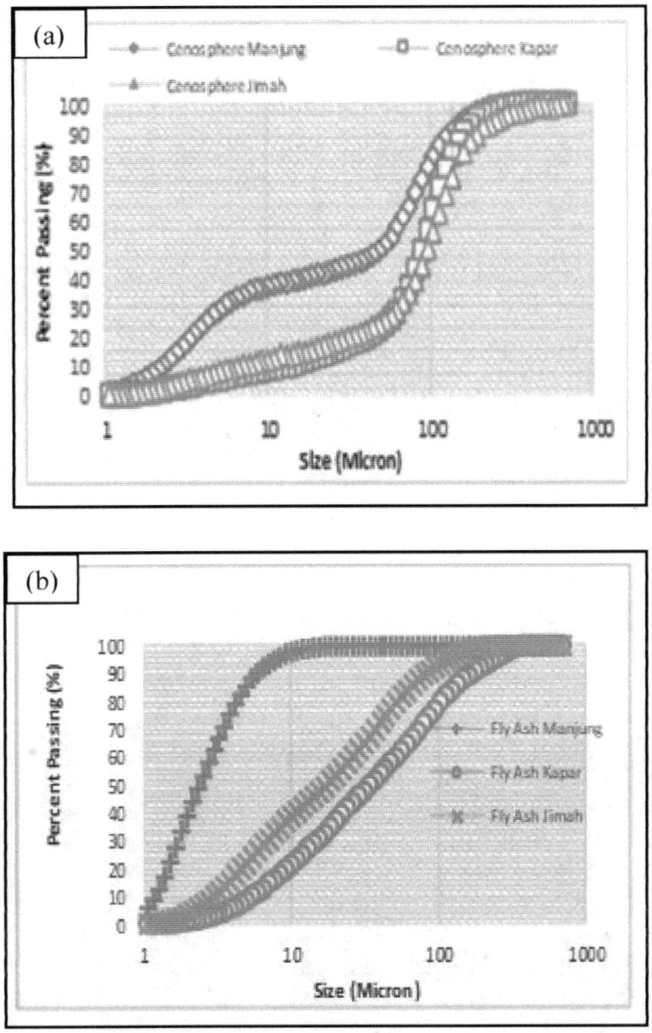

Figure 8 Particle size distribution of (a) cenospheres and (b) fly ash.

The extracted cenospheres sizes were ranged between 1 μm to 600 μm. However, fly ash of different sources could account for the different sizes of cenospheres. The particle size of the cenosphere is large compared with the particle size of the fly ash [6].

4. Conclusion

This research study is an effort to investigate the potential of cenospheres production from the existing power plants in Malaysia. The percentages of cenospheres production and the particles size of the extracted cenospheres were reported accordingly. The percentage of cenosphere extracted using wet methods for Jimah power plant sources, Kapar Power plant sources and Manjung Power Plant sources. It shows that the percentage of the cenosphere that is extracted for this study was around 1-2 % as stated by the previous researcher.

The percentages of cenospheres production and the particles size of the extracted cenospheres were reported accordingly. Sources from the Kapar power plant were found to have a larger size compared to the Jimah power plant while the finest particle size was observed at the Manjung power plant. The particle size of the cenospheres and fly ash for three separated power plants located in Malaysia showed that the particle size of the cenospheres was large compared to the particle size of the fly ash due to entangled air inside the cenospheres' rigid spher hence, raising the diameter.

Declaration of Competing Interest

The authors declare that they have no known competing financial interests or personal relationships which have, or could be perceived to have, influenced the work reported in this article.

Acknowledgments

The authors are thankful to Universiti Tenaga Nasional, Malaysia for providing financial supports (BOLD2020 Grant (RJO10517844/050) and YCU Grant (20190100101YCU/11) for this study.

References

[1] B. P. C. British Petroleum Company. (1981). BP Statistical Review of World Energy. London, "BP Statistical Review of World Energy. 67th edition, 2018," 2018.
[2] S. Beddu, M. Zainoodin, N. M. Kamal, D. Mohamad, S. Nabihah, and F. M. Nazri, "Material Characterization of Coal Combustion Product (Cenosphere) Generated from Coal–Fired Power Plant in Malaysia" In AIP Conference Proceedings, vol. 2030, No. 1, p. 020259, 2018.
[3] S. Jala and D. Goyal, "Fly Ash as a Soil Ameliorant for Improving Crop Production—a review" Bioresource Technology, vol. 97(9), pp. 1136-1147, 2016.
[4] K. H. Chuang, C. H. Lu, J. C. Chen, and M. Y. Wey, "Reuse of Bottom Ash and Fly Ash from Mechanical-Bed and Fluidized-Bed Municipal Incinerators in Manufacturing Lightweight Aggregates" Ceramics International, 44(11), pp.12691-12696, 2018
[5] M. D. A. Thomas, "Optimizing the Use of Fly Ash in Concrete," Portland Cement Association, vol. 5420, p. 24, 2007.
[6] V. B. Fenelonov, M. S. Mel'gunov, and V. N. Parmon, "The Properties of Cenospheres and The Mechanism of Their Formation During High-Temperature Coal Combustion at Thermal Power Plans" KONA Powder Part. J., vol. 28, pp. 189–208, 2010.
[7] N. Ranjbar and C. Kuenzel, "Cenospheres: A review" Fuel, vol. 207, pp. 1–12, 2017.
[8] M. Zyrkowski, R. C. Neto, L. F. Santos, and K. Witkowski, "Characterization of Fly-Ash Cenospheres from Coal-Fired Power Plant Unit" Fuel, vol. 174, pp. 49–53, 2016.
[9] J. Y. Wang, M. H. Zhang, W. Li, K. S. Chia, and R. J. Y. Liew, "Stability of Cenospheres in Lightweight Cement Composites in Terms of Alkali-Silica Reaction" Cement and Concrete Research, vol. 42, no. 5, pp. 721–727, 2012.
[10] X. Huang, R. Ranade, Q. Zhang, W. Ni, and V. C. Li, "Mechanical and Thermal Properties of Green Lightweight Engineered Cementitious Composites" Construction and Building Materials, vol. 48, pp. 954–960, 2013.
[11] S. Sen, "Physical Properties of Cenosphere" Physical properties of Cenosphere (Doctoral Dissertation), no. May, p. 63, 2014.
[12] E. V. Fomenko, N. N. Anshits, L. A. Solovyov, O. A. Mikhaylova, and A. G. Anshits, "Composition and Morphology of Fly Ash Cenospheres Produced from The Combustion of Kuznetsk Coal" Energy and Fuels, vol. 27, no. 9, pp. 5440–5448, 2013.
[13] J. Yu, X. Li, D. Fleming, Z. Meng, D. Wang, and A. Tahmasebi, "Analysis on Characteristics of Fly Ash from Coal Fired Power Stations" Energy Procedia, vol. 17, pp. 3–9, 2012.
[14] S. Ghosal and S. A. Self, "Particle Size-Density Relation and Cenosphere Content of Coal Fly Ash" Fuel, vol. 74, no. 4, pp. 522–529, Apr. 1995.
[15] A. Chávez-Valdez, A. Arizmendi-Morquecho, G. Vargas, J. M. Almanza, and J. Alvarez-Quintana, "Ultra-low thermal conductivity thermal barrier coatings from recycled fly-ash cenospheres" Acta Materialia, vol. 59, no. 6, pp. 2556–2562, 2011.

[16] T. Hirajima, H. T. B. M. Petrus, Y. Oosako, M. Nonaka, K. Sasaki, and T. Ando, "Hollow and Spherical Particles in Pulverized-Fuel Ash" International Journal of Mineral Processing, vol. 95, no. 1–4, pp. 18–24, 2010.
[17] Y. Li and H. Wu, "Ash Cenosphere from Solid Fuels Combustion. Part 1: An Investigation into Its Formation Mechanism Using Pyrite as a Model Fuel" Energy and Fuels, vol. 26, no. 1, pp. 130–137, 2012.
[18] P. K. Kolay and S. Bhusal, "Recovery of Hollowspherical Particles with Two Different Densities from Coal Fly Ash and Their Characterization," Fuel, vol. 117, pp. 118–124, 2014.

IOP Conf. Series: Materials Science and Engineering 1101 (2021) 012013 doi:10.1088/1757-899X/1101/1/012013

Mix proportion for medium grade concrete with silica fume as cement replacement for general purpose construction

A Rashidi[1*], M I S Mohammad Zain[1] and R Ahmadi[1]

[1] Department of Civil Engineering, Faculty of Engineering, Universiti Malaysia Sarawak, 94300 Kota Samarahan, Sarawak, Malaysia.
*Corresponding author: razida@unimas.my

Abstract. There are many demands of concrete in construction leading to the use of industrial waste as mineral additive or supplementary filler to improve the concrete performance and lower production cost. Silica fume is a distinct industrial waste. The making of silica fume concrete is made by case basis based on specific needs especially for the production of prescribed High Performance Concretes. In this study, the use of silica fume for low/medium grade concrete for general production is investigated [1]. It is an attempt to find a mix proportion with recommended silica fume replacement to produce a medium grade concrete for general purpose construction. Silica fume as cement replacement greatly increase the mechanical and durability properties of concrete. The concrete mix with 10% silica fume replacement gives the most practical replacement percentage resulting to a 32% strength increase compared to concrete with no silica fume. It is also possible to produce a normal grade concrete using a standard M40 mix proportion. Concrete samples with less than 10% cement replacement did not show significant differences in mechanical and durability properties compared to the reference sample.

1. Introduction

Concrete has a significant use in construction world that has made it became an important material in construction. The increasing demand of cement which requires production of cement at large scale consequently result in environmental problems, depletion of natural resources and increasing price [1-2]. To overcome these problems, numerous researches on the use of industrial waste as mineral additive or supplementary filler to improve the concrete performance to lower cost [3]. In Sarawak, most of these industrial waste included silica fume (SF), are categorized as schedule waste. As such it is expensive to dispose these waste materials. Latest advancement in new concrete almost invariably use mineral admixtures due to numerous advantages especially for the improvement of strength, durability, economy and ecological factors [4]. Despite its high price, silica fume's addition as one of the concrete admixtures seems to be promising when the aspect of very high strengths and durability are needed [4]. The silica fume is among the most popular industrial-by-product cementitious material from the smelting process in the silicon and ferrosilicon industry [1]. Silica fume is known to be a very reactive pozzolan because of the chemical content and physical properties. Due to its smaller size than cement, which is 100 to 150 times smaller, it acts as filler and contribute towards improving the properties of concrete. Silica fume is added to concrete mixtures for better workability, strength, high density, low permeability and resistance to chemical attack [3].

Extensive researches had been done to formulate silica fume as supplementary addition to cement in conventional concrete [1-4]. Bhanja et al [4] suggested a modified water-cementitious material ratio (cement plus additive) to be considered in concrete mix design incorporating silica fume instead of the Abrams' water-cement ratio formulation as normally used. Abrams' law is mainly associated with the strength of concrete mix and inversely correlate between the mass ratio of water and to cement [3-4]. Kumar [1] studied the optimum replacement percentages suitable under the Indian conditions by varying different replacement percentages of cement by silica fume on concrete grade 35 (M35). The water binder ratio (w/b) adopted was 0.32. It was found that the addition of silica fume in concrete has enhanced the strength and durability unlike to normal concrete. However, the compressive strength increased with up to 10% silica fume replacement but higher replacement of cement gave lower strength [1]. Kumar et al. [3] attempted to outline an exhaustive method of proportioning silica fume concrete for standard grade concretes (M20, M25, M30, M35 and M40). The extent of replacement of cement with silica fume was 13% with different w/c ratio. It was found that w/c of 0.45 gave the highest compressive strength to most of the concrete grades. Also, a 13% cement replacement gave a lower than expected compressive strength for most of the concrete grades.

There is still opportunity to investigate the use silica fume as cement replacement especially for low/medium grade concrete in general construction [1]. Hence, in this present study, an attempt to define a medium strength silica fume concrete mix proportion with recommended partial replacement is carried out. Measurement of properties such as concrete slump, compaction factor and compressive strength with different ratios of silica replacement were compared with corresponding values of conventional concrete of similar strength. An investigation on the durability of concrete with and without silica fume using the rapid-chloride penetration test was also considered.

2. Experimentation

2.1. Materials

In this study, a mix proportion is designed to achieve a M40 concrete is shown in table 1. The cementitious material used is the Ordinary Portland Cement (OPC). The particle size distribution for the silica fume ranges from 20 to 250 nm, with an average size of 150 nm. The particle size of cement is about 1nm. The fine and coarse aggregate used are sand and crushed granite respectively. Retarder (RS233) and superplasticizer (Sp) (RF611) are used in this experiment to improve the workability. Retarder is used to offset the accelerating effects of high temperature. Superplasticizer is used to make a high slump flowing concrete. Concrete mixes incorporating silica fume are casted into three different proportions of 0%, 5% and 10% by weight of cement; represented by SF0, SF5 and SF10 respectively. These mixes have a constant fixed water-cementitious materials ratio of 0.45 while keeping other mix design parameters constant. At these ratio, the concrete falls under the category of nominal weight concrete. The basis of these percentage replacement is to ensure the mix is economical and practical to be used. Too high percentage replacement would render the mix impractical for industry players to use.

2.2. Experimental Methods

The properties of fresh and hardened concrete due to some replacement of cement by silica fume are investigated. The workability of the fresh concrete is tested using the slump test based on BS1881-105:1984 [7]. Compressive strength test on 150x150mm harden concrete cubes were done to monitor the for 3, 7, 14 and 28 days strength development curve of the concrete as stated in BS 1881-116:1983 [8]. Flexural strength test on 100x100x500mm concrete prisms is done based on BS 1881-118: 1983 [9]. The rapid-chloride penetration test (RCPT) is done to evaluate the electrical conductance of the concrete samples. This will give a gauge indicator of their resistance to chloride ion penetration in accordance with ASTM C 1202 [10].

Table 1. Mix Proportion

Binder (Kg/m^3)	Water (Kg/m^3)	Sand (Kg/m^3)	Gravel (Kg/m^3)	Sp (x C%)	Retarder (x C%)
350	157	1200	700	3	5

3. Results and Discussion

3.1. Fresh properties

Slump tests obtained from the three mixes gave slump spread for SF0 and SF5 while slump height was obtained in SF10. Table 2 shows the result for slump test for the three mixes. For SF0 and SF5, the slump spreads are 37 cm and 33cm respectively. The width of spread for SF5 is smaller compared to SF0 replacement indicating that silica fume addition affect the workability of the concrete mix. The mix in SF10 was dry with poor workability and the slump height obtained is 3.7cm in height. Figure 1 shows the slump result for SF10.

In mixes SF5 and SF10, the silica fume particles have high surface area and absorbs most of the free water in the mix. Due to this effect, workability of the concrete is reduced with increase in amount of silica fume. The small particle size of the silica fume gave a ball bearing effect which enhances the thixotropic behavior of the concrete. The thixotropic behavior is the shear thinning behavior on the viscosity of concrete mix. The fresh mix with high silica fume content were more compact and cohesive.

Table 2. Slump Test Result

Mix Design	Slump Type	Measurement (mm)
SF0	Slump spread	37.1
SF5	Slump spread	33.0
SF10	Slump height	3.7

Figure 1. Slump height for 10% replacement (SF10)

3.2. Mechanical properties

Table 3 shows the rate of growth of compressive strength for 4, 7, 14 and 28 days for the three types of concrete mixes. It can be seen that SF10 experiences an improvement in compressive strength values from the initial days as compared to SF5. The 5% cement replacement in SF5 did not contribute much to strength enhancement and the rate of growth is almost the same as SF0. However, the rate of growth for SF5 becomes slightly significant after 7 days of age. This may be due to the secondary hydration process of OPC and silica fume in the concrete. When pozzolanic materials are combined with the concrete mix, the silica fume particles react with calcium hydroxide that is released during the hydration of cement to form additional calcium silicate hydrate (C – S – H) [11]. The silicate hydrate improves durability and the mechanical properties of concrete. The hydration of the silica fume in the concrete is rapid when the amount of replacement is increased as in SF10. From figure 2, it can be concluded that a 10% cement replacement in SF10 is an recommended cement replacement amount which achieve the maximum compressive strength with 32% increase compared to SF0. Mix SF5 shows that a small amount of replacement of the silica fume in the concrete does not produce a significant difference in the compressive strength of the concrete. It is worth to note in figure 2 that the change of strength behaviour of SF0 and SF5 are similar. It proves that any efforts to replace cement with any industrial waste must have a certain threshold value. Below this threshold value, any replacement percentage is deemed insignificant. Within the scope of this study for general construction purposes, it proves that the use of silica fume is possible provided the replacement factor reached 10%. Further increase in percentage replacement will indiscriminately affect the cost of construction, as the market price of silica fume is about 2.5 times more than OPC.

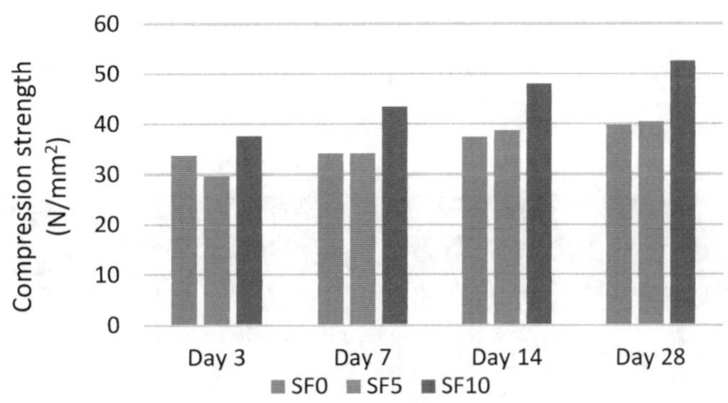

Figure 2. Compression strength test results

Table 3. Summary of Compressive Strength

Mix Design	3-Days Stress (N/mm^2)	7-Days Stress (N/mm^2)	14-Days Stress (N/mm^2)	28-Days Stress (N/mm^2)
SF0	33.7	34.1	37.5	39.8
SF5	29.7	34.1	38.6	40.4
SF10	37.6	43.5	47.9	52.6

Figure 3 shows the flexural strength results at 28 days for the three mixes. The SF5 and SF10 mixes exhibit better flexural strengths compared to zero cement replacement in SF0. The SF10 gives the best flexural strength. However, the flexural strength to compressive strength ratio in table 4 for SF5 and SF10 mix give similar ratio compared to SF0. This shows that silica fume as a filler contributes to the better flexural strength of the concrete. The tiny particles of the silica fume fills the voids in the concrete contributes to the strength of the concrete.

It became clear that addition of silica fume in the mix improved the flexure strength of concrete. SF0 gave a strength ratio of 12.2% while SF5 and SF10 showed a value above 14%. Thus, it showed that silica fume acts as filler in concrete voids and improves the ability to resist bending forces. This is a significant finding if the industry practitioner were to design and construct concrete pavement.

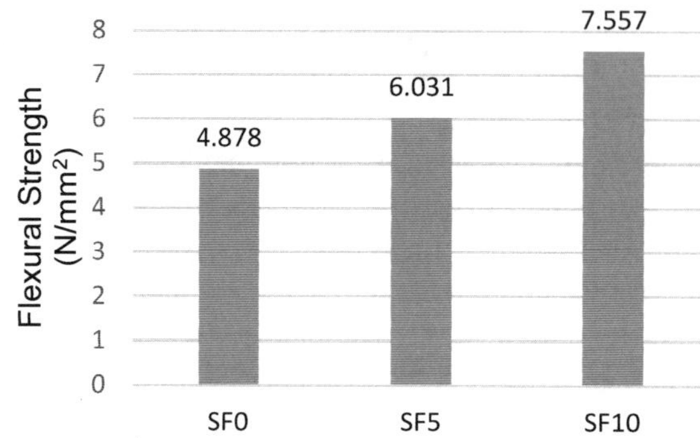

Figure 3. Flexural strength test results

Table 4. Summary of flexural and compressive strength at 28 days

Mix Design	Flexural Strength (N/mm^2)	Compressive strength (N/mm^2)	% Ratio
SF0	4.878	39.841	12.2%
SF5	6.031	40.464	14.9%
SF10	7.557	52.629	14.3%

3.3. Rapid Chloride Permeability Test (RCPT)

Table 5 shows the chloride ion penetrability scales and the related the average electrical charge. By comparing the criteria's in table 5 to table 6, the RCPT values for chloride ion penetrability in the mixes decreases as the amount of the replacement increases. Silica fume is able to reduce the permeability of the concrete against the chloride penetration, improve its resistance to chemical attack particularly chloride ions; thus enhance the durability of concrete in marine environment. The RCPT results indicates that as the amount of the replacement increases, the durability of the concrete increases.

Table 5. Chloride ion penetrability based on charge passed

Charge Passed (C)	Chloride Ion Penetrability Criteria
>4000	High
2000 - 4000	Moderate
1000 - 2000	Low
100 - 1000	Very Low
<100	Negligible

Table 6. RCPT result at 28 days

Mix Design	Electrical Charge (C)	Indicator
SF0	3366.33	Moderate
SF5	1932.33	Low
SF10	388.00	Very Low

The RCPT result is another data that supports the use of silica fume as fillers. The durability of SF10 and SF5 are much better than SF0. The addition of silica fume fills in the voids in the concrete matrix and cause it to be less permeable. For industry practitioners, the design of impermeable surfaces such as water tank or swimming pool can consider the incorporation of silica fume in their mixes.

4. Conclusion

In this study, a recommended cement replacement using silica fume for a medium strength concrete for general constructions is investigated. The mechanical properties and durability of the concrete mixes are compared. The results obtained confirm most findings that silica fume greatly enhanced the compressive strength of concrete. However, the replacement of silica fume must be above a threshold value to get a significant improvement in strength. In this study, a 10% replacement is commendable while a 5% replacement is deemed insignificant. The chloride ion diffusion resistance in concrete increases with the proportion of silica fume in mixes; thus indicating an improved durability for the concrete in marine environment. The RCPT result also proved that inclusion of silica fume, even at a low replacement percentage of 5% improved the durability of concrete significantly. As an overall, the concrete mix with 10% silica fume replacement gives the most recommended replacement percentage with the highest compressive strength of 52.6 kN/m^2 or 32% strength increase compared to concrete without silica fume. With the 10% replacement, it is possible to produce a high performance concrete using standard M40 mix proportion. Mix proportion with less than 10% cement replacement did not show significant mechanical and durability properties.

Acknowledgment
The experiments had been conducted with the financial support from the grant by Cahaya Mata Sarawak (CMS) Sdn. Bhd. at Universiti Malaysia Sarawak, 93400 Kota Samarahan, Sarawak, Malaysia.

References

[1] Kumar R and Dhaka J 2016 Review paper on partial replacement of cement with silica fume and its effects on concrete properties. *International Journal for Technological Research in Engineering,* **4**(1)
[2] Khan M and Ali M 2019 Improvement in concrete behavior with fly ash, silica-fume and coconut fibres. *Construction and Building Materials* **203** pp 174-187
[3] Kumar K D V and Paulson J 2019 Method of proportioning silica fume concrete *International Journal of Engineering and Advanced Technology (IJEAT)* ISSN: 2249 – 8958 Volume-8 Issue-3S February 2019
[4] Bhanja S, Dhir B S R K, Harrison T A, Newlands M D Water-cement ratio law and silica fume concrete mix design. *https://doi.org/10.1680/ccfdc.34013.0012*
[5] Types of concrete mix ratio design and their strengths *The Constructor Organization* https://theconstructor.org/concrete/types-of-concrete-mix-design/5984/
[6] Lee S T and Lee S H. Mechanical properties and durability of cement concrete incorporating silica fume *Journal of the Korean Ceramic Society* Vol. **4** No 5 pp 412~418 2010
[7] *British Standard N D* BS1881-105:1984. Slump Test
[8] *British Standard N D* BS 1881-116: 1983 Compressive Strength Properties.

[9] *British Standard N D* BS 1881-118: 1983 Flexural Strength Properties.
[10] *American Society for Testing and Materials N D* ASTM C1202 Rapid Chloride Penetration Test
[11] Neville A M *Properties of Concrete 5th Edition ISBN10:0273755803*
[12] Bhanja S and Sengupta B Influence of Silica Fume on the Tensile Strength of Concrete *Cement Concrete Res* **35** [4] 743-47 2005

A review on green pavement hydrological design and recommended permeable pavement with detention storage

N Bateni[1*], S H Lai[2], R Ahmad Bustami[1], M A Mannan[1] and D Y S Mah[1]

[1] Department of Civil Engineering, Faculty of Engineering, Universiti Malaysia Sarawak, 94300 Kota Samarahan, Sarawak, Malaysia.
[2] Department of Civil Engineering, Faculty of Engineering, University of Malaya, 50603 Kuala Lumpur, Malaysia.
*Corresponding author: bnorazlina@unimas.my

Abstract. This paper reviews innovation of green pavement technology for storm water management in an urban environment. This can be related with hydrological performance and assessments of the permeable pavement. Features of the typical permeable pavement are presented and discussed. Topics covered include recognizing important of permeable pavement, stormwater management benefit and detailed hydrological properties and design. The information in this paper provides stakeholders with an overview of research and development of green pavement. In particular, it discussed the benefit and advantages of the green pavement in current use. On the other hand, the permeable pavement with subsurface detention namely StormPav is presented. The hydrological design modification and innovation, as well as hydrological design and stormwater management benefits have been summarised. Therefore provided another option for green pavement infrastructure series to be used in roadworks.

Keywords: Stormwater management, green pavement infrastructure, permeable pavement, hydrological design, hydrological assessments

1. Introduction

Urban development affects land-use changes and thereby influences the natural hydrological cycle [1]. Percentage of land covered by impervious surfaces varied significantly with the changes of land use Arnold et al. [2] to cause a depletion in groundwater recharge, reduction in infiltration rate, creation of more runoff and thereby increasing the flood events [3,4]. Surface-water flooding in urban areas has become a critical issue due to the changing of precipitation patterns, congestion of the stormwater drainage system, expanding of urban areas and aging of drainage infrastructure [5]. The conventional approaches in stormwater management were mainly designed to reduce the peak flow but not the runoff volume. Consequently, it may lead to the occurrence of flood at downstream areas due to the excessive runoff. This is primarily a major issue in stormwater conveyance systems. It was normally addressed by enlarging the hydraulic capacity of the systems [6] and expanding existing urban drainage systems [4]. However, enlargement of the existing system had been proven its ineffectiveness in terms of economy and sustainability [4]. As a result, sustainability has appeared as the main concern while designing the stormwater management infrastructure since 1990s.

In past, there are numerous attempts to embrace sustainability in the design of stormwater intercourse [7-10] . Among the efforts, source control of stormwater with green infrastructure (GI) or

IOP Conf. Series: Materials Science and Engineering 1101 (2021) 012014 doi:10.1088/1757-899X/1101/1/012014

low impact development (LID) applications is a promising alternative for flood mitigation. GI practices had been established and widely employed as a promising solution to the problems, such as aging of water infrastructure, urbanisation, climate change and water shortage [11,12]. Barszcz [13] reported a reduction in runoff depth and peak flow rate, recorded at 50.0 and 38.5% respectively, in urbanized sub- catchment through their study on the characteristics of surface runoff/outflow with 4 different types of LID. Ahiablame and Shakya [14] found that, under a condition of 50-100% permeable pavement and 100% rain, the rate of flood reduction of the garden areas and the parking lot areas varied from 45.5 to 54.5% for major floods; 28.8–40.8% for action floods, and 36.4% for major floods; 21.6% for action floods, respectively. Rodríguez-Rojas et al. [15] showed an average water volume reduction of over 41% in a LID application. Meanwhile, Eaton [16] found a different reduction rate under various conditions, where it reaches 35-55% for individual land uses and 23-42% for the entire watershed of a low-density residential area in New York City. Meanwhile, via the recent study conducted by Matos et al. [17], it was reported a peak discharge reduction of 76% while applying the LID combinations with green roofs, as well as at parking lots, sidewalks, secondary roads and primary roads with permeable pavement

2. Green Pavement

The implementation of green infrastructure practices, especially green pavement, can significantly improve the rainfall-runoff responses and thereby minimise the occurrence of flood hazards, which may lead to uncountable economic losses [18]. Jaffe [19] reviewed the online economic model developed by the Center for Neighborhood Technology (CNT) through its application on construction costs, maintenance costs and component life spans (e.g. divert millions of gallons of stormwater). They found that the green infrastructures include permeable pavement, provided substantial economic benefits than the gray infrastructures. Permeable pavement (PP) has been widely used in stormwater management practices due to its environmental advantages, such as reducing the runoff [20], recharging the groundwater [21], mitigating the heat island effect and removing the pollutants [22].

Porous or pervious concrete, which was known and used for about 50 years ago in Europe and United States, is a type of concrete that contains little or no fines (i.e. sand).It is mostly composed of aggregate and cement paste. Its history started in the mid of 1940s as simple concrete turf blocks, which is a modular system to address flooding issues in the large cities of United States. In 1970s, due to the development of the turf blocks in plastic version, which provides the advantage in terms of cost-effectiveness, permeable pavement has become a more and more popular choice [23], while drawing the stormwater management strategies since 1980s in the United States, Canada, Europe and Japan. It was mostly applied to parking lots, low-density traffic lanes and pedestrian pathways [24]. Rowe [25] defined pervious and porous surfaces as "open to passage" and "full of openings (pores)", respectively. Meanwhile permeable surface was known as "capable of being passed through." In general, the term "pervious" is used to describe permeable concrete as it allows the water passage, while "porous" is used to label permeable asphalt because of the void spaces. Permeable pavement (PP) is a paving system that allows water to infiltrate to the underground. Therefore, it is suitable for a wide variety of residential, commercial and industrial applications.

PP can be generally categorised according to the surface type, such as pervious concrete (PC), porous asphalt (PA) and permeable interlocking concrete pavers (PICP). Figure 1 illustrates examples of different permeable pavement types and properties. PC and PA are permeable variations of concrete or asphalt where the binding agent coats the aggregate particles without filling the voids between the particles [26]. Meanwhile, PICP consists of modular units separated by the joints filled with open-graded aggregate [27]. PICP introduce numerous surface designs of pavers such as precast grid, concrete block paver, plastic grid, grid paver and etc. The PICP is growing in popularity due to the modularity of precast block paver design. It is well-established with the provision of design guidelines and various case studies [28-30]. Overall, the designed thickness of permeable pavement can be more than 200mm to 600mm with aggregate sizes of about 2 mm to 76mm, void capacity up to 15% to 40% and pore space between 0.5 mm to 50 mm.

Figure 1. Permeable pavement types and properties.

Table 1. PP development and research achievements.

Researcher	Outcome
	1990s
[23,35]	**Planning and Design** • Experimental and laboratory studies in small scale on the structural and hydrological design. • Determination on effectiveness in stormwater management. - e.g Studies on manufactured permeable pavers, field experiment to observe runoff subjected to different slopes and infiltration capacities.
	2000s
[20, 21, 26, 33, 36-69]	**Implementation and Application** • Full scale studies on hydrological performance and structural requirements. (e.g determination on optimal structure materials and thickness, preparation of the synthetic stormwater quality samples and selection of parameters for model verification) • Field and laboratory studies on stormwater quality, stormwater infiltration techniques and pollutant removal efficency. • Investigation of parameters establishment (structural and hydrological) with modelling tools to aid in assessment. (e.gdevelopment of stormwater management model and the code feasibility of the Storm Water Management Model (SWMM) • Introduction to various types of concrete grid pavers, laboratory monitoring, development and construction. • Case studies and application of the newest types of permeable pavers and implementation of software. • Published standard and guidelines of PPs.
	2010s
[70-97]	**Interpretation and Modification** • Many related studies on clogging and water quality performance of PP. • Wider application and model establishment (e.g. SWMM, PCSWMMPP, HYDRUS-2D). • Modification and introduction to new modification, technology and materials of PPs to enhance strength and durability for higher load, optimise hydrology benefit and stormwatermanagement and removal efficiency etc. • Field application, case studies and stormwater quantity and quality assessments.

3. PP research development

The PP has been widely used since the 1990s in Australia, Europe and Japan, and earlier in United States in 1984 [24]. The recent development of permeable pavement (PP) can be divided into three major time periods. In the 1990s, the PP system was introduced to the users. Most of the research studies were aimed to investigate the stormwater management benefits of the PP in reducing the peak runoff and its volume, as well as increasing the infiltration rate. Several prototypes had been constructed for the property and characteristic determinations via laboratory studies and field tests. The main concern of the analyses was the measurement of stormwater management benefits. Towards 2000s, the main focus of most of the research studies was on the applications aspect, which compared the designed PP with the existing road pavement or the pre- and post- conditions for the PP applications. Several modelling tools had been commonly used to assist the studies and applications of PP [31]. The permeable pavers, acting as the most successful pervious system had been seen for its wider application. This is because a number of design products, case studies benefits and standard guidance had been established [32,33]. From 2010s until recently, the main direction has changed to implementation and modification. Different modifications had been introduced, such as the adding of high strength materials with sufficient void to improve both structural and hydrological benefits, the modification of pavers surfaces with unique and aesthetic looks, etc.

Nevertheless, no matter how the modifications were implemented, the most important thing was the designed pavement must meet the standards and guidelines, especially to fulfill the structural and hydrological considerations. Of course, the point of view from the aspects of sustainability, economy and society have become more and more important. In addition, the optimal combination of low impact development (LID) practices or green infrastructure techniques has become the main concern in order to achieve the maximum benefit in stormwater management [34]. The chronology of PP development and researcher involved is simplified in table 1.

4. PP as green pavement

The PP has been proven its effectiveness in infiltrating stormwater runoff [13], reducing runoff volume [93,97] and peak flow [98]. Weiss et al. [84] discovered that the resulting variability on peak runoff and infiltration rate was mainly due to differences in PP, such as construction materials, mix designs, construction techniques, maintenance received, etc. As shown in table 2, many research on PP had been conducted. The PP had shown its potential to reduce the runoff to more than 40%, decrease the peak to at least 50-90% and improve the infiltration rate from 4-11 cm/h. Almost all studies agreed that permeable pavements, when constructed well and maintained regularly, could have the ability to reduce peak runoff and infiltrate a significant fraction of runoff volume. Ferguson [99] stated that, if the runoff coefficient that has been measured on the properly built porous pavement is zero, it meant there is no runoff due to the high surface permeability. On the other hand, numerous studies had been conducted on the new design materials used for different layers in permeable pavement system, with the objectives to find out the best combinations to minimise the clogging issues and optimise the permeability, whilst still can provide a good result in structural strength.

On the other hand, Bentarzi et al. [104] suggested a new eco-material, comprising a mixture of construction wastes (crushed concrete) and organic matter (compost) to produce permeable pavement. The crushed concrete was mainly for the structural support, while the compost was specifically for the retention and biological treatment of stormwater. Li et al. [105] used to overcome the issues such as low strength, high likelihood for clogging and inconvenient maintenance (as depicted in figure 1). Based on table 2, PP in PICP formed is the popular choice among the designers. This is due to the modular character, countless types of pavers block, surface materials, joints materials, layers and thicknesses, which provide opportunity of more research studies. Moreover, in term of the maintenance, Marchioni and Becciu [106] introduced a maintenance method to involved the removal of joint material and bedding layer. The other method for maintenance purposes is to remove the entire system, but the frequent replacement is impractical and expensive [102]. Thus, permeable pavement of PICP types becomes the preference, while selecting the type of pavement due to its modular interlocking concrete block characteristic that smoothens the remediation work [70,99].

Table 2. Researchers and types of PP used.

Sources	Types of PP used
[36,40,46,51,52,53,61,62,63,66,76,77,80,93,94,100,101]	PICP types: paving stones, pavers with infiltration joints, porous concrete pavers with filter layer, and greened porous pavers, interlocking concrete block pavers, Rocla-Ecoloc pavers). Concrete grid pavers (CGP), block paver, permeable pavement with impermeable paver blocks and enlarged joint spacing, porous concrete (PC), concrete grid pavers (CGP) filled with sand,and permeable interlocking concrete pavers (PICPs) with gravel fill , C&M Ecotrihex pavers and Atlantis turf, porous concrete block paving with different surface and layers and thicknesses, Impermeable rectangular concrete modules (Priora), grass paver, plastic geocells, Spaced modular, grass grid and gravel grid pavements, permeable interlocking concrete paving system (PICP), which comprised a maximum joint width
[24,50,56,88,93,100]	Porous Asphalt and Porous Concrete, porous asphalt and pervious concrete with different void contents
[96,102,103,104,105]	New design materials used for different layers in permeable pavement system: PICP at maximum joint width (5-6mm), a special aggregate, consisting of waste materials, new eco-material-(mixture of construction wastes (crushed concrete) and organic matter (compost)), reactive powder concrete with high strength pervious concrete as pores pavement.

5. PP design limitations

A typical PP is designed with layer of bases and connected through the pore spaces and interconnection between void. Permeable pavement invented with voids and porosity properties acts as stormwater management structure is prone to clog due to the blocking of debris, sediment and fine particles in the pore spaces, which disconnected the system [95,107, 108]. Besides, the layered characteristics with open-graded materials were opened for particle build-up block, which connected capillary pores, accumulated in void spaces of permeable pavements and clogged the system [65,90]. Clogging is a result of fine particles accumulating in void spaces and thus affecting the rate of infiltration of stormwater and causing the ponding of surface water over the pavements. Clogging can happen within three or four years of installation due to sediment in runoff water and collapsing pores resulting from vibrations caused by traffic [109]. Imaging studies showed that the top 25mm of pervious concrete is mostly subjected to clog. Numerous studies have reported clogging on permeable pavement with reduced permeability and infiltration rates over time [18,107,110].

On the other hand, it was sometimes for the permeable pavement to equip with underground detention storage. The concept of storing water in an underground carriageway could help to significantly reduce drain size and peak runoff, greywater usage and water harvesting. Such design has been applied in the University of South Australia, and in Gresham, Oregon. There also commercial products by integrating an underground retention and detention such as Permavoid® (Product of Permavoid Limited, Warrington, UK), StormCapture® and PermeCapture™ (Product of Oldcastle Infrastructure, UK), geo-cellular sub-base replacements and SingleTrap® and DoubleTrap® (Product of StormTrap, LLC, US). The products are large underground detention systems for water storage and groundwater recharge. This design might require big-scaled machinery and equipment for the construction. In addition, it faced numerous problems associated with the under road systems, such as water supply pipes, sewer pipes, cables and trees roots. Therefore, it is of interest to develop an innovative design for a micro-scale on-site detention pond permeable pavement to avoid potential conflict with the existing systems under roads besides having the ability to detain water with minimize clogging and recharge groundwater for sustainability development.

6. StormPav, innovation in green pavement

Previous researches on permeable pavement system were typically focused on its hydrological performance, the capability of handling traffic volume, materials and mix design, as well as handling, maintenance and safety measures. Nevertheless, among the recent studies that had been reviewed in this paper, it was found the absence of permeable pavement with a micro-detention pond system. The idea is mainly from the PICP design of the modular character, which allows quick and easy installation. To achieve improvement in terms of clogging the new system eliminates layer characteristics of PICP, where the large precast underground detention structure is replaced with a micro-detention hollow precast system that acted as base and water storage, figure 2a. This empty space enhances void storage for water retention and provides a longer time for infiltration to the ground.

StormPav as green pavement main features is a product of precast honeycomb structure that enhance self-interlocking, rain water holding micro-detention pond and self-drying through side and bottom seepage of water in comparison with conventional PP, figure 2b. It have features of modularity, adaptability, portability and self-interlocking. It is actually a hollow cylinder covered with the top and bottom plates, reinforced with two layers of steel and optionally covered with a geotextile layer. The system is constructed on the flat subgrade soil and dry-stacked to form an interlocking paver that has the ability to provide structural support and durability which in turn yielding benefits in terms of permeability in stormwater management. The permeability of the system is accomplished through the under drainage system of the hollow cylinder/vessel residing beneath the subsurface of the pavement. According to Shackel [70], a paver with dentate shape performed better than the rectangular paver. Thus, the StormPav is designed to have a hexagonal shape in order to achieve better performance from the perspective of hydrological design and structural durability. Acting as the open passage for stormwater inflow and infiltration respectively, both the top and bottom layers comprises a central hole with 40mm diameter and six grooves as interlocking keys.

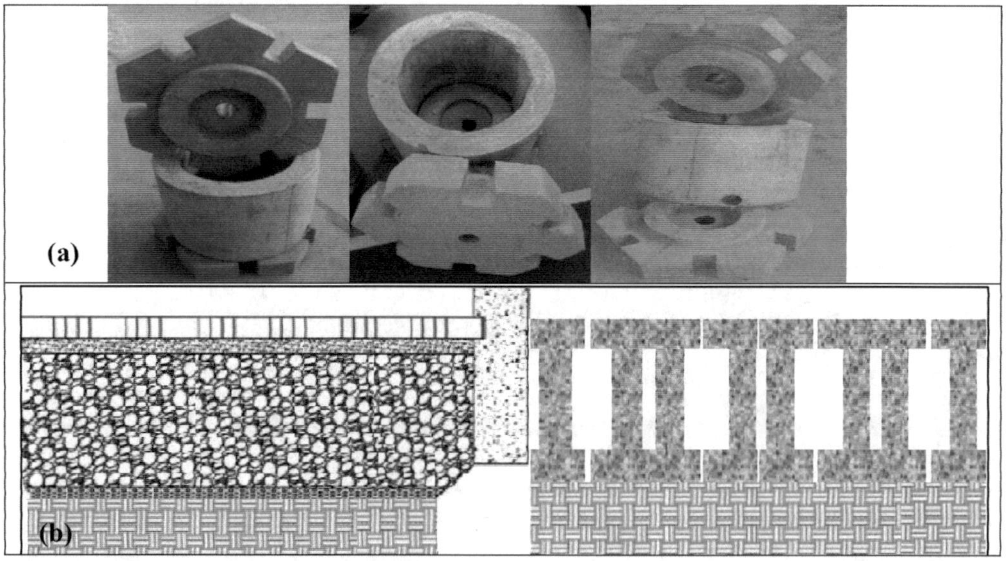

Figure 2. StormPav (a) single unit (b) cross-section of PICP with comparison to StormPav

This product ensures cost-effective as dry-stacked with reduce installation time and manpower. The aesthetics value is its enduring beauty as honeycomb structure and user-and environmental-friendly product. The new micro detention permeable pavement, StormPav consists of three layers of modular set: hexagonal shape top surface, micro detention pond hollow cylinder tank in between, hexagonal bottom and an optional geotextile layer, constructed on a flat subgrade native soil. The thickness of top and bottom layers and micro detention pond hollow cylinder tank is 75mm and 300mm respectively. The top and bottom layer of StormPav is a flat, concrete interlocking set. Each plate is reinforced by two layers of steel. A 0.04 m diameter hole in the centre and joint lines in the interconnection between sets characterize the open passage. The top cover functions as road pavement with an inflow to drain stormwater whilst the bottom cover functions as an outflow that allows infiltration. As for the horizontal flow, the micro detention pond hollow cylinder tank has two openings of 0.04 m diameter hole opposite to each other, which are placed close to the bottom of the cylinder. The micro structure promises easy and rapid installation which can be lifted manually, or using simple mechanical devices. In addition, the design comes with more percentage of an open surface area capable of providing a large amount of infiltration into the soil and with longer time of clogging. Instead of having a layer of open aggregates surface, base and sub-base, the StormPav is fabricated with a hexagon shape top and bottom layers with 40mm diameter hole and micro detention hollow cylinder tanks structure in between as a detention storage. The hollow cylindrical feature in StormPav has an empty space of 0.19 m^3/m^2 (30 L/unit) and about 70% void of the pavement area. It drains the surface water at a rate of 8400 mm/hr.

Table 3 presents a series of research studies on StormPav since the development in 2013. The development of StormPav is also following the chronology in table 1. The StormPav is design to apply as stormwater management infrastructure. The actual products are fabricated in 2015, which are then tested and investigated for various structural, roadworks, hydraulics and hydrological assessments. StormPav is a full precast product made using Grade 50 concrete. It's crushing load is 100kN/unit which is higher than 40kN/single wheel of 80kN Standard axle load. The StormPav can provide uniform settlement with interlocking and monolithic characters. With more manufactured and fabricated moulds to produce StormPav in 2016, the StormPav started it mass production to achieve more research accomplishment of the product.

Table 3. Research development of StormPav

Researcher	Research
[111-116]	**Planning and Pre-liminary Design (2013-2018)** • Prototype design and development of accessories and fabrication of testing accessories and devices (construction of rainfall simulator, formation of interlocking mould set, shear key for precast casting) • Preliminary studies on the structural and hydrological design properties, performances and characteristics. • Determination on effectiveness in stormwater management.
[117-128]	**Implementation and Application (2018-2020)** • Field and laboratory studies on the hydrological performance, hydraulics and flow regime, and structural requirements. (e.g determination on optimal structure materials and thickness, selection of parameters for model verification, stormwater quantity and quality, stormwater infiltration capacities and rate and pollutant removal efficiency) • Investigation of parameters establishment (structural and hydrological) with modelling tools to aid in assessment. (e.g case study application in various aspects and scenarios such as at different design lifespan, topographical features and identification of appropriate drainage areas.

StormPav is an innovative integrated building system, IBS green pavement with structural, environmental and economic advantages. For instance, the hydrological benefit provided by StormPav is capable of providing stormwater detention of 3 hours continuous rainfall intensity of 80mm/hr of 10-year ARI and met the hydrological design requirements for the events up to 10-yr ARI [124]. The system collect first flush of rainfall about 25mm and trap the pollutant as coating on large surface area inside StormPav and also at location of geotextile fabric placed onto subgrade. Case study application on StormPav resulted with runoff reduction to almost 60 to 70% with loss to infiltration of almost 90% [126]. Therefore, it is able to meet Environmental Protection Agency (EPA), USA stormwater regulations to reduce overall runoff and attenuate peak flow, whereby converting stormwater runoff through infiltration. Besides, StormPav, applying as road pavement, can utilise subgrade soil of hydrologic soil groups (HSG) A and B to control the flood for the rainfall events of 10-yr ARI by providing a detention period of 24 to 72 hr [126]. This also claims low risk of subsoil pollutant accumulation and also seasonal high water table should be at least 300 to 600 mm below of pavement formation level.

Furthermore, by comparing to the typical permeable pavement, StomPav appears as a more favourable choice in terms of cost-effectiveness. Due to its precast set with mini size, an easy and rapid installation can be expected. This is because it can be lifted manually or by using simple mechanical machines, and hence saving the installation time and manpower. In addition, the estimated production cost is around \$30/ m^2 or \$0.50/ ft^2, which is significantly lower if compared to the normal permeable pavement [124]. On the other hand, parking lot paved with this IBS acts as micro-detention pond thus it reduces the need for large detention basin. As the reducing runoff from parking lot paved with StormPav it allows the use of smaller drain. Thus, StormPav is able to reduce stormwater impact fee as imposed by authority on drainage system. In supporting local economy it is able to produce at temporary casting yard near construction site thus its use in low speed road especially parking lot, business centre and housing minimizes transport costs and related energy consumption. The light color of concrete pavement absorbs less heat from solar radiation and the large open space of $0.19m^3/m^2$ pavement area stores less heat, helping to lower heat island effects in urban areas. This pavement is suitable for protecting trees in a paved environment, sidewalks and landscaping, because it allows adjacent trees to receive more air and water. Therefore, the product shows a promising results as a new and innovative green pavement product.

Conclusion

StormPav is a green pavement with sustainability characteristics, which allow it to be incorporated into green infrastructure and stormwater management planning. In this paper, based on the recent studies, the proposed StormPav with current design can provide a promising result in terms of hydrological performance and assessments. Hence, it can be concluded that the StormPav appears as an alternative as a sustainable green pavement approach which can be served as a component in the road system and stormwater management.

Acknowledgments

The authors would like to express their gratitude for the financial support received from ERGS/TK03 (02)/1009/2013(06) awarded by the Ministry of Education, Malaysia. They would also like to thank the Universiti Malaysia Sarawak for the facilities provided for their experiments.

References

[1] H. Wheater and E. Evans, "Land use, water management and future flood risk," *Land use policy*, vol. 26, pp. S251–S264, Dec. 2009.

[2] J. Arnold, Chester L and C. J. Gibbons, "Impervious surface coverage : The emergence of a key environmental indicator," . *J. Am. Plan. Assoc.*, vol. 62, no. 2, pp. 243–258, 1996.

[3] Q. Li, F. Wang, Y. Yu, Z. Huang, M. Li, and Y. Guan, "Comprehensive performance evaluation of LID practices for the sponge city construction: A case study in Guangxi, China," *J. Environ. Manage.*, vol. 231, no. May 2018, pp. 10–20, 2019.

[4] Z. Zhu, Z. Chen, X. Chen, and G. Yu, "An assessment of the hydrologic effectiveness of low impact development (LID) practices for managing runoff with different objectives," *J. Environ. Manage.*, vol. 231, no. April 2018, pp. 504–514, 2019.

[5] J. L. Webber, M. J. Gibson, A. S. Chen, D. Savic, G. Fu, and D. Butler, "Rapid assessment of surface-water flood- management options in urban catchments," *Urban Water J.*, vol. 15:3, pp. 210–217, 2018.

[6] H. Haris, M. F. Chow, F. Usman, L. M. Sidek, Z. A. Roseli, and M. D. Norlida, "Urban Stormwater Management Model and Tools for Designing Stormwater Management of Green Infrastructure Practices," *IOP Conf. Ser. Earth Environ. Sci.*, vol. 32, no. 1, 2016.

[7] S. Beecham, "Water sensitive urban design, a technological assessment," *Waterfall J. Summer*, pp. 17, 5–12, 2003.
[8] T. D. Fletcher, W. Shuster, W. F. Hunt, R. Ashley, D. Butler, S. Arthur, S. Trowsdale, S. Barraud, A. Semadeni-Davies, J.-L. Bertrand-Krajewski, P. S. Mikkelsen, G. Rivard, M. Uhl, D. Dagenais, and M. Viklander, "SUDS, LID, BMPs, WSUD and more – The evolution and application of terminology surrounding urban drainage," *Urban Water J.*, no. October, pp. 1–18, Jul. 2014.
[9] (DID) Department of Irrigation and Drainage, *Urban Stormwater Management Manual for Malaysia*, 2nd Editio. Kuala Lumpur, Malaysia: Department of Irrigation and Drainage (DID) Malaysia, 2012.
[10] US EPA, "Low Impact Development (LID)," United States, 2000.
[11] B. L. Gordon, K. J. Quesnel, R. Abs, and N. K. Ajami, "A case-study based framework for assessing the multi-sector performance of green infrastructure," *J. Environ. Manage.*, vol. 223, no. June, pp. 371–384, 2018.
[12] C. Mei, J. Liu, H. Wang, Z. Yang, X. Ding, and W. Shao, "Integrated assessments of green infrastructure for flood mitigation to support robust decision-making for sponge city construction in an urbanized watershed," *Sci. Total Environ.*, vol. 639, pp. 1394–1407, 2018.
[13] M. Barszcz, "Influence of Applying Infiltration and Retention Objects to the Rainwater Runoff on a Plot and Catchment Scale – Case Study of Służewiecki Stream Subcatchment in Warsaw," *Pol. J. Environ. Stud.*, vol. 24, no. 1, pp. 57–65, 2015.
[14] L. Ahiablame and R. Shakya, "Modeling flood reduction effects of low impact development at a watershed scale," *J. Environ. Manage.*, vol. 171, pp. 81–91, 2016.
[15] M. I. Rodríguez-Rojas, F. Huertas-Fernández, B. Moreno, G. Martínez, and A. L. Grindlay, "A study of the application of permeable pavements as a sustainable technique for the mitigation of soil sealing in cities: A case study in the south of Spain," *J. Environ. Manage.*, vol. 205, pp. 151–162, 2018.
[16] T. T. Eaton, "Approach and case-study of green infrastructure screening analysis for urban stormwater control," *J. Environ. Manage.*, vol. 209, pp. 495–504, 2018.
[17] C. Matos, A. B. Sá, I. Bentes, S. Pereira, and R. Bento, "An approach to the implementation of Low Impact Development measures towards an EcoCampus classi fi cation," *J. Environ. Manage.*, vol. 232, pp. 654–659, 2019.
[18] K. Kumar, J. Kozak, L. Hundal, A. Cox, H. Zhang, and T. Granato, "In-situ infiltration performance of different permeable pavements in a employee used parking lot - A four-year study," *J. Environ. Manage.*, vol. 167, pp. 8–14, 2016.
[19] M. Jaffe, "Reflections on Green Infrastructure Economics," *Environ. Pract. (Environmental Rev. Case Stud.*, vol. 12, no. 4, pp. 357–365, 2010.
[20] C. F. Yong, A. Deletic, T. D. Fletcher, and M. R. Grace, "The clogging behaviour and treatment efficiency of a range of porous pavements," in *11th International Conference on Urban Drainage*, 2008.
[21] M. E. Dietz, "Low Impact Development Practices: A Review of Current Research and Recommendations for Future Directions," *Water. Air. Soil Pollut.*, vol. 186, no. 1–4, pp. 351–363, Sep. 2007.
[22] N. Xie, M. Akin, and X. Shi, "Permeable concrete pavements: A review of environmental benefits and durability," *J. Clean. Prod.*, vol. 210, pp. 1605–1621, 2019.
[23] D. B. Booth, J. Leavitt, and K. Peterson, "THE UNIVERSITY OF WASHINGTON PERMEABLE PAVEMENT DEMONSTRATION PROJECT – Background and First-Year Field Results by," Seattle, WA, 1991.
[24] J. Drake, "Performance and Operation of Partial Infiltration Permeable Pavement Systems in the Ontario Climate.," University of Guelph, 2013.
[25] A. Rowe, "Green Infrastructure Practices : An Introduction to Permeable Pavement," Rutgers, U.S, 2012.
[26] P. D. Tennis, M. L. Leming, and D. J. Akers, *Pervious Concrete Pavements*. Maryland, USA: Portland Cement Association & National Ready Mixed Concrete Association, 2004.

[27] H. Kuosa, E. Niemeläinen, and K. Loimula, "Pervious pavement systems and materials State-of-the-Art," Finland, 2013.
[28] Interpave, *Permeable pavements: guide to design construction and maintenance of concrete block permeable pavements*, 6th editio., no. January. Leicester, UK, 2010.
[29] Interlocking Concrete Pavement Institute (ICPI), "Construction of Permeable Interlocking Concrete Pavement Systems," Canada, 2013.
[30] Federal Highway Administration (FHWA), "Permeable Interlocking Concrete Pavement," U.S, Jan. 2015.
[31] W. R. C. James, W. James, and H. von Langsdorff, "Stormwater Management Model for Environmental Design of Permeable Pavement," *J. Water Manag. Model.*, vol. R207-26, 1996.
[32] Unilock, "Directions In Sustainable Design," USA, 2011.
[33] D. R. Smith, "Permeable Interlocking Concrete Pavements Permeable Interlocking Concrete Pavements," Washington DC, 2006.
[34] C. Li, C. Peng, P. C. Chiang, Y. Cai, X. Wang, and Z. Yang, "Mechanisms and applications of green infrastructure practices for stormwater control: A review," *J. Hydrol.*, vol. 568, no. September 2018, pp. 626–637, 2019.
[35] D. B. Booth and J. Leavitt, "Field Evaluation of Permeable Pavement Systems for Improved Stormwater Management," *J. Am. Plan. Assoc.*, vol. 65, no. 3, pp. 314–325, Sep. 1999.
[36] C. Dierkes, L. Kuhlmann, J. Kandasamy, and G. Angelis, "Pollution Retention Capability and Maintenance of Permeable Pavements," Hameln, Germany, 2000.
[37] C. W. Kipkie and W. James, "Feasibility of a permeable pavement option in SWMM for long-term continuous modeling," *Appl. Model. Urban Water Syst.*, vol. 8, no. 8, pp. 303–324, 2000.
[38] W. James and H. Von Langsdorff, "The use of permeable concrete block pavement in controlling environmental stressors in urban areas," in *Proceedings of the 7th International Conference on Concrete Block Paving (PAVE AFRICA 2003*, 2003, no. October, pp. 1–8.
[39] B. O. Brattebo and D. B. Booth, "Long-term stormwater quantity and quality performance of permeable pavement systems.," *Water Res.*, vol. 37, no. 18, pp. 4369–76, Nov. 2003.
[40] W. James, W. R. . James, and H. von Langsdorff, "Computer-Aided Design of Permeable Concrete Block Pavement for Reducing Stressors and Contaminants in an Urban Environment," in *Proceedings of the 7th International Conference on Concrete Block Paving (PAVE AFRICA 2003)*, 2003, vol. October, no. 7th.
[41] Z. He, "OPAC 2000: A New Pavement Design System," University of Waterloo, Waterloo, Ontario, Canada, 2000.
[42] J. Knapton, I. Cook, and D. Morrell, "A new design method for permeable pavements surfaced with pavers," *Highw. Transp.*, vol. 49, no. 1–2, pp. 23–26, 2002.
[43] T. Kuennen, "Voids Add Value To Pervious Concrete," *Road Science, Better Roads*, no. August, USA, Aug-2003.
[44] M. C. Adams, "Porous Asphalt Pavement With Recharge Beds : 20 Years and Still working," *Cahill Associates (CA)*, Philadelphia, USA, Apr-2003.
[45] J. Yang and G. Jiang, "Experimental study on properties of pervious concrete pavement materials," *Cem. Concr. Res.*, vol. 33, no. 3, pp. 381–386, Mar. 2003.
[46] B. Shackel and A. Pearson, "Permeable Concrete Eco-Paving as Best Management Practice in Australian urban road engineering," Australia, 2004.
[47] T. C. Ladd, "Water Quantity Study of A Porous Concrete Infiltration Basin Best Management Practice," Villanova University, 2004.
[48] Interlocking Concrete Pavement Institute (ICPI), "Permeable Interlocking Concrete Pavements," Virginia, USA, 2005.
[49] B. Shackel, "Design of permeable paving subject to traffic," in *8th International Conferce on Concrete Block Paving*, 2006, pp. 51–60.
[50] E. a Dreelin, L. Fowler, and C. Ronald Carroll, "A test of porous pavement effectiveness on clay soils during natural storm events.," *Water Res.*, vol. 40, no. 4, pp. 799–805, Feb. 2006.
[51] E. Z. Bean, W. F. Hunt, and D. A. Bidelspach, "Field Survey of Permeable Pavement Surface Infiltration Rates," *J. Irrig. Drain. Eng.*, vol. 133, no. 3, pp. 249–255, 2007.

[52] E. Z. Bean, W. F. Hunt, D. A. Bidelspach, and J. T. Smith, "Study on The Surface Infiltration Rate of Permeable Pavements," Norht carolina, 2004.
[53] K. Rankin and J. E. Ball, "A Review of the Performance of Permeable Pavers," Sydney, Australia, 2004.
[54] M. Offenberg, "Producing Pervious Pavements," *Concrete International*, no. March, Orlando, FL, pp. 50–54, Mar-2005.
[55] W. Boomsma and M. Huurman, "Permeable Paving Systems with Storing Capacity," in *8th International Conference on Concrete Block Paving*, 2006, no. November.
[56] J. K. Gilbert and J. C. Clausen, "Stormwater runoff quality and quantity from asphalt, paver, and crushed stone driveways in Connecticut.," *Water Res.*, vol. 40, no. 4, pp. 826–32, Feb. 2006.
[57] M. Wanielista, M. Chopra, J. Spence, and C. Ballock, "Hydraulic Performance Assessment of Pervious Concrete Pavements for Stormwater Management Credit," Orlando, FL, 2007.
[58] V. R. Schaefer, M. T. Suleiman, K. Wang, J. T. Kevern, and P. Wiegand, "An Overview of Pervious Concrete Applications in Stormwater Management and Pavement Systems," US, 2006.
[59] J. T. Kevern, "Advancements in pervious concrete technology," Iowa State University, 2008.
[60] A. S. Hager, "Sustainable Design of Pervious Concrete Pavements," University of Colorado Denver, 2009.
[61] J. Zhang, "A laboratory scale study of infiltration from Pervious Pavements," RMIT University, 2006.
[62] E. Fassman, S. Blackbourn, A. van Huyssteen, and D. Kettle, "Permeable Pavement Performance on the North Shore," in *2007 South Pacific Stomwater Conference*, 2007, no. December.
[63] N. Jayasuriya and N. Kadurupokune, "Impact of Pervious Pavements on Drainage Infrastructure," in *11th International Conference on Urban Drainage*, 2008, pp. 1–10.
[64] F. Straet, E. Beckers, and A. Degré, "Hydraulic behaviour of greened porous pavements : a physical study," *11th Int. Conf. Urban Drainage, Edinburgh, UK*, no. 1997, pp. 1–7, 2008.
[65] C. Yong, A. Deletic, T. D. Fletcher, and M. R. Grace, "The clogging behaviour and treatment efficiency of a range of porous pavements," *11th Int. Conf. Urban Drain.*, pp. 1–12, 2008.
[66] L. Hou, S. Feng, Z. Huo, Y. Ding, and S. Zhang, "Experimental study on rainfall-runoff relation for porous pavements," *Hydrol. Res.*, vol. 39, no. 3, p. 181, May 2008.
[67] UNI-GROUP, "Permeable Interlocking Concrete Pavement: Design Guide and Research Summary," U.S.A., 2008.
[68] Interpave, "Understanding Permeable Paving: Guidance for Designers, Planners and Local Authorities," UK, 2008.
[69] B. Shackel and D. Pezzaniti, "Development of Design Software for Permeable Interlocking Concrete Pavements," in *Paving, 9th. International Conference on Concrete Block*, 2009, pp. 1–8.
[70] B. Shackel, "The Design, Construction and Evaluation of Permeable Pavements in Australia," in *24th ARRB Conference – Building on 50 years of road and transport research*, 2010, pp. 1–14.
[71] S. Dierks and P. D. Associates, "Hydrologic and Water Quality Performance Evaluation of Porous Pavers & Vegetated Swales on Easy Street in Ann Arbor , Michigan," Ann Arbor, Michigan, 2009.
[72] L. . Rossman, *Storm Water Management Model User's Manual Version 5.0*. Cincinnati, Ohio: United States Environmental Protection Agency, 2010.
[73] Y. Wang and G. Wang, "Improvement of Porous Pavement.," US, 2011.
[74] X. . We, C. A. Pomeroy, and Y. . Sun, "Hydrological impacts evaluation of pervious pavement based on a storm water management model," in *ISWREP 2011 - Proceedings of 2011 International Symposium on Water Resource and Environmental Protection*, 2011, vol. 1, pp. 614–617.
[75] C. T. Agouridis, J. A. Villines, and J. D. Luck, "Permeable Pavement for Stormwater

Management," Lexington, Kentucky, 2011.

[76] E. Gomez-Ullate, J. R. Bayon, S. Coupe, and D. Castro-Fresno, "Performance of pervious pavement parking bays storing rainwater in the north of Spain.," *Water Sci. Technol. t*, vol. 62, no. 3, pp. 615–21, Jan. 2010.

[77] E. Gomez-Ullate, a V Novo, J. R. Bayon, J. R. Hernandez, and D. Castro-Fresno, "Design and construction of an experimental pervious paved parking area to harvest reusable rainwater.," *Water Sci. Technol.*, vol. 64, no. 9, pp. 1942–50, Jan. 2011.

[78] H. Jia, Y. Lu, S. L. Yu, and Y. Chen, "Planning of LID-BMPs for urban runoff control: The case of Beijing Olympic Village," *Sep. Purif. Technol.*, vol. 84, no. February, pp. 112–119, 2012.

[79] H. . Qin, Z. . Li, and G. Fu, "The effects of low impact development on urban fl ooding under different rainfall characteristics," *J. Environ. Manage.*, no. August, pp. 1–9, 2013.

[80] M. Carbone, G. Brunetti, and P. Piro, "Hydrological Performance of a Permeable Pavement in Mediterranean Climate," in *14th SGEM GeoConference on Water Resources. Forest, Marine And Ocean Ecosystems, Book 3*, 2014, no. SGEM2014 Conference Proceedings, ISBN 978-619-7105-13-1 / ISSN 1314-2704.

[81] M. Korkealaakso, J., Kuosa, H., Niemeläinen, E. & Tikanmäki, "Review of pervious pavement dimensioning, hydrological models and their parameter needs," Finland, 2014.

[82] H. Kim, M. Jung, K. J. B. Mallari, G. Pak, S. Kim, S. Kim, L. Kim, and J. Yoon, "Assessment of porous pavement effectiveness on runoff reduction under climate change scenarios," *Desalin. Water Treat.*, pp. 1–6, May 2014.

[83] M. Kayhanian, P. T. Weiss, J. S. Gulliver, and L. Khazanovich, "The Application of Permeable Pavement with Emphasis on Successful Design , Water Quality Benefits , and Identification of Knowledge and Data Gaps," California, US, 2015.

[84] P. T. Weiss, M. Kayhanian, L. Khazanovich, and J. S. Gulliver, "Permeable Pavements in Cold Climates: State of the Art and Cold Climate Case Studies," Minnesota, 2015.

[85] L. A. Rossman and W. C. Huber, *Storm Water Management Model Reference Manual*, Volume I –., vol. I. Cincinnati, Ohio, 2016.

[86] A. Palla and I. Gnecco, "Hydrologic modeling of Low Impact Development systems at the urban catchment scale," *J. Hydrol.*, vol. 528, pp. 361–368, 2015.

[87] Z. Zahmatkesh, S. J. Burian, M. Karamouz, H. Tavakol-Davani, and E. Goharian, "Low-impact development practices to mitigate climate change effects on urban stormwater runoff: Case study of New York City," *J. Irrig. Drain. Eng.*, vol. 141, no. 1, 2015.

[88] M. Brugin, M. Marchioni, G. Becciu, F. Giustozzi, E. Toraldo, and V. C. Andrés-Valeri, "Clogging potential evaluation of porous mixture surfaces used in permeable pavement systems," *Eur. J. Environ. Civ. Eng.*, vol. 8189, no. December, pp. 1–11, 2017.

[89] M. Alsubih, S. Arthur, G. Wright, and A. Deonie, "Experimental study on hydrological performance of a permeable pavement," *Urban Water J.*, vol. 14, no. 4, pp. 427–434, 2017.

[90] A. Kia, H. S. Wong, and C. R. Cheeseman, "Clogging in permeable concrete : A review," *J. Environ. Manage.*, vol. 193, pp. 221–233, 2017.

[91] K. Zhang, F. Yong, D. McCarthy, and A. Deletic, "Predicting long term removal of heavy metals from porous pavements for stormwater treatment," *Water Res.*, vol. 142, pp. 236–245, 2018.

[92] M. Razzaghmanesh and M. Borst, "Investigation clogging dynamic of permeable pavement systems using embedded sensors," *J. Hydrol.*, vol. 557, pp. 887–896, 2018.

[93] A. Timm, B. Kluge, and G. Wessolek, "Landscape and Urban Planning Hydrological balance of paved surfaces in moist mid-latitude climate – A review," *Landsc. Urban Plan.*, vol. 175, no. March, pp. 80–91, 2018.

[94] V. Ioannidou and S. Arthur, "Hydrological Response of a Permeable Pavement Laboratory Rig for Stormwater Management," *Proc. MDPI*, vol. 2, no. 11, p. 607, 2018.

[95] M. Kayhanian, H. Li, J. T. Harvey, and X. Liang, "Application of permeable pavements in highways for stormwater runoff management and pollution prevention : California research experiences," *Int. J. Transp. Sci. Technol.*, 2019.

[96] H.M. Imran, S. Akib, and M. R. Karim, "Permeable pavement and stormwater management systems: a review.," *Environ. Technol.*, vol. 34, no. 18, pp. 2649–2656, 2013.
[97] W. Lin, S. Ryu, and Y. Cho, "A case study of flow characteristics of permeable pavements by time and space model," *Can. J. Civ. Eng.*, vol. 41, pp. 660–666, 2014.
[98] K. Lee, H. Kim, G. Pak, S. Jang, L. Kim, C. Yoo, Z. Yun, and J. Yoon, "Cost-effectiveness analysis of stormwater best management practices (BMPs) in urban watersheds," *Desalin. Water Treat.*, vol. 19, no. 1–3, pp. 92–96, Jul. 2010.
[99] B. K. Ferguson, "Porous Pavements Q & A," *Stormwater-The J. Surf. Water Qual. Prof.*, no. September, 2009.
[100] K. . Collins, "A fied evaluation of four types of permeable pavement with respect to water quality improvement and flood control," North Carolina State University, 2007.
[101] Z. Niu, Z. Lv, Y. Zhang, and Z.-Z. Cui, "Stormwater infiltration and surface runoff pollution reduction performance of permeable pavement layers," *Env. Sci Pollut Res*, no. 23, pp. 2576–2587, 2016.
[102] T. Lucke and C. Dierkes, "Addressing the demands of the new German permeable pavement design guidelines and the hydraulic behaviour of a new paving design," *J. Eng. Sci. Technol.*, vol. 10, no. Spec. Issue on Advances in Civil and Environmental Engineering ACEE 2015, August 2015, pp. 14–28, 2015.
[103] T. Lucke, "Using Drainage Slots in Permeable Paving Blocks to Delay the Effects of Clogging: Proof of Concept Study," *Water*, pp. 2660–2670, 2014.
[104] Y. Bentarzi, A. Ghenaim, A. Terfous, A. Wanko, F. Feugeas, J. B. Poulet, and R. Mosé, "Hydrodynamic behaviour of a new permeable pavement material under high rainfall conditions," *Urban Water J.*, vol. 13, no. 7, pp. 687–696, 2015.
[105] J. Li, Y. Zhang, G. Liu, and X. Peng, "Preparation and performance evaluation of an innovative pervious concrete pavement," *Constr. Build. Mater.*, vol. 138, pp. 479–485, 2017.
[106] M. Marchioni and G. Becciu, "Experimental results on permeable pavements in urban areas: A synthetic review," *Int. J. Sustain. Dev. Plan.*, vol. 10, no. 6, pp. 806–817, 2015.
[107] J. Sansalone, X. Kuang, G. Ying, and V. Ranieri, "Filtration and clogging of permeable pavement loaded by urban drainage.," *Water Res.*, vol. 46, no. 20, pp. 6763–74, Dec. 2012.
[108] C. F. Yong, A. Deletic, T. D. Fletcher, and M. R. Grace, "The drying and wetting effects on clogging and pollutant removal through porous pavements," in *NOVATECH 2010*, 2010, pp. 1–10.
[109] M. Scholz and P. Grabowiecki, "Review of permeable pavement systems," *Build. Environ.*, vol. 42, no. 11, pp. 3830–3836, Nov. 2007.
[110] A. Al-Rubaei, M. Engström, M. Viklander, and G. Tobias-Blecken, "Long-term hydraulic performance of stormwater infiltration systems : a field survey," *Novatech 2013*, pp. 1–7, 2013.
[111] D.Y. S. Mah, *Potential of Road Subsurface On-Site Stormater Detention System*. Sarawak, Malaysia: UNIMAS Publisher, 2016.
[112] D. Y. S. M. and C. H. J. B. J O K Ngu, "Flow characteristics of individual lot stormwater detention," *Water Pract. Technol.*, vol. 11, no. 4, pp. 721–727, 2016.
[113] E. E. W. Tang, D. Y. S. Mah, and O. S. Selaman, "Improving drowned drainage outlets with submerged orifices," *Water Pract. Technol.*, vol. 11, no. 4, pp. 849–856, 2016.
[114] D. Y. S. Mah, B. Y. Koh, F. J. Putuhena, and N. Rosli, "Modelling of MSMA components: Porous pavement with detention system underneath for low traffic roads," *J. Civ. Eng. Sci. Technol.*, vol. 7, no. 1, pp. 30–38, 2016.
[115] D. Y. S. Mah, S. L. Loh, G. S. Poul, and W. H. Wan Ibrahim, "Road Intersection As Stormwater Detention Basin," in *Proceedings of the 37th IAHR World Congress*, 2017, vol. 6865, no. 1, pp. 2088–2091.
[116] N. Bateni, S. H. Lai, F. Putuhena, Y. S. Mah, and A. Mannan, "A Rainfall Simulator Used for Testing of Hydrological Performances of Micro-Detention Permeable Pavement," *Int. J. Eng. Technol.*, vol. 7, pp. 44–48, 2018.
[117] D. Y. S. Mah, S. W. Loh, M. A. Mannan, and W. H. Wan Ibrahim, "Modelling of surface permeation in multiple-orifice permeable road," *Int. J. Eng. Technol.*, vol. 7, no. 3, pp. 40–43,

2018.

[118] D. Y. S. Mah, J. O. K. Ngu, V. Liew, and W. H. Wan Ibrahim, "Augmenting drainage system in the old town of Kuching, Sarawak, Malaysia," *Int. J. Eng. Technol.*, vol. 7, no. 3, pp. 36–39, 2018.

[119] H. Maseri, O. S. Selaman, M. A. Mannan, and S. N. L. Taib, "Performance of cylindrical detention pond (CDP) as depression storage under fully saturated condition," *Int. J. Eng. Technol.*, vol. 7, no. 4, pp. 5996–6000, 2018.

[120] D. Y. S. Mah, C. V Liow, and A. Mannan, "Application of Storm Pav Green Pavement System in a Government Buildings Redevelopment Scheme," *Int. J. Innov. Technol. Explor. Eng.*, vol. 8, no. 9, pp. 723–727, 2019.

[121] C. V Liow, D. Y. S. Mah, and M. A. Malek, "Modelling of stormpav green pavement system using storm water management model and solidworks flow simulation," *Int. J. Recent Technol. Eng.*, vol. 8, no. 2, pp. 4673–4679, 2019.

[122] C. V Liow, D. Y. S. Mah, and M. R. R. Mohd Arif Zainol, "Modelling of Stormpav green pavement: Inlet and outlet of integrated permeable road and stormwater detention system," *Int. J. Civ. Eng. Technol.*, vol. 10, no. 2, pp. 966–976, 2019.

[123] S. H. Lui, D. Y. S. Mah, and Y. T. Fang, "Stormwater detention in road shoulder using StormPav green pavement system," *Int. J. Civ. Eng. Technol.*, vol. 10, no. 3, pp. 1555–1564, 2019.

[124] N. Bateni, S. H. Lai, F. J. Putuhena, D. Y. S. Mah, M. A. Mannan, and R. J. Chin, "Hydrological Performances on the Modified Permeable Pavement with Precast Hollow Cylinder Micro detention Pond Structure," *KSCE J. Civ. Eng.*, vol. 23, pp. 3951–3960, 2019.

[125] H. Maseri, O. S. Selaman, M. A. Mannan, and S. N. L. Taib, "Relationship Performance of Cylindrical Detention Pond (CDP) as Depression Storage and Runoff Reduction Under Fully Saturated Condition," *J. Comput. Theor. Nanosci.*, vol. 17, no. 2, pp. 1539–1546, 2020.

[126] N. Bateni, S. H. Lai, F. Putuhena, D. Y. S. Mah, M. A. Mannan, and R. J. Chin, "Hydrological Impact Assessment on Permeable Road Pavement with Subsurface Precast Micro-Detention Pond," *Water Environ. Journal*, pp. 1–10, 2020.

[127] N. Bateni, L. S. Hin, A. Mannan, J. H. Adam, Kipli K., and Bustami R.A. (2021) "Performance of Permeable Pavement with Subsurface Micro Detention Storage as Rainwater Harvesting Device," in *ICCOEE 2021: Lecture Notes in Civil Engineering, vol 132. Springer, Singapore.*

[128] N. Bateni, "Permeable Road Pavement with Subsurface Precast Micro Detention Storage: A Green Pavement Practice.," University of Malaya, 2020.

IOP Conf. Series: Materials Science and Engineering 1101 (2021) 012015 doi:10.1088/1757-899X/1101/1/012015

Numerical investigation on the performance of stone columns under raft foundation in soft clayey soils

Danish Ahmed[1,2] Siti Noor Linda bt Taib[1*] Tahar Ayadat[2] and Alsidqi Hasan[1]

[1] Faculty of Engineering, Universiti Malaysia Sarawak, Malaysia.
[2] Engineering Department, Prince Mohammad bin Fahd University, Saudi Arabia.
*Corresponding author: tlinda@unimas.my

Abstract. Stone columns are the most common and effective technique used for enhancing the overall strength and performance of soft soils. They are more effective for moderately lightweight structures. This investigation presents a parametric study of stone columns embedded in ground to strengthen the soft clayey soil under stiff raft foundation. This research is based on a computational analysis by creating a foundation finite element model consisting of a group of stone columns that are mounted under rafting using PLAXIS 3D. A certain range of parameters (for example spacing, diameter and angle of friction of stone columns) is considered and it is concluded that the increase in diameter and angle of friction of stone columns can improve the ultimate load bearing capacity of the foundation system. Furthermore, the failure mechanism of this foundation scheme is analysed and a semi-empirical model for the determination of the bearing capacity ratio *(BCR)* is developed from the results of the parametric analysis.

1. Introduction

Foundations for structures on soft soil usually involve excessive settlement and insufficient load bearing capacity. Among all the soil improvement techniques, stone columns are treated as the most common and economical practice. Granular material compacted and compressed in cylinder-shaped hole in the ground to reinforce and enhance the behaviour of soft soils. Stone columns are very effective for moderately lightweight constructions. It was introduced first in in 1830 in France as stated by Hanna et al [1]. Hughes and Withers [2] are considered in the first few researchers who conducted the experimental tests on the single granular columns their finding includes that the carrying capacity of a group of stone columns can be considered as the sum of carrying capacity of individual column. They also have concluded that the stone columns failure occurs separately by bulging. In addition, they presented the concept of a unit cell system for estimating a stone column's carrying capacity.

Since then, a lot of works have been carried out to examine the efficiency of stone columns under raft base. Some of these investigations were based on laboratory experimental works [2-6]. Whereas, others have investigated the performance of stone columns under raft using numerical and analytical methods [7-20].

The principle objective of this work is to present a parametric study of stone columns embedded in ground to strengthen the soft clayey soil under stiff raft foundation. This study is based on a numerical investigation by developing a finite element model of the foundation system consists of a group of stone columns installed under raft using PLAXIS 3D. A certain range of parameters (for example spacing,

IOP Conf. Series: Materials Science and Engineering 1101 (2021) 012015 doi:10.1088/1757-899X/1101/1/012015

diameter and angle of friction of stone columns) is considered. The failure mechanism of this foundation scheme is analysed. Furthermore, from the results of the parametric study a semi-empirical model is developed for the determination of bearing capacity ratio *(BCR)*.

2. Numerical Modeling

PLAXIS 3D Software has been used to simulate a finite element model for numerical study using nonlinear elasto-plastic element to reflect the condition of assembly of stone columns under raft in soft clayey soil. The model consists of 20 m square raft foundation either resting on the soft clay layer treated with group of stone columns or resting on untreated layer of soft clay underlined by thick layer of sand.

In the present study all the soil elements were modelled as 10-noded tetrahedral elements for 3D meshing, globally medium mesh was used for all the soil domain including stone columns, whereas it was refined to fine mesh and very fine mesh in the region where critical conditions (high stresses and displacements) were expected. In the model, soft soils were reinforced with stone columns and both were modelled as the soil, separately with their material properties and there were no additional elements provided as an interface between the soil and stone columns. For this reason, similar kind of settlements were expected in both stone columns and soils in the surrounding region and therefore no differential settlement was expected. Boundary conditions were selected for this model after many trials to eliminate the effect of vertical and horizontal stresses due to confinement. Fixed supports were provided at the base in figure 1 and figure 2 and roller supports were presumed on the model's sides (vertical boundaries).

For all the soil elements including soft clay, sand and stone columns were modelled with the elasto-plastic Mohr-Coulomb model which consists of five basic parameters, cohesion *(c)*, friction angle (φ), dilatancy angle (ψ), Young's modulus *(E)*, and Poisson's ratio (v).

Experimental test results were used from the literature [5] to validate the present numerical model, the experimental test setup consists of large test box (120 cm x 120 cm x 90 cm) carrying the bed of clay in it with maximum depth of 500 mm and reinforced with a single stone column 200 mm in diameter, also a stiff steel plate with 200 mm in diameter and 30 mm in thickness were used for loading on the stone column. Figure 3 shows promising match between the results obtained from numerical model and the experimental data.

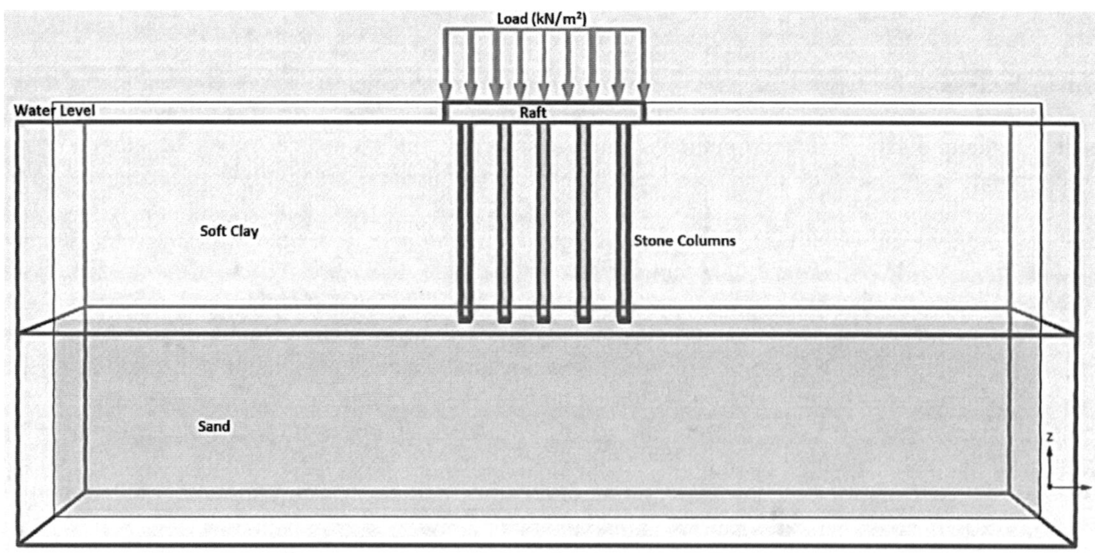

Figure 1. Side view of the finite element model

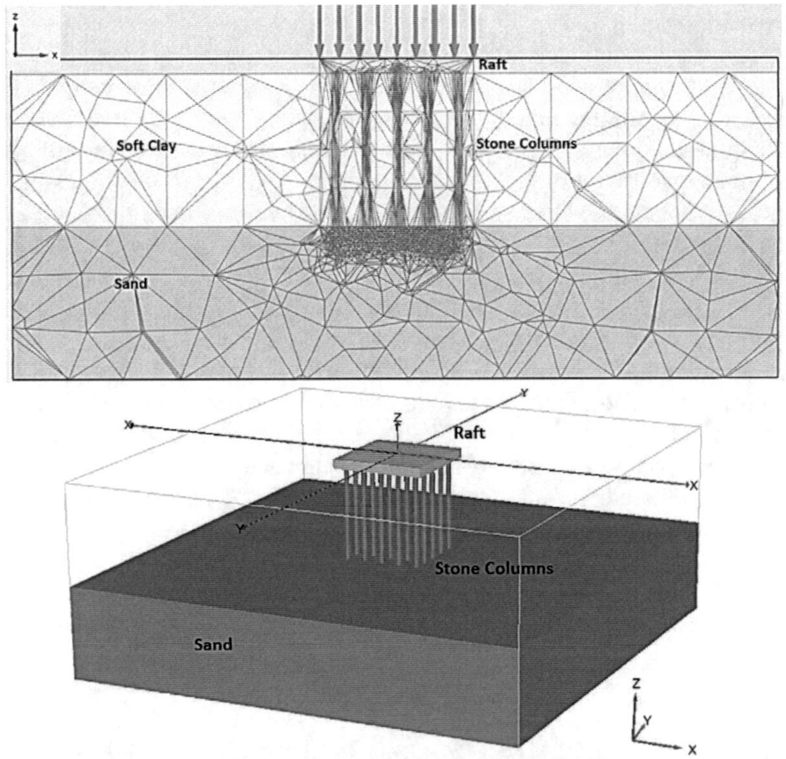

Figure 2. Perspective of the finite element model.

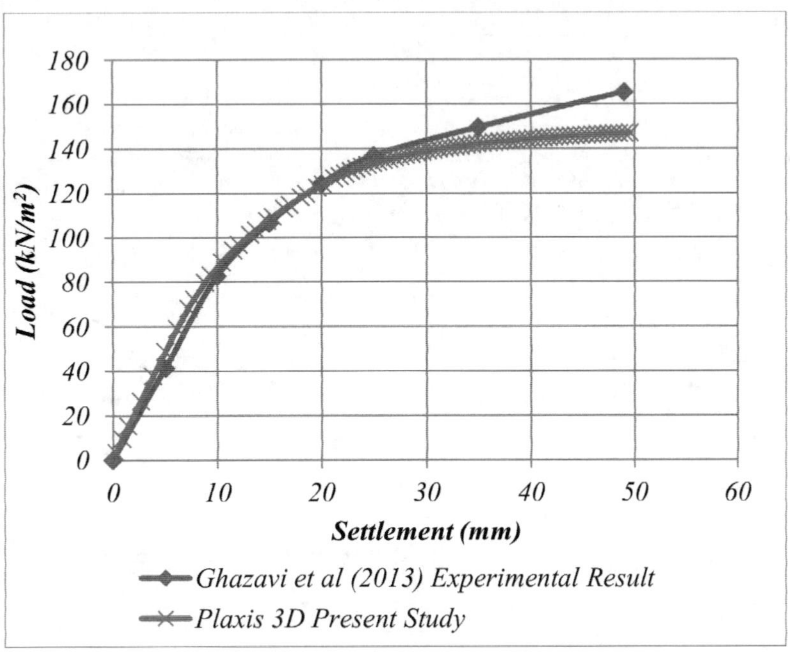

Figure 3. Load vs settlement curves for validation of the present numerical model

3. Parametric Investigation

Parameters used in the present study are taken from the studies reported in literature ([1], [14], and [15]) as presented in table 1. The numerical investigation has been conducted to investigate the effect of these parameters on the load bearing capacity of raft foundation on untreated and treated soft soil (strengthened with stone columns). Figure 4(a) shows stress strain relationship of raft foundation on both untreated and treated soil with different spacing-diameter ratio *(S/D)* of the stone columns. It can be observed that the raft on untreated soft soil has lowest load bearing capacity, whereas the load bearing capacity of raft on treated soft soil is increasing with decrease in *(S/D)* of the stone columns. Since stress-strain curves could not show any sign of failure, Chin's method [21] was applied to estimate the ultimate load bearing capacity of the raft foundation under different conditions. In this method each strain value first must be divided by its corresponding stress and the resulting values must be plotted against the strains as shown in figure 4(b). When the plotted values fall in straight lines it indicates the load approaches fault tension and the ultimate load efficiency (q_u) can be determined by measuring the reverse slope of this straight line. The obtained values of the ultimate load bearing capacity (q_u) of raft foundation system under various cases are plotted in figure 5 against spacing-diameter ratio *(S/D)* of the stone columns. These curves clearly show the influence of *(S/D)* ratio on the ultimate load bearing capacity of the foundation system, and it can be detected that (q_u) increase with decrease in the ratio *(S/D)*.

Table 1. Material Properties (Parameters for Numerical Model).

Properties	Soft Clay	Sand	Stone Column	Raft
Unit weight, γ_{unsat}, γ_{sat}	17, 18 kN/m^3	17.5, 19.5 kN/m^3	19, 20 kN/m^3	-
Young's modulus, E_s	4 MPa	28 MPa	80 MPa	27.8 GPa
Poisson's ratio, v_s	0.5	0.3	0.3	0.15
Angle of friction, φ	0°	42°	42°, 45°, 48°	-
Cohesion, c	25 kPa	0	0	-

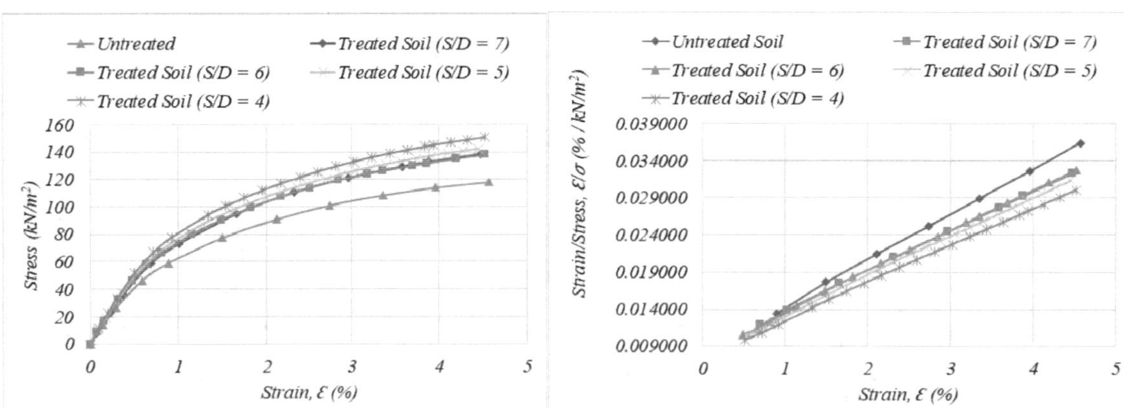

Figure 4. (a) Stress against Strain for the cased of $\varphi_c = 42^o$ & (b) Strain/Stress against Strain (Chin [21]) for raft and stone columns plus raft foundation for the case of $\varphi_c = 42^o$

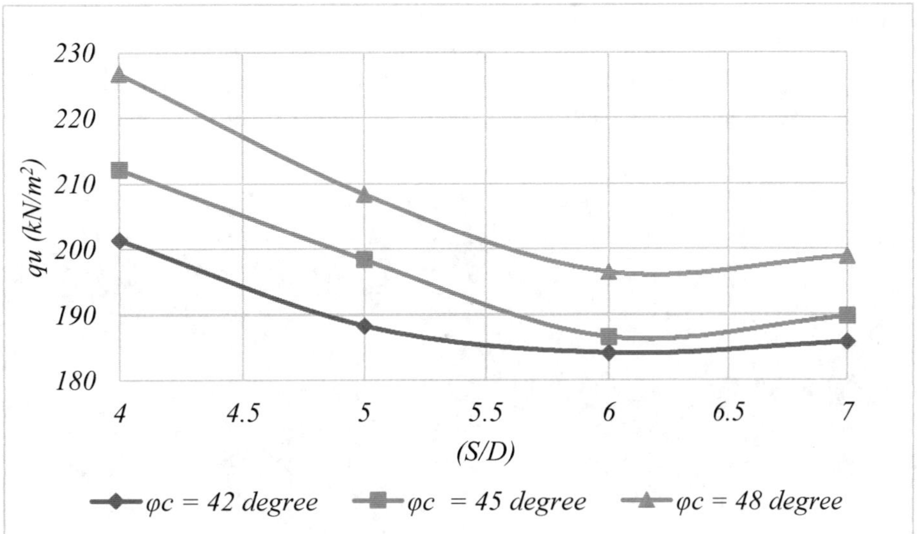

Figure 5. Change in ultimate load bearing capacity *(q_u)* with the ration *(S/D)*

The term bearing capacity ratio *(BCR)* was introduced to clearly understand the effect of ratio *(S/D)* and angle of friction of stone columns (φ_c), bearing capacity ratio is described as the ratio between the ultimate load bearing capacity of raft foundation on treated soft soil to that of raft foundation on untreated soft soil. It can be observed from figure 6 that bearing capacity ratio *(BCR)* and ratio *(S/D)* has linear relation. It is clearly indicating that the ratio *(S/D)* of stone columns have no influence on the bearing capacity when *(S/D)* is greater than 7, whereas the bearing capacity ratio *(BCR)* can be improved when the ratio *(S/D)* is 4 or less. Figure 6 also shows the positive influence of angle of friction (φ_c) on the bearing capacity ratio *(BCR)*, increase in the angle of friction of stone columns (φ_c) can increase the ultimate load bearing capacity of the foundation system on treated soft clayey soil. This can be described by the fact that the stone columns with higher angle of friction have good interlocking between their granular particles and have good shear strength.

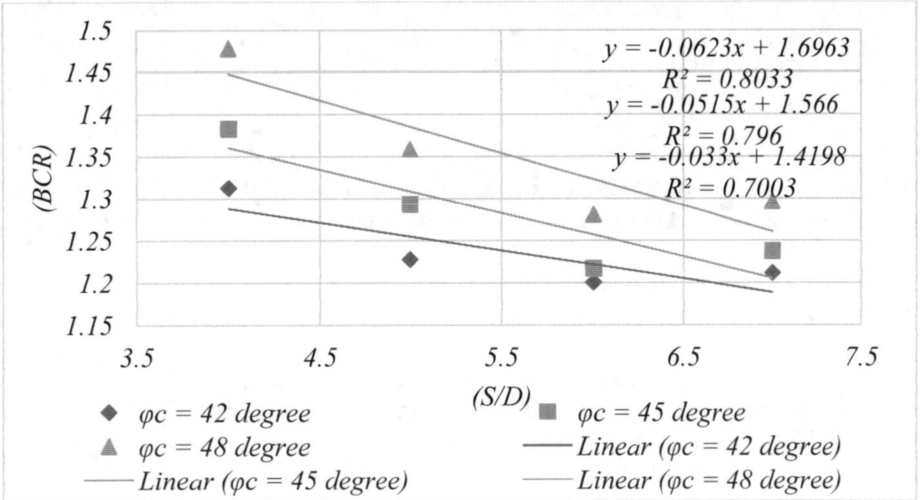

Figure 6. Change in the bearing capacity ratio *(BCR)* with the ration *(S/D)*

The outcomes of the present study (numerical model) was also utilized to observe the failure modes of stone columns with raft foundation under loading. It was deduced that the foundation systems were failed mostly because of shear both in columns and the soil in the surrounding regions, large shear failure can be seen in figure 7. This findings concord well with the results reported by Hanna et al [1].

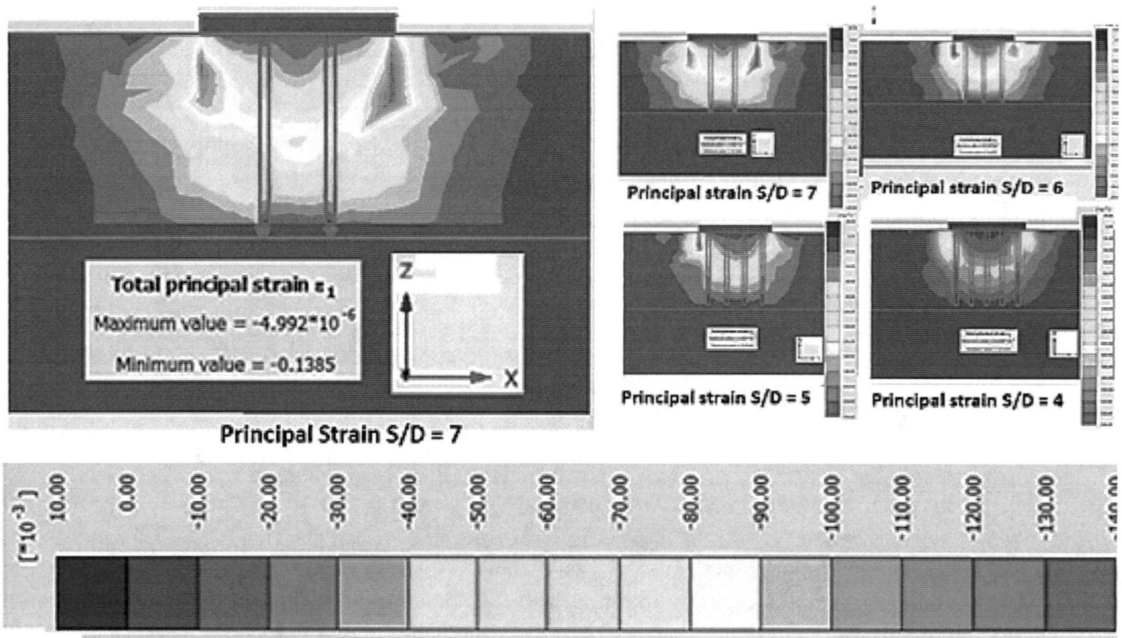

Figure 7. Variation of the improvement factor *(BCR)* with the ratio *(S/D)*

As discussed earlier, the findings of this parametric study were also used to establish an empirical model to estimate the of bearing capacity ratio *(BCR)* of the foundation system (stone columns under raft foundation). Based on the results presented in figure 6 least square regression was used to establish expression in equation (1) for estimating the bearing capacity ratio of such foundation system:

$$BCR = (0.14 \tan \emptyset_c - 0.1894)\, S/D - (1.3169 \tan \emptyset_c - 2.882) \quad (1)$$

4. Conclusion

Parametric study was conducted by developing a finite element model to study the performance of assembly of stone columns installed under a rigid raft foundation in soft clayey soil. In this parametric investigation the effect of certain parameters on the behaviour of this foundation system was investigated.

The main conclusions were drawn are summarized as follows:

- The ultimate load bearing capacity (q_u) of the foundation system increases with the decrease in ratio *(S/D)*.
- The bearing capacity ratio *(BCR)* and the ratio *(S/D)* has linear relation, whereas the ratio *(S/D)* of stone columns have no influence on the bearing capacity when *(S/D)* is greater than 7, on the other hand the bearing capacity ratio *(BCR)* improves when the ratio *(S/D)* is 4 or less.
- The increase in the angle of friction of stone columns (φ_c) increases the ultimate load bearing capacity (q_u) of the foundation system on treated soft soil which also indicates that *(BCR)* is higher for the higher angle of friction of stone columns.
- The assembly of stone columns installed under a rigid raft foundation in soft clayey soil were failed mostly because of shear both in columns and the soil in the surrounding regions.

- A semi-empirical model was developed from the parametric study to estimate the ultimate load bearing capacity of assembly of stone columns embedded in soft clayey soil under raft foundation.

Acknowledgments
The authors would like to extend appreciation to Universiti Malaysia Sarawak (Grant: F02/PGRG/1947/2020) and Prince Mohammad bin Fahd University.

References

[1] Hanna M, Etezad M, and Ayadat T 2013 Mode of failure of a group of stone columns in soft soil *Int. J. of Geomechanics* **13** 87-96
[2] Hu W, Wood D M and Stewart W 1997 Ground improvement using stone column foundations: result of model tests *Int. Conf. on Ground Improvement Tech.*247-256
[3] Black J, Sivakumar V and Madhav M 2007 Reinforced stone columns in weak deposits: laboratory model study *J. Geotech. Geoenviron. Eng. ASCE* **133 No 9** 1154-1161
[4] Black J, Sivakumar V and McKinley J 2007 Performance of clay samples reinforced with vertical granular columns *Can. Geotech. J.* **44 No 1** 89-95
[5] Black J A, V Sivakumar and A Bell 2011 The Settlement Performance of Stone Column Foundations *Geotech.* **61 (11)** 909-922
[6] Ghazavi M and Afshar J N 2013 Bearing capacity of geosynthetic encased stone columns *Geotextiles. and Geomembranes.* **10 no 2** pp 47-53
[7] Wehr W 1999 Schottersäulen das verhalten von einzelnen säulen und säulengruppen *Geotech.* **22 (1)** 40-47
[8] Wood D M, Hu W and Nash D F T 2000 Group effects in stone column foundations: model tests *Geotech.***50 (6)** 689-698
[9] Bae W S, Bang W S and Byung C A 2002 Behavior of foundation system improved with stone columns *Proc. 12th Int. Offshore and Polar Eng. Conf.* Kitakyushu Japan 675-678
[10] Nazari Afshar J and Ghazavi M 2014 A simple analytical method for calculation of bearing capacity of stone-column *Int. J. of Civil Eng.* **12 No 1**
[11] Ziaie Moayed R and Mohammadi Haji B 2016 Circular raft footings strengthened by stone columns under static loads *Int. J. of Civil Environ. Struct. Const. and Arch. Eng.* **10 No 9**
[12] Fattah M Y, Al Neami M A and Al Suhaily A S 2017 Estimation of bearing capacity of floating group of stone columns *Eng. Science and Technology an Int. J.* **20** 1166-1172
[13] Castro J 2014 Numerical modelling of stone columns beneath a rigid footing *Journal paper Computers and Geotech.* **60** 77–87
[14] Killeen M M and McCabe B A 2014 Settlement performance of pad footings on soft clay supported by stone columns: a numerical study *Soils and Foundations* **54 4** 760 776
[15] Elsawy M and El-Garhy B 2016 Behavior of raft foundation resting on improved soft soil with conventional granular piles *J. of Scientific and Eng. Research* **3 4** 428 434
[16] Lajevardi S H, Shamsi H R, and Hamidi M, 2018 Numerical and Experimental Studies on Single Stone Columns *Soil Mech Found Eng* **55** 340 345.
[17] Etezad M, Hanna A M, and Ayadat T 2015 Bearing Capacity of a Group of Stone Columns in Soft Soil *International Journal of Geomechanics* **15** Issue 2.
[18] Debnath P and Dey A K 2018 Prediction of Bearing Capacity of Geogrid-Reinforced Stone Columns Using Support Vector Regression *International Journal of Geomechanics* **18** Issue 2.
[19] Ong D E L, Sim Y S, and Leung C F 2018 Performance of Field and Numerical Back-Analysis of Floating Stone Columns in Soft Clay Considering the Influence of Dilatancy *International Journal of Geomechanics* **18** Issue 10.

[20] Etezad M, Hanna A M, and Khalifa M 2018 Bearing Capacity of a Group of Stone Columns in Soft Soil Subjected to Local or Punching Shear Failures *International Journal of Geomechanics* **18** Issue 12.

[21] Chin F K 1972 The inverse Slope as a Prediction of Ultimate Bearing Capacity of Piles *Proc. 3rd Southeast Asian Conf. on Soil Eng.* Hong Kong pp 83-91

IOP Conf. Series: Materials Science and Engineering 1101 (2021) 012016 doi:10.1088/1757-899X/1101/1/012016

An evaluation of highway crash-prone areas: A case study on Pan Borneo Highway in the state of Sarawak

S L Tan[1*], M R Kabit[1] and O A Johnson[2]

[1] Department of Civil Engineering, Universiti Malaysia, Sarawak, Malaysia
[2] School of Energy, Geoscience, Infrastructures and Society Heriot Watt University, Malaysia
* Corresponding author: tslcivil@yahoo.com.my

Abstract. Road traffic crashes is one of the major causes of death that needs to be addressed globally. Many studies have been conducted to identify the contributing factors to traffic crashes and to determine the required preventive measures. This study evaluates the causes of 164 traffic crashes along a 10 kilometres section of Pan Borneo Highway Sarawak. The study investigated the main causes of traffic crashes within this section of road through multiple sources: based on expert on site evaluation, analysis of site incident report and police accident record. The finding reveals that most of the accidents (93%) occurs during the daytime, weather contributes a combined 69% and private car owners (80%) cause most of the traffic crashes. The study also indicates that the road condition is a significant factor to the occurrence of traffic crashes in that area contributing 64% of the total crashes.

1. Introduction

Road transportation system plays a major part on the development of socio-economic status in a country. The mobilisation of people or goods from their origin to destination depend on its accessibility. However, frequent travelling on road causes increasing of traffic crash. The World Health Organization reported in 2016 that road accident is eighth leading cause of death and it is expected to be the fifth cause of death in year 2030 worldwide [1-3]. In Malaysia, road traffic crash was ranked at fourth (4.6%), after Ischemic heart diseases (13.9%), Pneumonia (12.7%) and Cerebrovascular diseases (7.1%) for the causes of death according to government's statistic report in year 2017 [4]. An average of 6,150 deaths were reported yearly due to road traffic crashes from 2007 to 2018 in Malaysia [5], this has made it imperative to investigate the causes and to determine the required preventives measures [6-7]. The probable causes of traffic crash can be identified based on the crash patterns [8], according to Malaysia Institute of Road Safety Research (MIROS), the possible causes of road traffic crashes are but not limited to human careless (80.6%), unsafe road conditions (13.2%) and vehicle condition (6.2%) [5].

Jesna and Anjaneyulu [1] reported the uncertainty in the behaviour of drivers to the same situation, and the geometry of road is a significant factor in the cause of accident [1]. Dudziak et al [6] identify driver behaviour as a major factor in road accident [8]. The reliability of different drivers or vehicles when reaching an unsafe road condition will be different, and the failure in controlling the vehicle may cause the traffic crash. While human errors appear to be the major cause of traffic crash, poor road conditions could lead to traffic crashes. This article presents a case study finding based on an evaluation conducted at crash-prone areas along Km 421 to Km 431 (Jalan Betong / Sarikei / Meradong) of Pan Borneo Highway in the state of Sarawak. Pan Borneo Highway is a major route connecting major towns in the state of Sarawak from the southern part in Teluk Melano to the northern part to Sabah through Brunei.

2. Methodology

This study was divided into three parts. The first part was the evaluation of the road traffic incident's reports recorded by the Works Package Contractor's Site Safety Officer. The evaluation to the road traffic incident's reports including determination of the actual crash-prone areas based on each kilometre. The evaluation was done according to four (4) major factors: time of incidents (daytime from 07:00am to 07:00pm or night-time from 07:00pm to 07:00am), weather conditions (fine, raining or after rain), type of vehicles (private cars, small lorries and vans, large lorries, or motorcycles) and cause of traffic crashes (human errors, unsafe road conditions or due to vehicle).

The second part was on site evaluation of the accident areas. Site evaluation was conducted to determine the hazard areas (improper geometry design and unsafe pavement surface condition) with driving on the studied area during daytime, night-time and raining time with normal speed (80 km/hr) and slow speed (40 km/hr). And the third part is the examination of traffic crashes report from Sarikei's Police Station. The police's reports evaluated based on the accident description as stated in the reports. Each individual traffic crash was evaluated to determine the cause whether it is human careless, unsafe road conditions or mechanical conditions of vehicles.

3. Data analysis and discussion

3.1. *Evaluation to the road traffic incident's reports*

Figure 1 illustrates the actual crash-prone areas based on kilometre. From the analysis, it can be seen that crash-prone areas were in Km 430 to Km 431 (34 cases or 21%), Km 423 to Km 424 (33 cases or 20%), and Km 426 to Km 427 (27 cases or 16%) which accounted for merely half of the total numbers of incidents reported.

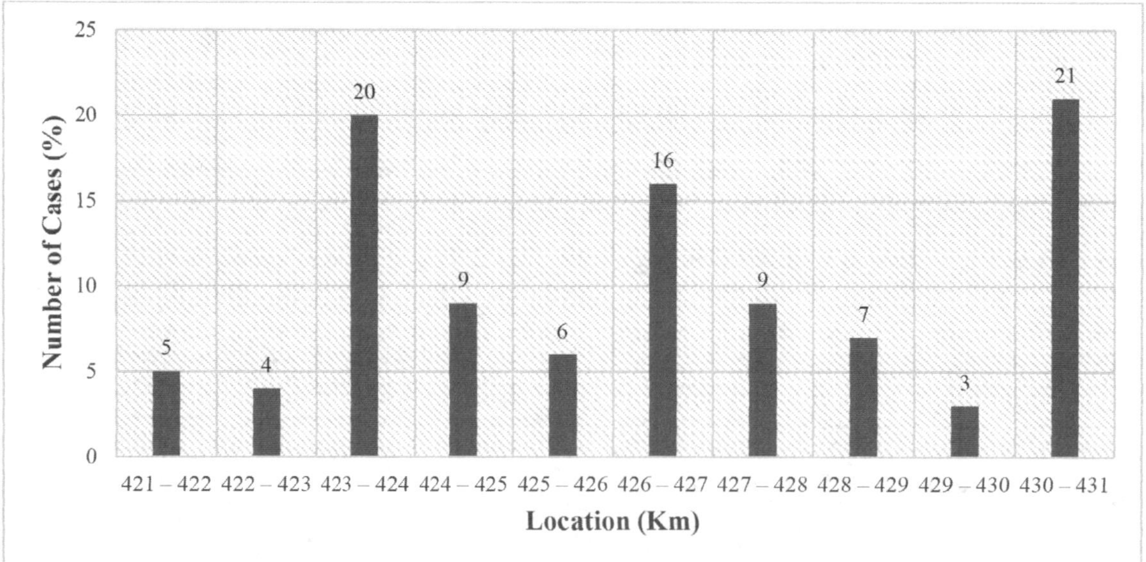

Figure 1. Actual crush-prone areas based on Km

The analysis of the incident time indicates that most of the accidents (93%) occurs during the daytime, as shown in figure 2, it could be concluded that poor night vision may not have been the cause of accident along the study area. This is quite different from what most studies has established that the

IOP Conf. Series: Materials Science and Engineering 1101 (2021) 012016 doi:10.1088/1757-899X/1101/1/012016

number of night-time crashes generally declines when days are longer, or when there is enough lighting on the road [9-12].

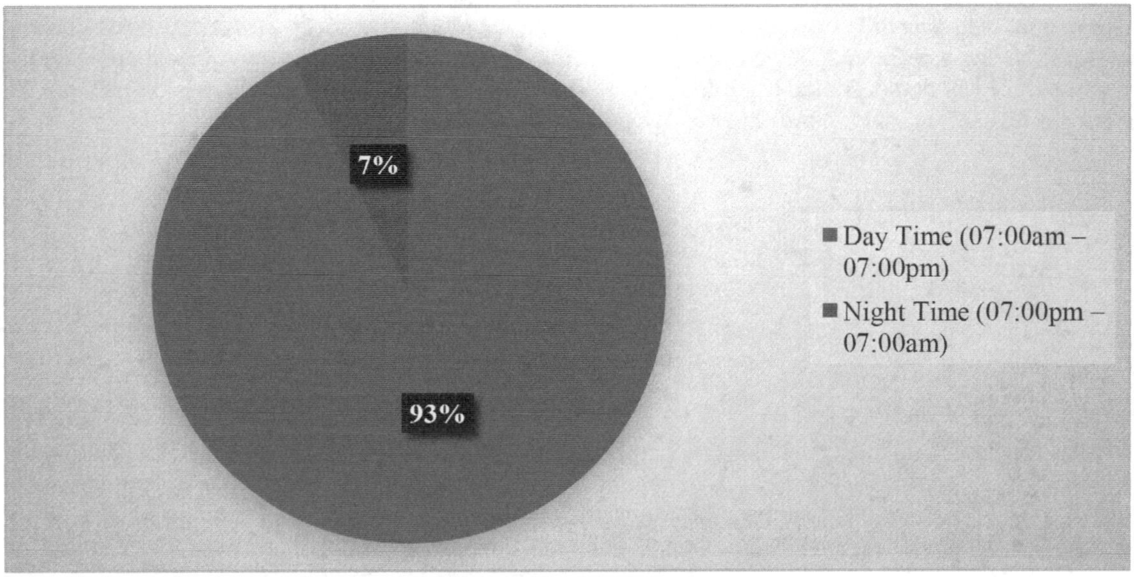

Figure 2. Total Traffic Crashes Based on Time

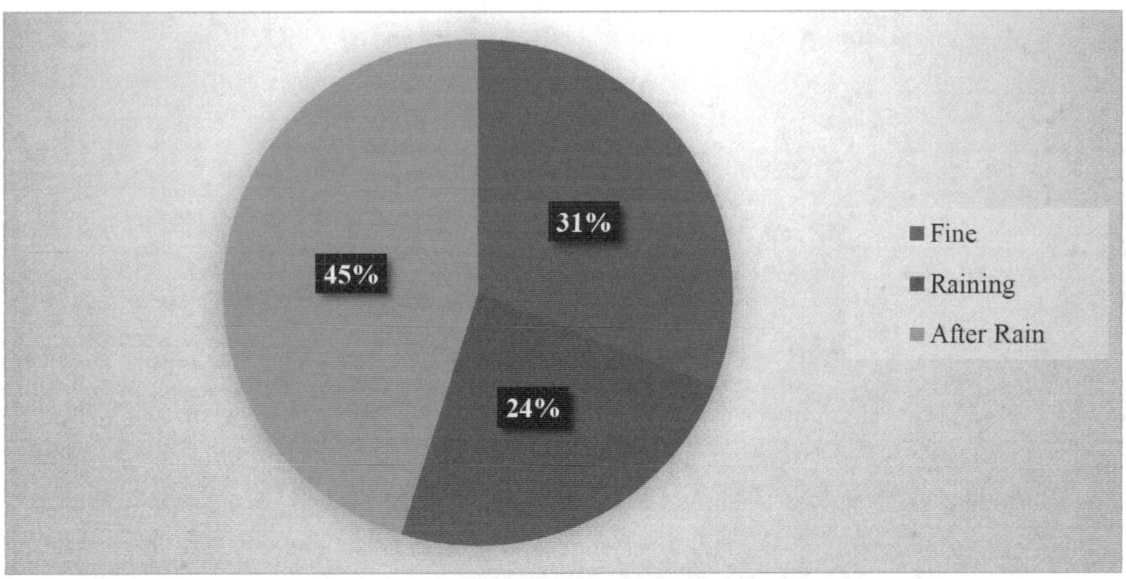

Figure 3. Total Traffic Crashes Based on Weather

It has been established weather is one of the major causes of accident along the study area. As shown in the figures 3, raining time or immediately after rain fall (which might be due to flooding of the roadway) with a combined 69% of incident occurrence. Rain or flooding on the roadway could easily affect the visual reaction of the driver and the visual reaction times will consequently be affected [13]. When the visual reaction time, which is the estimate of the time of vehicular collision with an object under a specific condition is affected, then the critical stopping distance (which is a key factor of safety) [14] will be jeopardized, thus, leading to accident.

Figure 4 shows that private car owners (80%) cause most of the traffic crashes. The high dependence on private cars has really resulted in the high density of cars. There exists less effective public transport in most rural areas, which has led to a greater dependence on private cars. However, it was not measure or there is no data related to the effect of age as one of the factors in this study. However, most accident that occur during the day may have been causes due to age and caused by drivers older than 65 years old. Literature has demonstrated that older drivers have a raised crash risk on a per kilometer driven basis, which is second only to the very youngest age groups [15].

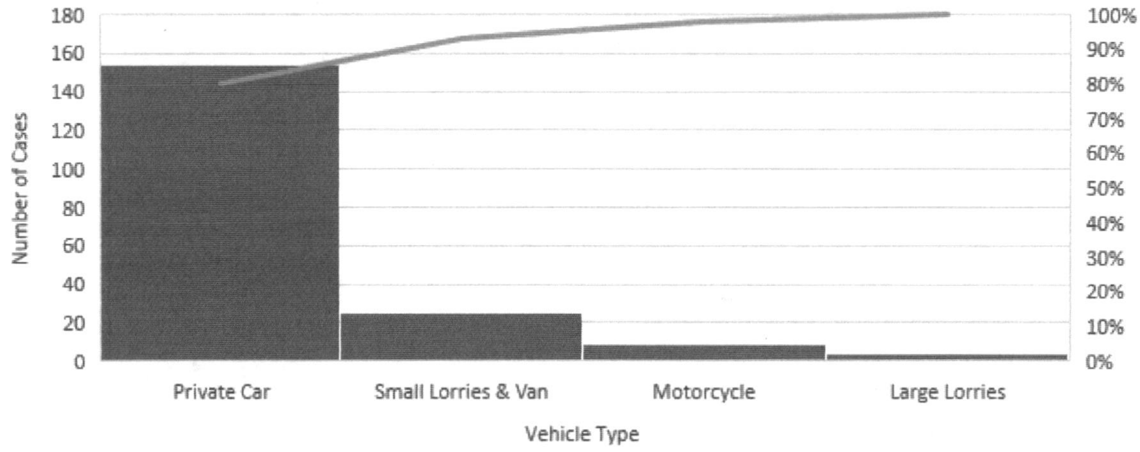

Figure 4. Total Traffic Crashes Based on Type of Vehicle

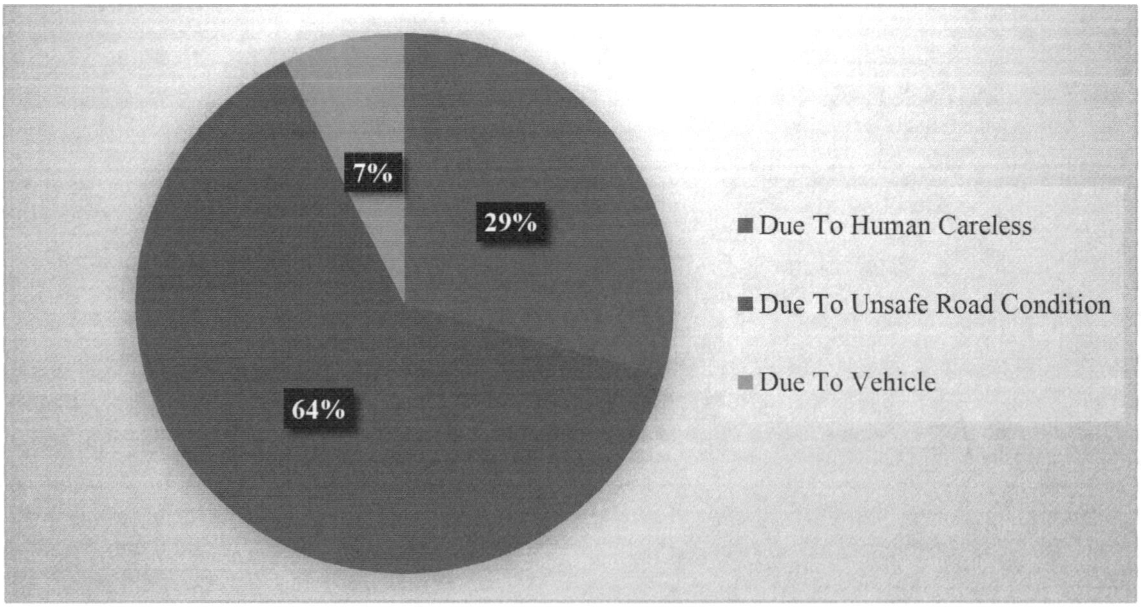

Figure 5. Total Traffic Crashes Based on Cause of Incident (Analysed from Incident's Reports)

Figure 5 shows total traffic crashes based on cause of incident. From the results, human errors contributed to 29% of the traffic crashes which is significantly lowered than the national recorded

IOP Conf. Series: Materials Science and Engineering 1101 (2021) 012016 doi:10.1088/1757-899X/1101/1/012016

statistics. Interestingly, unsafe road condition was found as the major contributing factor (64%) to traffic crashes at the study area.

3.2. Site evaluation
From the site evaluation conducted, Km 421 until Km 431 of Jalan Betong / Sarikei / Meradong constructed at mountainous terrain area as shown in figure 6.

In addition, figures 7 to figures 10 shows the existing site conditions that is unsafe to the road users due to uneven surface and improper geometry design like sharp curve, continuing horizontal and vertical curve, horizontal curve in sag vertical curve, and horizontal curve in crest vertical curve. The reliability of different drivers or vehicles when reaching these unsafe road conditions will be different, and the failure in controlling the vehicles may cause the traffic crash.

This existing R3 single dual carriageway Pan Borneo Highway will be upgraded to R5 two dual carriageway under the Proposed Development and Upgrading of Pan Borneo Highway in the state of Sarawak. From site observation, the sight distance along this studied area is sufficient, as the obstructions along roadside have now been cleared for road construction works. However, it is dark during night-time in most of the areas. Temporary warning signs and longitudinal yellow crossbars have been provided in advance of the crash-prone areas. Moreover, the safety blinker lights are functioning during the night-time in this studied area. The drainage system of the studied area was found to be insufficient. Water ponding observed on the road surface in many places during rain and after rain.

Figure 6. Mountainous Terrain

Figure 7. Uneven Road Surface

Figure 8. Sharp Curve

(a) (b)

Figure 9. Continuing Horizontal and Vertical Curves

(a) (b)

Figure 10. Horizontal Curve in Sag or Crest Vertical Curve

3.3. Police's Report Analysis

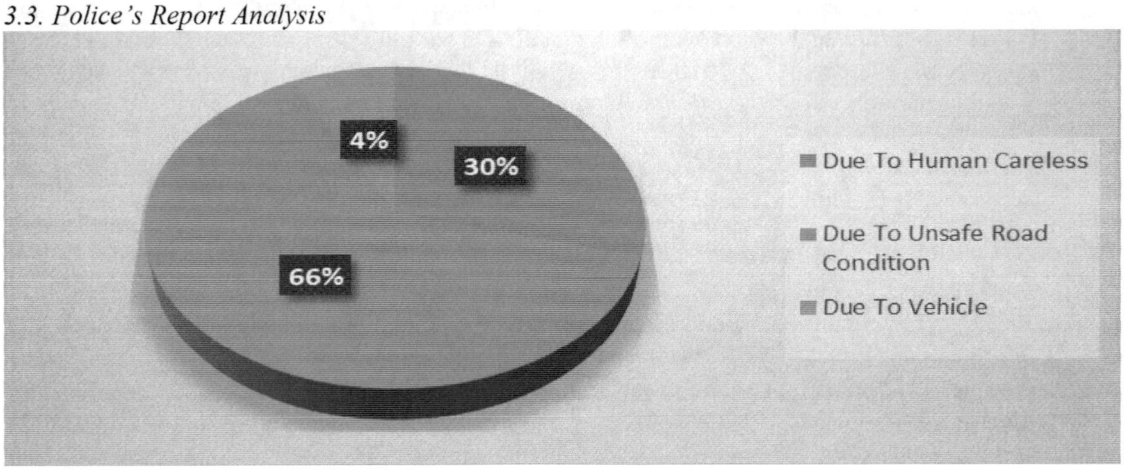

Figure 11. Total Traffic Crashes Based on Cause of Incident (Analysed from Police's Reports)

Based on description stated in the police's reports, the type of crashes has been identified to determine probable causes with respect to drivers, road, and vehicle. The collisions which have involved more than one vehicle, for example head-on collision and rear-end collisions will consider as due to human careless. The individual vehicle crash to the road side furniture or tree will consider as due to unsafe road condition. And the crash due to vehicle breakdown on road will consider as due to vehicle.

The cause of road traffic crashes based on the police reports are 30% due to human careless, 66% due to unsafe road condition and 4% due to vehicle as shown in figure 11. It was significantly similarity between the percentages in each causes of incident compared with the finding recorded by Safety Officer as shown in figure 5.

4. Conclusion and recommendations

Road crashes have significantly increased in the recent years and it has been identified as the eight causes of death. The effect of road crashes cannot be underestimated, it could significantly affect the socio-economic condition of a nation. To address this important factor, attempt has been made by researchers through study to reduce accidents and thus alleviate the negative effect on the environment. The road section in this study has been identified as the most dangerous along the Pan Borneo Highway Sarawak in terms of the number of traffic crashes. The present study examined road crash factors of crash-prone areas of Km 421 to Km 431 Jalan Betong / Sarikei / Meradong of Pan Borneo Highway in the state of Sarawak. The results showed that majority of the road crashes occurred during daytime and during inclement weather condition involving primarily private cars. It was also found that unsafe road surface conditions have significantly contributed traffic crashes for the studied road segment.

Thus, it could be recommended that the unsafe surface conditions along this existing road should be rectified with new geometric alignment design in the proposed development and upgrading of Pan Borneo Highway in the state of Sarawak. This is important as the road quality of the existing road, especially along the studied road segment has significantly reduced and unsafe for the road users on this segment of road, a road safety audit should be carried out more frequently to ensure sufficient safety precaution on site. Road traffic crash can be avoided with proper and effective implementation of the required preventives measures, such as proper planning system, high standard of safety, education (proper public awareness) and prevention methods. Further study should be conducted on most of these preventive measures and its application on the area.

References

[1] Jesna N M and Anjaneyulu M V L R 2016 Reliability analysis of horizontal curves on two lane highway *Transportation Research Procedia (Mumbai)* vol 17 (India: 11th Transportation Planning and Implementation Methodologies for Developing Countries 2014) pp 107-115

[2] Khademi N and Choupani A A 2018 Investigating the road safety management capacity: Toward a lead agency reform, *International Association of Traffic and Safety Sciences Research.* **42** 105-120

[3] World Health Organization 2018 *The Top 10 Cause Of Death* Retrieved on 24 May, 2018 from http://www.who.int/news-room/fact-sheets/detail/the-top-10-cause-of-death

[4] Mohamad R 2018 *Statistics on causes of death, Malaysia, 2018* Department of Statistic Malaysia. Pp 1-6 (31 October 2018)

[5] JKJR 2019 *Road Safety Statistic Book* (Malaysia: Road Safety Department of Malaysia) pp 1-34

[6] Dudziak M, Lewandowski A and Sledzinski M 2017 Uncommon road safety hazards *XXI International Polish-Slovak Conference "Machine Modelling and Simulations 2016" (Poland)* **177** (Poland: Procedia Engineering) pp 375-380

[7] Wegman F 2017 The future of road safety: A worldwide perspective, *International Association of Traffic and Safety Sciences Research.* **40** 66-71

[8] Nicholas J Garber and Lester A Hoel 2009 *Traffic and highway engineering* (United States of America: Cengage Learning) pp 151-208

[9] Regev S, Rolison J J and Moutali S 2018 Crash risk by driver age, gender, and time of day using a new exposure methodology, *Journal of Safety Research.* **66** 131-140

[10] Elvik R 1995 *Meta-analysis of evaluation of public lighting as an accident countermeasure* Transportation Research Record No. 1485 (Washington: Transportation Research Board) pp 112-123

[11] Sullivan J M and Flannagan M J 2002 The role of ambient light level in fatal crashes: inferences from daylight saving time transitions, *Accident Analysis and Prevention.* **34** 487-498

[12] Fergusons S A, Preusser D F and Lund A K et al. 1995 Daylight saving time and motor vehicle crashes: the reduction in pedestrian and vehicle occupant fatalities, *American Journal of Public Health.* **85** 92-96

[13] Plainis S and Murray I J 2002 Reaction times as an index of visual conspicuity when driving at night, *Ophthalmic and Physiological Optics.* **22** 409-415

[14] Leibowitz H W, Owens D A and Tyrrell R A 1998 The assured clear distance ahead rule: implications for nighttime traffic safety and the law, *Accident Analysis & Prevention.* **30** 93-99

[15] Casado-Sanz N, Guirao B and Attard M 2020 Analysis of the risk factors affecting the severity of traffic accidents on Spanish crosstown road: The driver's perspective, *Sustainability.* **12** 1-26

Trend analysis of seasonal rainfall patterns in Peninsular Malaysia

Y X Tan[1*]

[1] Department of Civil Engineering, Faculty of Engineering, Technology and Built Environment, UCSI University, 56000 Cheras, Kuala Lumpur, Malaysia.

Abstract. Long records of rainfall data with good quality is a significant component that triggers the entire chain of responses in a hydrological cycle by providing useful information to hydrologist and related parties. In this study, the temporal trend analysis of seasonal rainfall data was carried out in Peninsular Malaysia using the non-parametric Mann-Kendall test. Then, the magnitude of the seasonal rainfall trend line was determined using Sen's Slope Estimator. Historical rainfall data with a minimum duration of 25 years were checked using four homogeneity tests and the missing values in the data were infilled using multiple imputation method. According to the output of the homogeneity tests, 64.3%, 7.1% and 28.6% of the rainfall time series can be categorised as useful, doubtful and suspect, respectively. From the trend test results, almost all the rainfall data were detected to have an increasing trend except Bayan Lepas for Inter-Monsoon 1, Southwest Monsoon and Inter-Monsoon 2. In addition to that, Bayan Lepas, Hospital Seremban, Ipoh and KLIA Sepang were detected to have a significant increasing trend for Northeast Monsoon as well as Batu Pahat and KLIA Sepang for Inter-Monsoon 2. In conclusion, Peninsular Malaysia exhibited increasing rainfall trend from year 1984 to 2019.

1. Introduction

Hydrological cycle is the repetition of the water flow from the ocean to the atmosphere and returns to earth by rainfall where it accumulates in the rivers and runs back into the ocean. This circulation is endless and repetitive. Rainfall has become a key element in the movement of water flow since it fluctuates not only spatially but also temporally as a result of the climate variability. The analysis of variability in the rainfall patterns are beneficial and vital towards the adequate allocation of water supply, development of a sustainable water resources and management as well as drought and rainfall forecasting for current and future climate conditions. According to [12], the rainfall-runoff, soil moisture and groundwater storage will severely influence the frequency of drought and flood occurrence depending on the variation of rainfall patterns. Thus, it is crucial to evaluate the variations of the historical rainfall patterns to understand the impact of climate change on agricultural, economic, environmental and societal sectors.

In the recent years, a number of researches has been performed to explore the spatio-temporal trends of the rainfall using both parametric and non-parametric test. In India, a study on the rainfall trend and variability with the available record of 102 years were employed using analysis of linear regression and the non-parametric tests such as Mann-Kendall (MK) test and Spearman Rho as well as Sen's Slope to determine the magnitude of the trend pattern [13]. Based on the results, decreasing trend was observed at all stations except two stations for annual rainfall time scale. Similarly, [12] employed Kendall Rank Correlation (KRC) test, MK test and its modified version at multiple significance level to conduct spatio-temporal trends of the rainfall. Additionally, the simple linear regression (SLR) test, Theil-Sen's Slope (TSS) were adopted to assess the magnitude of the rainfall trend pattern. Both significant increasing and decreasing trends were discovered at all the districts for monthly, seasonal and annual time scales. Besides that, the non-parametric test such as MK test, Spearman's Rank Correlation test and Sen's Slope was applied to evaluate the homogeneity, trend and shifting for the rainfall data at the arid regions of India [16]. The results indicated that six rainfall stations showed significant upward trend in the rainfall data during the summer season. On the contrary, significant downward trend were observed at two rainfall stations for each monsoon season.

Meanwhile, assessments on the distribution of the trend for the rainfall patterns were also carried out in Malaysia. The MK Test was adopted to analyze the trend for various time scales with several different hydro-climatic data in Selangor [5]. Based on the outputs generated, the rainfall at the upstream flow experienced a positive trend whereas the downstream flow displayed a negative trend. A combination of both increasing and decreasing trends were detected at the middle streamflow. In addition to that, similar trend were also detected for other time scales studied. The spatial trend for rainfall indices taking consideration of the long term persistence in Peninsular Malaysia was also investigated using the MK test and its modified version as well as the Sen's Slope to determine the magnitude of the secular trend [9].

Although Malaysia is located in a region which is less vulnerable to the calamities of natural disasters, the high variability in the rainfall patterns may expose the country to the risk of natural disasters [15]. The vulnerability of climate change is not uniformly distributed across the country [22]. In Peninsular Malaysia, the east coast region receives higher amount of rainfall as compared to the west coast region. Flooding has become an annual issue in the east coast region due to the high amount of rainfall received [14,20]. Therefore, the study on the rainfall trend analysis and the magnitude of the seasonal rainfall trend in Peninsular Malaysia may provide useful insights for flood mitigation and water resources management. In contrast to that, the states of Malacca, Klang Valley and Selangor which are located at the west coast region encountered severe dry weathers back in year 1991, 1998 and 2014, respectively [6,7]. The impact of El-Nino effect has caused the country to experience an extensive loss in both agricultural and economic sectors. As such, the trend analysis on the rainfall patterns is essential and necessary to mitigate the impacts of the natural disaster. With the availability of the long and good quality historical rainfall records, it has become a motivation to carry out an assessment on the trend for the seasonal rainfall patterns in Peninsular Malaysia. This study aims to evaluate the temporal trend of the seasonal rainfall data using Mann-Kendall (MK) test and to determine the magnitude of the seasonal rainfall trend line using Sen's Slope Estimator.

2. Study area and materials

The area for this case study was applied in the Peninsular Malaysia. Since Peninsular Malaysia is located near the equator, the region experiences hot and humid weather with high amount of rainfall and uniform temperature throughout the year [8]. The distribution of the rainfall patterns at the study area of interest is influenced by two major monsoon seasons namely the Southwest Monsoon (SWM) and the Northeast Monsoon (NEM). There are also two inter-monsoon seasons i.e. the Inter-Monsoon 1 (INM1) and Inter-Monsoon 2 (INM2).

Table 1. Description of rainfall stations selected.

Station Name	State	Available Period	Duration (year)
Batu Pahat	Johor	1992 - 2019	28
Bayan Lepas	Pulau Pinang	1984 - 2019	36
Cameron Highlands	Pahang	1994 - 2016	33
Hospital Seremban	Negeri Sembilan	1975 – 2012	37
Ipoh	Perak	1984 – 2019	37
KLIA Sepang	Selangor	1984 – 2019	37
Senai	Johor	1984 – 2019	36
Subang	Selangor	1984 – 2019	37

IOP Conf. Series: Materials Science and Engineering 1101 (2021) 012017 doi:10.1088/1757-899X/1101/1/012017

Daily rainfall dataset with a minimum historical record of 25 years were obtained from the Malaysian Meteorological Department (MMD). The details of the selected rainfall stations are presented in table 1. Data quality checking was first carried out to ensure the rainfall dataset are suitable for further analysis and to check for missing rainfall values. If missing rainfall values are present, the multiple imputation method is applied to fill up the missing gaps.

3. Homogeneity test

For this study, the Buishand Range (BR) Test, Pettitt Test (PT), Standard Normal Homogeneity Test (SNHT) and Von Neumann Ratio (VNR) Test were employed to test for the homogeneity of the acquired dataset.

3.1. Buishand Range (BR) test

The BR test can identify the shift in the intermediate for the applied time scales [4]. The adjusted partial sums are defined as follow:

$$S_0^* = 0 \; and \; S_k^* = \sum_{i=1}^{k} (P_i - \bar{P}) \; where \; k = 1, 2, 3, \ldots, n. \tag{1}$$

where P_i is the time series to be tested and $\overline{P}$ is the average. There is no deviations detected for the values of P_i with regards to the average if the series is homogeneous. The values of S_k^* will vary around zero if the series is homogeneous. The significance of the change in the average can be assessed using the value of R as expressed below:

$$R = \frac{(\max_{0 \le k \le n} S_k^* - \min_{0 \le k \le n} S_k^*)}{s} \tag{2}$$

3.2. Pettitt Test (PT)

The non-parametric Pettitt Test evaluates the homogeneity of a time scale in accordance with the ranks of an element [17]. The statistic can be defined as follow:

$$X_k = 2\sum_{i=1}^{k} r_i - k(n+1) where \; k = 1, 2, 3, \ldots, n. \tag{3}$$

where $r_1, r_2, r_3, \ldots, r_n$ are the ranks of the tested variables arranged according to ascending order. The value of the statistic will reach its highest or lowest near the year k = E if any shift is discovered in the mean of year E. Maximum value of the statistic is given as follow:

$$X_E = \max_{1 \le k \le n} |X_k| \tag{4}$$

3.3. Standard Normal Homogeneity Test (SNHT)

The SNHT assumes that the variates in the test are normally distributed. The SNHT is able to detect inhomogeneity near the starting and the ending of a given time scale [1]. The statistic can be expressed as follow:

$$T_k = k\overline{z_1^2} + (n-k)\overline{z_2^2} \; where \; k = 1, 2, 3, \ldots, n. \tag{5}$$

and

$$\overline{z_1} = \frac{1}{k}\sum_{i=1}^{k}\frac{P_i - \bar{P}}{s} \text{ and } \overline{z_2} = \frac{1}{n-k}\sum_{i=k+1}^{n}\frac{Y_i - \bar{Y}}{s} \tag{6}$$

where $\overline{z_1}^2$ and $\overline{z_2}^2$ are the arithmetic averages of the z_i sequence before and after the shift. P_i and Y_i are the time series to be tested. $\overline{P}$ and $\overline{Y}$ are the average and s is the standard deviation. T_k reaches the peak value close to the year k = K if the shift is located at the year K. The statistic can be defined as follow:

$$T_0 = \max_{1 \le k \le n} T(k) \tag{7}$$

3.4. Von Neumann Ratio

The VNR test evaluates where the time series is randomly distributed. The test is sensitive towards the departure of homogeneity within a given time series. The test applied the quotient between the average square successive (year to year) differences to the variance [3]. The statistic can be defined as follow:

$$N = \frac{\sum_{i=1}^{n-1}(P_i - P_{i+1})^2}{\sum_{i=1}^{n}(P_i - \bar{P})^2} \tag{8}$$

where P_i is the time series to be tested and $\overline{P}$ is the average.

4. Trend analysis

For this study, the Mann-Kendall (MK) test and the Sen's Slope Estimator were used to analyze and to determine the temporal trend of the rainfall dataset. The time series used in this study is only based on the seasonal scale.

4.1. Mann-Kendall (MK) test

The Mann-Kendall test is proposed to identify the existence of trend in rainfall data. Climatic data such as the temperature and rainfall are among the few parameters used for trend analysis. The null hypothesis was checked against the alternative hypothesis to detect the increasing trend or decreasing trend using the MK Test which is non-parametric [19]. The MK test statistics, S, can be defined as follow:

$$S = \sum_{i<j} sgn\,(x_j - x_i), \text{ where } sgn\,(x) = \begin{pmatrix} 1 \text{ if } x > 0 \\ 0 \text{ if } x = 0 \\ -1 \text{ if } x < 0 \end{pmatrix} \tag{9}$$

where x_i and x_j are the rank for the *i*th and *j*th observations respectively. The standard deviation can be obtained using:

$$Z(var)\begin{cases} \frac{s-1}{\sqrt{Var(S)}} & when\ S > 0, \\ 0 & when\ S = 0, \\ \frac{s-1}{\sqrt{Var(S)}} & when\ S < 0. \end{cases} \tag{10}$$

Positive value of *Z(var)* indicates an upward trend while negative value of *Z(var)* indicates a downward trend.

4.2. Sen's Slope Estimator

For Sen's Slope Estimator [18], the magnitude between all the consecutive data points can be determined using:

$$\beta = med(\frac{x_j - x_k}{j - k}) \tag{11}$$

where β is the magnitude among two points x_j and x_k calculated at the time j and k. The rate of change over the entire period is given by the median of all the magnitudes (slopes) which is also known as the Sen Slope.

5. Results and discussion

5.1. Homogeneity test

For homogeneity test, the daily rainfall dataset obtained from the MMD were aggregated into monthly scale to be analyzed for NEM, SWM, INM1 and INM2. Then, data quality checking for the rainfall dataset were validated using the four homogeneity tests namely the BR, PT, SNHT and VNR. If the null hypothesis is accepted with a p-value larger than 0.05, the time scale of rainfall data is homogeneous at the significance level of 95% [11]. The critical values at significance level of 95% are summarized in table 2 for the four homogeneity tests applied as a function of n, number of years. The results of all the homogeneity test for INM1, SWM, INM2 and NEM are presented in table 3. Hospital Seremban rejected the null hypotheses for most of the homogeneity tests for all the seasons analyzed. Figure 1 shows the result of PT for INM1 rainfall at Hospital Seremban. It is noticeable that there is a break in the rainfall data for the year 1998 and 2012. Similar results are also drawn from the BR, SNHT and VNR tests where all three null hypotheses were rejected. Therefore, the rainfall data at Hospital Seremban were said to be inhomogeneous since significant break was detected. Besides, the rainfall data at KLIA Sepang were also inhomogeneous for all the seasons analyzed except during the SWM. The null hypotheses were rejected for all four homogeneity tests during INM1, SWM and INM2. On the contrary side, Bayan Lepas accepted the null hypotheses for all the homogeneity tests for all the time series analyzed. Figure 2 shows the result of BR for SWM rainfall at Bayan Lepas. In addition to that, Senai accepted all null hypotheses except PT during the INM2. However, the rainfall data for these two rainfall stations were still considered to be homogeneous for all the season.

Table 2. The 95% critical values for the four homogeneity tests.

n	*PT*	*SNHT*	*BR*	*VNR*
20	57	6.95	1.43	1.30
30	107	7.65	1.50	1.42
40	167	8.10	1.53	1.49
50	235	8.45	1.55	1.54

The time series for the rainfall data can undergo further classifications into several categories namely Class I (useful), Class II (doubtful) and Class III (suspect) based on the outputs from the homogeneity test conducted [21]. A summary on the classification of the outputs from the homogeneity tests is presented in table 4. For Class I, rainfall data for all the time scales were homogeneous and they can be applied for further analysis. Next, for Class II, rainfall data for INM1 and INM2 showed inhomogeneity and they should be carefully handled if they were to be considered for further analysis. For Class III, rainfall data for all the time scales were inhomogeneous and they cannot be applied for further analysis. Homogeneity tests are carried out to ensure that the rainfall data are measured under similar influence by the environmental conditions using the same instrument. Relocations of rainfall stations and different measurement techniques or methods could lead to inhomogeneities in rainfall data i.e. in class of II and III [20].

IOP Conf. Series: Materials Science and Engineering 1101 (2021) 012017 doi:10.1088/1757-899X/1101/1/012017

Table 3. Results for homogeneity tests.

Station Name	**Inter-Monsoon 1**			
	PT	***SNHT***	***BR***	***VNR***
Batu Pahat	0.5230	0.0415[a]	0.2477	0.0483[a]
Bayan Lepas	0.5981	0.4447	0.5716	0.0550
Hospital Seremban	0.0021[a]	0.0172[a]	0.0058[a]	0.0398[a]
Ipoh	0.0509	0.1897	0.0608	0.6805
KLIA Sepang	< 0.0001[a]	0.0044[a]	<0.0001[a]	0.0009[a]
Senai	0.5350	0.8281	0.5811	0.6257
Subang	0.3337	0.5044	0.2502	0.1817
Station Name	**Southwest Monsoon**			
	PT	***SNHT***	***BR***	***VNR***
Batu Pahat	0.7031	0.0423[a]	0.5448	0.7457
Bayan Lepas	0.8703	0.9319	0.9862	0.4490
Hospital Seremban	0.0033[a]	0.0005[a]	0.0012[s]	0.0029[a]
Ipoh	0.5349	0.8023	0.4852	0.5246
KLIA Sepang	0.4283	0.5688	0.3961	0.8339
Senai	0.6729	0.7958	0.5987	0.2982
Subang	0.5156	0.6210	0.6418	0.4537
Station Name	**Inter-Monsoon 2**			
	PT	***SNHT***	***BR***	***VNR***
Batu Pahat	0.1322	0.1601	0.1046	0.4056
Bayan Lepas	0.6239	0.7927	0.4962	0.3045
Hospital Seremban	0.1102	0.0014[a]	0.0318[a]	0.0188[a]
Ipoh	0.9215	0.7341	0.9557	0.3324
KLIA Sepang	0.0001[a]	0.0001[a]	0.0001[a]	0.0003[a]
Senai	0.0430[a]	0.4798	0.2418	0.6772
Subang	0.0407[a]	0.0487[a]	0.0763	0.0894
Station Name	**Northeast Monsoon**			
	PT	***SNHT***	***BR***	***VNR***
Batu Pahat	0.5312	0.0319[a]	0.4366	0.0726
Bayan Lepas	0.1719	0.3764	0.1750	0.1923
Hospital Seremban	0.0052[a]	0.0115[a]	0.0066[a]	0.1361
Ipoh	0.0064[a]	0.0125[a]	0.0091[a]	0.0324[a]
KLIA Sepang	< 0.0001[a]	< 0.0001[a]	<0.0001[a]	<0.0001[a]
Senai	0.0877	0.3171	0.1666	0.1086
Subang	0.3612	0.0279[a]	0.1917	0.5122

[a] These values indicate that the time series is inhomogeneous.

Table 4. Classification of homogeneity tests.

Class	Season			
	INM1	SWM	INM2	NEM
I – Useful	57.14%	85.71%	57.14%	57.14%
II – Doubtful	14.29%	0.00%	14.29%	0.00%
III – Suspect	28.57%	14.29%	28.57%	42.86%

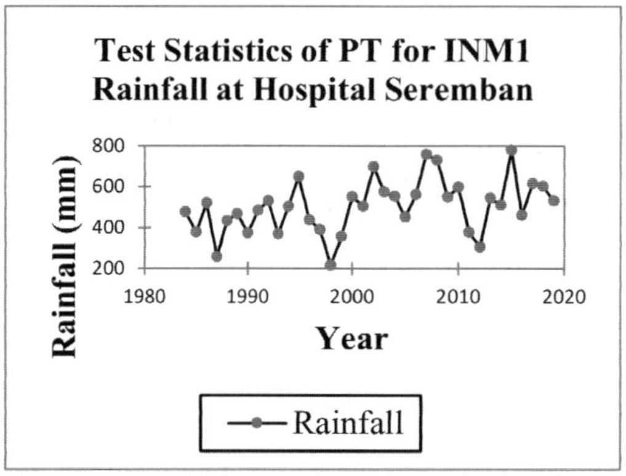

Figure 1. Test statistics of PT for INM1 rainfall at Hospital Seremban.

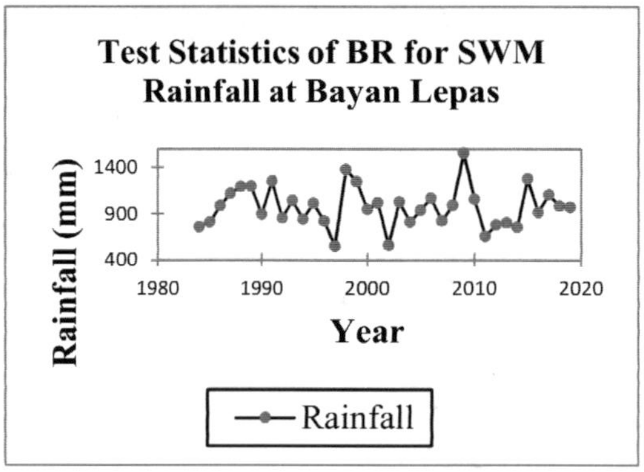

Figure 2. Test statistics of BR for SWM rainfall at Bayan Lepas.

5.2. Trend analysis

For trend analysis, the MK test was employed and an indication of an increasing or decreasing trend in the rainfall data will be provided based on the value of Z as shown in Equation 10. Similarly, the p-value of the MK test is obtained b on the significance level of 95%. As for the Sen's Slope Estimator, the value obtained indicates the magnitude of the trend in the rainfall data. In contrast to that, there is no trend observed in the rainfall data if the value of Sen's Slope Estimator obtained is zero. The results for MK test are presented in table 5 for indicating that all the rainfall stations experienced increasing trends except for Bayan Lepas during INM1, SWM and INM2. However, Bayan Lepas exhibited significant increasing trends for the NEM. It was observed that the p-values were lesser than 0.05 and the values for Z as well as Sen's Slope Estimator were significantly higher during the NEM than all other monsoon seasons. Figure 3 shows the trend line of the rainfall data during NEM at Bayan Lepas.

The presence of trend in the rainfall data may be contributed by the changes in the long term rainfall which are influenced by several factors such as geographic location of the rainfall stations and period of the rainfall data analyzed [20]. Besides that, significant increasing trends were observed at Bayan Lepas and Ipoh during INM2 and NEM, respectively. In addition to that, Hospital Seremban detected significant increasing trend during INM1, SWM and NEM. Similarly, significant increasing trends were also detected for KLIA Sepang for INM1, INM2 and NEM. The results obtained from the trend test are consistent with the findings in which Peninsular Malaysia encountered significant increasing trends especially during the NEM [2,10,22]. The reason significant increasing trends were detected at these rainfall stations could be due to the localized rainfall in which higher amount of rainfall are received at only one particular station.

Table 5. Results for trend tests.

Station Name	Inter-Monsoon 1		
	p-value	Z	Sen's Slope
Batu Pahat	0.8912	0.0212	0.3978
Bayan Lepas	0.4405	-0.0921	-2.6188
Hospital Seremban	0.0111[a]	0.2952[a]	4.7683[a]
Ipoh	0.0790	0.2063	4.3320
KLIA Sepang	<0.0001[a]	0.4476[a]	7.0569[a]
Senai	0.8606	0.0222	0.4521
Subang	0.4246	0.0952	1.7158
Station Name	Southwest Monsoon		
	p-value	Z	Sen's Slope
Batu Pahat	0.7694	0.0423	1.9113
Bayan Lepas	0.7766	-0.0349	-1.1869
Hospital Seremban	0.0006[a]	0.3937[a]	9.6899[a]
Ipoh	0.6361	0.0571	1.5379
KLIA Sepang	0.6361	-0.0571	-1.5566
Senai	0.7152	0.0444	1.1393
Subang	0.2171	0.1460	3.5936[a]
Station Name	Inter-Monsoon 2		
	p-value	Z	Sen's Slope
Batu Pahat	0.0248[a]	0.3016[a]	5.6827[a]
Bayan Lepas	0.4903	-0.0825	-2.5825
Hospital Seremban	0.2606	0.1333	2.5315
Ipoh	0.8183	0.0286	0.4181
KLIA Sepang	0.0005[a]	0.3968[a]	10.2713[a]
Senai	0.0889	0.2000	3.9074
Subang	0.0342[a]	0.2476[a]	5.4759[a]
Station Name	Northeast Monsoon		
	p-value	Z	Sen's Slope
Batu Pahat	0.7221	0.0503	1.7035

Bayan Lepas	0.0237[a]	0.2653[a]	7.3124[a]
Hospital Seremban	0.0025[a]	0.3542[a]	12.9125[a]
Ipoh	0.0050[a]	0.3288[a]	13.1481[a]
KLIA Sepang	<0.0001[a]	0.4877[a]	19.0418[a]
Senai	0.0565	0.2240	6.0696
Subang	0.2414	0.1382	5.8501

[a] These values indicate that trend is significant.

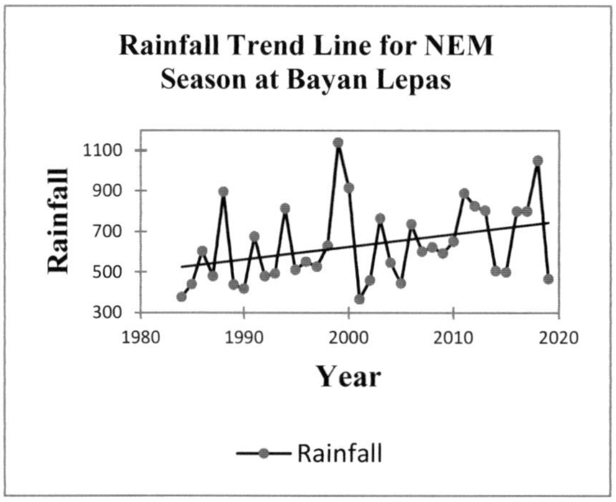

Figure 3. Rainfall trend line for SWM rainfall at Bayan Lepas.

6. Conclusion

In a nutshell, the temporal trend of the rainfall data and the magnitude of the rainfall trend line in Peninsular Malaysia were obtained using Mann-Kendall (MK) test and Sen's Slope Estimator. Rainfall dataset were obtained from the MMD and data quality checking were performed using the homogeneity tests namely BR, PT, SNHT and VNR. Based on the result of the homogeneity test, there were four, six, four and four 'useful' rainfall stations for the INM1, SWM, INM2 and NEM, respectively. Besides that, there were two 'doubtful' rainfall stations, one for INM1 and the other for INM2. As for the 'suspect' rainfall stations, there were one for SWM, two each for INM1 and INM2, and three for NEM. In general, rainfall stations that are 'useful' can be used for further hydrological analysis whereas the doubtful' class should be handled with care and the 'suspect' class should be discarded.

For trend analysis, the results of MK test indicated that most of the rainfall stations showed positive value which is an upward trend except for Bayan Lepas during the INM1, SWM and INM2. However, a significant increasing trend was detected at Bayan Lepas during the NEM. On the other hand, significant increasing trends were also detected at stations such as Batu Pahat, Hospital Seremban, Ipoh and KLIA Sepang. Overall, an increasing trend of the seasonal rainfall was observed for Peninsular Malaysia. With the results obtained, it may be able to provide pertinent guidance and as an indication for the planners to prepare for the repercussions of the natural catastrophes such as drought and flood. Beholding the knowledge of the trend in the rainfall data, this can ensure an adequate allocation of water supply and the development of sustainable water resources.

Although this study was performed for Peninsular Malaysia, however, the result could not reflect on the entire study area and a full indication of the trend for the rainfall data. Further studies should include more rainfall stations in each state to cover the entire region in order to provide more insight and accurate information. Besides that, this scope of this study did not cover the adjustment towards the rainfall data which were classified under the class of 'doubtful'. As such, it is recommended that future studies may include investigation on the inhomogeneity present in the rainfall data and the appropriate methods to improve the quality of the rainfall data.

Acknowledgements
The study was supported by UCSI University through Pioneer Scientist Incentive Fund (PSIF) under project code Proj-2019-In-FETBE-065 and the authors are grateful for the climatic data provided by Malaysian Meteorological Department (MMD).

References

[1] Alexandersson H and Moberg A, 1997 *Int. J. Climatol.* **17** 25–34
[2] Amanda S P L 2017 Assessment of climate change impacts on rainfall series in Peninsular Malaysia using statistical methods
[3] Bingham C amd Nelson L S 1981 *Technometrics* **23** 285–288
[4] Buishand T A 1982 *J. Hydrol.* **58** 11–27
[5] Ebrahimian M, Nuruddin A A, Soom M A M, Sood A M, Neng L J and Galavi H 2018 *Singap. J. Trop. Geogr.* **39** 192–214
[6] Fung K F, Huang Y F and Koo C H 2018 *E3S Web of Conferences* **65** 07007
[7] Hong D and Hong K A 2016 *Int. J. u- e- Serv. Sci. Technol.* **9** 53–66
[8] Hui-Mean F, Yusof F, Yusop Z and Suhaila J 2019 *Theor. Appl. Climatol.* **138** 657–671
[9] Khan N, Pour S H, Shahid S, Ismail T, Ahmed K, Chung E, Nawaz N and Wang X 2019 *Meteorol. Appl.* **26** 655–670
[10] Lian C Y, Huang Y F, Ng J L, Mirzaei M, Koo C H and Tan K W 2019 *J. Water Clim. Chang.*
[11] Lin N J, Aziz S A, Feng H Y, Wayayok A and Kamal M R 2015 *J. Teknol.* **76** 1–6
[12] Malik A and Kumar A 2020 *Theor. Appl. Climatol.* **140** 183–207
[13] Meshram S G, Singh V P and Meshram C *Theor. Appl. Climatol.* **129** 729–744
[14] Ng J L, Abd Aziz S, Huang Y F, Wayayok A and Rowshon M K 2017 *III International Society for Horticultural Science 1152* 11–17.
[15] Ng J L, Yap S Y, Huang Y F, Noh N I F M, Al-Mansob R A and Razman R 2020 *IOP Conf. Ser. Earth Environ. Sci.* **476** 012118.
[16] Patakamuri S K, Muthiah K and Sridhar V 2020 *Water* **12** 211
[17] Pettitt A N 1979 *Appl. Stat.* **28** 126
[18] Sen P K, 1968 *J. Am. Stat. Assoc.* **63** 1379–1389
[19] Sun Y, Liang X, Xiao C and Fang Z 2019 *Water Sci. Technol. Water Supply* **19** 19–29
[20] Tan Y X, Ng J L and Huang Y F 2020 *Carpathian J. Earth Environ. Sci.* **15** 103-112
[21] Wijngaard J B, Klein Tank A M G and Können GP 2003 *Int. J. Climatol.* **23** 679–692
[22] Wong C L, Yusop Z and Ismail T 2018 *Int. J. GEOMATE* **14** 65–72

IOP Conf. Series: Materials Science and Engineering 1101 (2021) 012018 doi:10.1088/1757-899X/1101/1/012018

Mechanical properties of reactive powder concrete under distinct curing environment

A Vijay[1*], S Senthilselvan[1]

[1] Department of Civil Engineering, SRM Institute of Science and Technology, Kattankulathur, Tamilnadu 603203. India

*Corresponding author: vijaya@srmist.edu.in

Abstract. Reactive Powder Concrete (RPC) is composed of fine powders (cement, sand, quartz powder and silica fume), steel fibres (optional) and superplasticizer. This gives good compaction and an impenetrable matrix is achieved, and this strong concentration gives RPC, an ultra-high strength and durability properties. In this research work, focus was aimed to produce reactive powder concrete (RPC) composite material with compression strength up to 100 N/mm^2. Components for RPC mixture are carefully selected to achieve optimal mixture. Detailed concrete mix proportions are given in this research paper. Preparation and testing of material were produced in laboratory belonging to the Department of Civil Engineering in SRM Institute of Science and Technology. Casting of concrete cubes and beams was done for both M60 and M100 grades of concrete. This paper investigates mechanical and fresh concrete properties of reactive powder concrete for both the design mix. Laboratory investigations like; Compressive test, split tensile test and flexural strength were conducted to evaluate strength characteristics of RPC, under different curing conditions like oven, steam, and hot water with respect to time factor of T1 (60 min), T2 (61-90 min) and T3 (91-120 min). In every aspect of testing, the steam curing at 100°C shows improved results of 15.71-17.40% in compressive strength, 24.66-25.93% in tensile strength and 20% in flexural strength for both grades of concrete

1. Introduction

Concrete is an extensively used building material that governs the building engineering all over the world. In the last two decades, the development in the concrete industry resulted in numerous types of concrete with significant results for the reactive powder concrete. The application of reactive powder concrete (RPC) extensively used for military establishment and other tough applications like blast proofing, etc. In 1990s, RPC was formed by the addition of ancillary material, eradication of coarse aggregates, exceptionally low water/binder ratio, with calculated amount of superplasticizer, and if required, fine steel fibre reinforcement is also added for further reinforcement. Literature studies states that the ingredients may be set to experience heat curing and application of pressure before and during setting of concrete. RPC is an ultra-high-performance concrete (UHPC) capable of reaching compressive strength more than 200MPa and corresponding flexural strength more than 40MPa, as stated by its makers Richard and Cheyrezy [1-6]. There is an expanding use of RPC due to the remarkable mechanical properties and robustness.

Table 1. Properties of materials used

MATERIALS	CEMENT	RIVER SAND	SILICA FUME	QUARTZ POWDER
Chemical composition [%]				
Silica (SiO_2)	20.79	51.00	99.5%	99.886%
Alumina (Al_2O_3)	4.71	6.83	0.08%	0.043%
Titanium oxide (TiO_2)	2.21	0.58	0.04%	0.001%
Calcium oxide (CaO)	60.20	0.48	0.01%	0.001%
Magnesium oxide (MgO)	2.74	-	0.01%	0.000%
Sodium oxide (Na_2O)	0.23	-	1.23	0.003%
SO_3	2.71	-	0.64	-
Mineral composition [%]				
C_3S	55.4	-	-	-
C_2S	19.7	-	-	-
C_3A	6.6	-	-	-
C_4AF	9.3	-	-	-
Physical and Mechanical properties				
Specific gravity	3.15	2.65	2.3	2.72
Density g/cm3	3.13	-	-	2.63
Initial setting time [min]	140	-	-	-
Final setting time [min]	250	-	-	-

2. Materials and Methods

Ultra-high performance is especially valuable attribute of RPC. The markets in which high-performance concrete products will be challenging are where resilience characteristic leads: any concrete composition that will have to deal with harsh environmental situations will be made with a high-performance concrete to enhance its working life. In addition, the lower maintenance obligations effect in significant economic gain. Furthermore, in the creation of RPC, a partial replacement of cement by silica fume (which is a waste by-product of silicon alloy) results in less cement utilization and therefore less greenhouse gas release. The use of RPC allows not only improved mechanical performance of the structural elements, but it also guarantees a substantially prolonged service life due to its fundamental material properties. The primary tests of materials such as cement, river sand, silica fume, quartz powder as well as chemical admixture (superplasticizer) are discussed in this section. Insights into research methodology were also discussed briefly.

2.1. Materials and mix proportions
In this research study, the properties of various materials used to constitute RPC was tested and based on those outcomes, the experiments were conducted. Ordinary Portland Cement (OPC) of grade-53 is utilized throughout the work whose specific gravity was found to be 3.15. Fine aggregate was acquired from nearby obtainable river sand which passed through 4.75 mm sieve. In the reactive powder concrete sand plays the foremost role. Because of its high particle size its function is adds more strength to the concrete mix. Silica fume, which is a by-product of silicon metal or ferrosilicon alloys was also utilized in the production of RPC. Because of its finer particles, large surface area, and the high SiO_2 content, high strength and durability be able to be attained in concrete. Silica fume in concrete does not bleed and is more vulnerable in plastic shrinkage cracking. Segregation is also reduced since concrete mix is more cohesive. Another main element quartz powder is the very well used mineral admixture in high strength concrete. Combining it to the concrete mix will significantly boost the workability, strength & impermeability of concrete mixes whilst producing the concrete resilient to chemical strikes, abrasion & reinforcement oxidization, and increasing the comprehensive strength. Quartz powder is a very reflex and effective pozzolanic material due to its fine particle size and high purity of SiO_2 content. The main function of quartz powder is to give highest endurance to the concrete against heat. The properties of different materials employed in this research is tabulated in the table 1.

3. Testing and results
The RPC design mix is obtained from various trial mix based on previous literature studies and conducted experiments on M60 and M100 grade of concrete. Higher properties such as higher compressive and split tensile strength were achieved with the use of finer particle such as silica fume and quartz. Testing of hardened concrete is performed to ascertain the compressive strength for cube specimen, split tensile strength for cylindrical specimen and flexural strength test for beam specimen for 7 day, 14 days, 28 days of RPC grade M60 and M100. Their results are discussed in this paper with respect to time factor from the mixing minutes.

3.1. Compressive strength
Compression test is done to find the characteristic strength of concrete specimen and it depends on cement strength, water cement ratio, worth of concrete mix. The cube specimen of size 70.6 mm × 70.6 mm is casted and tested for 7 day, 14 days, 28 days under varied curing environments, such as normal water curing (room temperature), oven curing, steam curing and hot water curing. The cubes casted at time T1 are shown in figure 1. The graphical representations of compressive strength of M60 and M100 concrete are illustrated in figure 2 (a) and (b), correspondingly.

Figure 1. Sample cubes casted

IOP Conf. Series: Materials Science and Engineering 1101 (2021) 012018 doi:10.1088/1757-899X/1101/1/012018

The cubes casted are varied by means of delayed time where T1 indicates casting time not greater than 60 minutes, T2 indicates 61-90 minutes, T3 indicates 91-120 minutes. This variation in time for casting cubes were indicated in figure 2 (a) and (b) for compressive strengths of M60 and M100 grades of RPC, respectively.

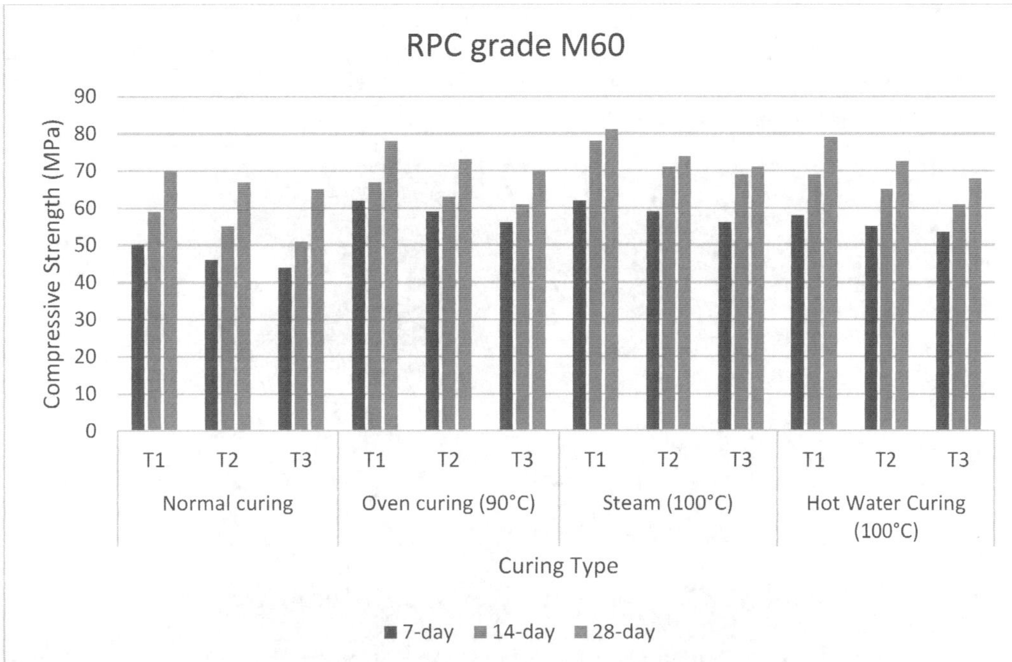

Figure 2. (a) Compressive strength of M60 grade RPC

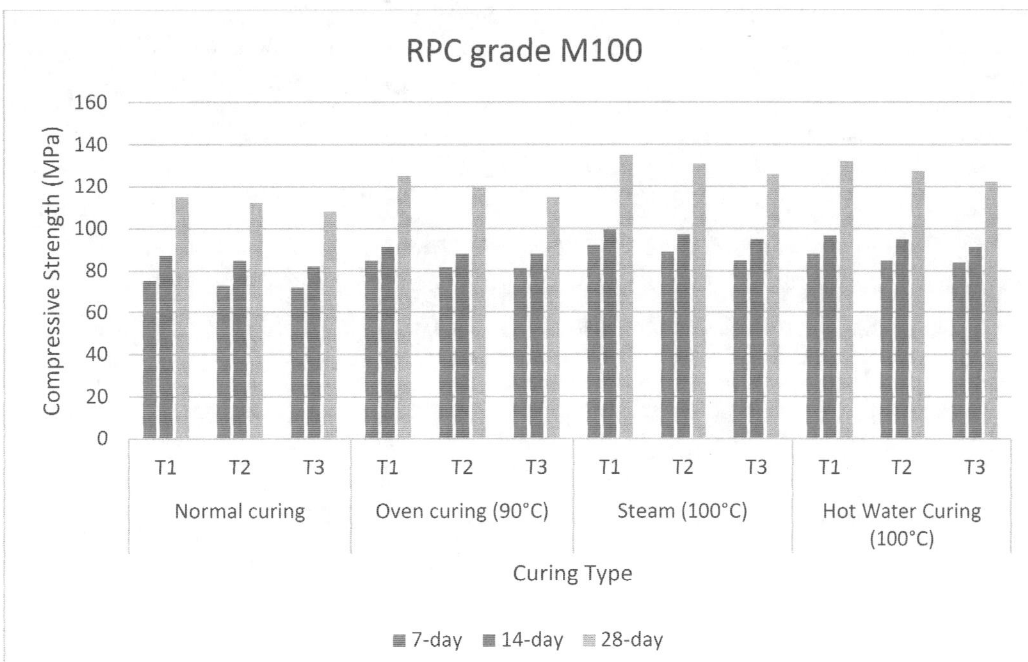

Figure 2. (b) Compressive strength of M100 grade RPC

It is observed that the RPC grades irrespective of curing, at T1 timing has achieved higher strength than the delayed ones.

3.2. Split tensile strength

Split tensile potency analysis is performed for cylindrical specimen of size 100 mm × 50 mm are casted and tested for 7 days, 14 days, 28 days under varied curing environments, such as normal water curing (room temperature), oven curing, steam curing and hot water curing. The tested cylindrical specimen is displayed in figure 3. The graphical representations of tensile strength of M60 and M100 concrete are illustrated in figure 4 (a) and (b), respectively.

Figure 3. Testing of cylinder specimen

The cylinders casted are also varied by means of delayed time where T1 indicates casting time not greater than 60 minutes, T2 indicates 60-90 minutes, T3 indicates 90-120 minutes. This variation in time for casting were indicated in figure 4 (a) and (b) for tensile strengths of M60 and M100 grades of RPC, respectively.

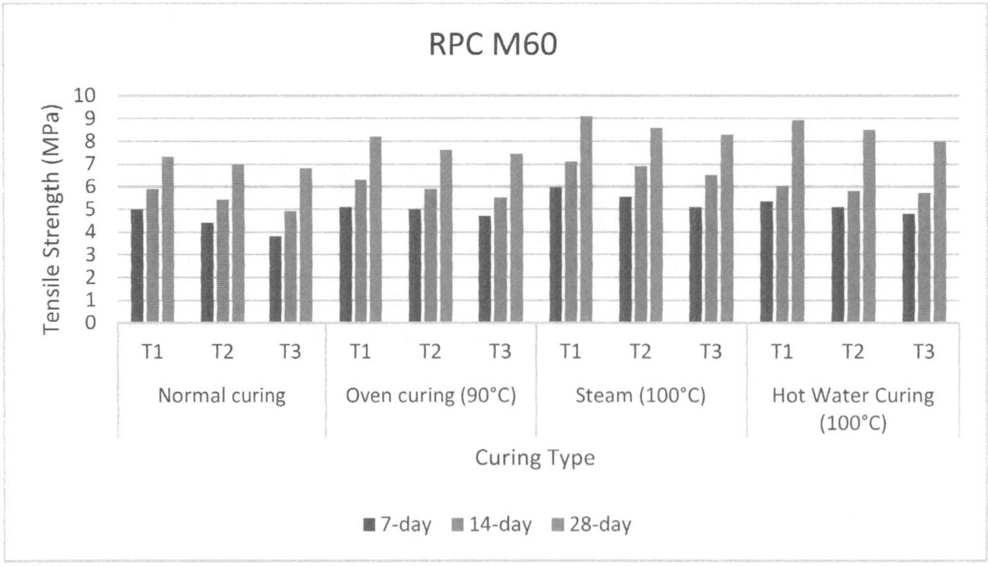

Figure 4. (a) Split tensile strength of M60 grade RPC

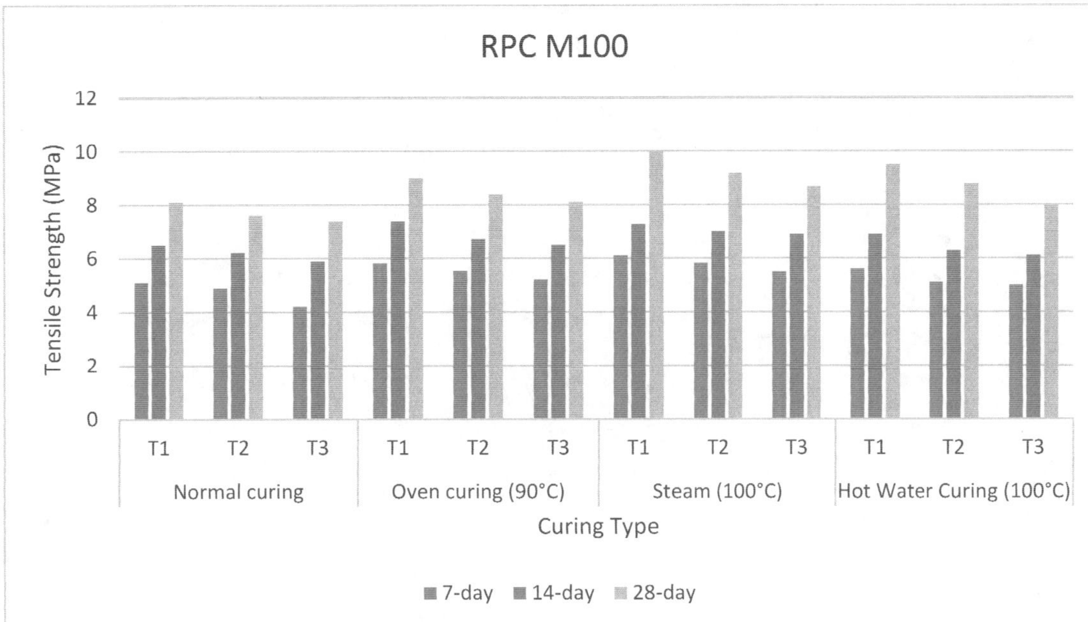

Figure 4. (b) Split tensile strength of M100 grade RPC

3.3. Flexural strength

The flexure strength of beam specimen of size whose width and depth are 40 mm and length 160 mm are tested for 14 days in addition to 28 days under varied curing environments, such as normal water curing (room temperature), oven curing, steam curing and hot water curing. Figure 5 shows the flexural testing of beams at SRMIST Strength of Materials laboratory. The graphical description of flexural strength of M60 as well as M100 concrete are shown in figure 6.

Figure 5. Flexural strength of M60 and M100 grade RPC

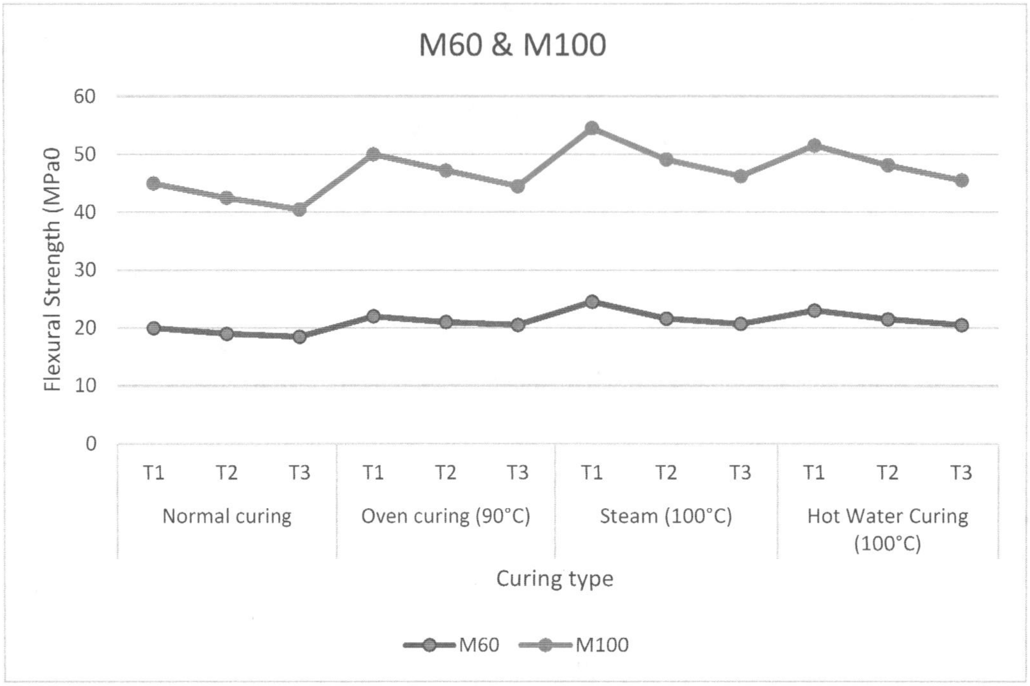

Figure 6. Flexural strength of M60 and M100 grade RPC

4. Inference

The mechanical properties for instance, compression, tension and flexure of the cube, cylinder and beam specimen are mainly concentrated. The specimens were cured at varied curing environments for 7, 14 and 28 days, with respect to time of casting from the mixing of ingredients to produce RPC compound. A sample of mixing method involved in preparation of RPC is indicated in figure 7.

Figure 7. Formulation of RPC

The specimens casted within 60 minutes (T1) of mixing all the necessary ingredients of RPC yields good results in all the curing methods.

4.1 M60 grade RPC results

With respect to normal curing, considering the time T1, the RPC of grade M60 has attained a compressive strength of 50 MPa, 59 MPa and 70 MPa in 7, 14 and 28 days, correspondingly. For time T2, it has attained a compressive strength of 46 MPa, 55 MPa and 67 MPa in 7, 14 and 28 days, correspondingly. For time T3, it has attained a compressive strength of 44 MPa, 51 MPa and 65 MPa in 7, 14 and 28 days, correspondingly.

With respect to oven curing at 90°C, considering the time T1, the RPC of grade M60 has attained a compressive strength of 62 MPa, 67 MPa and 78 MPa in 7, 14 and 28 days, correspondingly. For time T2, it has attained a compressive strength of 59 MPa, 63 MPa and 73 MPa in 7, 14 and 28 days, correspondingly. For time T3, it has attained a compressive strength of 56 MPa, 61 MPa and 70 MPa in 7, 14 and 28 days, correspondingly.

With respect to steam curing at 100°C, considering the time T1, the RPC of grade M60 has attained a compressive strength of 62 MPa, 78 MPa and 81 MPa in 7, 14 and 28 days, correspondingly. For time T2, it has attained a compressive strength of 59 MPa, 71 MPa and 74 MPa in 7, 14 and 28 days, correspondingly. For time T3, it has attained a compressive strength of 56 MPa, 69 MPa and 71 MPa in 7, 14 and 28 days, correspondingly.

With respect to hot water curing at 100°C, considering the time T1, the RPC of grade M60 has attained a compressive strength of 58 MPa, 69 MPa and 79 MPa in 7, 14 and 28 days, correspondingly. For time T2, it has attained a compressive strength of 55 MPa, 65 MPa and 72.5 MPa in 7, 14 and 28 days, correspondingly. For time T3, it has attained a compressive strength of 53.5 MPa, 61 MPa and 68 MPa in 7, 14 and 28 days, correspondingly.

The tensile strength of specimens of RPC grade M60, under normal curing has attained 7.3 MPa, 7 MPa and 6.8 MPa in 28 days for the period T1, T2, T3, correspondingly. Under oven curing at 90°C it has attained 8.2 MPa, 7.6 MPa and 7.45 MPa in 28 days for the period T1, T2, T3, correspondingly. Under steam curing at 100°C it has attained 9.1 MPa, 8.6 MPa and 8.3 MPa in 28 days for the period T1, T2, T3, correspondingly. Under hot water curing at 100°C it has attained 8.9 MPa, 8.5 MPa and 8 MPa in 28 days for the period T1, T2, T3, correspondingly.

The flexural strength of specimens of RPC grade M60, under normal curing has attained 20 MPa, 19 MPa and 18.5 MPa in 28 days for the period T1, T2, T3, correspondingly. Under oven curing at 90°C it has attained 22 MPa, 21 MPa and 20.50 MPa in 28 days for the period T1, T2, T3, correspondingly. Under steam curing at 100°C it has attained 24.5 MPa, 21.6 MPa and 20.7 MPa in 28 days for the period T1, T2, T3, correspondingly. Under hot water curing at 100°C it has attained 23 MPa, 21.5 MPa and 20.5 MPa in 28 days for the period T1, T2, T3, correspondingly.

4.2 M100 grade RPC results

With respect to normal curing, considering the time T1, the RPC of grade M100 has attained a compressive strength of 75 MPa, 87 MPa and 115 MPa in 7, 14 and 28 days, correspondingly. For time T2, it has attained a compressive strength of 73 MPa, 85 MPa and 112 MPa in 7, 14 and 28 days, correspondingly. For time T3, it has attained a compressive strength of 72 MPa, 82 MPa and 108 MPa in 7, 14 and 28 days, correspondingly.

With respect to oven curing at 90°C, considering the time T1, the RPC of grade M100 has attained a compressive strength of 85 MPa, 91 MPa and 125 MPa in 7, 14 and 28 days, correspondingly. For time T2, it has attained a compressive strength of 81.5 MPa, 88 MPa and 120 MPa in 7, 14 and 28 days, correspondingly. For time T3, it has attained a compressive strength of 81 MPa, 88 MPa and 115 MPa in 7, 14 and 28 days, correspondingly.

With respect to steam curing at 100°C, considering the time T1, the RPC of grade M100 has attained a compressive strength of 92 MPa, 99.5 MPa and 135 MPa in 7, 14 and 28 days, correspondingly. For time T2, it has attained a compressive strength of 89 MPa, 97 MPa and 131 MPa in 7, 14 and 28 days,

IOP Conf. Series: Materials Science and Engineering 1101 (2021) 012018 doi:10.1088/1757-899X/1101/1/012018

correspondingly. For time T3, it has attained a compressive strength of 85 MPa, 95 MPa and 126 MPa in 7, 14 and 28 days, correspondingly.

With respect to hot water curing at 100°C, considering the time T1, the RPC of grade M60 has attained a compressive strength of 88 MPa, 96.5 MPa and 132 MPa in 7, 14 and 28 days, correspondingly. For time T2, it has attained a compressive strength of 85 MPa, 95 MPa and 127 MPa in 7, 14 and 28 days, correspondingly. For time T3, it has attained a compressive strength of 84 MPa, 91 MPa and 122 MPa in 7, 14 and 28 days, correspondingly.

The tensile strength of specimens of RPC grade M100, under normal curing has attained 8.1 MPa, 7.6 MPa and 7.4 MPa in 28 days for the period T1, T2, T3, correspondingly. Under oven curing at 90°C it has attained 9.0 MPa, 8.4 MPa and 8.1 MPa in 28 days for the period T1, T2, T3, correspondingly. Under steam curing at 100°C it has attained 10 MPa, 9.2 MPa and 8.7 MPa in 28 days for the period T1, T2, T3, correspondingly. Under hot water curing at 100°C it has attained 9.5 MPa, 8.8 MPa and 8 MPa in 28 days for the period T1, T2, T3, correspondingly.

The flexural strength of specimens of RPC grade M100, under normal curing has attained 25 MPa, 23.5 MPa and 22 MPa in 28 days for the period T1, T2, T3, correspondingly. Under oven curing at 90°C it has attained 28 MPa, 26.2 MPa and 24 MPa in 28 days for the period T1, T2, T3, correspondingly. Under steam curing at 100°C it has attained 30 MPa, 27.5 MPa and 25.5 MPa in 28 days for the period T1, T2, T3, correspondingly. Under hot water curing at 100°C it has attained 28.5 MPa, 26.6 MPa and 25 MPa in 28 days for the period T1, T2, T3, correspondingly.

5. Conclusions

The mechanical properties of RPC grades of M60 and M100 are experimentally verified for different curing conditions with specimens casted under set of batches where T1 refers to casting time not greater than 60 minutes, T2 refers to 60-90 minutes and T3 refers to 90-120 minutes.

- The results of both the grades of concrete in any curing conditions shows higher strength under curing period T1, which is not greater than 60 minutes.
- Considering three time periods, the compressive strength of M60 grade of RPC with oven curing at 90°C gains strength in the range of 7.70 – 11.43%, the range increases to 7.70 – 12.86% under hot water curing at 100°C and the range further increases to 9.23 – 15.71% for steam curing at 100°C.
- Considering three time periods, the compressive strength of M100 grade of RPC with oven curing at 90°C gains strength in the range of 8.04 – 11.30%, the range increases to 12.96 – 14.78% under hot water curing at 100°C and the range further increases to 16.67 – 17.40% for steam curing at 100°C.
- Considering three time periods, the tensile strength of M60 grade of RPC with oven curing at 90°C gains strength in the range of 8.57 – 12.33%, the range increases to 17.65 – 21.92% under hot water curing at 100°C and the range further increases to 22.06 – 24.66% for steam curing at 100°C.
- Considering three time periods, the tensile strength of M100 grade of RPC with oven curing at 90°C gains strength in the range of 10.81 – 13.58%, the range increases to 15.80 – 18.52% under hot water curing at 100°C and the range further increases to 19.0 – 25.93% for steam curing at 100°C.
- Considering three time periods, the flexural strength of M60 grade of RPC with oven curing at 90°C gains strength in the range of 10.81 – 12.50%, the range increases to 13.51 – 17.50% under hot water curing at 100°C and the range further increases to 13.51 – 20.00% for steam curing at 100°C.
- Considering three time periods, the flexural strength of M100 grade of RPC with oven curing at 90°C gains strength in the range of 9.09 – 12.80%, the range increases to 13.64 – 15.20% under hot water curing at 100°C and the range further increases to 15.91 – 20.00% for steam curing at 100°C.

The scope of research on reactive powder concrete is vast and the key lies in the right mix proportion of ingredients with right amount of time. The micro elementary level analysis on tested specimen of different curing methods are under progress of this study.

Acknowledgments
The authors bid to show gratefulness to the management, pro vice chancellor of faculty of engineering and technology, the dean of college of engineering and technology and the head of the civil engineering department for the moral support and assistance rendered in carrying out the experiments at the laboratory facilities of SRM Institute of Science and Technology.

References

[1] Richard P and Cheyrezy M 1995 Composition of Reactive Powder Concrete *Cement and Concrete Research* **25** 1501-1511

[2] Aitcin PC 2000 Cements of yesterday and today, Concrete of tomorrow *Cement and Concrete Research* **30** 1349-1359

[3] Harish KV, Dattatreya JK, Sabitha D and Neelamegam M 2008 Role of ingredients and curing regime in ultra-high strength concrete *Journal of Structural Engineering* **34** 421-428

[4] Cwrizen A 2007 The effect of heat-treatment regime on the properties of reactive powder of concrete *Advances in Cement Research* **19** 25-33

[5] Matte V and Moranville M 2000 Simulated microstructure and transport properties of ultra-high performance cement-based materials *Cement and Concrete Research* **30** 1947-1954

[6] Matte V and Moranville M 1999 Durability of reactive powder composites: Influence of silica fume on the leaching properties of very low water/binder pastes *Cement and Concrete Research* **21** 1-9

IOP Conf. Series: Materials Science and Engineering 1101 (2021) 012019 doi:10.1088/1757-899X/1101/1/012019

A study on topography versus sediment yield under simulated rainfall

L K Yong[1*], P L Law[1], S N L Taib[1], J O K Ngu[1], D Y S Mah[1] and S L G Law[2]

[1] Department of Civil Engineering, Universiti Malaysia Sarawak, 93000, Kota Samarahan, Sarawak, Malaysia
[2] Faculty of Engineering, Computing and Sciences, Swinburne University of Technology, Kuching, Sarawak, Malaysia
*Corresponding author: yongleongkong@gmail.com

Abstract. This study investigates the effects of topography on the amount of sediment yield under simulated rainfall. The slope gradient and length would affect the runoff depth (V) and peak flow volume (Q_p) and thus the amount of surface runoffs. In this study, the simulated 150mm/hour rainfall intensity was applied on triangular prism-shaped, cone-shaped and pyramid-shaped models for determination of the amount of respective sediment yields (tons/storm event). It was observed that the sediment yields of the triangular prism-, cone- and pyramid-shaped amounted to 0.144, 0.143 and 0.125 tons/storm event, respectively. The triangular prism-shaped topography has the highest sediment yield amount as it experiences highest runoff depth and highest surface runoff velocity at downslope. Based on the experimental outcomes, it was shown that MUSLE could over-estimate sediment yield as much as 3.6 times for areas characterized by hilly landscape.

1. Introduction

The term "Sediment Yield" can be defined as the amount of eroded soils collected at the outflow end of an observation plot, field, channel, or watershed [1]. In a watershed, sediment yield refers to the amount of eroded material at a designated point remote from the origin of the detached particles; includes the soil loss from slopes and channels, mass wasting and deposition of eroded material on both slopes and in channels [1,2]. The quantity of sediment yield and sediment's size characteristics are important in analyses of the scale of pollution in a watershed and indicate the decision-making for best management practices [2].

The Modified Universal Soil Loss Equation (MUSLE) has been widely used to predict sediment yield amount for a watershed or a catchment area [3,4,5]. The MUSLE is a modified version of Universal Soil Loss Equation (USLE) used to predict the amount of sediment yield on a storm basis from the outlet of watershed based on the runoff characteristics [5,6,7,8]. Williams and Berndt [5] developed a runoff factor $11.8(V\times Q_p)^{0.56}$ by investigating 778 individual storm events in 18 catchments with areas ranging from 15 to 1500 ha, and subsequently replaced the Rainfall Erosivity Factor (R) in the Universal Soil Loss Equation (USLE) to become MUSLE [5,7]. The MUSLE was recognized by United States of America, Department of Agriculture (USDA) as the best single indicator for sediment yield prediction, the amount of sediment yield in a watershed to a specific location for a specific storm as expressed in equation 1 [7,8]. Equation 1 was later revised by CPESC,

Inc. (2010) and rewritten as in equation 2 [9]. Table 1 illustrates some of the researchers who adopted the MUSLE to estimate sediment yield at different regions 1.

$$T = 11.8(V \times Q_p)^{0.56} \times K \times LS \times C \times P \quad \text{(tons/storm event)} \tag{1}$$

$$T = 95(V \times Q_p)^{0.56} \times K \times LS \times C \times P \quad \text{(tons/storm event)} \tag{2}$$

where,

V = Runoff Depth (inch)
Q_p = Peak flow rate for the storm event (cfs)
K = Soil erodibility factor (tons•ac•hr/10^2 ft•tonf•in)
LS = Topographical factor, unitless
C = Cover management factor, unitless
P = Support practice factor, unitless

Table 1. Sediment Yield Assessments at Different Regions

Researchers	Study Area	Sediment Yield (tons/storm event)
Sadeghi & Mizuyama, 2007 [7]	Khanmirza watershed, Iran	7,495.8
Fu et al., 2006 [10]	Pataha Creek Watershed, USA	3.1
Santhi et al., 2006 [11]	West Fork Watershed, USA	3,160 .0
Hao et al., 2004 [12]	Yellow River Basin, China	770.0
Jain & Kothyari, 2000 [13]	Nagwa Watershed, India	4,187.6
	Karso Watershed, India	191.6
Meyer et al., 1997 [14]	Mississippi, USA	30.0

Generally, higher rainfall intensity comprises of higher percentage of larger raindrops and thus generate higher volume of surface runoffs [15]. Previous studies show that the rainfall intensities and raindrops size in equatorial region are comparatively higher than other regions [16-23]. Besides, the terrains in equatorial region comprise of comparatively more cone-shaped and pyramid-shaped topography characterized with steeper slopes and shorter slope lengths as compared to agricultural lands in the eastern USA [24].

Hence, in equatorial areas, higher rainfall intensity coupled with steeper and shorter hill slopes would result in higher volume and velocity of surface runoff (as compared to RUSLE's 37 experimental sites in eastern USA). Obviously, higher surface runoff has higher runoff depth (V) and higher peak runoff volume (Q_p) that would produce higher amount of sediment yield at downslope and in a catchment. In addition, MUSLE LS-Factor only consider inclined plane-surface slope, while triangular prism-, cone- and pyramid-shaped terrains with steeper and shorter horizontal slope length are commonly in equatorial region. Thus, the adoption/application of RUSLE's LS-Factor (inclined plane-surface) in the MUSLE used to predict the sediment yield in equatorial region would undoubtedly been underestimated. Figure 1 shows the differences between MUSLE and current research.

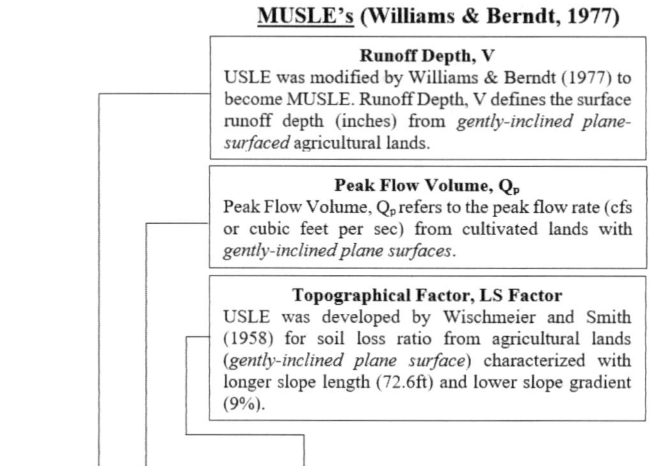

$$A = 95(V \times Q_p)^{0.56} \times K \times (L \times S) \times C \times P$$

Figure 1. A Comparison of Development of MUSLE's and Equatorial's (LS) Factors.

2. Methods and Materials

Experimental trails on sediment yield under simulated rainfall on three different topographical soil models (pyramid-, triangular prism- and cone-shaped) were carried out at Geotechnical Engineering laboratory of Universiti Malaysia Sarawak (UNIMAS). Soil samples were put inside the designed moulds and well compacted in accordance to Standard Protocol Test (SPT). A rainfall simulator was fabricated to simulate high rainfall intensity of approximately 150 mm/hr. Surface runoffs were collected for 30-minute rainfall duration, and the collected surface runoffs were weighed and recorded. One liter (1-L) of water sample from each topographical shape was also taken to dry for determination of Total Suspended Solids (TSS).

2.1. Rainfall Simulator to Simulate Equatorial Rainfall Intensity

The experimental rainfall simulator consists of a water tank, pump, hose, valve, steel structure and sprinkler as shown in figure 2. The mobility of the rainfall simulator is high because the structure is lightweight and provided with rollers. Supporting equipment such as toolbox, extension wire and hose could be stored at the drawer of the structure of simulator rainfall after use to keep the experimental site clean and tidy. The rainfall simulator produces a rainfall intensity of approximately 150 mm/hr. The raindrop falling height was setup at 10 meters above the peak of soil sample to ensure that the largest possible raindrop (5mm) would achieve terminal settling velocity (V_{ts}) [25].

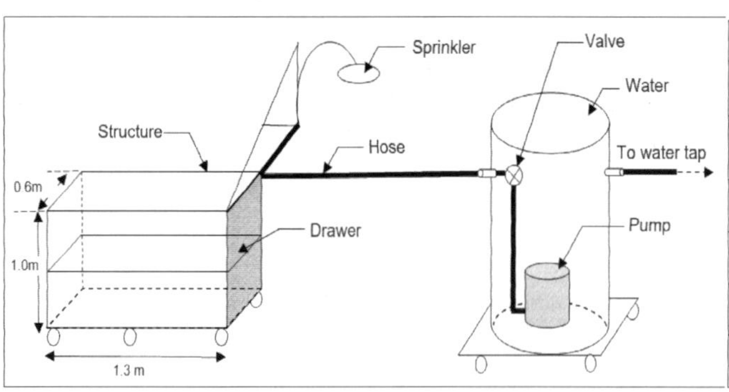

Equipment	Description
Mounting Structure	Steel structure mounted on rollers (Light-weight & mobility)
Size of Mounting Structure	Length : 1.3 m Width : 0.6 m Height : 1 m
Rainfall Intensity	High-to-Extremely High, approximately 150 mm/hr
Raindrop size	0.1–5 mm in diameter
Raindrops Falling Height	10-m high from the peak of soil sample
Water Tank	200 Liters

Figure 2. Rainfall Simulator

IOP Conf. Series: Materials Science and Engineering 1101 (2021) 012019 doi:10.1088/1757-899X/1101/1/012019

2.2. Soil Samples

Two types of mineral soils were used in this study to examine the amount of sediment yield after each rainfall event. The soil samples were taken at Kuching, Sarawak: Sample A was collected at a field near Santubong River and Sample B was collected from downslope of a hill located at Kota Samarahan, Kuching as shown in figure 3. The Grain size distribution and hydrometer analysis were performed to determine the soil particle sizes and subsequently categorized the soil groups by using the Soil Textural Triangle. Grain size distribution curves of the Sample A and Sample B are shown in figures 4 and 5, respectively. Based on the results of grain size distribution analysis, Sample A recorded 17.05% gravel, 82.85% sand and 0.1% silt and clay, and thus Sample A can be classified as Poorly-graded Sand (S-P). The soil compositions of Sample B are 17.2% gravel, 81.3% sand and 1.5% silt and clay, classified as Poorly-graded Sand (S-P) based on the Soil Textural Triangle.

Figure 3. Soil Samples A and B

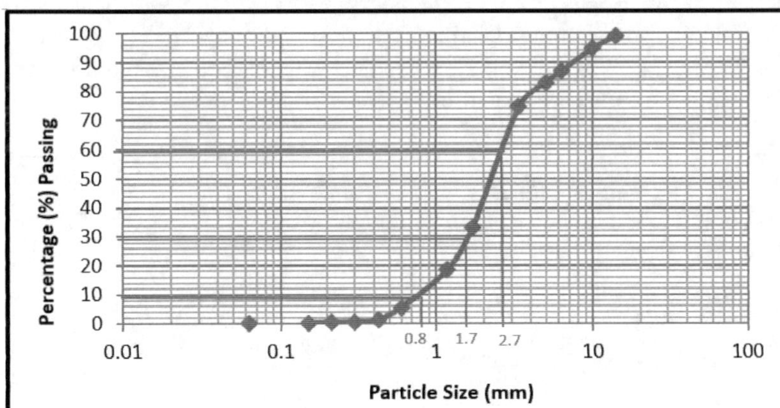

Particle Size (mm)	% Passing
14	99.10
10	94.75
6.3	87.05
4.75	82.95
3.35	74.45
1.7	33.10
1.18	18.45
0.6	5.60
0.425	1.40
0.3	0.60
0.212	0.50
0.15	0.30
0.063	0.10

Gravel	**Sand**	**Silt**	**Clay**
17.05	**81.82**	**0.1**	
Soil Classification			**S-P**

Figure 4. Soil Sample A - Grain Size Distribution of Soil

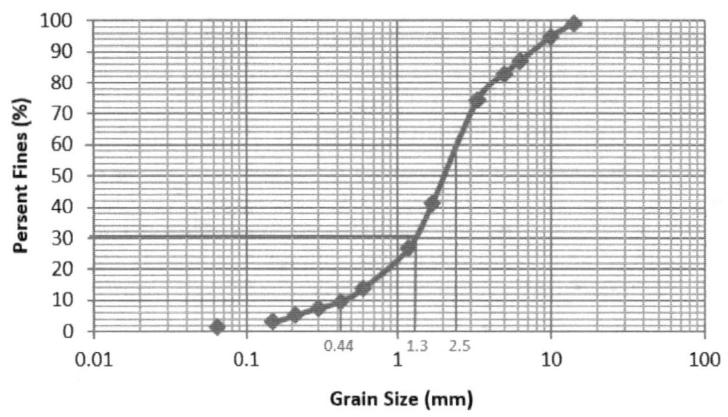

Particle Size (mm)	% Passing
14	100.00
10	99.70
6.3	91.85
4.75	82.80
3.35	65.25
1.7	44.90
1.18	36.95
0.6	27.30
0.425	23.40
0.3	20.80
0.212	17.70
0.15	9.60
0.063	1.50

Gravel	Sand	Silt	Clay
17.2	81.3	1.5	
Soil Classification			S-P

Figure 5. Soil Sample B - Grain Size Distribution

2.3. Laboratory Analysis of Total Suspended Solids (TSS)

The total suspended solids (TSS) measures the amount of solid materials content in the runoff waters. In this study, one liter (1-L) of runoff water sample was taken from sediment basin after a 30-minute simulated rainfall event; the water in the basin was agitated to make sure the water was completely mixed before sampling. The water sample was transferred to tray, weighed and recorded. After weighing, the tray was put into the oven and dried for 24 hours (to make sure the oven was on and oven temperature was set at 104°C). After oven-dried at 104°C for 24 hours, the tray was removed from the oven and left to cool. The tray was put on the scale and waited until the reading stabilized. The TSS can be computed by using the following equation:

$$TSS\ (mg/L) = \frac{Oven\ dried\ mass\ (mg) - Empty\ tray\ (mg)}{Sample\ volume\ (L)} \tag{3}$$

2.4. Topographical Shape and Soil Compaction

The Revised Universal Soil Loss Equation (RUSLE) is based on the unit plot design of 72.6ft slope length and slope steepness of 9%. Normalization of slope length and slope steepness are based on the unit plot [26,27]. The Modified Universal Soil Loss Equation (MUSLE) was developed by Williams and Berndt in 1977 by replacing the rainfall erosivity, R to a runoff factor $11.8(V\times Q_p)^{0.56}$. The runoff factor was later revised by CPESC, Inc. and rewritten as $95(V\times Q_p)^{0.56}$ in 2010. The other sub-factors such as Soil Erodibility (K), Crop Management (C) and Support Practice (P) were remained. Experimental plots were designed in three topographical landscapes (triangular prism-, cone- and pyramid-shaped) with base area of 1.0 m^2 as shown in table 2. The horizontal slope length of soil sample is 0.5 m and the slope steepness is 100% (angle of slope 45°).

The experimental soil compaction method and number of blows in accordance with Standard Proctor Test (SPT). A 2.5kg rammer was used to compact the soil samples. However, there were three compartments of experimental moulds and the surface areas are larger than Proctor mould; therefore, each compartment was compacted separately in three layers to ensure that soil samples were well compacted. The number of Rammer blows for triangular-, cone- and pyramid-shape moulds are shown in table 2.

Table 2. Experimental Moulds of Different Topographical Landscapes and Number of Rammer Blows [24]

Topographical Shape	Moulds' Design	Mounting Moulds	Number of Rammer Blows
Triangular Prism	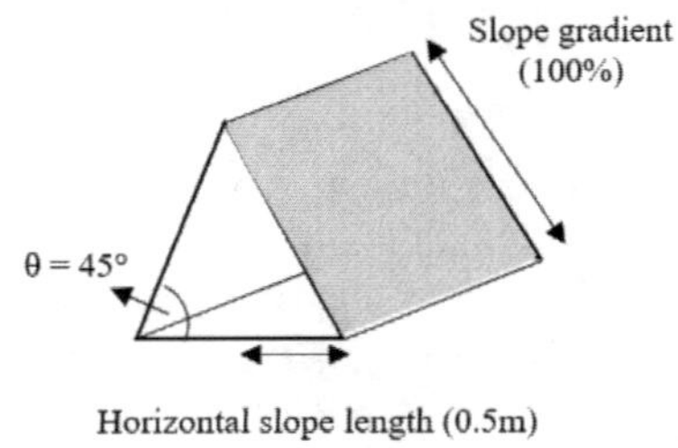	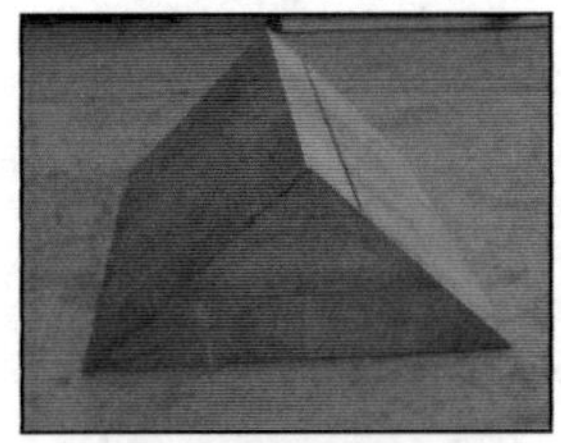	**Top part:** 25 blows for 1 layer. **Middle part:** 621 blows per layer for 3 layers. **Bottom part:** 1863 blows per layer for 3 layers.
Cone	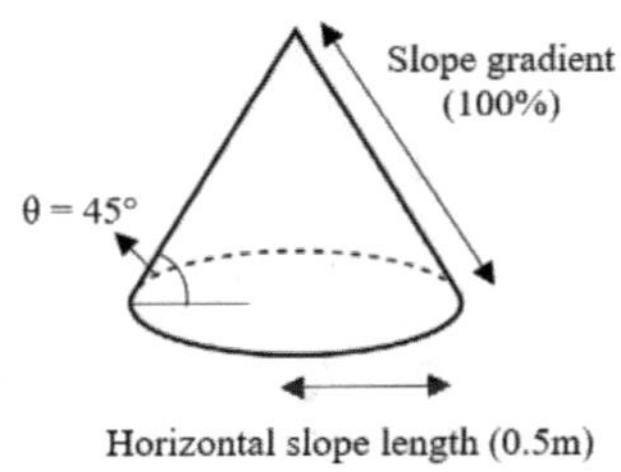		**Top part:** 25 blows for 1 layer. **Middle part:** 108 blows per layer for 3 layers. **Bottom part:** 891 blows per layer for 3 layers.
Pyramid	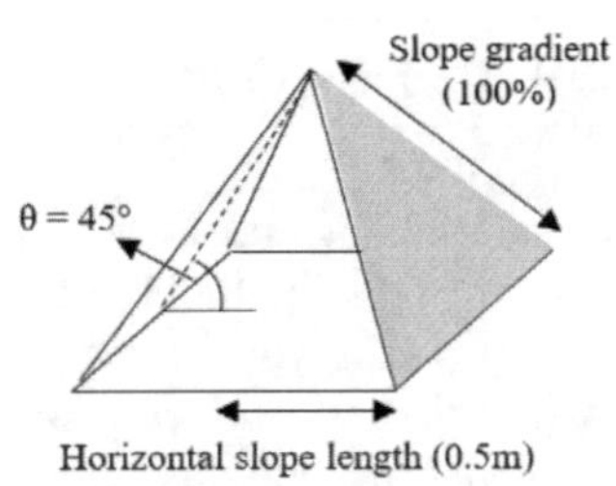		**Top part:** 25 blows for 1 layer. **Middle part:** 135 blows per layer for 3 layers. **Bottom part:** 1134 blows per layer for 3 layers.

2.5 Surface Runoff Patterns

The amount of surface runoff and sediment yield can be related to the surface runoff pattern of different topographical features. Figure 6 shows the surface runoff pattern for pyramid-, triangular prism- and cone-shaped soil samples. It was observed that surface runoffs of triangular prism-shaped would flow in an evenly dispersed pattern downslope along two sides of plane surfaces, while the pyramid-shaped would flow onto the four sides, and cone-shaped model presented even flow in all directions.

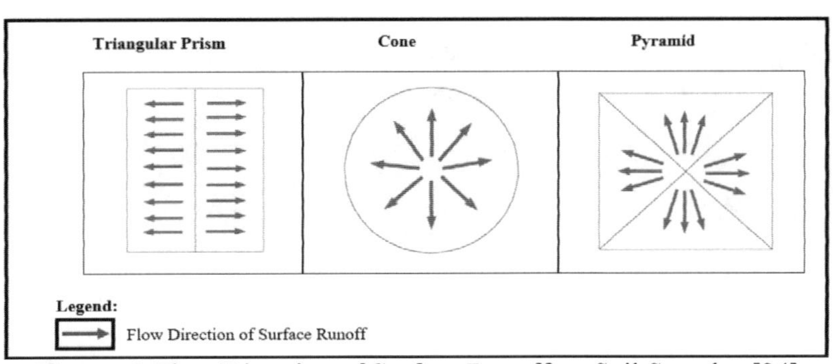

Figure 6. Flow Direction of Surface Runoff on Soil Samples [24]

IOP Conf. Series: Materials Science and Engineering 1101 (2021) 012019 doi:10.1088/1757-899X/1101/1/012019

3. Results and Analysis

The Modified Universal Soil Loss Equation (MUSLE) is commonly used to estimate sediment yield from a watershed [9]. In this research, the Total Suspended Solids (TSS) can be treated as sediment yield from a watershed as the eroded soil particles (mainly silt and clay particles) are predominantly transported by surface runoffs into sediment basin. Runoff samples were collected from the sediment basin and the dry-mass of sediments are the Total Suspended Solids (TSS) and the results are presented in table 3.

Table 3. Total Suspended Solids in Runoff Waters.

Topographical Landscape	Soil Sample	Oven-dried Sample Mass (g)	Empty Tray Mass (g)	TSS (g/L)	Sediment Yield (tons/storm event per m^2)	Sediment Yield (tons/storm event/ac)	Average Sediment Yield (tons/storm event)
Triangular Prism	A	272.39	237.24	35.15	35.15×10^{-6}	0.142	0.144
	B	273.64	237.24	36.40	36.40×10^{-6}	0.146	
Cone	A	271.18	237.24	33.94	33.94×10^{-6}	0.136	0.143
	B	274.31	237.24	37.07	37.07×10^{-6}	0.150	
Pyramid	A	267.93	237.24	30.69	30.69×10^{-6}	0.121	0.125
	B	269.25	237.24	32.01	32.01×10^{-6}	0.129	

In table 3, the total suspended solids (TSS) of soil samples A and B after each rainfall event at rainfall intensity of 150mm/hr were recorded 35,150 mg/L and 36,400 mg/L for triangular prism-shaped, 33,940 mg/L and 37,070 mg/L for cone-shaped, and 30,690 mg/L and 32,010 mg/L for pyramid-shaped. The sediment yields of triangular prism-shaped and cone-shaped topography are almost identical, recorded 0.144 and 0.143tons/storm event, respectively. Pyramid-shaped recorded the lowest sediment yield of 0.125tons/storm event, which is 13.2% lower than triangular prism-shaped and 12.6% lower than cone-shaped. Sediment yield is the amount of soil particles been washed by surface runoff and deposit at downslope. Thus, the amount of sediment yield depends on the surface runoff depth and velocity. The surface runoff depth and velocity are different for experimental triangular prism-shaped, cone-shaped and pyramid-shaped topography. Surface runoff velocity can be computed by knowing the Surface runoff velocity (RV, m/s), Volume of surface runoff (Q, m^3/s), and Slope or gradient of hydraulic (S, m/m). Equation 4 can be used to determine the height of surface runoff on slope (h). Figure 7 shows the cross-sectional length and surface runoff velocity at downslope locations a, b, c, d and e.

$$h^3 = \frac{Q^2}{2gL_n{}^2} \qquad (n=1,2,3,4,5) \qquad (4)$$

IOP Conf. Series: Materials Science and Engineering 1101 (2021) 012019 doi:10.1088/1757-899X/1101/1/012019

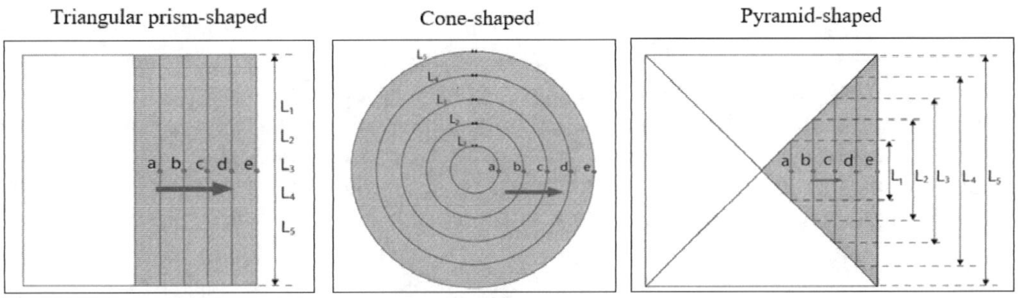

Triangular prism-shaped	Cone-shaped	Pyramid-shaped
L_1 = 1.00 m	L_1 = 0.71 m	L_1 = 0.20 m
L_2 = 1.00 m	L_2 = 1.42 m	L_2 = 0.40 m
L_3 = 1.00 m	L_3 = 2.13 m	L_3 = 0.60 m
L_4 = 1.00 m	L_4 = 2.84 m	L_4 = 0.80 m
L_5 = 1.00 m	L_5 = 3.55 m	L_5 = 1.00 m

	Runoff Depth (mm)			Runoff Velocity (m/s)		
Location	Triangular Prism	Cone	Pyramid	Triangular Prism	Cone	Pyramid
a	0.089	0.066	0.058	0.042	0.036	0.034
b	0.141	0.104	0.092	0.053	0.045	0.043
c	0.184	0.136	0.121	0.060	0.052	0.049
d	0.223	0.165	0.146	0.066	0.057	0.054
e	0.259	0.192	0.170	0.071	0.061	0.058

Figure 7. Experimental Runoff Depth and Runoff Velocity at Locations "a" to "e"

In figure 7, triangular prism-shaped recorded highest surface runoff depth of 0.259mm and highest surface runoff velocity of 0.071m/s at location "e" (downslope of experimental plot), and thus the surface runoff would carry more soil particles such as silt and clap as compared to cone-shaped and pyramid-shaped topography.

3.1 Experimentally Observed versus MUSLE Predicted Sediment Yield

The MUSLE was developed using gentle-inclined plane surface with 72.6ft slope length, 6ft width and 9% slope steepness [5,24,27]. The triangular prism-shaped topography has similar shape (inclined plane surface) as MUSLE plot, thus it can be compared with MUSLE prediction. Table 4 shows the MUSLE predicted sediment yield of 0.449tons/storm event, which is found approximately 67.9% higher than triangular prism-shaped topography.

Table 4. MUSLE Predicted Sediment Yield

Volume of runoff for the storm event, V (ac-ft)	0.246
Peak flowrate for storm event, Q_p (cfs)	6.53×10^{-4}
Topography factor, T (unitless)	0.63
Cover-management factor, C (unitless), bare soil condition	1.0
Support practice factor, P (unitless), bare soil condition	1.0
Sediment Yield (tons/storm event)	0.449

4. Conclusions

The sediment yield amount is affected by surface runoff depth and surface runoff velocity, whereby higher surface runoff depth and velocity would generate higher sediment yield amount. Experimental results show that the triangular prism-shaped topography has highest sediment yield amount as compared to cone-shaped and pyramid-shaped topography. A comparison of the experimental outcomes with the MUSLE, it was shown that the MUSLE predicted sediment yield could over-estimate by 3.6 times in hilly areas that comprise of mostly triangular prism-, cone- and pyramid-shaped terrains.

References

[1] Zachar D 1982 *Soil Erosion (Development in Soil Science 10)*. Bratislava, Slovakia: Elsevier.
[2] Renard K G and Foster G R 1983 Soil Conservation: Principles of Erosion by Water. Dryland Agriculture. *Agronomy Monograph,* **23** 156-176
[3] Simons D B and Şentürk F 1992 *Sediment Transport Technology*. Littleton, USA: Water Resources Publications.
[4] Renard K G, Foster G R, Weesies G A, McCool D K and Yoder D C 1997 *Predicting Soil Erosion by Water: A Guide to Conservation Planning with the Revised Universal Soil Loss Equation (RUSLE)*. Agriculture Handbook No. 703. Washington, DC: US Government Printing Office.
[5] Williams J R and Berndt H D 1977 Sediment Yield Prediction based on Watershed Hydrology. *Transaction of the American Society of Agricultural Engineers* **20** 1100-1104
[6] Sadeghi S H R 2004 Application of MUSLE in Prediction of Sediment Yield in Iranian Conditions. In: *Proceedings International Erosion Control Association, ISCO2004* (Brisbane, Australia, 4–8 July 2004) **998** 1–4
[7] Sadeghi S H R and Mizuyama T 2007 Applicability of the Modified Universal Soil Loss Equation for Prediction of Sediment Yield in Khanmirza Watershed, Iran. *Hydrological Sciences Journal* **52** 1068-107
[8] Hrissanthou V 2005 Estimate of Sediment Yield in a Basin without Sediment Data. *Catena* **64** 333–347
[9] Certified Professional in Erosion and Sediment Control (CPESC, Inc.) 2010 Section 4: Certified Professional in Erosion and Sediment Control Exam Review.
[10] Fu G, Chen S and McCool D K 2006 Modeling the Impacts of No-till Practice on Soil Erosion and Sediment Yield with RUSLE, SEDD, and ArcView GIS. *Soil and Tillage Research,* **85** 38-49
[11] Santhi C, Srinivasan R, Arnold J G and Williams J R 2006 A Modeling Approach to Evaluate the Impacts of Water Quality Management Plans Implemented in a Watershed in Texas. *Environmental Modelling & Software,* **21** 1141-1157
[12] Hao F B, Zhang X S and Yang Z F 2004 A Distributed Non-point Source Pollution Model: Calibration and Validation in the Yellow River Basin. *Journal of Environmental Sciences* **16** 646-650
[13] Jain M K and Kothyari U C 2000 Estimation of Soil Erosion and Sediment Yield using GIS. *Hydrological Sciences Journal* **45** 771-786
[14] Meyer L D, Dabney S M, Murphree C E, Harmon W C and Grissinger E H 1997 Effects of Crop Land Management Practices on Runoff and Erosion. *Proceedings of the Conference on Management of Landscapes Disturbed by Channel Incision*.
[15] Norman J M, Kustas W P and Humes K S 1995 Source Approach for Estimating Soil and Vegetation Energy Fluxes in Observations of Directional Radiometric Surface Temperature. *Agricultural and Forest Meteorology* **77** 263-293.
[16] Alvares C A, Stape J L, Sentelhas P C, Goncalves J L D M and Sparovek G 2013 Koppen's Climate Classification Map for Brazil. *Meteorologische Zeitschrift* **22** 711-728

[17] Dijk V A I J M, Bruijnzeel L A and Rosewell C J 2002 Rainfall Intensity-Kinetic Energy Relationships: A Critical Literature Appraisal. *Journal of Hydrology* **261** 1-23
[18] Kottek M, Grieser J, Beck C, Rudolf B and Rubel F 2006 World Map of the Köppen-Geiger Climate Classification Updated. *Meteorologische Zeitschrift* **15** 259-263
[19] Lal R 1985 Soil Erosion and Sediment Transport Research in Tropical Africa. *Hydrological Sciences Journal* **30** 239-256
[20] Maxwell J B 1981 Climatic Regions of the Canadian Arctic Islands. *Arctic*: 225-240
[21] Ohmura A and Reeh N 1991 New Precipitation and Accumulation Maps for Greenland. *Journal of Glaciology,* **37** 140-148
[22] Oliver Y M, Lefroy E C, Stirzaker R and Davies C L 2005 Deep-drainage Control and Yield: The Trade-off between Trees and Crops in Agroforestry Systems in the Medium to Low Rainfall Areas of Australia. *Australian Journal of Agricultural Research* **56** 1011-1026
[23] Peel M C, Finlayson B L and Mcmahon T A 2007 Updated World Map of the Koppen-Geiger Climate Classification. Hydrology and Earth System Sciences Discussions. *European Geosciences Union,* **4** 439-473
[24] Yong L K, Law P L, Taib S N L, Mah D Y S and Johari A H 2018 Refinement of Topographical Factor for Estimating Soil Loss and Sediment Yield in Equatorial Regions. *Journal of Civil Engineering, Science and Technology,* **9** 98-116
[25] Gunn R and Kinzer G D 1949 The Terminal Velocity of Fall for Water Droplets in Stagnant Air. *Journal of Meteorology,* **6** 243-248
[26] Wischmeier W H and Smith D D 1978 *Predicting Rainfall Erosion Losses – A Guide to Conservation Planning.* Agriculture Handbook No. 537. Washington, DC: United States Department of Agriculture (USDA).
[27] Renard K G, Foster G R, Weesies G A, McCool D K and Yoder D C 1997 *Predicting Soil Erosion by Water: A Guide to Conservation Planning with the Revised Universal Soil Loss Equation (RUSLE).* Agriculture Handbook No. 703. Washington, DC: US Government Printing Office.

IOP Conf. Series: Materials Science and Engineering 1101 (2021) 012020 doi:10.1088/1757-899X/1101/1/012020

Response spectra for moderate seismic area – application to Miri district of Sarawak, Malaysia

R Ahmadi[1]*, I A Najar[2]*, A F Abdullahi[3] and T Galin[4]

[1, 2] Department of Civil Engineering, Universiti Malaysia Sarawak, 94300, Kota Samarahan, Sarawak, Malaysia
[3] Department of Civil and Environmental Engineering, Dhofar University, Salalah, P.O. Box 2509, Sultanate of Oman
[4] Department of Mineral and Geoscience Malaysia, Kuching, Sarawak, Malaysia
*Corresponding authors: araudhah@unimas.my, imtiyaznajar999@gmail.com

Abstract. The aim of this paper is to estimate the response spectra for moderate seismic area in Sarawak, Malaysia. In the present study, the response spectra for Miri district, in Sarawak, has been obtained by using 1-D equivalent linear site response analysis at 114 borehole locations. All the borehole sites are classified based on average shear wave velocity (Vs30) as per Malaysia National Annex to Eurocode 8 (MNA-EC8) site classifications using the geotechnical relationship. In this study, the input motion was selected from Pacific Earthquake Engineering Research (PEER-NGA) online database compatible with target response spectra of Sarawak at engineering bedrock by considering the seismic hazard map for 2475 years as return period and by following the 475 years return period as shown in (MNA-EC8). From the results of this study, it was found that the maximum response spectral acceleration (response spectra with 5% damping) is 2.25 g, 3.25 g and 1.7 g at the ground surface for MNA-EC8 site classes C, D and E respectively. It was also found that the results are under the provisions of seismic design code of Malaysia. The suggested results in this study can be used as a reference in Malaysia to support the MNA-EC8.

1. Introduction

A response spectrum is a useful tool available to engineers designing earthquake resistant structures. Site response analysis is used to measure ground surface motions, amplification potential and the development of design response spectra. Shear wave velocity (Vs) is an important parameter for conducting ground response analysis [1]. In general, the following three different approaches can be implemented to conduct ground response analyses: (1) use of geotechnical data (i.e. SPT blow counts obtained from bore logs), (2) use of microtremor data (i.e. ambient noise measurement) and (3) use of a geophysical method (spectral analysis of surface wave tests, SASW) [2].

Amirsardari et al. [3] investigated the seismic site response analysis for Australia which has a history of low-to-moderate earthquakes. An equivalent linear program was used to analyze the impacts of the depth of rock/soil to engineering bedrock and the shear wave velocity (Vs30) on the site amplification factors for sand and clay with a plasticity index of 30%. The analysis was completed using the SHAKE2000 program. On the other hand, analysis of the site response was carried out by [4-6] by considering the vertical and horizontal components of ground motions. The analyses were performed using both nonlinear and equivalent linear analyses utilizing the DEEPSOIL program. Furthermore, site response analysis was carried out by using the software SHAKE and then integrated with probabilistic seismic hazard assessment using MATLAB. Also the matching of response spectrum was carried out using RSPMATCH software [7]. Tsang et al. [8] studied the design spectrum for flexible soil sites in regions of low-to-moderate seismicity. An alternative design model was introduced to consider the soil resonance phenomenon without having the need to conduct computational site response analysis of the subsurface model.

Pappin et al. [9] used the Oasys SIREN program to produce the response spectra by using the one-dimensional site response analysis in the moderate seismicity region of Hong Kong and found that the results obtained were more realistic compared to the SHAKE program. One-dimensional site response analysis was carried using three industry-standard programs, the time domain nonlinear program, DEEPSOIL, and the frequency domain equivalent linear program, SHAKE. Four soil locations, which are deemed suitable for nuclear safety-related building, varying from hard rock to stiff soil were selected. Comparing the forecasts for the various available programs showed that standard programs for locations with stiff soil and low-intensity soil movements are in excellent agreement, however, for analyses involving greater strains and at greater frequencies, there is wide variation [10].

From the above literature review, it is evident that researchers have used several types of data, softwares and methodologies in order to produce the response spectra. In the present study, geotechnical data will be used and evaluated using the 1-D equivalent linear response analysis by utilizing the DEEPSOIL software.

2. Study Area

Miri is a coastal area in northeast of Sarawak, Malaysia. It is situated on the island of Borneo, near the border with Brunei. The study area is located within coordinates longitudes-113° 58′ 46.4″ E to 113°49′ E and latitudes 04° 21′ N to 04° 07′ N. The study area is shown in figure 1.

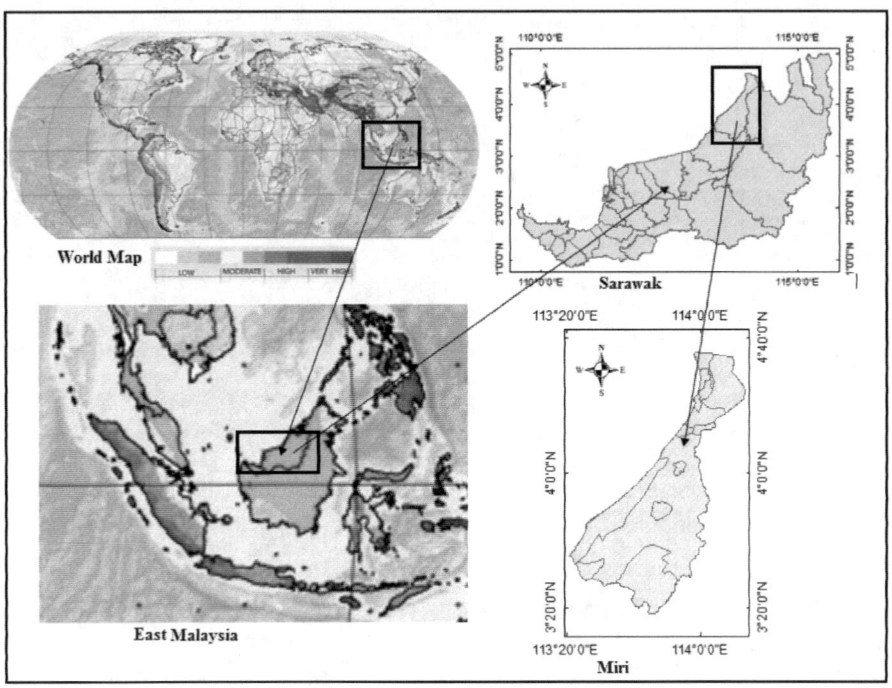

Figure 1. Miri district study area

3. Methodology

The response spectra for various site class profiles for the Miri district were determined with DEEPSOIL software using a 1-D equivalent linear response approach.

3.1 Geotechnical dataset utilized in site response analysis

For this research, standard penetration tests were carried out by various government agencies and geotechnical engineering consultancies of Malaysia, and penetration values (N) were calculated. Data was taken from 114 boreholes, each with a depth not exceeding 30 m. The borehole locations are shown in figure 2.

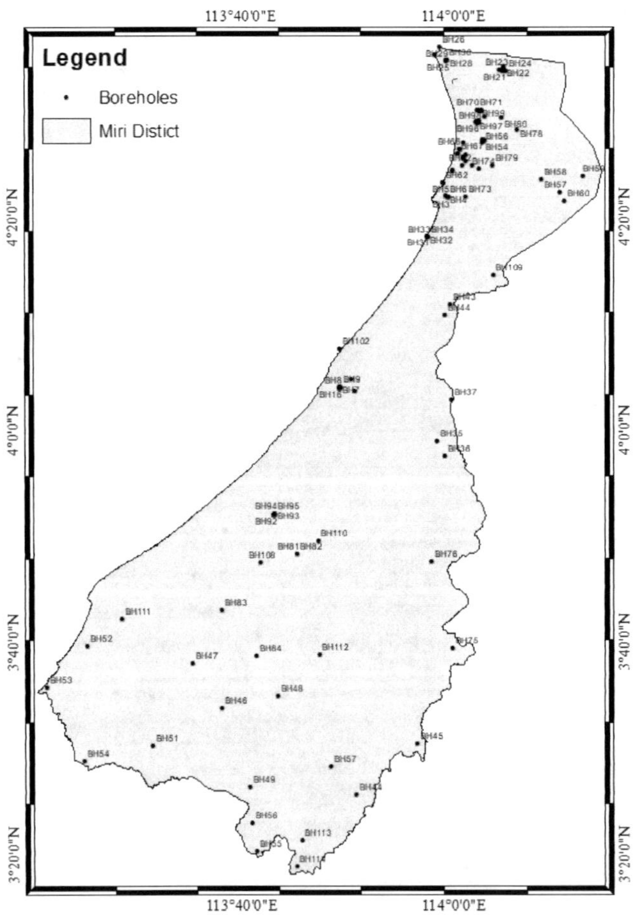

Figure 2. Location of boreholes across the Miri district

3.2. MNA-EC8 shear wave velocity (Vs30) site classification

The acquired SPT-N values provide an estimate of the shear strength of the soil segment under consideration. To decide the dynamic properties of soil, the achieved SPT-N values are used to assess the shear wave velocity up to a depth of 30 m (Vs30) by utilizing soil type and additionally area explicit connections. The parameter Vs is used by the MNA-EC8 [11] for site classification up to a depth of 30 m to design seismic-resistant infrastructure. The recommended site classification according to MNA-EC8 [11] based on average shear wave velocity up to the depth of 30 m is shown in table 1. This research calculated Vs30 of soil segments utilizing the relationships between SPT-N values and the Vs30. Imai and Tonouchi [12] relationship is used to calculate Vs30. This was introduced by Looi et al. [13] by using it in the Kuching area of Sarawak, Malaysia. The relationships between SPT-N values and Vs30 have been studied by many researchers since the 1960s. Imai and Tonouchi [12] have investigated the largest dataset, including 1654 data pairs from 386 borings at 250 destinations in Japan. It is applicable for all geological ages and all types of soils [13-15]. Imai and Tonouchi [12] created a relationship for Vs, which depends on N-value. The applicable range of N-value is from less than one blow per feet (bpf) to nearly 400 bpf. The relationship used to calculate Vs30 is given in equation 1.

$$Vs = 97 \times N^{0.314} \tag{1}$$

where,
Vs = shear wave velocity
N = SPT-N value

Table 1. Classification of borehole sites according to MNA-EC8 [11] on the basis of Vs30 up to a depth of 30 m

MNA-EC8 site class	Rock/soil type	Vs30 (m/s)
A	Hard rock	>1500
B	Rock	760-1500
C	Dense soil/soft soil	360-760
D	Stiff soil	180-360
E	Soft soil	<180

The shear wave velocity up to a depth of 30 m (Vs30) was determined for the soil types available in the Miri district based on 114 SPT boreholes data. The obtained Vs30 values showed that the site classes C, D and E of the MNA-EC8 [11] sites are the dominant classes in the Miri district as shown in figure 3. Site class D is the predominant while site classes C and E covers a small area of the Miri district.

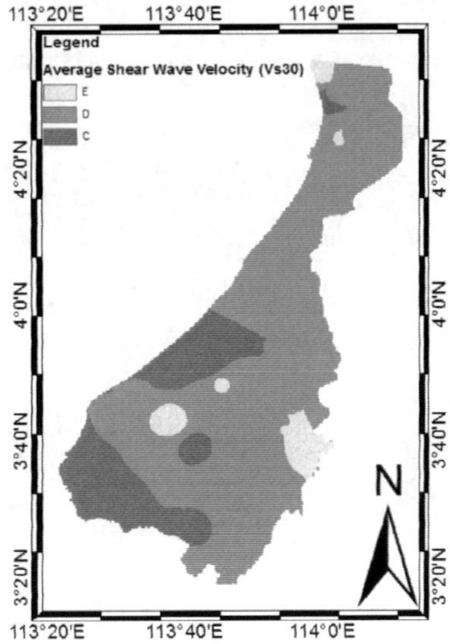

Figure 3. Distribution of Vs30 in Miri district

IOP Conf. Series: Materials Science and Engineering 1101 (2021) 012020 doi:10.1088/1757-899X/1101/1/012020

3.3. Input motion

Due to the unavailability of records of input ground motions for low-to-moderate seismicity regions, this research used an earthquake record from other locations well matched from a target response spectrum with similar characteristics. The record was obtained from the Pacific Earthquake Engineering Research Center (PEER) online database (PEER NGA) [16]. The seismogram of the ground motion is shown in figure 4.

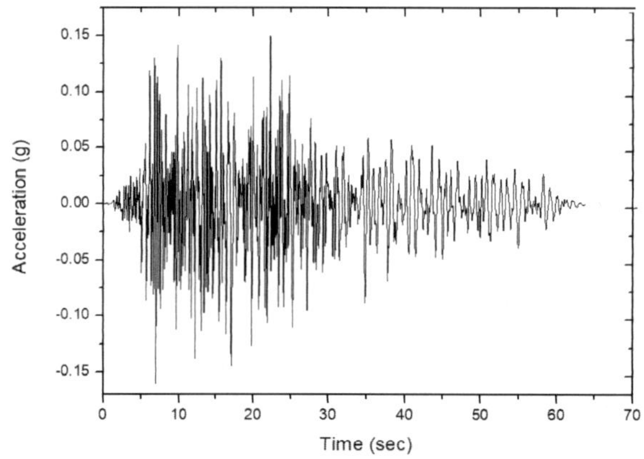

Figure 4. Input motion for the study area, compatible with target response spectra having moment magnitude 6.0 with PGA 0.15 g

3.4. DEEPSOIL

In the present research, the equivalent linear method was used to calculate the ground response analysis by using DEEPSOIL, a 1-D site response platform. The equivalent linear methodology introduced by Seed and Idriss [17] accounts for the non-linearity-of-the-shear-modulus and the damping-ratio (%), using an iterative process to obtain modulus values and damping compatible with the actual strains in each layer. By utilizing the small-strain shear modulus reduction ($^{G}/_{Gmax}$), and damping ratio (%) versus shear strain curves, each layer is modeled according to corresponding soil properties. 'Gmax' is the initial shear modulus and 'G' is the shear modulus. The effective shear strain ratio was taken as"0.65 during the analysis. The bedrock was assumed to be of a rocky nature with a shear wave velocity of 760 m/sec. This equivalent linear method is more suitable for low to moderate levels of earthquakes and is commonly used in engineering.

There are various active approaches to determine the response of a single degree of freedom (SDOF). The Duhamel integral solutions, the frequency-domain solution and the Newmark β method can all be used to analyse the response spectra. For the present study, the frequency domain solution approach has been used to estimate the response of SDOF systems and to solve the dynamic equilibrium equation defined in [18-19] by using the DEEPSOIL software, as shown in equation 2.

$$m\ddot{u} + c\dot{u} + ku = -m\ddot{u}g \tag{2}$$

where

m = mass

c = viscous damping

k = system stiffness of SDOF system

$\ddot{u}$ = nodal accelerations

$\dot{u}$ = relative velocities

u = relative displacements

$\ddot{u}g$ = the exciting acceleration at the base of SDOF.

In the frequency-domain solution, the Fourier Amplitude Spectra (FAS) input motion is modified by a transfer function as defined in equation 3:

$$H(f) = \frac{-f_n^2}{(f^2 - f_n^2) - 2i\xi f f_n} \tag{3}$$

where

f_n = natural frequency of the oscillator which is calculated as in equation 4:

$$fn = \frac{1}{2\pi} \sqrt{k/m} \tag{4}$$

ξ = damping ratio which is calculated as in equation 5:

$$\xi = \frac{c}{2\sqrt{km}}). \tag{5}$$

The-frequency-domain-solution-requires the use of FFTs (Fast Fourier Transforms)-to-move between the frequency-domain, where the oscillator transfer function is applied, and the time-domain, where the peak-oscillator-response is-estimated. Over-the-frequency-range-of the-ground-motion, the frequency-domain solution is exact.

4. Results and Discussion

The-PGA-of soil is not enough to describe the ground-motion since the -frequency parameter of the motion is not taken into account. The response of the building-depends on the frequency-content and it is standard practice to describe ground-motion in terms of response-spectra [20]. Response spectra provide the peak structural responses in the linear range, which can be utilized for acquiring lateral forces created in structures in response seismic activity. This knowledge is essential in designing earthquake-resistant structures. Engineers and designers need a response spectrum of accelerations in order to design a seismic resistant building structure. Ground response analysis has been carried out for 114 boreholes across the Miri district. From the results of this study, it was found that the maximum response spectral acceleration (response spectra with 5% damping) is 2.25 g, 3.25 g and 1.7 g at the ground surface for MNA-EC8 site classes C, D and E respectively. The spectral accelerations (response-spectra-with 5% damping) at the ground surface are shown in figure 5, figure 6 and figure 7 respectively. They were compared to the response spectra at surface according to MNA-EC8. It was found that the computed surface acceleration obtained from site-specific ground response analysis corresponding to response spectrum of published MNA-EC8 [11] are in close agreement with the target (surface measurements) for Sarawak, Malaysia.

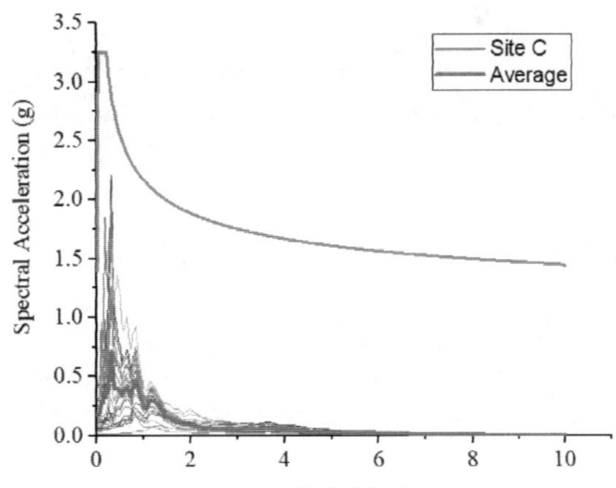

Figure 5. Response spectra for 5% damping at the ground surface of the Miri district for site class C as per MNA-EC8

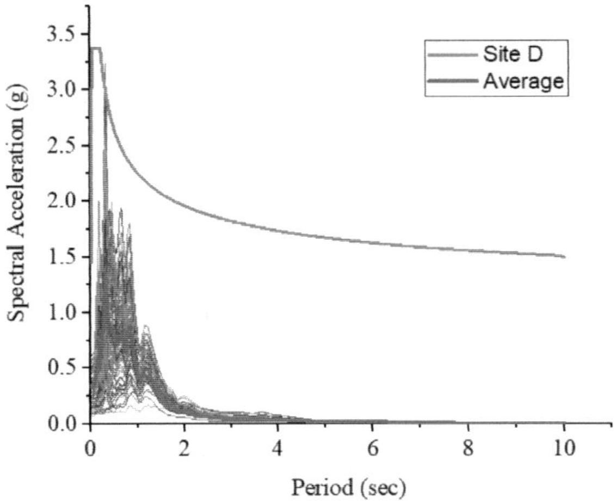

Figure 6. Response spectra for 5% damping at the ground surface of the Miri district with respect to site class D as per MNA-EC8

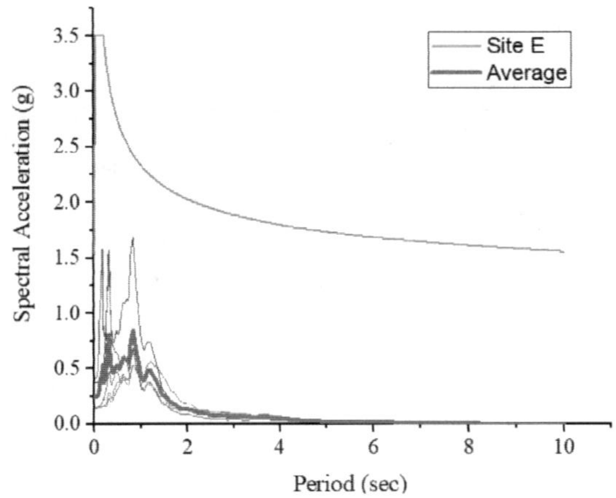

Figure 7. Response spectra for 5% damping at the ground surface of the Miri district with respect to site class E as per MNA-EC8

5. Conclusions

In this study, response spectra of Miri district have been developed by using a 1-D-equivalent-linear-response-analysis under frequency-domain-approach using DEEPSOIL software corresponding to 2475 years of return period. For the study, the input-motion was selected by considering the seismic hazard map for 2475 years as return period by following the Malaysia National Annex to Eurocode 8 (MNA-EC8). In this study, the ground response spectra have been evaluated for all locations and for all the soil types of Miri district and these results were compared with MNA-EC8 and showed that results of spectral acceleration of this study are in close agreement with the design response spectra provided by MNA-EC8.

Acknowledgments
Authors wishing to acknowledge the financial support from Cross Disciplinary Research Grant, F02/CDRG/1817/2019. Acknowledgement is also given to the Department of Mineral and Geoscience, Sarawak Branch for this research collaboration and financial support.

References

[1] Nazri F M, Tan C G and Rashid M A 2018 Investigation of Shear Wave Velocity by using Multi-Channel Analysis of Surface Wave Method for Microzonation Map Development and its Application to Industrial Frame Structures *Soil Mechanics and Foundation Eng.* **55** 2 pp 110–114

[2] Sitharam T G and Anbazhagan P 2007 Seismic Hazard Analysis for the Bangalore Region. *Natural Hazards* **40** 2 pp 261–278

[3] Amirsardari A, Lumantarna E and Goldsworthy H M 2014 Seismic site response analysis for Australia. *In Proceedings of the Australian Earthquake Engineering Society, Lorne, Victoria, Australia*, 21–23 November 2014

[4] Tsai C C and Liu H W 2017 Site Response Analysis of Vertical Ground Motion in Consideration of Soil Non-linearity *Soil Dynamics and Earthquake Engineering* **102** 2 pp 124–136

[5] Sana H, Nath S K and Gujral K S 2019 Site Response Analysis of the Kashmir valley during the 8 October 2005 Kashmir Earthquake (Mw 7.6) using a Geotechnical Dataset *Bulletin of Engineering Geology and the Environment* **78** 4 pp 2551–2563

[6] Afshari K and Stewart J P 2019 Insights from California Vertical Arrays on the Effectiveness of Ground Response Analysis with Alternative Damping Models. *Bulletin of the Seismological Society of America,* **109** 4 pp 1250–1264

[7] Yunatci A A and Çetin K O 2008 Site Specific Ground Response Analysis Integrated Within Probabilistic Seismic Hazard Framework *In Proceedings of the 8th International Congress on Advances in Civil Engineering, North Cyprus* 15-17 Sep 2008

[8] Tsang H H, Wilson J L, Lam N T K and Su R K L 2017 A Design Spectrum Model for Flexible Soil Sites in Regions of Low-to-Moderate Seismicity *Soil Dynamics and Earthquake Engineering* **92** pp 36-45

[9] Pappin J W, Free M W, Bird J and Koo R 2004 Evaluation of Site Effects in a Moderate Seismicity Region, Hong Kong. In *Proceedings of the 13th World Conference on Earthquake Engineering* Vancouver, British Columbia, Canada 01–06 August 2004

[10] Bolisetti C, Whittaker A S, Mason H B, Almufti I and Willford M 2014 Equivalent Linear and Nonlinear Site Response Analysis for Design and Risk Assessment of Safety-Related Nuclear Structures *Nuclear Eng. and Dsgn.* **275** pp 107–121

[11] MNA-EC8 2017. NA-2017 to MS EN1998-1, Eurocode 8: Design of Structures for Earthquake Resistance – Part 1: General Rules, Seismic Actions and Rules for Buildings. Malaysia.

[12] Imai T 1982 Correlation of N-value with S-wave Velocity and Shear Modulus. *In Proceedings of the 2nd European Symposium of Penetration Testing* Amsterdam, Netherlands, 24-27 May 1982 Balkema Publishers

[13] Looi D T W, Tsang H H, Hee M C and Lam N T K 2018 Seismic Hazard and Response Spectrum Modelling for Malaysia and Singapore *Earthquakes and Structures* **15** 1 pp 67-79

[14] Wair B R, DeJong J T and Shantz T 2012 *Guidelines for Estimation of Shear Wave Velocity Profiles.* Available through: University of California, Berkeley. Available at:https://peer.berkeley.edu/sites/default/files/webpeer-2012-08-bernardrwairjas.pdf. [Accessed on 10 August 2018].

[15] Kumar S S, Krishna A M and Dey A 2014 Nonlinear Site-Specific Ground Response Analysis: Case Study of Amingaon, Guwahati. In *Proceedings of the 15th Symposium on Earthquake Engineering*, Roorkee, 11–13 December 2014. IIT Roorkee.

[16] PEER (Pacific Earthquake Engineering Research Center) 2019 *PEER ground motion database.* Available at: https://ngawest2.berkeley.edu/spectras/new?sourceDb_flag=1 [Accessed on 15 September 2018]
[17] Seed H B and Idriss I M 1969 The Influence of Soil Conditions on Ground Motions during Earthquakes. *Journal of Soil Mechanics and Foundation Engineering* **94** 1 pp 93–137
[18] Newmark N M 1959 A method of computation for structural dynamics. *Journal of the Engineering Mechanics Division* **3** pp 67-94
[19] Chopra A K 1995 *Dynamic of Structures, Theory and applications to Earthquake Engineering Englewood Cliffs,* New Jersey: Prentice Hall
[20] Housner G W 1952 Spectrum Intensities of Strong-Motion Earthquakes. *In Proceedings of the Symposium on Earthquake and Blast Effects on Structures*, Los Angeles, California, 15 June 1952. Earthquake Engineering Research Institute.

IOP Conf. Series: Materials Science and Engineering 1101 (2021) 012021 doi:10.1088/1757-899X/1101/1/012021

Treatment of septic tank effluent using sequencing batch reactor along with the incorporation of rice husk and Bael pericarp as a natural adsorbent in reducing BOD and COD

Prasanna K[1]*, Annadurai R[1], Godson M D[1], Aravind Murali[1], Induja Ashok[1], Krishnan M V[1]

[1]Department of Civil Engineering, SRM Institute of Science and Technology, Kattankulathur – 603 203, Tamil Nadu, India.
*Corresponding author: prasannk@srmist.edu.in

Abstract. The generation of liquid forms of waste has rapidly shown its hike in the presently expanding world. This quantification hugely includes the day today's secretion of the human excreta into the septic tanks that are commonly found in every household. Proper treatment and disposal of faecal waste, once the septic tank is full, becomes challenging for every resident and local bodies. The primary contaminants found in the Septic Tank Effluents (STE) include many disease-causing pathogens, faecal coliform bacteria, detergents and toxic gases. The Sequencing Batch Reactor (SBR) takes its position when treating any waste, which contains a high amount of organic content. In this study, the characteristics, namely COD and BOD of STE samples collected from different locations of the semi-urban area at the outskirt of a town were analyzed. The STE samples treated using SBR mainly to reduce COD and BOD concentration found to be in higher concentration beyond the permissible limit. The same samples subjected to contact with adsorbents such as rice husk and powdered form of dried Bael (*Aegle marmelos*) outer shell. Results showed a good reaction with the adsorbents and found a significant reduction in COD and BOD concentration.

1. Introduction

The drastic increase in population, industrialization and urbanization is inevitable in developing countries like India. Nowadays, it becomes challenging for every resident and local government to identify and pick out the proper treatment and disposal of faecal waste once the septic tank is filled. The improper monitoring and maintenance of these wastes have led to the contamination of various resources, especially India's water resources mainly in the monsoon season. Most contaminants found in septic tank effluent include pathogens (anything that can produce diseases), faecal coliform bacteria (bacteria from human waste), detergents (including chlorine and bleach) and toxic gases (like methane and hydrogen sulphide). These contaminants have led to the outbreak of severe diseases, including typhoid and cholera. In urban areas, septic tank effluents' disposal has to be done through well-designed sewer lines and treatment plants. However, in many rural areas, the effluents are disposed of without proper treatment in the nearby railway tracks or lakes. Considering this situation of rural areas, simple and effective treatment plants have to be adopted [1][2][3].

SBR is the best idea for this scenario. SBR is a reactor in which specific steps are carried out sequentially through a fill-and-draw process [4] in which the effluent is put on to a single batch reaction tank, treated for the removal of unwanted components and then released [5][6]. It is a modified system of an Activated Sludge Process (ASP) [7]. The main feature that separates SBR from ASP is that in SBR all the process, including equalization, aeration and clarification, with a controlled time sequence in a single tank. The main advantage of using SBR is that the effluent released from it is low in organic compounds (as it effectively reduces the value of COD and BOD); it also meets effluent discharge standards. SBR can also be used as part of a more extensive system when nutrients such as nitrogen and phosphorus have to be removed [8][9].

Moreover, SBR requires a smaller installation area and the operation time is flexible and feasible [10][11]. Since this system has a small footprint, it is relatively easy to expand this system by adding additional reactors [12]. Importance of protecting the environment is increasing day by day. Therefore, there is a need to develop this SBR technology for the effective disposal of these STEs [13]. Among these modern innovations, adsorption would be an efficient and user-friendly technique for removing substrate present in the STEs [14][15]. Various researchers widely use the adsorption process to remove suspended particles and heavy metals present in the wastewater. The need for economical and readily available adsorption material is essential towards the commercially available activated carbon. In SBR, adsorbents can be added quickly and effortlessly [16][17]. In this study, the characteristics, namely COD and BOD of STE samples collected from different semi-urban areas at the outskirts of a Chengalpattu district in Tamil Nadu, India, were analyzed. Ten samples collected from different places within the district to analyze its common behaviours. Then, STE samples treated using SBR mainly to reduce BOD and COD, which are found to be in higher concentration beyond the permissible limit. The same samples then subjected to contact with adsorbents such as rice husk and powdered form of dried Bael (*Aegle marmelos*) outer shell, which are mainly available in these areas.

2. Materials and methods

2.1. Sample Collection

The septic tank effluent samples collected in a 5-litre jar for SBR treatment. The jar cleaned with water and naturally dried in the sunlight. Ten samples each of 20 litres collected and stored in these sun-dried airtight containers. Figure 1 shows the collection of samples from the septic tank. The samples were collected from different Chengalpattu district locations, as shown in Figure 2, whose latitude and longitude are given in Table 1.

Figure 1. Collecting STE samples

IOP Conf. Series: Materials Science and Engineering 1101 (2021) 012021 doi:10.1088/1757-899X/1101/1/012021

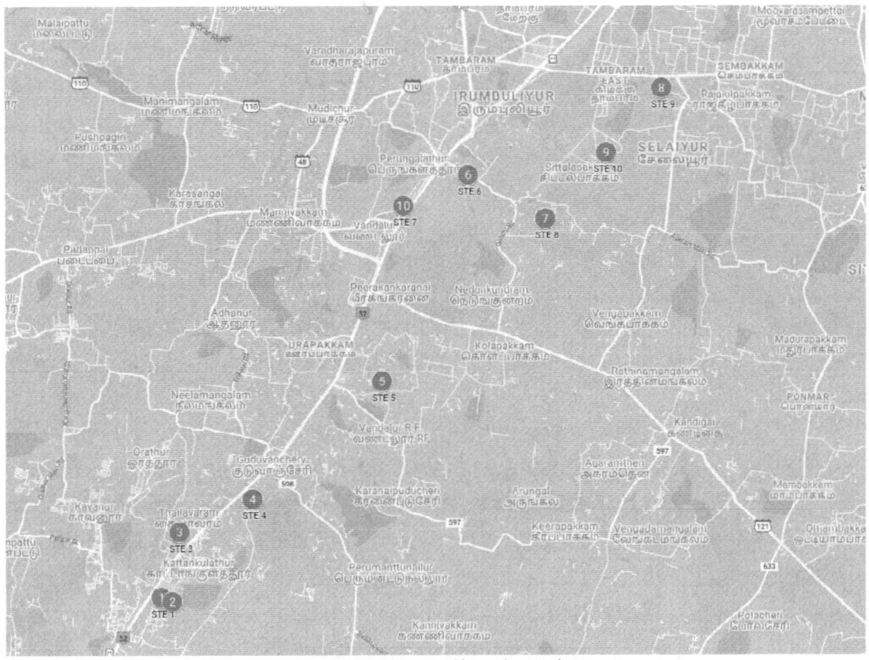

Figure 2. Sampling locations

Table 1. Characteristics of STE samples

Sample designation	Latitude (°)	Longitude (°)
STE 1	12.816	80.037
STE 2	12.816	80.039
STE 3	12.83	80.041
STE 4	12.836	80.056
STE 5	12.86	80.082
STE 6	12.902	80.1
STE 7	12.893	80.11
STE 8	12.92	80.14
STE 9	12.907	80.128
STE 10	12.896	80.086

2.2. Preparation of absorbent
Rice husk

Rice husk is profusely accessible waste material, extensively produced as a by-product from the rice mills and available in excess all around significant rice-producing countries [18][19]. This material's primary content was carbon, about 30 – 60% of the total constituents [20]. The rice husk is collected from the rice mill as a raw form, as shown in Figure 3 (a). These collected rice husks are then sun-dried and ground into fine form, as shown in Figure 3 (b). The powdered form was used as the adsorbent in this study. The subtle form will accelerate the boost up the reaction rate compared to concrete rice husk due to increase in the adsorbing particles, thus increasing the area of contact of the adsorbent with the suspended particles present in the effluent to be treated.

Figure 3. Rice husk (a) Raw form (b) powdered form

Bael (Aegle marmelos)

The fruit of *Aegle marmelos* is mainly available in the selected study areas and is collected from its vicinity and dried under the sun for 4 to 5 days. After drying, the inner pulpy part removed and the outer shell crushed into smaller fragments, as shown in Figure 4 (a). This follows oven drying of those fragments after cleaning with water at 50°C. The crushed outer shells are subjected to grinding to make it a semi-fine powder, as shown in Figure 4 (b), used for the study. The same concept of increasing the area of contact to accelerate the treatment process in the SBR was used, which paved the way to grind the Bael pericarp into powdery form and utilize it.

Figure 4. Bael (a) crushed pieces (b) powdered form

2.3. Experimental setup

The SBR consists of a holding tank in which the collected sample retained reaction tank where all the treatment performed. An outflow pipe attached to the main tank for collecting the treated effluent. Two air diffusers attached to the main tank to induce the aeration process. The mixing is done using a mechanical stirrer. Valves provided in the tanks to adjust the flow of the septic tank effluent. A schematic diagram showing the SBR design and its components shown in Figure 5.

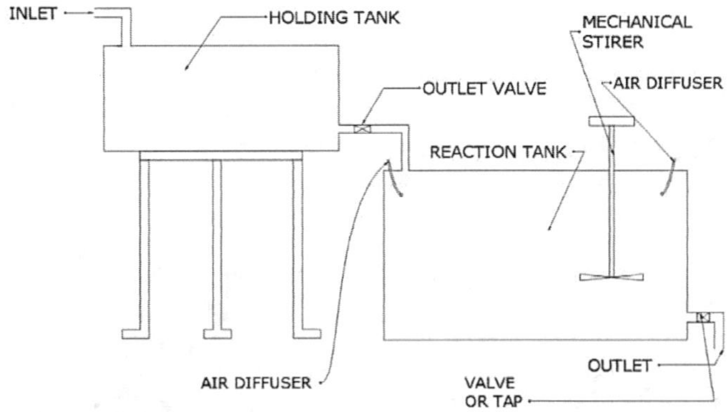

Figure 5. Experimental setup of SBR

The SBR specification and design include:

STE holding tank – specifications

Holding tank has a full capacity of 15.75 L with a dimension of 35 x 15 x 30 cm. It has a working volume of 13 L.

Reaction tank – specifications

The tank's total capacity is 24 L with a working capacity of 20 L having a dimension of 40 x 30 x 20 cm.

Operating conditions

The reactor is operated for 8-hour cycle for about three days (filling - 30 minutes, mixing - 30 minutes, aeration - 120 minutes, settling - 120 minutes, drawing - 30 minutes and idle - 150 minutes).

2.4. SBR treatment process

The treatment of SBR occurs mainly in five different phases performed in order to achieve higher efficiency. The following sections will elaborately discuss each phase.

2.4.1 Phase 1 – Filling and Mixing. The septic tank effluent poured into the effluent holding tank according to its volume. The valve opened and effluent filled into the main vessel for treatment process under gravity. This gravity dropping helps settle the heavier particles in the bottom and leave only the rarer particles in suspension with the water. The filling process is shown in Figure 6, and it follows by mixing process, as shown in Figure 7. The mixing is done mechanically using **a stirring machine**. This phase took about 30 minutes, and the SBR operated in static fill phase.

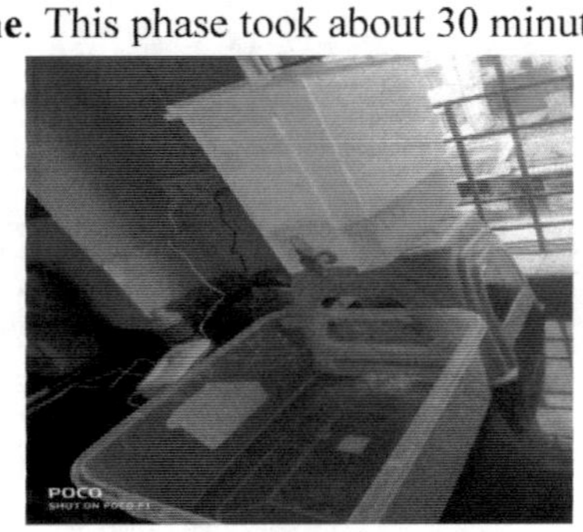

Figure 6. Phase 1 – Fill

Figure 7. Phase 1 – Mixing

2.4.2 Phase 2 – Adding Natural Adsorbent and Aeration. The second phase is the addition of natural adsorbent, followed by aeration. In this phase, crushed rice husk and powdered form of dried Bael (**Aegle marmelos**) add to the effluent and a porous air diffuser immersed into the reactor tank to induce aeration. This aeration and the added natural adsorbent will help in the decomposition of organic matter present inside the reactor. The aeration phase was done for about 150 minutes. The process of aeration, along with the added adsorbent shown in Figure 8.

Figure 8. Phase 2 - Aeration using air diffusers

Figure 9. Phase 3 - Sludge settled in the main tank

2.4.3 Phase 3 – Settling. In this phase, effluent neither added nor removed and SBR kept undisturbed for the microbial reaction and the natural adsorbent present. The microbe will exploit the remaining substrate present in the tank, and when these substrates decrease, the microbe starts to die due to lack of food. The perished microbes along with the added substrate, settle down as sludge inside the reactor. Simultaneously natural adsorbent present in the tank will absorb the remaining substrates and settle as sludge after its potential. This phase took about 120 minutes to complete. The sludge settled in the main tank after this phase is shown in Figure 9.

2.4.4 Phase 4 – Drawing. The settling phase followed by a drawing phase where the treated effluent removed from the reactor. During this phase, the reactor kept undisturbed so that the settled sludge did not mix up.

2.4.5 Phase 5 – Idle. The idle phase was the last stage in the treatment of effluent using SBR. The reactor kept undisturbed for 150 minutes by not withdrawing or adding any effluent in the idle phase. This phase is not mandatory in all SBR operating conditions if the sludge completely separates.

3. Results and discussion

The STE characteristics were analyzed based on various parameters. A total of 25 parameters taken to understand the typical behaviour of STE samples. Characteristics of STE samples are as shown in Table 2.

Table 2. Characteristics of STE samples

Parameter (units)	STE1	STE2	STE3	STE4	STE5	STE6	STE7	STE8	STE9	STE 10
pH	7.28	6.92	7.02	7.51	6.98	7.16	6.88	7.35	7.10	7.24
Turbidity (NTU)	149	258	154	195	165	208	269	165	186	164
Electrical conductivity	3.1	2.4	3.6	2.8	3.1	2.6	2.4	3.0	2.8	2.9

(µs/cm)										
TSS ×10³(mg/L)	10.3	20.8	11.5	16.2	12.03	16.8	21.1	11.9	15.2	11.9
TDS (mg/L)	3.6	2.4	3.54	3.0	3.4	2.9	2.3	3.3	3.0	3.4
Ammonium (mg/L)	109.3	133.9	110.6	124.2	113.2	128.3	135.5	115.7	121.0	114.1
BOD (mg/L)	22	44	24	35	27	40	49	35	29	31
COD (mg/L)	80	224	86	156	113	172	232	105	210	96
Lead (mg/L) LOQ-0.1	-- BLQ --									
Copper (mg/L) LOQ-0.05	BLQ	0.53	BLQ	0.1	BLQ	0.28	0.51	BLQ	BLQ	BLQ
Zinc (mg/L) LOQ-0.01	BLQ	0.06	BLQ	BLQ	BLQ	BLQ	0.07	BLQ	BLQ	BLQ
Nickel (mg/L) LOQ-0.1	BLQ	BLQ	BLQ	BLQ	BLQ	BLQ	BLQ	BLQ	BLQ	BLQ
Boron (mg/L) LOQ-0.1	BLQ	BLQ	BLQ	BLQ	BLQ	BLQ	BLQ	BLQ	BLQ	BLQ
Arsenic (mg/L) LOQ-0.005	BLQ	BLQ	BLQ	BLQ	BLQ	BLQ	BLQ	BLQ	BLQ	BLQ
Mercury (mg/L) LOQ-0.01	BLQ	BLQ	BLQ	BLQ	BLQ	BLQ	BLQ	BLQ	BLQ	BLQ
Cadmium (mg/L) LOQ-0.01	BLQ	BLQ	BLQ	BLQ	BLQ	BLQ	BLQ	BLQ	BLQ	BLQ
Selenium (mg/L) LOQ-0.005	BLQ	BLQ	BLQ	BLQ	BLQ	BLQ	BLQ	BLQ	BLQ	BLQ
Cyanide (mg/L) LOQ-0.01	BLQ	BLQ	BLQ	BLQ	BLQ	BLQ	BLQ	BLQ	BLQ	BLQ
Total phosphorous (mg/L)	10.95	37.8	20.75	31.3	15.59	26.7	38.6	20.6	127	10.29
Potassium (mg/L)	9.0	5.0	2.0	10.2	5.6	4.8	7.8	5.0	9.2	2.8
Total Nitrogen (mg/L)	198.3	206	168.4	192	175.2	216	159	135	178	196.23
Nitrates (mg/L)	22.35	31.2	28.64	29.5	33.26	24.0	16.5	29.7	19.1	23.56
E-coli (Mpn/100ml)	1600	1600	1600	1700	1700	1600	1500	1400	1400	1600
Total coliform (Mpn/100ml)	1600	1600	1500	1500	1700	1600	1400	1400	1400	1600
Bacterial count ×10⁵(Cfu/ml)	2	1.8	2	2	1.8	1.9	2	1.8	1.8	1.8

It was found that some parameters were higher than the prescribed standards as given IS code provisions. For this study, reduction in BOD and COD took for analysis. The concentration of BOD before and after treatment with both the adsorbent shown in Figure 10. The dosage for the addition of adsorbent fixed for both the adsorbent and dosage is 20 g/L.

It is seen that the reduction of the concentration of BOD is more in case of powdered rice husk with a maximum reduction of 50 % than the powdered Bael fruit with a maximum reduction of 48%. This may be due to the higher amount of Silicon present in the rice husk with the content being approximately 96.34%, which act as a medium in accelerating the rate of adsorption, based on the XRD analysis done by various researchers [21] [22] [23] [24]. The concentration of COD before and after treatment with both the adsorbent shown in Figure 11.

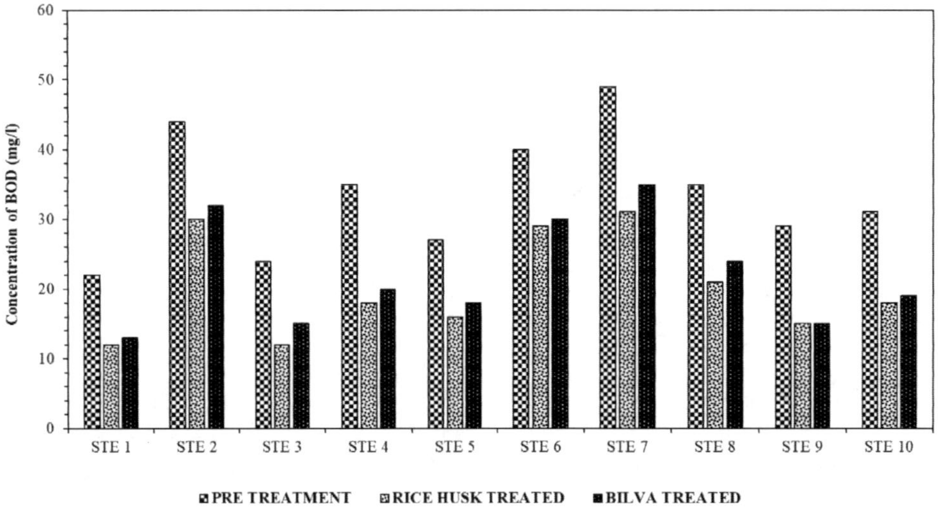

Figure 10. Graphical representation of the concentration of BOD

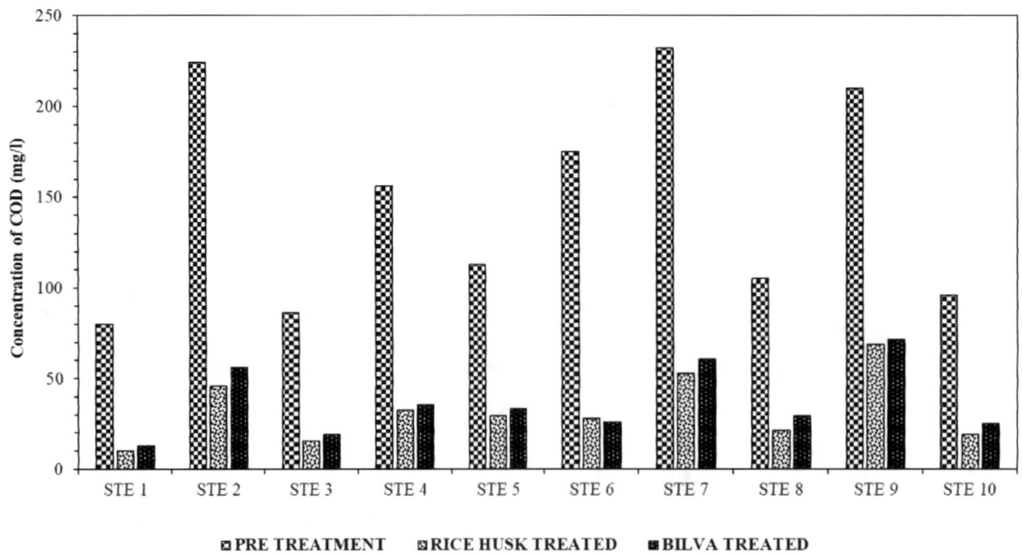

Figure 11. Graphical representation of the concentration of COD

From Figure 11, the same pattern observed as that of Figure 10. SBR effectively reduces COD concentration when subjected to rice husk with a maximum reduction in the percentage of 87.25%

than that of powder outer shell of Bael fruit, which has a maximum reduction percentage of 84%. One reason for reducing COD and BOD may be due to the accelerated aeration process given to the effluent through SBR [25], and the other may be due to the Oxygen content available in the adsorbents. Many studies based on XRF analysis reveal that the amount of oxygen present in the rice husk and Bael are about 25.982% and 20% respectively [26] [27], which diminishes both the oxygen demands that are found in the

4. Conclusion

The septic tank effluent has been collected from different places of Chengalpattu district, Tamil Nadu and its essential characteristics were analyzed. From the analysis of the septic tank effluent, it was found that the samples have a high content of COD, BOD, and bacterial Coliforms. From the observation, metals like lead, zinc, nickel, copper, boron, arsenic, mercury, cadmium, selenium and cyanide are below the limit of qualifications as per IS 3025. Hence, the paper's main aim was curated to reduce COD and the BOD contents from the septic tank waste. The treatment was carried out with the help of SBR. The natural adsorbents were added to the SBR after grinding it into fine form. The adsorbent which included rice husk powder and powdered outer shell of Bael fruit, are added with a dosage of 20 mg/L. The effluent has been subjected to a physical and chemical process to reduce BOD and COD concentration, high in concentration. The results revealed that COD and BOD's removal efficiency was between 73 -87% and 27 - 50% respectively in rice husk and 65 – 87% and 25 – 50% when subjected with powdered Bael shell as the adsorbent. It was observed that SBR was adequate for the treatment of septic tank effluent for the reduction of COD as well as BOD with the help of added adsorbent and proving rice husk to be slightly effective than the powdered outer shell of Bael fruit in reducing the different oxygen demands.

5. References

[1] MeghaGidde, Raveena Chavan, Yash Chaudhari, Sudhir Ghule, Datta Chate,(2018), "Sequential Batch Reactor" *GRD Journals - Global Research and Development Journal for Engineering,* **3**, *Issue 4.*

[2] Janani, T., Prasanna, K., & Annadurai, R. (2019, June). Impact of solid waste dumping in Pallavaram Dumping site, India. In *AIP Conference Proceedings* (Vol. **2112**, No. 1, p. 020145). AIP Publishing LLC.

[3] Le, H. T., Jantarat, N., Khanitchaidecha, W., Ratananikom, K., & Nakaruk, A. (2015). Development of sequencing batch reactor performance for nitrogen wastewater treatment. *Microbial and Biochemical Technology*, **7**, 363-366.

[4] Pannirselvam, R., Ibrahim, Y., & Gopalakrishnan, A. N. (2015). Simultaneous carbon and nutrient removal from dairy wastewater in sequencing batch reactor (SBR). *Global NEST Journal,* **17**(3), 628-636.

[5] K Lim, C. K., Seow, T. W., Neoh, C. H., Nor, M. H. M., Ibrahim, Z., Ware, I., & Sarip, S. H. M. (2016). Treatment of landfill leachate using ASBR combined with zeolite adsorption technology. 3 *Biotech*, **6**(2), 195.

[6] Dutta, A., & Sarkar, S. (2015). Sequencing batch reactor for wastewater treatment: recent advances. *Current Pollution Reports*, **1**(3), 177-190.

[7] Dohare, D., & Kawale, M. (2014). Biological Treatment of Wastewater Using Activated Sludge Process and Sequential Batch Reactor Process-A Review. *Int J Eng Sci Res Technol*, **3**(11), 728-736.

[8] Abu Bakar, A., Daud, Z., Ahmad, Z., Othman, M., & Rashid, A. (2014). Treatment of leachate using sequencing batch reactor (SBR).

[9] Patil, P. G., Kulkarni, G. S., Kore, S. S., & Kore, V. S. (2013). Aerobic sequencing batch reactor for wastewater treatment: a review. *Int J Eng Res Technol*, **2**(10), 534-550.

[10] ZA, N. F., Norli, I., Norfariha, S., Aisyah, S., & Renuka, R. (2014). Ammoniacal Nitrogen Removal in Semiconductor Wastewater by Sequence Batch Reactor Using Bacteria Inoculum from WormTea. *Journal of Medical and Bioengineering* Vol, **3**(4).

[11] Jungles, M. K., Campos, J. L., & Costa, R. H. R. (2014). Sequencing batch reactor operation for treating wastewater with aerobic granular sludge. *Brazilian Journal of Chemical Engineering*, **31**(1), 27-33.

[12] Mane, S. S., & Munavalli, G. R. (2012). Sequential batch reactor-Application to wastewater–A Review. *In Proceeding of International Conference* SWRDM (pp. 121-128).

[13] Mahvi, A. H. (2008). Sequencing batch reactor: a promising technology in wastewater treatment.

[14] Spagni, A., Marsili-Libelli, S., & Lavagnolo, M. C. (2008). Optimization of sanitary landfill leachate treatment in a sequencing batch reactor. *Water Science and Technology*, **58**(2), 337-343.

[15] Sarti, A., Garcia, M. L., Zaiat, M., & Foresti, E. (2007). Domestic sewage treatment in a pilot-scale anaerobic sequencing batch biofilm reactor (ASBBR). *Resources, conservation and recycling*, **51**(1), 237-247.

[16] Daifullah, A. A. M., Girgis, B. S., & Gad, H. M. H. (2003). Utilization of agro-residues (rice husk) in small waste water treatment plans. *Materials Letters*, **57**(11), 1723-1731.

[17] Farrokhi, M., Ashrafi, D., Roohbakhsh, E., & Yoonesi, A. (2014). Hospital wastewater treatment by integrated fixed film activated sludge, using rice husk as fixed media. *Advances in Life Sciences*, **4**(3), 178-183.

[18] Nhapi, I., Banadda, N., Murenzi, R., Sekomo, C. B., & Wali, U. G. (2011). Removal of heavy metals from industrial wastewater using rice husks. *The Open Environmental Engineering Journal*, **4**(1).

[19] Masoud, M. S., El-Saraf, W. M., Abdel-Halim, A. M., Ali, A. E., Mohamed, E. A., & Hasan, H. M. (2016). Rice husk and activated carbon for waste water treatment of El-Mex Bay, Alexandria Coast, Egypt. *Arabian Journal of Chemistry*, **9**, S1590-S1596.

[20] Shalaby, N. H., Ewais, E. M., Elsaadany, R. M., & Ahmed, A. (2017). Rice husk templated water treatment sludge as lowcost dye and metal adsorbent. *Egyptian journal of petroleum*, **26**(3), 661-668.

[21] Habeeb, G. A., & Mahmud, H. B. (2010). Study on properties of rice husk ash and its use as cement replacement material. Materials research, **13**(2), 185-190.

[22] Mansaray, K. G., & Ghaly, A. E. (1997). Physical and thermochemical properties of rice husk. Energy sources, **19**(9), 989-1004.

[23] Ahmaruzzaman, M., & Gupta, V. K. (2011). Rice husk and its ash as low-cost adsorbents in water and wastewater treatment. Industrial & Engineering Chemistry Research, **50**(24), 13589-13613.

[24] Yalcin, N., & Sevinc, V. (2001). Studies on silica obtained from rice husk. Ceramics international, **27**(2), 219-224.

[25] K Prasanna, R Annadurai, S Vimal Kumar, M Aravind (2018). Removal of COD and Ammoniacal Nitrogen Present in Leachate Using Sequencing Batch Reactor, Jour of Adv Research in Dynamical & Control Systems, Vol. **10**, 08-Special Issue.

[26] Kaur, A., & Kalia, M. (2017). Physico chemical analysis of bael (Aegle Marmelos) fruit pulp, seed and pericarp. Chemical Science Review and Letters, **6**(22), 1213-1218.

[27] Charoensiddhi, S., & Anprung, P. (2010). Characterization of bael fruit (Aegle marmelos [L.] correa) hydrolysate as affected by enzyme treatment. Journal of food biochemistry, **34**(6), 1249-1267.

IOP Conf. Series: Materials Science and Engineering 1101 (2021) 012022 doi:10.1088/1757-899X/1101/1/012022

Effect of aggregate-binder proportion and curing technique on the strength and water absorption of fly ash-based one-part geopolymer mortars.

S Haruna[1, 2, *], B S Mohammed[1], M M A Wahab[1], A Al-Fakih[1]

[1] Department of Civil Engineering, Universiti Teknologi PETRONAS, Malaysia.
[2] Department of Civil Engineering, Bayero University Kano, Nigeria.

Abstract. This paper This paper investigated the effects of binder-aggregates proportion on the performance of one-part geopolymer mortars. High calcium fly ash together with sodium metasilicate have been utilized as the binder, the powdered sodium metasilicate activator was kept at 12% by weight of the fly ash. Three types of mortars were produced with a different binder to fine aggregates proportions (B: A) of 1: 0.5, 1:1, 1:2. The strength properties of the one-part geopolymer mortars (OPGM) which comprises compressive, flexural, splitting tensile strength and water absorption have been investigated. At 28 days of outdoor curing, the OPGM exhibited compressive strength of 50 MPa and 43 MPa at ambient curing. The flexural strength of the OPGM represents 16 – 17 % of its compressive strength. Regardless of the curing techniques, the strength properties of the OPGM is almost the same. The optimum OPGM was found to be at 1: 0.5 binder-aggregates proportions at outdoor curing.

1. Introduction

Over the past three decades geopolymer has been developed and its utilizations in civil infrastructures begin to gain reputation by replacing cement materials with the greener material [1, 2]. Geopolymer is regarded as an encouraging sustainable and environmentally favorable material substitute to Ordinary Portland cement (OPC) in building applications. The polymerization system of geopolymer constitute of an intensely accelerated chemical reaction under alkaline medium on Si-Al minerals that yields a 3-dimensional polymeric sequence as well as a ring framework composed of Si-O-Al-O bonds [1, 3, 4].

Geopolymer has tremendous ability to reduce CO_2 emissions and the use of natural resources linked to conventional concrete OPC production. In geopolymers, the reaction mechanism is distinct from a well-known OPC hydration reaction. The source materials used to produce geopolymers are triggered in an alkaline condition and the major components of the geopolymers dissolve first in the alkaline environment and then undergo speciation, gelation, reorganization and polymerization till they form a binder with cementing characteristics [5]. Despite this potentiality exhibited by geopolymer, it is applications has been restricted to precast construction due to the need of curing at elevated temperature and low strength development when cured at ambient temperature [6-8]. Moreover, dealing with large volumes of highly corrosive and viscous alkaline solutions may be awkward and time consuming for bulk production of geopolymer concrete, and therefore limits its large-scale applications [9]. However, alkaline solution-activated geopolymers do not appear to be a viable substitute for OPC in the construction industry, despite its enormous environmental benefits and excellent engineering properties owing to its complexity in mixing and handling. In order to address these problems, dry mixture is needed in such a way that only water will be added to the materials similar to that of OPC binders [10, 11]. Therefore, one-part geopolymers can be well suited for both in-situ and precast applications.

Ke et al. [12], have produced one-part geopolymers using thermally treated mud as the source materials. The red mud was calcined with NaOH pellets at 800 °C to promote the development of fresh hydraulic compounds, along with a partially ordered aluminosilicate structure C_3A and C_2S calcium-rich phases. They have reported a compressive strength of 7.5 – 10 MPa at 7 days curing. However, the strength drastically deteriorates after 1 week, at 4 weeks, the compressive strength was reduced to less than 2 MPa. Ye et al. [13], have reported increase in compressive strength of OPG binder using Bayer red mud blended with 20 – 30 % of silica fume (SF). Their reason for the increased in strength with the addition of SF was due enhancement in Si/Al ratio which result to a more stable structure particularly at lower w/b ratio. They conclude that, their developed one-part geopolymer could be suitable for construction purposes, it could potentially replace ordinary clay bricks for interior wall or back bricks and other building materials that are non-load bearing. Sturm et al. [14], have produced one-part geopolymer from rice husk ash (RHA). They obtained a compressive strength of 30 MPa after 1 day of 80 °C oven curing. Hajimohammadi et al. [6], have also produced one-part geopolymer binder using low calcium fly ash. They have achieved high compressive strength of 65 MPa after 3 weeks of curing at 40 °C. However, the requirements for oven curing make it unsuitable for cast-in-situ applications.

Panda et al. [15], have reported increase in strength of fly ash/ slag based OPG activated with solid potassium silicate for 3D concrete printing. The developed binder exhibited orthotropic mechanical properties. The developed one-part geopolymer can be used for load bearing applications. Due to the layer-wise manufacturing strategy used in concrete printing, the printed geopolymers showed anisotropic mechanical efficiency relative to the mould cast samples. The findings is in agreement with that of [16]. Because of the advantages of one-part geopolymer binders and GGBS as a calcium-rich aluminosilicate material, numerous attempts have been made to produce one-part alkali-activated slag binders [17, 18]. GGBS based OPG binders however have some weaknesses, which include volumetric instability and high shrinkage compared to OPC based binders. The high shrinkage of GGBS based OPG contributes to the development of cracks on its surface which negatively influence its long-term performance [19]. Based on the extensive literature search, there is limited studies on the use of C-class fly ash in the production of one-part/ in-situ geopolymer mortars. In view with that the paper is aimed to determine the influence of binder-aggregate proportion and curing methods on the strength and water absorption behaviours of the one-part geopolymer mortars.

2. Experimental methods

2.1 Materials

Class C fly ash conforming with requirement of ASTM 618-10 has been utilized as the source material in this investigation. The oxide content of fly ash was obtained by X-Ray fluorescence (XRF) and presented in Table 1. Powdered anhydrous sodium metasilicate (50 % of Na_2O, 46% of SiO_2 and 4 % of H_2O) has been utilized as the activator. The granular activator has been used at 12% by mass of the fly ash according to previous work [3, 20]. In this investigation, natural river sand was utilized as the fine aggregate.

Table 1. Oxide Composition of the source materials

Oxide	(%)
CaO	17.1
Al_2O_3	14.9

SiO_2	37.3
Fe_2O_3	16.5
MgO	3.72
P_2O_5	2.59
SO_3	2.56
Na_2O	1.74
K_2O	1.66
TiO_2	1.07
Others	0.69
LOI	0.17

2.2 Preparation of one-part geopolymer mortar mixtures

Three types of one-part geopolymer mortars (OPGM) have been produced with a different binder to aggregates proportions (B: A) of 1: 0.5, 1:1, and 1:2 respectively. The preparation of one-part geopolymer mortars (OPGM) involve blending of granular sodium metasilicate and fly ash with the aid of a mechanical Hobart mixer for about 3 minutes to obtain a uniform dry geopolymer cement. River sand was then added and mix again for about 3 minutes, water was then poured to the dry mixture and continue mixing for additional 3 minutes until a consistent mixture of OPGM was formed. The fresh mixture of the OPGM mixture has been placed into the moulds of 50 mm cube and compacted using vibrating table. The freshly formed OPGM were then removed from the moulds after 24 hours and then subjected to two curing methods, namely, ambient and outdoor curing till the testing periods of 3, 7, 28 and 90 days respectively. The mixture proportion of the OPGM was shown in Table 2.

Table 2. Experimental design parameters

Mixture Identification	Fly ash Kg/m^3	Dry Na_2SiO_3 activator (%)	Fine aggregate (Kg/m^3)	w/b ratio	B: A
OPGM1	1235.43	0.12	617.71	0.25	1:0.5
OPGM2	960.89	0.12	960.89	0.25	1:1
OPGM3	665.23	0.12	1330.46	0.25	1:2

3. Result and discussion

3.1 Compressive strength of OPGM

The influence of binder aggregate proportion on the compressive strength of the OPGM is illustrated in Figure 1. As expected, the compressive strength of the OPGM increases with age of curing. Figure 1(a) illustrates the compressive strength development of laboratory cured OPGM. At early age of curing, the strength of OPGM1 and OPGM2 is almost the same while that of OPGM3 increases by 23.3% and 13.4% at 3 and 7 days respectively. At 28 and 90 days, the compressive strength of OPGM1 significantly increases compared to OPGM2 and OPGM3. The improvement in the strength is attributed to the continued pozzolanic reaction of the binder at later ages and enhancement of bonding between the geopolymer paste and the river sand particles. Figure 1(b) illustrate the compressive strength growth of outdoor cured OPGM. It is interesting to note that there is no significance difference between the ambient and outdoor cured OPGMs particularly at early age. All the developed OPGMs specimens achieved more than 20 MPa at 7 days and almost 50 MPa was achieved at 28 days of outdoor curing. At 28 days of outdoor curing, the compressive strength of OPGM1 (1: 0.5) was significantly higher than that of OPGM2 and OPGM3 by 24.1 % and 26.8 % respectively. Moreover, it should be noted that the compressive strength of the developed OPGM demonstrated continuous strength development over a prolonged curing period. This finding is consistent with the reported literatures [3, 21, 22]. It is interested to note that at 90 days of curing, there is insignificant difference between the two curing regimes considered.

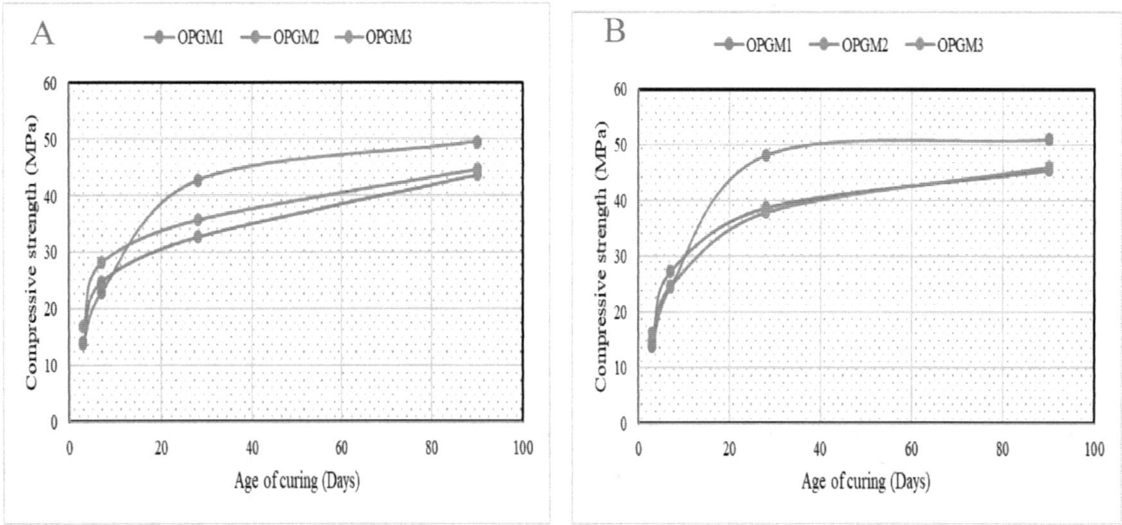

Figure 1. Compressive strength of one-part geopolymer mortar (a) ambient cured (b) outdoor cured.

3.2 Splitting tensile strength of OPGM

Figure 2 represent the effect of binder aggregate proportion on the splitting tensile strength of the OPGM. As expected, the tensile strength of the developed OPGM follows similar pattern with the compressive strength that is it increased with age of curing. Figure 2(b) illustrated the splitting tensile strength of ambient cured OPGM. As shown in Figure 2(b), the split tensile strength of almost 3 MPa and 3.25 MPa was obtained at 28 and 90 days for OPGM2 sample. For OPGM1 and OPGM3, the splitting tensile strength of 2.45 MPa and 2.72 MPa was realized at 28 days of ambient curing. Similarly, at 90 days, the tensile strength of 3.22 and 2.86 MPa was achieved. The tensile strength of the OPGM2 samples outperformed their OPGM1 and OPGM3 counterpart. Moreover, Figure 2(b) shows the strength growth of the outdoor cured OPGM. It is worth to mention that there are no significant discrepancies between the ambient (indoor) and outdoor curing and the strength growth follows the same trend.

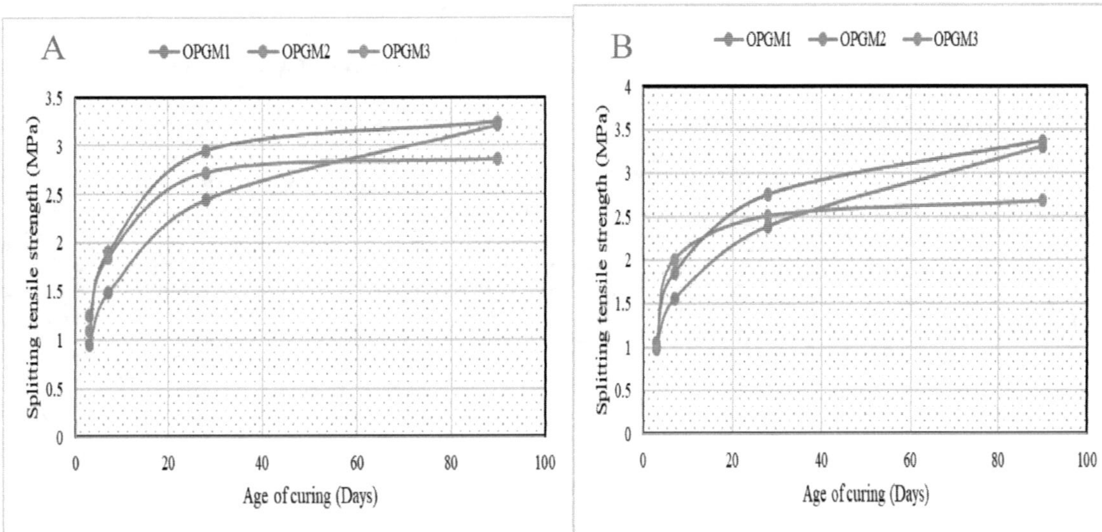

Figure 2. Splitting tensile strength of one-part geopolymer mortar (a) ambient cured (b) outdoor cured

3.3 Flexural strength

The flexural strength of the OPGM samples could also be utilized as its tensile strength. Nevertheless, the flexural strength usually indicates a higher strength result than the split tensile strength. As presented in Figure 3, the flexural strength of the OPGM follows similar trend with that of the compressive strength. Figure 3(a) and Figure 3(b) shows the flexural strength growth of ambient and outdoor cured OPGM. The highest flexural strength of 5.83 MPa and 6.11 MPa was achieved at 28 days for ambient and outdoor curing methods respectively. At 90 days, the strength continues to increase gradually. The flexural strength of the developed OPGM represents 16 – 17 % of its compressive strength values. These values are high compare to the OPC based mortars. It is noted that the flexural strength decreased with the increased in the fine aggregate proportion in the OPGM. The strength enhancement at lower aggregate content could be attributed to the sustained pozzolanic reaction of fly ash at a later age by generating more calcium alumina silicate hydrate (C-A-S-H), thus enhancing its strength properties [23]. Regardless of the curing methods, the flexural strength of the OPGM is almost the same, even though the outdoor cured OPGM exhibited slightly higher flexural strength than the ambient cured one-part geopolymer mortar.

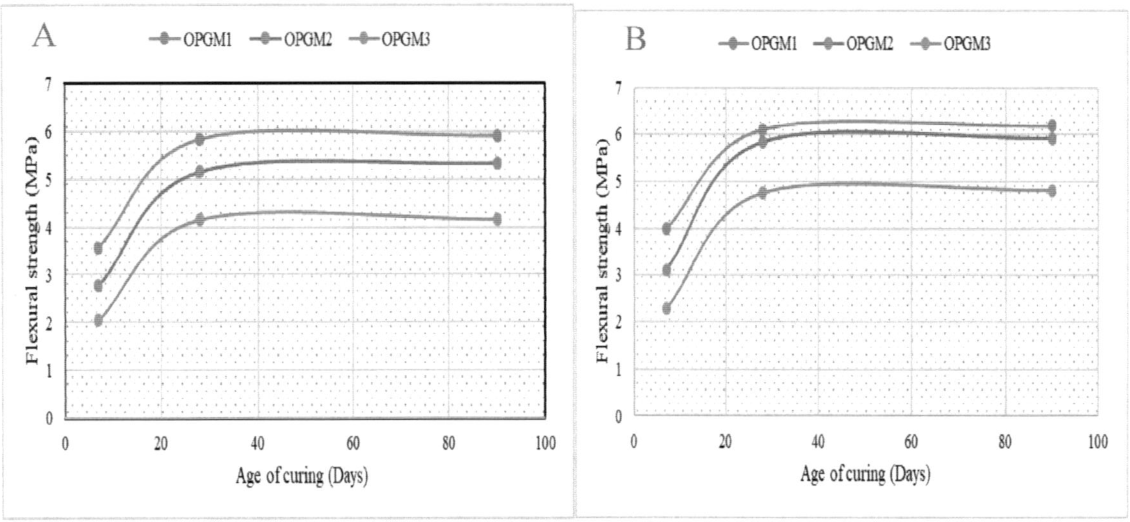

Figure 3. Flexural strength of one-part geopolymer mortar (a) ambient cured (b) outdoor cured.

3.4 Water absorption of OPGM

The water absorption behaviour of OPGM was presented in Figure 3. The water absorption values of OPGM obtained in this investigation were obtained in the range of 6.19 – 3.94 % at 28 and 90 days of ambient curing. It has been revealed from Figure 3 that the water absorption decreased with increased in the aggregate content and vice versa. This explained that as the aggregate content in the mixes increases the water absorption decreased. Regardless of the curing condition, the water absorption of the OPGM is almost the same. The high-water absorption could be attributed to the high binder content within the OPGM mixtures [23]. At higher paste content, there exist a multiple number of micro-cracks on the surface of the concrete which is associated to the drying shrinkage mechanism of the OPGC. The micro-cracks tend to absorb extra water resulting in the increase of water absorption of the OPGC at higher paste volume.

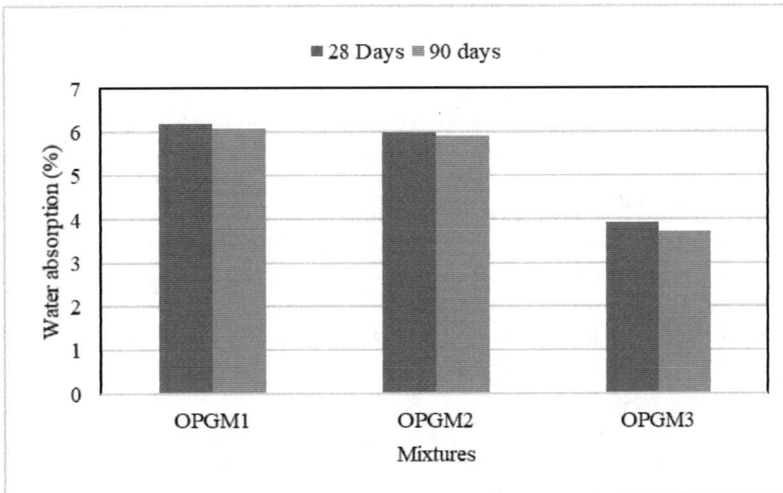

Figure 4. water absorption of OPGM

4. Conclusion

The study reported experimental investigation on the influence of binder-aggregate proportion and curing methods on the strengths and water absorption behaviours of the one-part geopolymer mortars. Based on the experimental work performed in this investigation, the following findings can be outlined:

1. The compressive strength of the one-part geopolymer mortars decreases with the increase in the aggregate proportion similar to OPC-based mortars. At early age of curing, there is no significance difference in strength between the ambient and outdoor cured OPGM.
2. The flexural strength of the developed OPGM are very promising in comparison with the OPC based mortars. The obtained flexural strength of the OPGM represents 16 – 17 percent of their compressive strength values. The outdoor cured samples exhibited higher strength than the ambient cured samples.
3. The water absorption of the developed OPGM1 and OPGM2 is almost the same while that of OPGM3 decreases significantly by 34.3% and 37.1% respectively.
4. The optimum OPGM was found to be at 1: 0.5 binder-aggregates proportions at outdoor curing.

Acknowledgement

The authors would like to thank Universiti Teknologi PETRONAS (UTP) and Yayasan UTP (YUTP 015LC0-097) for the support provided for this research.

References

[1] Davidovits, J., *Geopolymer chemistry and applications*. 2008: Geopolymer Institute.

[2] Duxson, P., Fernández-Jiménez, Ana, Provis, John L, Lukey, Grant C, Palomo, Angel, van Deventer, Jannie SJ, *Geopolymer technology: the current state of the art.* Journal of materials science, 2007. **42**(9): p. 2917-2933.

[3] Mohammed, B.S., Haruna, Sani, Wahab, M. M. A. Liew, M. S. Haruna, Abdulrahman, *Mechanical and microstructural properties of high calcium fly ash one-part geopolymer cement made with granular activator.* Heliyon, 2019. **5**(9): p. e02255.

[4] Haruna Sani, M.B.S., Liew, M.S, Alaloul Wessam Salah, Haruna Abdulrahman, *Effect of water-binder ratio and NAOH molarity on the properties of high calcium fly ash geopolymer*

mortars at outdoor curing. International Journal of Civil Engineering & Technology, 2018. **9**(10): p. 1339-1352.

[5] Hajimohammadi, A., Ngo, Tuan, Kashani, Alireza, *Sustainable one-part geopolymer foams with glass fines versus sand as aggregates.* Construction and Building Materials, 2018. **171**: p. 223-231.

[6] Hajimohammadi, A., van Deventer, Jannie SJ *Characterisation of one-part geopolymer binders made from fly ash.* Waste biomass valorization, 2017. **8**(1): p. 225-233.

[7] Neupane, K., *"Fly ash and GGBFS based powder-activated geopolymer binders: A viable sustainable alternative of portland cement in concrete industry".* Mechanics of Materials, 2016. **103**: p. 110-122.

[8] Nurruddin, M.F., Haruna, Sani, Mohammed, Bashar S, Sha'aban, Ibrahim Galal, *Methods of curing geopolymer concrete: A review.* International Journal of Advanced and Applied Sciences 2018. **5**(1): p. 31-36.

[9] Tennakoon, C.K., *Assessment of properties of ambient cured geopolymer concrete for construction applications.* 2015, Ph. D. Thesis), Swinburne University of Technology Melbourne, Australia.

[10] Duxson, P. and J.L. Provis, *Designing precursors for geopolymer cements.* Journal of the American Ceramic Society, 2008. **91**(12): p. 3864-3869.

[11] Nematollahi, B., J. Sanjayan, and F.U.A. Shaikh, *Synthesis of heat and ambient cured one-part geopolymer mixes with different grades of sodium silicate.* Ceramics International, 2015. **41**(4): p. 5696-5704.

[12] Ke, X., Bernal, Susan A, Ye, Nan, Provis, John L, Yang, Jiakuan *One-Part Geopolymers Based on Thermally Treated Red Mud/NaOH Blends.* Journal of the American Ceramic Society, 2015. **98**(1): p. 5-11.

[13] Ye, N., Yang, Jiakuan, Liang, Sha, Hu, Yong, Hu, Jingping, Xiao, Bo, Huang, Qifei, *Synthesis and strength optimization of one-part geopolymer based on red mud.* Construction and Building Materials, 2016. **111**: p. 317-325.

[14] Sturm, P.G., G. J. G, Brouwers, H. J. H, Kühne, H. C., *Synthesizing one-part geopolymers from rice husk ash.* Construction and Building Materials, 2016. **124**: p. 961-966.

[15] Panda, B., Singh, G. V. P. Bhagath, Unluer, Cise, Tan, Ming Jen, *Synthesis and characterization of one-part geopolymers for extrusion based 3D concrete printing.* Journal of Cleaner Production, 2019. **220**: p. 610-619.

[16] Nematollahi, B., Xia, Ming, Bong, Shin Hau, Sanjayan, Jay. *Hardened Properties of 3D Printable 'One-Part' Geopolymer for Construction Applications.* in *First RILEM International Conference on Concrete and Digital Fabrication – Digital Concrete 2018.* 2019. Cham: Springer International Publishing.

[17] Hajimohammadi, A., Ngo, Tuan, Mendis, Priyan, Kashani, Alireza, van Deventer, Jannie S. J., *Alkali activated slag foams: The effect of the alkali reaction on foam characteristics.* Journal of Cleaner Production, 2017. **147**: p. 330-339.

[18] Haruna, S., Mohammed, Bashar S, Wahab, MMA, *Effect of GGBS Slag on Setting Time and Compressive Strength of One-Part Geopolymer Binders.* Journal of Infrastructure & Facility Asset Management, 2020. **2**(2).

[19] Abdollahnejad, Z., Mastali, M., Luukkonen, T., Kinnunen, P., Illikainen, M., *Fiber-reinforced one-part alkali-activated slag/ceramic binders.* Ceramics International, 2018. **44**(8): p. 8963-8976.

[20] Mohammed, B.S., Haruna, Sani, Mubarak bn Abdul Wahab, Mohamed Liew, M. S. , *Optimization and characterization of cast in-situ alkali-activated pastes by response surface methodology.* Construction and Building Materials, 2019. **225**: p. 776-787.

[21] Ban, C.C., Ken, Part Wei, Ramli, Mahyuddin, *Effect of sodium silicate and curing regime on properties of load bearing geopolymer mortar block.* Journal of Materials in Civil Engineering, 2017. **29**(3): p. 04016237.

[22] Sani Haruna, B.S.M., Mohamed Mubarak bn Abdul Wahab, Abdulrahman Haruna *Compressive strength and workability of High Calcium One-Part alkali activated mortars using response surface methodology* IOP Conference Series: Earth and Environmental Science 2nd International Conference on Civil and Environmental Engineering, Langkawi, Kedah, Malaysia. , 2020. **476**.
[23] Haruna, S., Mohammed, Bashar S., Wahab, M. M. A., Liew, M. S., *Effect of paste aggregate ratio and curing methods on the performance of one-part alkali-activated concrete.* Construction and Building Materials, 2020. **261**: p. 120024.

Generating Landslide Susceptibility Map using Airborne Lidar Derived Parameters and Geological Mapping Factors for Canada Hill, Miri, Sarawak

E Muol[1]* and G T Noweg[2]

[1] Jabatan Mineral dan Geosains Malaysia Sarawak, Jalan Wan Abdul Rahman Kenyalang Park, 93658 Kuching, Sarawak, Malaysia
[2]Institute of Biodiversity and Environmental Conservation,Universiti Malaysia Sarawak, 94300 Kota Samarahan, Sarawak, Malaysi
Email: edwardm@jmg.gov.my

Abstract. Landslide Susceptibility Mapping using GIS software and remote sensing data have been conducted in several location involving geological and geomorphological sensitive at Canada Hill, Miri. The previous researcher has conducted quantitative analyses using different statistical methods with different parameters in the same study area. The mapping of landslides using high-resolution Airborne LiDAR data is a valuable effort. All of this play important role, in the analysis and development of landslide susceptibility map. High-resolution Airborne LiDAR data has the ability to penetrate thick forest cover and produce Digital Terrain Model. Using Digital Terrain Model, the landslide parameter can be generated and extracted. The main objective of this study was to produce landslide susceptibility map using the Probability Frequency Ratio Model method. This study involved the delineating of causative factors from Digital Terrain Model generated by Airborne LiDAR data as well as the data collected from the field. Apart from parameters derived from LiDAR, parameters from filed and site investigation were included into the mapping process. This study was different from the previous studies in the same area in terms of various analytical approaches and samples used. The results of the landslide susceptibility map were verified via randomly selected landslides samples using two different methods. The landslide susceptibility map produced is more refined and is able to predict more effectively compared to the existing map. The landslide susceptibility map produced in this study could be used for land use planning and management by decision makers and land use planners.
Keywords: LiDAR, landslide susceptibility map, digital terrain model, probability frequency ratio, geology

1. Introduction

Sarawak has recorded numerous tragedies due to this geological hazard. Since the occurrence of landslides in Punda Ruan, Simunjan which killed 16 longhouse inhabitants, the Sarawak government and Department of Mineral and Geosciences Malaysia has seriously monitored the areas with risk of landslide. Canada Hill, which stretches 8 km striking northeast-southwest and formed the backbone of Miri City is a prominent landslide risk area. In the span of 5 years since 2009, a total of four (4) was reported in Miri City with two (2) fatalities and damages to infrastructures [1,2,3].

IOP Conf. Series: Materials Science and Engineering 1101 (2021) 012023 doi:10.1088/1757-899X/1101/1/012023

The slopes, anthropogenic activities, vegetation cover, climate, its geomorphological and geological features play important roles in triggering the process. Locating the distribution of these features are substantial as predicting and planning of any development is crucial.

2. Location and Accessibility

The study area is located in the middle of Miri City. It is bounded by Borneo Rectified Skew Orthomorphic (BRSO) 475000mE to 481000mE and 481000mN to 489000mN. The overall accessibility to the study area is excellent. The area is connected by the flight from majority of international airport in Malaysia, major town in Sarawak and rural air service (Figure 1). Miri City is well served by Pan Borneo Highway.

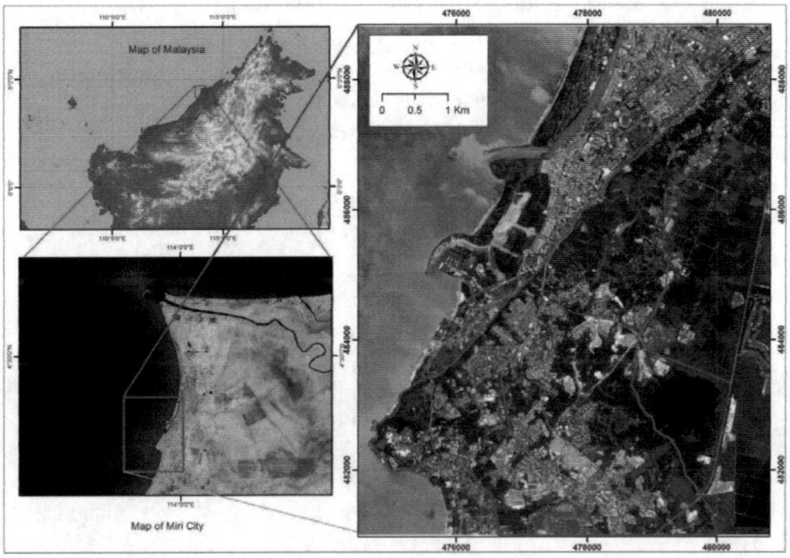

Figure 1. The location map of Canada Hill, Miri, Sarawak.

3. Material and Method

In this investigation, a probabilistic approach was carried out to reveal the correlation between several landslide causative factors. Frequency ratio method (Fr) as referred to equation (1) was applied to see the spatial relationship between the landslide locations and each landslide related factors. In the relation analysis, the ratio is that of the area where landslides occurred to the total area.

The formula for the calculation of frequency ratio is as follows;

$$Fri = \frac{\left(\frac{Npix(Si)}{Npix(Ni)}\right)}{\left(\sum Npix(Si)/\sum pix(Ni)\right)} \quad (1)$$

Rating of each factor's type or range *Fri*
Where,
Npix(Si) = The number of pixels containing slides in class (i)
Npix(Ni) = Total number of pixels having a class (i) in the whole study area
$\sum$ *Npix(Si)* = Total number of pixels containing landslides and
$\sum$*pix(Ni)* = Total number of pixels in the whole study area

By taking 1 as an average value, if the value of Fri is greater than 1 it means a higher correlation, while the value of Fri lower than 1 means a lower correlation. Factor ratings were summed up to yield Landslide Susceptibility Index (LSI) according to equation (2).

$$LSI = \sum Fr(i\ldots m) \tag{2}$$

Where FR is the rating of each factor's type or range.

3.1. Data Acquisition

The first step for the production of a Landslide Susceptibility Map (LSM) is to gather the available data in an effort of the selection of the effective parameters in preparation of thematic digital maps. All such data, maps used in this study are raster based on a cell size of 0.5 m x 0.5 m. An important step for the production of a landslide susceptibility map is the construction of an inventory map based on previously occurred landslides and the preparation of the causative data layers by extracting them from reliable sources. These data areathen converted to raster for the analysis.

Light detection and ranging (LiDAR) is a remote sensing technology that emits intense, focused beams of light. These laser beams travel at the speed of light and the time taken for the laser to return back to the position of the target can be determined accurately. It is similar to radar (radio detecting and ranging), except that it is based on discrete pulses of laser light. However, accurate digital terrain models (DTM) are necessary for this study. DTM accuracies achieved by Shuttle Radar Topographic Mapping (SRTM; 30 m), and the Advanced Spaceborne Thermal Emission and Reflection Radiometer (ASTER; 30 m) [4] compared to LiDAR (±0. 5 m). The LiDAR data were acquired In 2011 using Optech ALTM 3100 sensors.

The spot (Satellite Pour-1'Observation de la Terre) satellites in orbit are Spot 2, Spot 4 and Spot 5 provide images with a large choice of resolutions (2.5 m – 1 km). Spot 5 is a high- resolution satellite that was launched in 2002. It is owned and operated by Airbus Defence & Space. The satellite of Spot 5 captures both panchromatic (black and white) and multispectral (colour) digital imagery. The spot 5 imagery used in the study was acquired on the 1st August 2011 and was purchased from MACRES in 3 bands mode (BGR) after undergoing atmospheric correction with the pixel resolution of 2.5 m.

The Airborne Laser Scanning (ALS) data were processed to create Digital Terrain Models (DTM) out of ALS ground points and Digital Surface Models (DSM) out of the adjusted first return ALS data with a 0.5 m resolution. The ground point data are interpolated with different ground routines such as binning interpolation type, simple nearest neighbour or complex kriging techniques to produce hillshades together with slope gradient map and topographic contours. Hillshades which are calculated from DTMs is a grayscale 3D representation of the surface, with the sun's relative position taken into account for shading the image. This function uses the altitude and azimuth properties to specify the sun's position. This will produce a 3D false image of the landscape that creates illumination and shadowing to emphasize and highlight geomorphic features.

High resolution DTM and high resolution of derivative maps enables a high level of accuracy and visibility of main landslide features. Landslide body is clearly visible, including main scarp as well as the zone of accumulation and zone of depletion on topographic derivatives, hillshades and slope maps (Figure 2 & Figure 3). These landslides are relict rotational slides which have been covered by thick overgrowth and hampered visualisation by satellite images. Slope and contour maps were the two that proved to be the most useful maps of all three topographic derivative map types used to identify landslides. Hillshade maps can provide a very powerful visual interpretation when trying to identify landslides, but their usefulness depends on the sun angle and the azimuth with which the derivative map was processed (Figure 3 & Figure 4). A total number of 35 landslides were detected and mapped either by visual inspection of LiDAR imagery or field investigation, or both (Figure 4 & Figure 5). Some of these landslides were abstract from previous study and are confirm using hillshades and field observation.

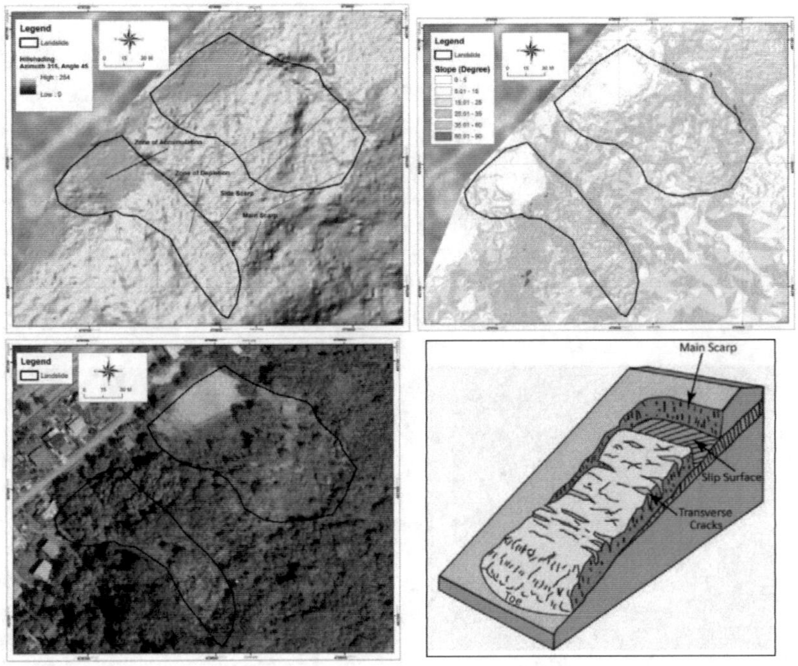

Figure 2. Composite display of three different maps (hillshades, slope and orthophoto) of the landslide at Kg. Lereng Bukit.

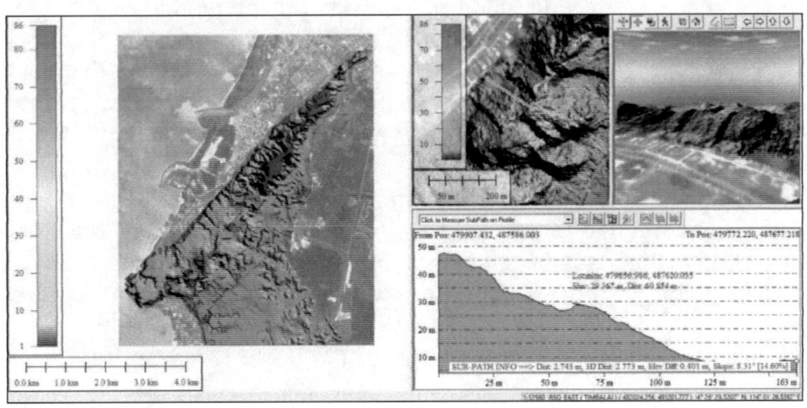

Figure 3. The cross-section of landslide's relict with its location and 3D image.

3.2. LiDAR Derived Parameters

3.2.1. Slope Gradient. The slope gradient map of the study area was created in ArcMap in degrees from the DTM which itself was processed from LiDAR data (Figure 6). The slope map is reclassify manually based on the slope data used by [5] for the hillside development procedure.

IOP Conf. Series: Materials Science and Engineering 1101 (2021) 012023 doi:10.1088/1757-899X/1101/1/012023

Figure 4. Photographs' combination of landslides in the study area.

3.2.2. Slope Aspect. A map of aspect direction was created for the field area from the high-resolution DEM by using the aspect tool from the spatial analyst's toolbox, in ArcGIS. The aspect tool uses an algorithm similar to the slope to fit a plane and then determines the direction that this plane faces [6]. The cell values (0° to 360° and -1 for flat cells) were rounded up to the nearest integer. The aspect direction for each hillslope failure was determined by using the majority value of aspect direction cells.

3.2.3. Curvature. The Spatial Analyst Tools Curvature creates the raster curvature, profile curvature, and plan curvature by inputting the DTM of the study area. The curvature is the combination of plan and profile curvature, which separates the concavity of the slope in different directions. A negative value equals to the concave slope, a positive value indicates a convex slope and a value zero is linear. Curvature analysis also allows areas to be identified on a surface where this plan features are more or less localized and consequently, could help to identify zones prone to landslides.

IOP Conf. Series: Materials Science and Engineering 1101 (2021) 012023 doi:10.1088/1757-899X/1101/1/012023

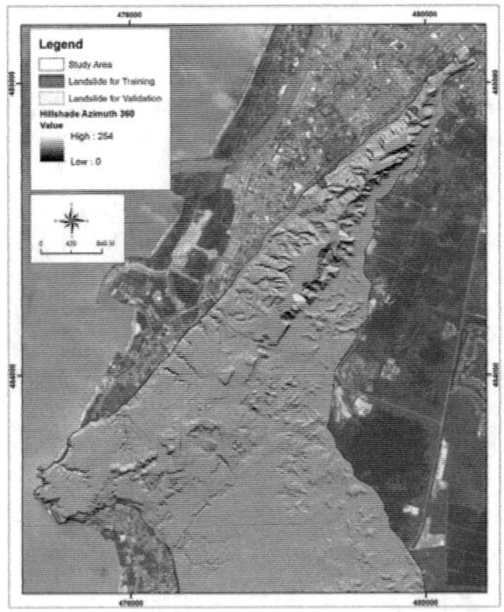

Figure 5. The distribution of mapped landslides both for preparing the model (training) and validation .

3.2.4. LS Factor. The Slope Length and Steepness Factors (LS) factors represent erodibility due to a combination of slope length and steepness relative to a standard unit plot. It expresses the effect of topography, specifically hill slopes length and steepness, on soil erosion. An increase in hillslope length and steepness results in an increase in the LS factor which may contribute to the slope failure with another triggering factor. The LS factor which, influenced by slope length and steepness were analysed using the equation below (Equation 3):

$$LS = (\lambda/22.1)^m(65.41 \sin^2 \theta + 4.56 \sin \theta + 0.065) \quad (3)$$

Where λ is the horizontally measured plot length,
θ is the slope angle and m is a variable plot exponent adjustable to match terrain and soil variants. m varies between 0.5 (slopes of 5% or more) and 0.2 (slopes of < 1%).

The flow accumulation raster was calculated using Spatial Analyst Tools hydrology based on the DTM LiDAR data. Finally, the LS factor was analyst using a raster calculator in Spatial Analyst Tool of ArcMap.

3.2.5. Lineament Density. Many authors have noted the correlations between faulting and the distribution of landslides [7]. Since the intensity of fault-related rock fracturing in the zone around the fault increase with the density of landslides, the lineament or fault density is a good indicator of rock mass strength [8]. The lineaments of the study area are extracted from previous studies, satellite images, aerial photographs and high-resolution DTM LiDAR data. Lineament density of a lineament is the total line length of the fault network divided by the area. A high density indicated one or more lineament exist in that particular area.

3.2.6. Distance to Drainage. The stream networks of the study area were extracted from the high-resolution DTM LiDAR using Spatial Analyst Tools. The study area contains more than 10 streams which are closely related to the geological structure in the study area. The drainage of the study area was analysed from the DTM LiDAR data using hydrology in the Spatial Analyst Tools. Most of the stream networks are parallel to each other and only the stream network in the southern part of the study area eventually merges together. The buffers were created, using 3D Analyst, around each of the drainages in the following increments in meters, 0-25, 25-50, 50-75, 75-100 and >100. The ranges for the buffers were chosen in part because these ranges were typical in similar studies and due to how close the drainage channels were to each other.

3.2.7. Topographic Wetness Index. TWI describes, how do topographical and hydrological features influence the location and size of saturated areas which act as initiation points for the runoff. Wood et al. [9] predict that steady-state conditions and uniform soil properties for their equation (4) to calculate TWI,

$$TWI = \ln (A / (\tan \beta)) \tag{4}$$

A represents the flow accumulation raster and the slope gradient is β (in degrees).

Wood et al. [9] explains that steady state and uniform soil properties can be assumed because the variation of topography often far exceeds soil transitivity.

3.3. Satellite Image and Field Derived Parameters

3.3.1. Soil. The soil map of the study area was extracted from the 1:50,000 scales which might not be very accurate in comparison with the study that used the high-resolution LiDAR data to develop Landslide Susceptibility Map. However, eight (8) soil series were used to analyse whether this factor contributed to the stability of the slopes which led to landslides in the study area (Figure 7). The soil series were Tatau, Similajau, Rajang, Nyalau, Merit, Miri1, Miri and Igan.

3.3.2. Lithology. The study area is underlain by two (2) geological formations, namely: The Tukau Formation and Miri Formation, and alluvium of the quaternary deposit. However, the lithology is crucial as the types of rocks are one of the factors that influence the stability of the slope. A total of five (5) different lithological were extracted from the previous studies.

3.3.3. Landuse. The land cover of the area is a thematic classification of the Spot 5 images which is also counter check on the ground. Spot 5 images with resolution 2.5 m are used to identify the high vegetation, low vegetation, built up and barren. Field checking is important in order to demarcatethe extent of the land cover to be digitized to vector format for analysis.

3.3.4. Biomass. The plot mapping of biomass together with Canopy Height Model (CHM) LiDAR data and Spot 5 images have made the study possible. The plot mapping and spot 5 with a resolution of 2.5 m are successful in delineating the biomass based on the spot 5 images.

4. Result and Discussion

4.1. Landslide Susceptibility Analyses
For calculation of the landslide susceptibility index (LSI) of the study area, the frequency ratios of the each of the class or type of each factor (Table 1, Figure 8 & Figure 9) are calculated based on the formula (1).

The overall Landslide Susceptibility Index (LSI) of the area is calculated as follows;

LSI = Fr (Aspect) + Fr (Curvature) + Fr (Slope) + Fr (LS Factor) + Fr (Lineament Density) +Fr (Lithology) + Fr (Soil) + Fr (Distance to Drainage) + Fr (Topographic Wetness Index) + Fr (Land use) + Fr (Biomass).

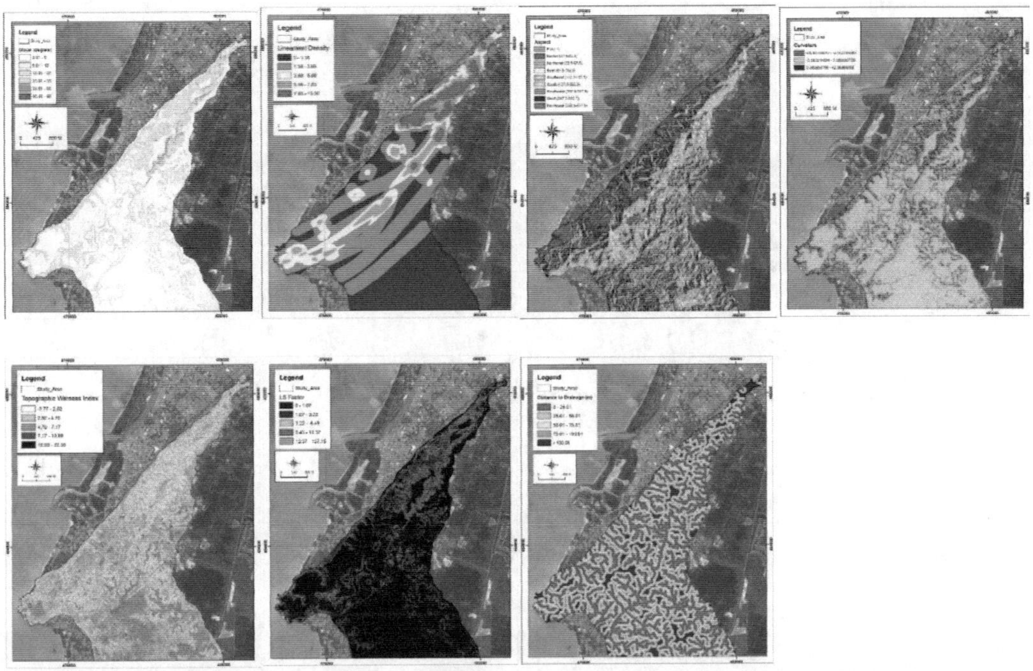

Figure 6. The causative factor maps derived from LiDAR (Slope, Aspect, Curvature, LS Factor, Lineament Density, Distance to Drainage and Topographic Wetness Index).

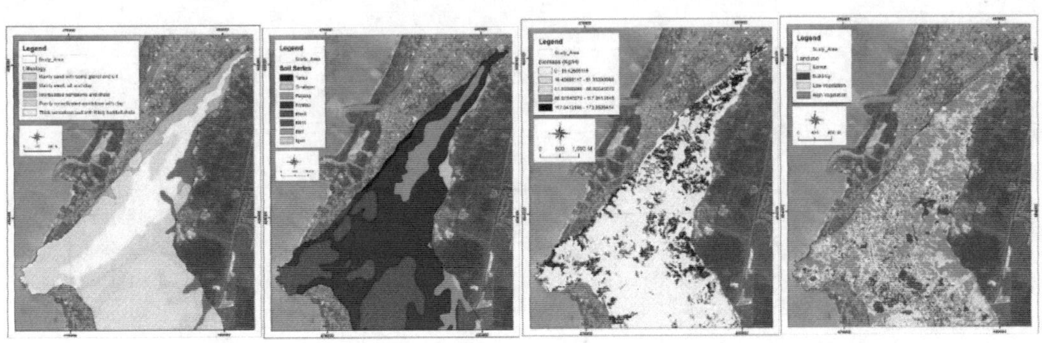

Figure 7. The causative factor maps derived from satellite image and field mapping. (Lithology, Soil, Biomass and Land Use)

Table 1. Frequency Ratio of Causative Factors

Parameters		Pixel of Occurrences	% Pixel of Occurrences	Pixel of Landslides	% pixel of Landslides	Frequency Ratio
Aspect	Flat	139	0.0002	0	0.0000	0.0000
	North	8341842	12.7558	47944	19.1472	1.5011
	Northeast	7678637	11.7416	15356	6.1327	0.5223

	East	8238603	12.5979	18625	7.4382	0.5904
	Southeast	8236733	12.5950	18095	7.2265	0.5738
	South	8227661	12.5812	19463	7.7729	0.6178
	Southwest	7549745	11.5446	24197	9.6635	0.8371
	West	8252807	12.6196	35555	14.1995	1.1252
	Northwest	8870433	13.5641	71162	28.4197	2.0952
Curvature	-5,760.2 - -29.3	3999042	6.1151	31833	12.7130	2.0790
	-29.3 - 49.1	59321009	90.7096	202703	80.9526	0.8924
	49.1 - 4,249.1	2076549	3.1753	15861	6.3343	1.9949
Slope	0 – 5	25892944	39.5933	19439	7.7633	0.1961
	5 – 15	20726345	31.6930	39643	15.8321	0.4995
	15 – 25	9723458	14.8683	77129	30.8027	2.0717
	25 – 35	5664446	8.6616	71218	28.4420	3.2837
	35 – 60	3266853	4.9954	42041	16.7897	3.3610
	60 – 90	122554	0.1874	927	0.3702	1.9755
Ls_Factor	0 - 1.07	43008030	65.7642	36192	14.4538	0.2198
	1.07 - 3.22	11805779	18.0524	65023	25.9680	1.4385
	3.22 - -6.45	7985592	12.2109	103900	41.4941	3.3981
	6.45 - 12.37	2417637	3.6968	41668	16.6408	4.5013
	12.37 - 137.15	179562	0.2746	3614	1.4433	5.2566
Lineament Density	0-1.38	24369949	37.2644	9923	3.9629	0.1063
	1.38 - 3.60	23762902	36.3362	40996	16.3724	0.4506
	3.60 - 5.56	8303470	12.6970	57878	23.1145	1.8205
	5.56 - 7.83	5278734	8.0718	70853	28.2963	3.5056
	7.83 - 13.50	3681545	5.6295	70747	28.2539	5.0189
Lithology	Mainly sand with some gravel and silt.	29052537	44.4278	27479	10.9751	0.2470
	Mainly sand, silt and clay	3746517	5.7293	0	0.0000	0.0000
	Interbedded sandstone and shale.	5807369	8.8808	161510	64.5072	7.2637
	Poorly consolidated sandstone with clay.	9973651	15.2519	0	0.0000	0.0000
	Thick sandstone bed with thinly bedded shale.	16816526	25.7162	61408	24.5264	0.9537
Soil	Nyalau	37575484	57.4572	109938	43.9055	0.7641
	Igan	1696396	2.5940	0	0.0000	0.0000
	Miri	5156964	7.8856	105765	42.2389	5.3565

IOP Conf. Series: Materials Science and Engineering 1101 (2021) 012023 doi:10.1088/1757-899X/1101/1/012023

	Similajau	192284	0.2940	0	0.0000	0.0000
	Rajang	2128122	3.2541	0	0.0000	0.0000
	Merit	1186455	1.8142	0	0.0000	0.0000
	Miri1	16846362	25.7600	29638	11.8364	0.4595
	Tatau	614533	0.9397	5056	2.0192	2.1488
Distance to Drainage	0 - 25m	23221411	35.5087	48920	19.5370	0.5502
	25 – 50	17983086	27.4986	83474	33.3367	1.2123
	50 – 75	13017174	19.9051	68834	27.4899	1.3811
	75 – 100	7304873	11.1702	32897	13.1379	1.1762
	> 100m	3870056	5.9178	16272	6.4985	1.0981
Topographic Wetness Index	-3.77 - 2.52	20095621	30.7285	103676	41.4046	1.3474
	2.52 - 4.70	22083529	33.7682	95003	37.9409	1.1236
	4.70 - 7.17	14528514	22.2158	39299	15.6947	0.7065
	7.17 - 10.89	6936525	10.6067	10947	4.3719	0.4122
	10.89 - 22.56	1752411	2.6796	1472	0.5879	0.2194
Land Use	Barren	12142921	18.5763	52160	20.8309	1.1214
	Build-Up	11569868	17.6997	4599	1.8367	0.1038
	High Vegetation	20230889	30.9494	123983	49.5146	1.5999
	Low Vegetation	21452922	32.8188	69655	27.8178	0.8476
Biomass	0 -16.43	42841417	65.5360	107192	42.8128	0.6533
	16.43 - 51.33	3949365	6.0415	34595	13.8173	2.2871
	51.33 - 86.92	4100837	6.2732	31599	12.6207	2.0119
	86.92 - 117.04	5696373	8.7139	39264	15.6821	1.7997
	117.04 - 173.85	8808608	13.4748	37747	15.0762	1.1188

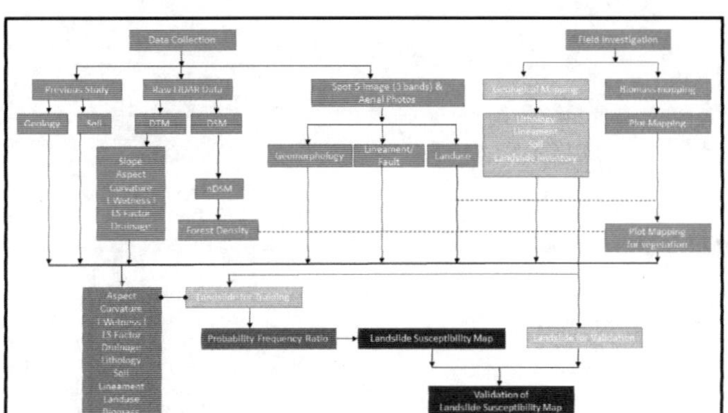

Figure 8. Conceptual Framework of data capture, processing, and validation of the study.

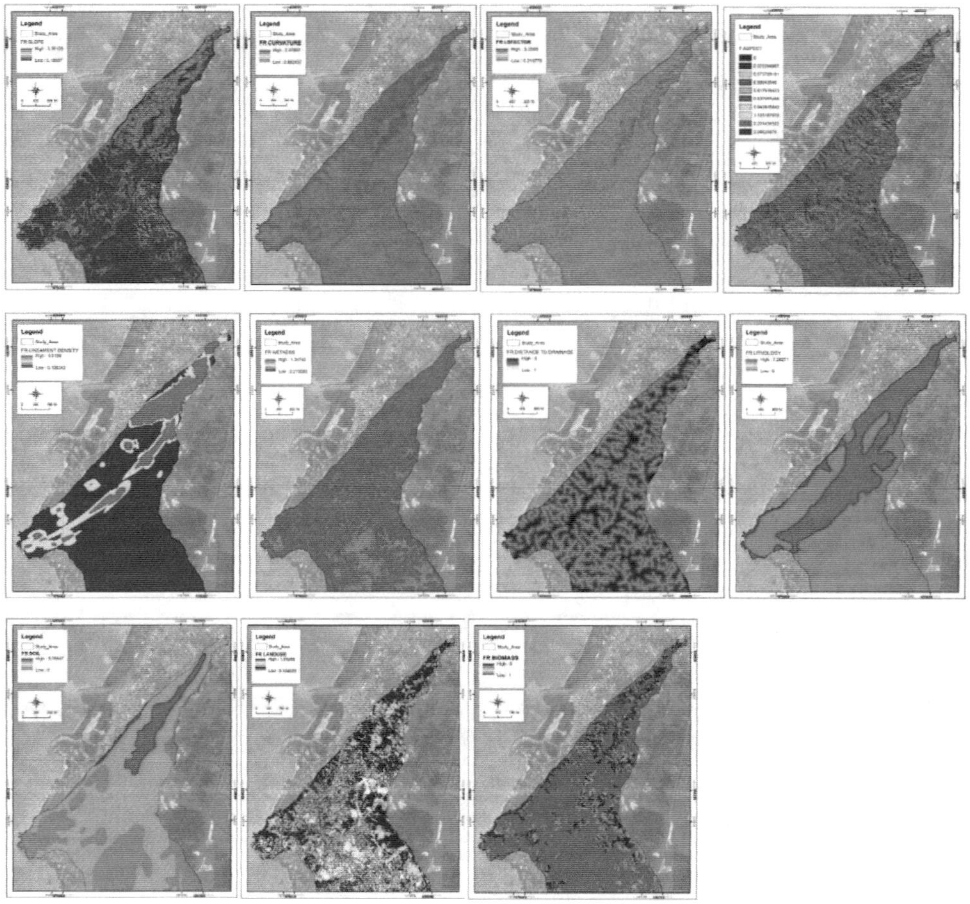

Figure 9. Frequency ratio maps of all the causative factor.

The calculation was carried out by using the raster calculator in Spatial Analyst Tool in ArcGIS (Figure 10). The LSI then was classified using the natural breaks methods because it allows to gather similar values and to maximize the differences between classes (Figure 11).

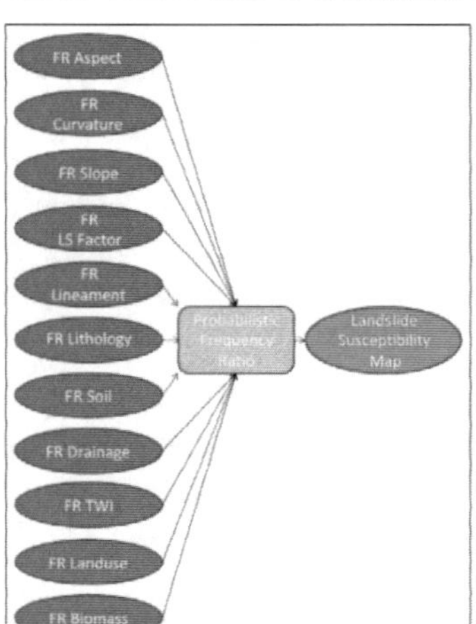

Figure 10. The Model Builder of calculating LSI using raster calculator in spatial analyst tool.

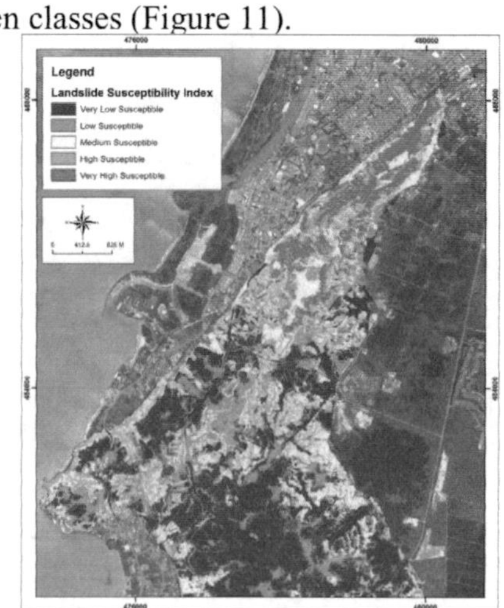

Figure 11. Landslide Susceptibility Map of the study area overlay on Miri City.

4.2. Validation.
To check susceptibility model developed by the probabilistic frequency ratio of bivariate statistical, the randomly assigned landslide inventory was plotted on LSM for validation.

For the validation of the output from our analysis, the receiver operating characteristics (ROC) curve was drawn, and the area under curve (AUC) value was calculated for the proposed model (Figure 12). The AUC value ranges from 0.5 to 1.0 where an AUC value close to 1.0 reflect model fitness and a value below 0.5 indicates model's inaccuracy and unacceptability. The ROC plot assessment results of the validation data set (Figure 8) show that the AUC for the landslide susceptibility map using the frequency ratio model is 0.971, which corresponds to a prediction accuracy of 97.1% (Table 2). According to the decision rules suggested by [10] when the susceptibility value of 0.5 is taken into consideration as the cut-off value, 97.1% of observed landslides are located on the areas having high susceptibility values. Therefore, the landslide susceptibility maps produced using these sets are considered as spatial effective.

Another method to validate the landslide susceptibility map is to check using the landslide data from the study area. The validation process posits that an active landslide must be triggered in those areas showing at least moderate values of susceptibility (≥14. 665 and <20.573, class 3 in Table 3), and will be more probable for higher susceptibility values (Class 4 and 5 in Table 3). Most of the landslides in the entire study area (90.02%) are included in the areas showing susceptibility values higher than 20.574 or class 4 and 5.

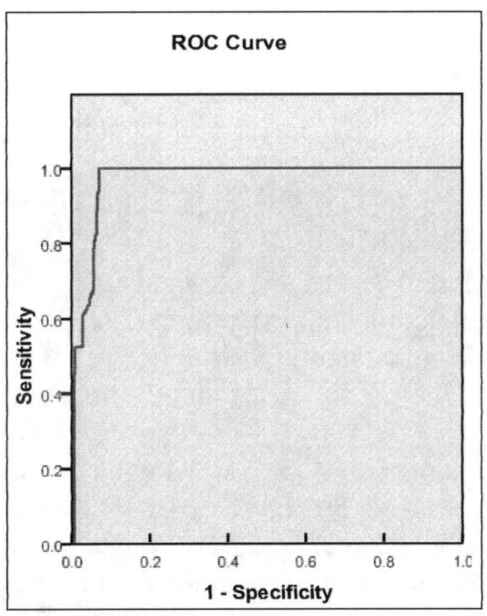

Figure 12. Prediction rate curve for the landslide potential maps using the Frequency Ratio model.

Table 2. Result of the ROC for the LSI using Frequency Ratio model

Area Under the Curve

Test Result Variable(s):Value

Area	Std. Error[a]	Asymptotic Sig.[b]	Asymptotic 95% Confidence Interval	
			Lower Bound	Upper Bound
.971	.006	.000	.960	.982

[a] Under the nonparametric assumption
[b]Null hypothesis: true area = 0.5

IOP Conf. Series: Materials Science and Engineering 1101 (2021) 012023 doi:10.1088/1757-899X/1101/1/012023

Table 3. Validation of susceptibility zones

Reclassified Value	Susceptibility Class	Pixel Number	% Area Covered	Landslide Pixel	% Landslide area covered
4.738 – 10.016	Very Low	16598211	25.41	1	0.00
10.016 – 14.655	Low	23612456	36.14	8804	0.62
14.655 – 20.574	Medium	13339317	20.42	12228	9.36
20.574 – 27.612	High	8254028	12.63	66357	50.81
27.612 – 45.525	Very High	3523620	5.39	51202	39.21
	TOTAL	65327632		130592	

4.3. Discussion

In the calculation of frequency ratio of the causative factors that influence the slope stability in the study area, a number of the frequency ratio is greater than 1 which means that the respective frequency ratio has a higher correlation. The aspect with the slope facing west, northwest has the FR value of 1.125, 2.09, and 2.02 respectively. The slope aspect has a significant impact on landsliding because it controls the exposure to sunlight, winds, rainfall [11] and vegetation cover [12]. However, in this particular case, the aspect is controlled by the bedding plane as it covers a total area of 32.8%. As in curvature the concave accounts for 6.1%, while convex which covers 3.2% have higher correlations with FR values of 2.07 and 1.9. Slope gradient plays significant factors to slope stability with any s lopes more than 35° are considered steep and could cause the slope to be unstable, but in this case slope of more than 15° are having FR of more than 1. These slopes covered 28% of the tot al area. The LS Factor, which controls by slope length and steepness are having 34.3% of the area with FR more than 1. This is due to the long stretch of slopes, especially to the western flank.

Different lithologic units have different sensitivities to landslides [13]. As a result, the wide variation of the lithological types from different geological formation in the Canada Hill area makes the lithology as a major predictive factor, which controls the landslide occurrences. The more stable units are the quaternary alluvium and Tukau Fm. These con sist of mainly poorly consolidated sand, silt and clay, underlying the flat area, mostly on top of the ridges. Thick bedded sandstone exhibit FR nearly 1 as the outcrops are found along the ridges and the slopes at Tanjong Lobang. The interbedded shale and sandstone exhibit FR of 7.26 which cover 8.8% of the area. These beds are dipping to the west and northwest causing FR of aspect and slopes having greater FR. The present soil series of Miri and Tatau which are comprised of fine sand and sandy loem with poorly drained conditioned are the reason that it has FR more than 1. Beside the lithology which greatly shaped the geomorphology of Canada Hill, the presence of the few sets of lineaments also detrimental in affecting the slope stability. The lineament density with FR 3.6 to 13.50 covered 26.5% of the study area.

In any slope failure, runoff plays an important role as that of a triggering factor for a landslide. Surface water flows and accumulates to form a network of stream which was once a fault plane. In hydrologic factor, distance to stream and Topographic Wetness Index (TWI) are taken into account as the causative factors in a landslide susceptibility index. A distance of landslides more than 25 m to drainage shows that the value of the FR is more than 1 and the values decrease in cases where the distance is more than 100 m. This covers a total area of 64.5%. TWI on the other hand, shows high FR in the class of 1 and 2 with FR values 1.34 and 1.12 respectively with 64.4% coverage. TWI is predicting areas which are susceptible to saturation and are prone to overflows. FR values are lower along the stream channel indicating that lower FR values are the opposite of TWI.

Causative factors such as Land use, which has four categorical classes shows the classes of barren and high vegetation have FR values greater than 1 indicating that high vegetation (FR value 1.60) is also prone to landslides beside the barren area (FR value 1.12) with a coverage area for both classes is 49.4%. The recent landslides body is without vegetation cover while the relict landslides are mostly found in the area with biomass of more than 16.43 Kg/Ha. Two classes of biomass showing FR values more than 2; class 2 and 3 with FR values of 2.28 and 2.01 respectively. Relict of landslides is having trees of height lower than surrounding after being disturbed by movement. Landslides with FR values below 2 but with biomass nearly 100Kg/Ha indicate that the relicts are much older. The soil condition in the landslide area is not compacted and loosely colluvium with high water content and this causes the plants to grow rapidly.

Landslide distribution is mapped via DTM and fieldwork shows that it varies depending on the location and types of debris. Landslides of planar failure are found only in the south-west slopes whereas, toppling and circular failures are in the southeast. These data are not mentioned in the previous study by [14] and only manage to outline three (3) landslide without considering the past landslide hidden under the canopy. Moreover, little were covered during the field mapping, especially, under the forest cover which geological formation is the key to the slope stability in the study area. This was due to geological factors that contribute to the types of failure. The role of topographic factor is also closely related to geology because thick massive sandstone that shaped the slope aspect and slope gradient especially in the southwest. Similarly, the LS factor and curvature formed as a result of the weathering effect of shale and sandstone beds. JKR [14] also failed to mention and relate the geological factors such as lithology and lineament in the developing landslide susceptibility map, hazard map and risk map.

5. Conclusion

There are eleven (11) factors generated that are associated with the distribution of landslides in order to produce the landslide susceptibility map. The factors were selected based on the previous studies in which it is mentioned that these factors are main causative factors of landslides. Landslide inventory is significant in this study as the distribution of the landslides are related to the causative factor, especially the topographic and geologic factor. The delineation of topographic factors is also appropriate as it is closely related to the geological structure, especially the bedding plane which has similarities with the slope aspect. Anthropogenic factors of biomass are taken into account for the analyses of frequency ratio (FR), however the value is not substantial to prove that it greatly affected the stability of the slopes. This is due to the geological conditions of the study area as well as other causative factors. Geological factors of lithology and lineament density is a major factor that affects stability in the region other than the influence of human on the environment especially the vegetation cover.

This study proved the geological factor acted as one of the main causative factors that affected the stability of slopes. Mapping of landslide inventory was crucial in using the frequency ratio method as it was related to the landslide and the causative factors. The geomorphology of Canada Hill was formed by the geological structure, especially the anticline and the main Miri fault. These geological structures also shaped the topographical features such as the aspect and slope gradient which in turn also cause the failures in the western flank of Bukit Canada. The Landslide Susceptibility Map produced can be used for optimum management by decision makers and land use planners, and also avoidance of landslide susceptible regions in the study area.

6. References

[1] Kaluni, L. (2006). Gelinciran Tanah di Kolej Tuanku Datu Haji Bujang, Miri. Jabatan Mineral dan Geosains Malaysia, Sarawak. Unpublished.

[2] Muol, E. (2009). Laporan Tanah Runtuh di Kg. Lereng Bukit, Bukit Kanada, Miri. Jabatan Mineral dan Geosains Malaysia, Sarawak. Unpublished.

[3] Daud, Z. (2009). Geologi Kejuruteraan Kawasan Kg. Lereng Bukit, Bukit Kanada, Miri. Jabatan Mineral dan Geosains Malaysia, Sarawak. Unpublished. 28pp.

[4] Tighe, M. L., & Chamberlain, D. (2009, November). Accuracy comparison of the SRTM, ASTER, NED, NEXTMAP® USA digital terrain model over several USA study sites. In ASPRS/MAPPS 2009 Fall Conference.

[5] KPKT. (1994). Garis Panduan Pembangunan di Kawasan Bukit, Jabatan Kerajaan Tempatan. Kementerian Perumahan dan Kerajaan Tempatan. Unpublished.

[6] Burrough, P. A., & McDonnell, R. A. (1998). 1998: Principles of geographical information systems. Oxford: Oxford University Press.

[7] Zhou, S., Chen, G., & Fang, L. (2016). Distribution pattern of landslides triggered by the 2014 Ludian earthquake of China: Implications for regional threshold topography and the seismogenic fault identification. ISPRS International Journal of Geo-information, 5(4), 46.

[8] Sarkar, S., Kanungo, D. P., Patra, A. K., & Kumar, P. (2008). GIS based spatial data analysis for landslide susceptibility mapping. Journal of Mountain Science, 5(1), 52-62.

[9] Woods, R. A., Sivapalan, M., & Robinson, J. S. (1997). Modelling the spatial variability of subsurface runoff using a topographic index. Water Resources Research, 33(5), 1061-1073.

[10] Can, T., Nefeslioglu, H. A., Gokceoglu, C., Sonmez, H., & Duman, T. Y. (2005). Susceptibility assessments of shallow earthflows triggered by heavy rainfall at three catchments by logistic regression analyses. Geomorphology, 72(1), 250-271.

[11] Yalcin, A., Reis, S., Aydinoglu, A. C., & Yomralioglu, T. (2011). A GIS-based comparative study of frequency ratio, analytical hierarchy process, bivariate statistics and logistic regression methods for landslide susceptibility mapping in Trabzon, NE Turkey. Catena, 85(3), 274-287.

[12] García-Rodríguez, M. J., Malpica, J. A., Benito, B., & Díaz, M. (2008). Susceptibility assessment of earthquake-triggered landslides in El Salvador using logistic regression. Geomorphology, 95(3), 172-191.

[13] Ozdemir, A., & Altural, T. (2013). A comparative study of frequency ratio, weights of evidence and logistic regression methods for landslide susceptibility mapping: Sultan Mountains, SW Turkey. Journal of Asian Earth Sciences, 64, 180-197.

[14] JKR. (2011). Slope Hazard and Risk Assessment for Penang Island and Bukit Kanada, Miri Area. Jabatan Kerja Raya Malaysia, Kuala Lumpur. Unpublished.

IOP Conf. Series: Materials Science and Engineering 1101 (2021) 012024 doi:10.1088/1757-899X/1101/1/012024

Assessment of Subsurface Condition of Peat Soil due to Brackish Groundwater Seepage

H D Daud[1], R May[1], R Rajali[2], J Bujang[2], E Muol[2]

[1] Department of Civil Engineering, Faculty of Engineering, Universiti Malaysia Sarawak, 94300 Kota Samarahan, Sarawak, Malaysia
[2] Jabatan Mineral dan Geosians Sarawak (JMG), Jalan Simpang Tiga, Bangunan Sultan Iskandar, 93300 Kuching, Sarawak, Malaysia
Corresponding author: rmay@unimas.my

Abstract. The purpose of this assessment is to assess the subsurface condition of peat soil due to brackish groundwater seepage along drainages using a numerical method. Visual MODFLOW is applied to conceptualize a condition in the study area and to produce outcomes related to the objectives of the study. The results of the simulation show there is only a small portion of high salt concentration along the drainage. The migration of salt concentration is mostly affected by the groundwater head and velocity vectors of the aquifer materials. Meanwhile, most of the other places in the study area receive a low salt concentration. Therefore, the impact of brackish groundwater on the environment is not significant since the salt concentration is low in most of the study area.

1. Introduction

Haze pollution is one of the most prominent disasters in Southeast Asia. This disaster happens almost every year in the fourth quarter of the year. This disaster is mostly due to open burning by plantation landowners in Indonesia to clear off their land. The purpose of land clearance is for farming and palm oil plantation. This has also become a trend in Sarawak.

Meanwhile, haze pollution in Kuala Baram, Miri occurs frequently during the dry season. Peat soil covers most of the plantation land in Kuala Baram. Peat soil is an unconsolidated superficial deposit with high non-crystalline colloid (humus) content in which forming the subsurface of wetland systems [1]. The dry peat from open burning activities burns easily during the dry season, and this creates smouldering peat [2]. The smouldering peat will cause a longer period of haze pollution.

Hence, groundwater is the only water supply in the study area that helps to reduce the smouldering effect. However, this groundwater highly concentrates Sodium Chloride (NaCl) since the study area is near to the coast. Figure 1 shows the location of the study area. Therefore, the concern of this assessment is about the migration of brackish groundwater in the aquifer materials. The salt intrusion in the aquifer system is investigated in Brunei, a neighbouring country of Malaysia and close to the study area [3]. The study area mostly covers the plantation land in Miri. Therefore, this research concern about the migration of brackish groundwater in the aquifer materials through existing drainages. The salinity of groundwater must consider to ensure the crops can survive under the present condition. The main objective of this assessment is to assess the subsurface condition of peat soil due to brackish groundwater seepage through drainages by using a numerical simulation.

Figure 1. Location of Study Area.

2. Methodology

Several tasks are included in the research methodology. Visual MODFLOW is used to prepare the conceptual model and to simulate the groundwater flow and salt migration [4,5,6]. Input parameters of aquifer characteristics, sources of salt movement, and salt properties are databased in the software. A 3D visualization of simulation results is impressively produced. Besides, the software has a good user interface where the program is simple to use, understand, enhance and modify [7]. Figure 2 shows the sequent activity of the research study. Figure 3 shows the conceptual aquifer system representing the condition and parameter of aquifer materials in the study area. There are five layers of soil in the study area, namely peat soil, fined sand, clay, medium-grained sand, and silty clay. Table 1 shows the hydraulic conductivity (K), specific storage (S_s), specific yield (S_y), effective porosity (n_e), and porosity (n) for different aquifer materials. Besides, this study also involves two rivers, namely Baram River, and Baong River. Besides that, the recharge value used in the simulation is 11% of the annual rainfall in the study area, which is 675.90 mm/year.

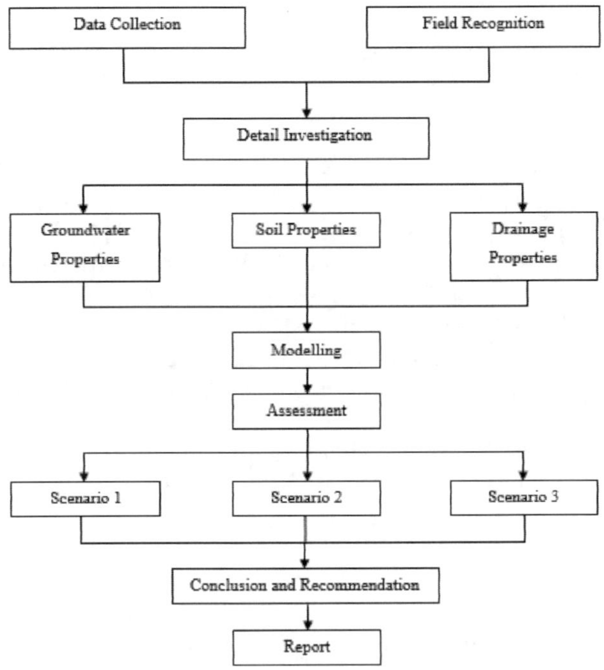

Figure 2. Activity chart.

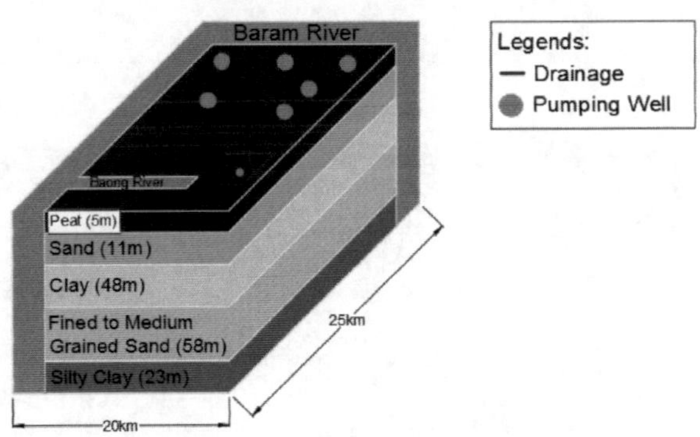

Figure 3. Conceptual aquifer system.

Table 1. Aquifer material properties.

Types of Soil	K (m/s)	S_s (m^{-1})	S_y	n_e	n
Peat	3.23×10^{-4}	9.6×10^{-6}	0.44	0.4	0.83
Fined Sand	2×10^{-4}	4.96×10^{-4}	0.33	0.2	0.40
Clay	9×10^{-6}	4.95×10^{-3}	0.06	0.11	0.46
Medium Grained Sand	5×10^{-3}	4.97×10^{-4}	0.32	0.23	0.39
Silty Clay	9.5×10^{-6}	4.95×10^{-3}	0.06	0.11	0.46

3. Results and discussion

The results include the calibration result, groundwater head, velocity vectors, and salt concentration migration in the aquifer materials, mainly peat soil.

3.1. Calibration result

Calibration is to minimize the root-mean square (RMSE) and maximize the correlation coefficient (R^2). The best RMSE value is less than 1 m, while the best R^2 value is closer to 1.

The calibration result shows a comparison between the observed head from observation in the study area and the calculated head from the simulation. The observation head represents the elevation of groundwater level in each pumping well. Table 2 shows groundwater heads of pumping wells for the calibration process. The calibration result in figure 4 shows a slight difference between the observed and the calculated head. The figure shows that the RMSE value is 1.071 m, while R^2 is equal to 0.75. Hence, the boundary condition and parameters of aquifer materials in table 1 are acceptable.

Table 2. Groundwater head of pumping wells.

Well No.	Coordinate (m) X	Coordinate (m) Y	Head (m ASL)*
KB3	171178.71	503597.25	4.5
KB5	174509.55	503251.60	7.53
KB7	172837.59	501819.51	6.9
KB9	172277.26	500604.26	7.87
KB15	169287.63	503672.68	3.526
KB18	172446.28	495859.50	7
KB19	169268.74	500582.73	5.3

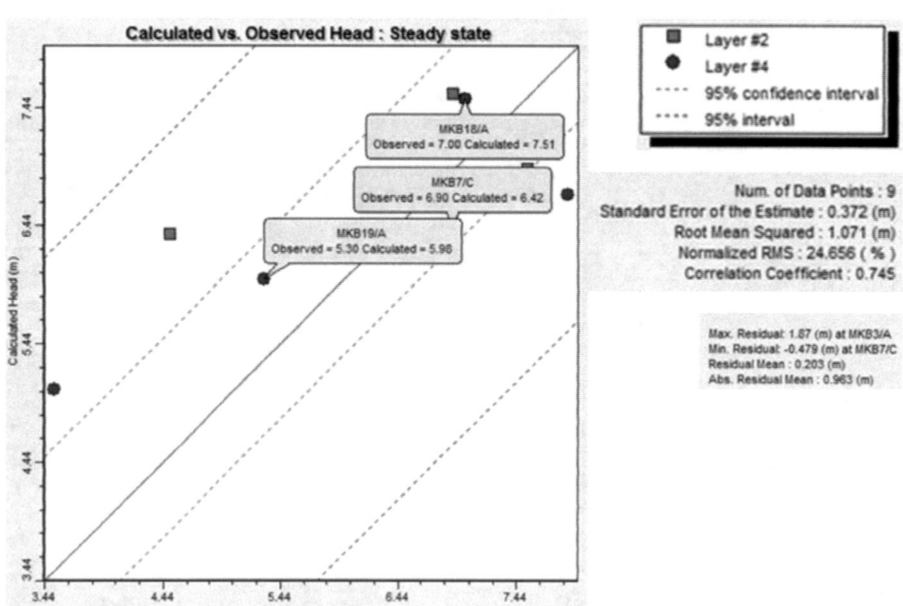

Figure 4. Calibration result of the simulation.

IOP Conf. Series: Materials Science and Engineering 1101 (2021) 012024 doi:10.1088/1757-899X/1101/1/012024

3.2. Groundwater head

Figure 5 shows the movement of groundwater head in peat soil. The groundwater head flows from 10 m ASL at the east of the study area to 0 m ASL downstream of Baong River at the south. Besides, the groundwaterheads are toward downstream of the Baram River at the north.

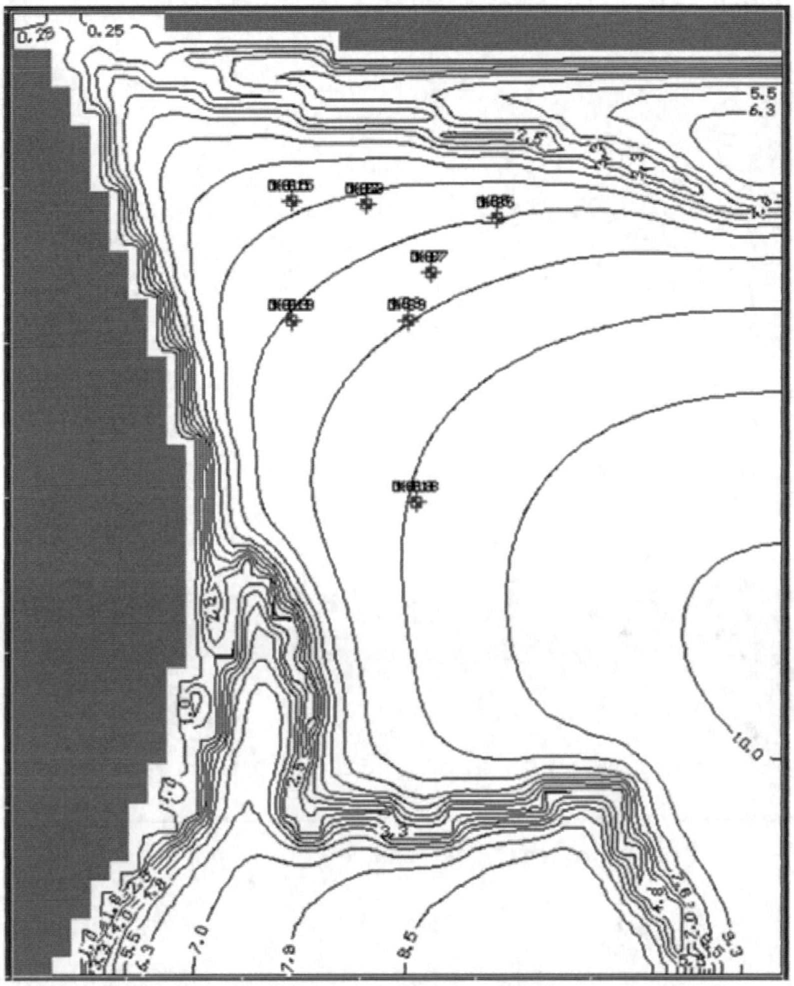

Figure 5. Groundwater head distribution in the peat soil.

3.3. Groundwater flow

Figure 6 shows the velocity vectors in the peat soil. The vectors move from the east of the study area to the north and northwest. Besides, the vectors also move to the west and southwest of the study area near Baong River. The maximum velocity magnitude is 7.5×10^{-5} m/s.

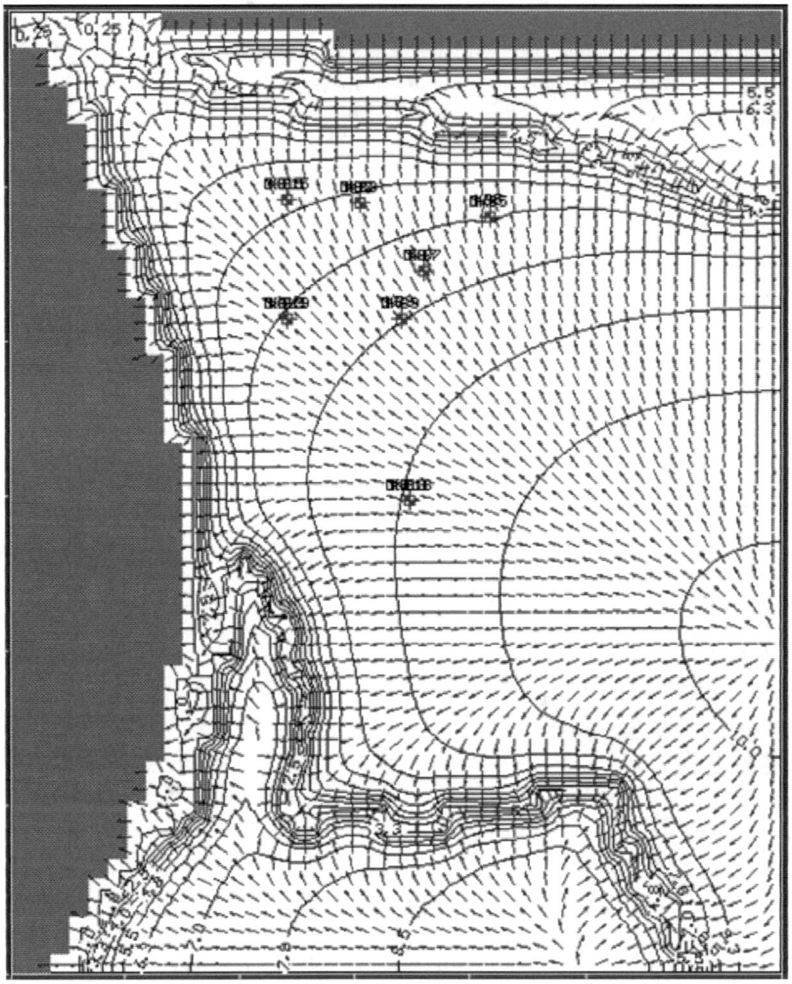

Figure 6. Distribution of groundwater flow vector in the peat soil.

3.4. Salt concentration distribution

The salt concentration results are estimated based on two conditions, natural and development conditions. The development condition is crucial for the research. The development condition is a condition where it involves pumping activities. The pumping activities is an abstraction of groundwater, and the pumped water fills the drainage system that helps to extinguish peat fire in the study area. Hence, the brackish groundwater from the pumping activities is expected to contribute to the migration of salt concentration in the study area.

Figure 7 shows the formation of a small plume below the drainage. The high concentration is possibly seen in the peat soil layer (layer 1) and fined sand layer (layer 2). Figure 8 shows the salt concentration of brackish water moves further down until the end of the clay layer (layer 3) after 2 years of pumping activities. The salt concentration is low as it moves down from the fine sand layer. Figure 9 illustrates the low salt concentration migrates further down until the middle of the medium-grained sand layer (layer 4) after 10 years of groundwater abstraction. The salt concentration stops migrating downwards as the times go on. However, figure 10 shows that salt concentration moves towards the coastal zone after 20 years. This is due to the directions of the velocity vectors as seen in figure 6. In figure 11, the high salt concentration moves down until the end of the clay layer, and the low salt concentration moves slightly more horizontally after 50 years. The salt concentration along the coastline is consistent from

Day 1 until 50 years of pumping activities. The clay layer shows a high salt concentration migrating inland until it reaches a certain point. The salt concentration for every layer becomes less as it migrates further inland. However, there is a formation of punching holes as the times go on.

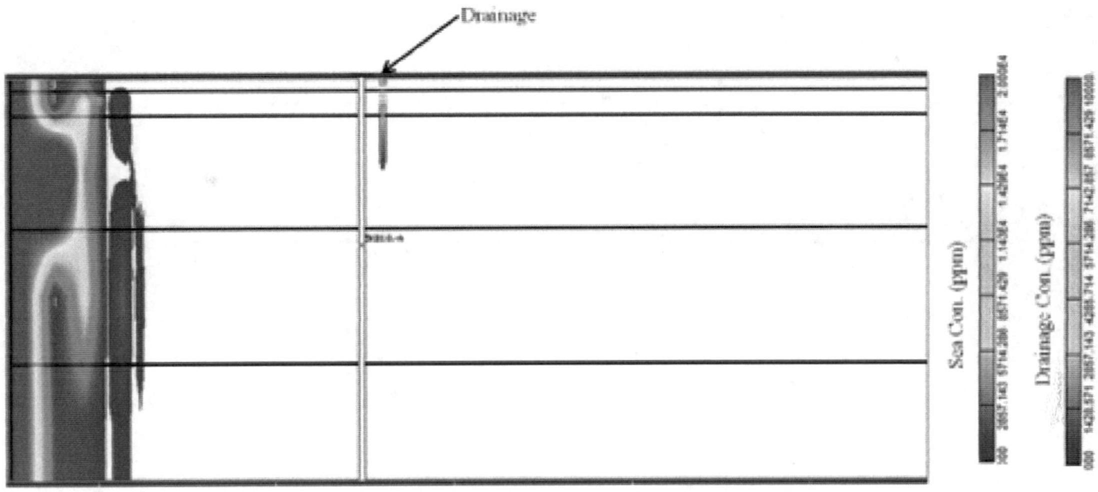

Figure 7. Vertical distribution of salt concentration from the drainage in day 1.

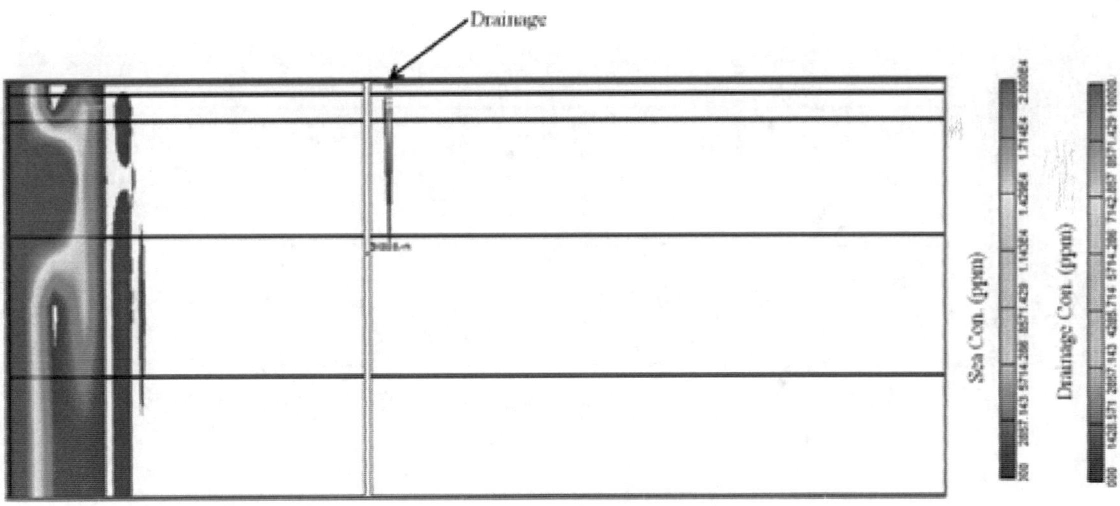

Figure 8. Vertical distribution of salt concentration from the drainage in 2 years.

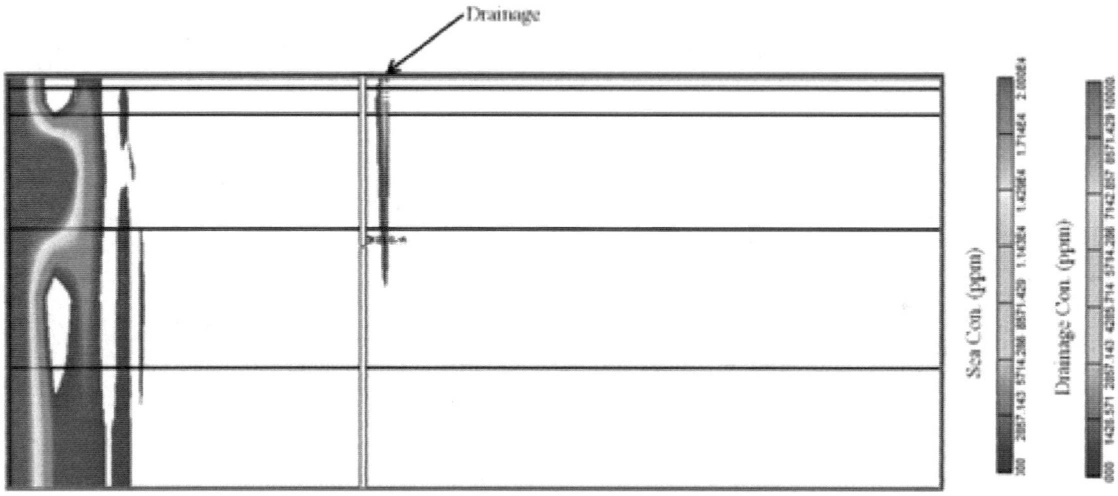

Figure 9. Vertical distribution of salt concentration from the drainage in 10 years.

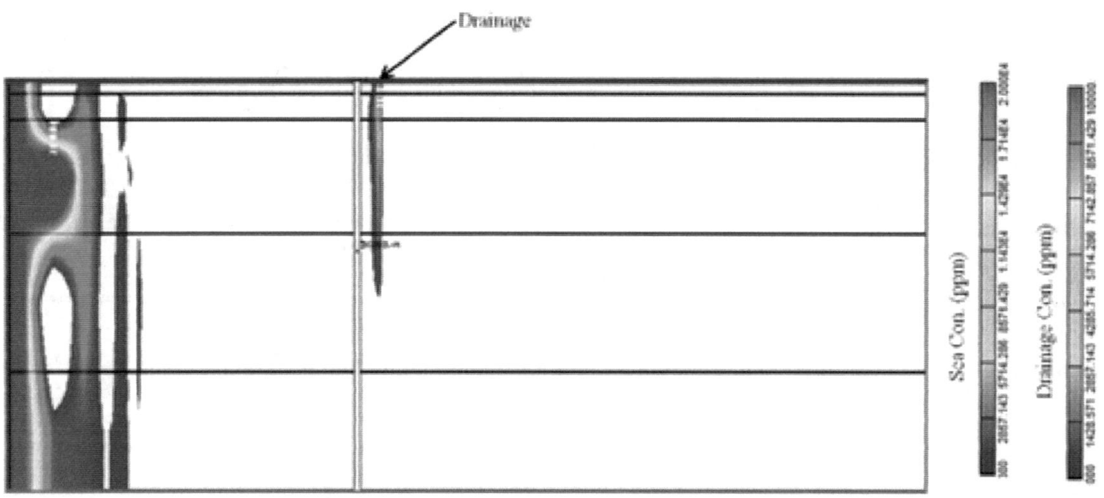

Figure 10. Vertical distribution of salt concentration from the drainage in 20 years.

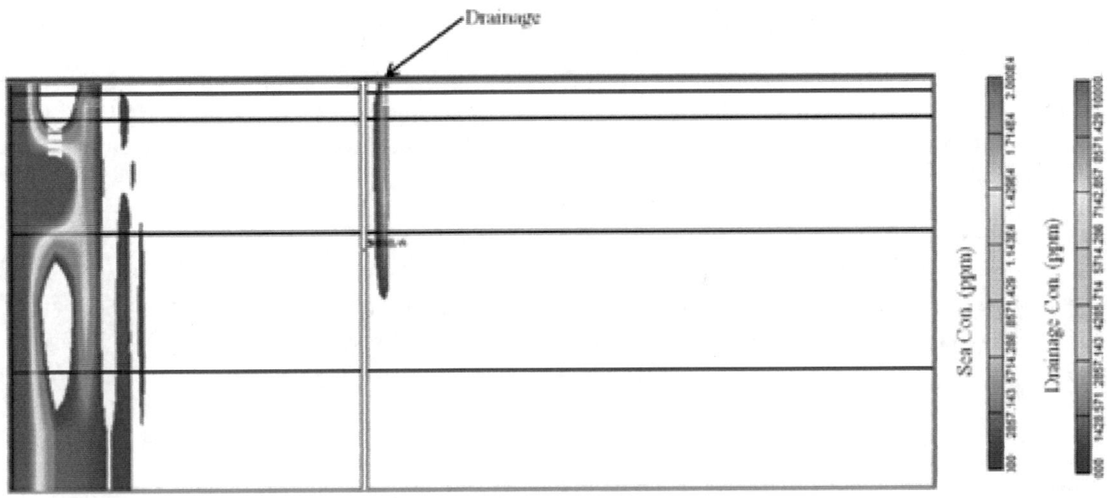

Figure 11. Vertical distribution of salt concentration from the drainage in 50 years.

4. Conclusion

This study assesses the subsurface condition of the peat soil due to brackish groundwater seepage along the existing drainages by using a numerical method. Using Visual MODFLOW in this assessment has brought to the achievement of the objective. The results show there is only a small portion of high salt concentration along the drainage. The furthest that the salt concentration can migrate downwards is until the middle of the medium-grained sand layer as the times go on. Then, the salt concentration will move horizontally towards the coast, where the concentration decrease. The results show there is only a slight movement in the horizontal direction. The flow of brackish groundwater depends on the velocity vectors in every layer of soils. Meanwhile, the high salt concentration from the sea only concentrates along the coastline until it reaches a certain point. Then, the salt concentration reduces as it moves further inland. Therefore, the impact of brackish groundwater on the environment is not significant since the salt concentration is low in most of the study area.

5. Recommendation

A further study on the impact of salt concentration on the crops is one of the best recommendations since most of the study area is for plantation. This will help the plantation owner decides the best location to plant their crops. Besides, a study on designing drainage that able to reduce the salt concentration before it migrates into the aquifer materials is also one of the best recommendations. The selection of proper materials is crucial when designing the drainage system, so it will not pollute the environment.

6. References

[1] Huat B B, Prasad A, Asadi A, Kazemian S 2014 *Geotechnics of Organic Soils and Peat* (London: CRC Press/Balkema)

[2] Rajali R B 2015 *Tube Well Constructions and Peat Management to Combat and Prevent Peat Fires in Kuala Baram, Miri, Sarawak. Miri, Sarawak* (Sarawak: Minerals and Geoscience Malaysia)

[3] Azhar A S, Abdul Latiff A H, Gödeke S H 2019 Groundwater investigation of a coastal aquifer in Brunei Darussalam using seismic refraction. *Environmental Earth Sciences* **78** p 220

[4] Sajeena S, Swathy P S, Abdul Hakkim V M 2020 Studies on saline water intrusion in the coastal stretch of Kadalundi river basin, Malappuram district, Kerala using visual MODFLOW – A case study. *Research Square* pp 1-17.

[5] Kotb A, Hagra M, Hassan N 2019 Impact of groundwater exploitation on saltwater intrusion in coastal aquifer case study: Derna-Libya *Proc. Twenty-Second International Water Technology Conference, IWTC22* pp 52-63
[6] Sajeena S and Kurien E K 2019 Studies on groundwater resources using visual MODFLOW-A case study of Kadalundi river basin, Malappuram, Kerala. *Indian Journal of Soil Conservation* **47** (1) pp 21 – 9.
[7] Waterloo Hydrogeologic 2015 *Visual Modflow 2011.1 User's Manual.* Waterloo Hydrogeologic.

Acknowledgments
The authors gratefully acknowledge field and data supports from Jabatan Mineral dan Geosain (JMG) Sarawak. The authors would also like to thank anonymous reviewers for their constructive comments.

Gallivanting with Architecture

J Wahid[1], B Saleh[2], J Chuloh[2], A Abdullah[3]

[1] Architecture Department, Faculty of Built Environment, University Malaysia Sarawak (UNIMAS), Malaysia
[2] Architecture Engineering Department, College of Architecture, Art and Design, Ajman University, U.A.E.
[3] Department of Architecture, Faculty of Architecture and Ekistics, University Malaysia Kelantan (UMK), Malaysia.
Corresponding author: wjulaihi@unimas.my; abac.saleh.b@ajman.ac.ae; c.jung@ajman.ac.ae; azli.ab@umk.edu.my

Abstract. Lately, it has been observed that buildings are becoming daring in their design, with forms and structures that appears to defy the law of gravity. The architecture discipline has been deployed to create eye-catching, even startling forms isolated in space that serves multiple purposes from authoritarian state-building to symbols of development, to "must-see" objects driving tourist to a particular place. These dramatic, monumental assemblies are not always well-received, however, with public outrage frequently directed toward their undesirable impacts on or lack of consideration toward the built context of a place. The building is not only a functional structure in any urban area but rather an intervention in the public space that remain there for many years or became an iconic landmark for someone to remember. The design of these buildings is not entirely based on the skill of the architect. Possibly influenced by the desires and tastes of the clients somewhat influenced by the practical constraints of the budget. Likewise, within those parameters, the creativity and even personality of the architect often shines through. The paper employed an exploratory techniques coupled with a descriptive analysis from printed materials as the framework to gather the written criticism of the projects that has been built. Through the pictorial and repertory grid applied in the investigation, this examines the essence of architecture design from the modern era until the present day. The findings from pictorial analysis showed that resemblance of the façade of the building symbolize the common objects and simple form found in daily life.

Keywords. *architectural design principles, iconic building, gallivanting.*

1. Introduction

Architecture is based on the creation of individual skill, and there is no definite rule to say it is wrong or right. After the subject of architecture became a discipline in the higher institution, it is guided by the principles of design and theory [1]. Subsequently, those who follow the course of architecture are trying to show that they are different and proving to the world that their creations are unique, pleasant and admired by the participants. Alongside that pre-conceived ideas, the so-called architect is working very hard to find a way to produce what is said to be a good design. The principles of design in architecture are to achieve aesthetic through compositional unity. The definition of compositional unity itself cannot be expressed in a single word. It is a composition of form, shape, texture, scale, rhythm, sequence, colour, balance, proportion, and craftsmanship. These characteristics employed in the design is theoretically accomplished comprehensively in order to produce a good architecture.

How does the architect derive to the conclusion that this is the final product of design? The definition of concept and approach in design sometimes misleading when presented by the architect. Architects preferred to relate these buzzy words to describe their work and possibly they derived it from nature or from observing a certain image in the environment. Sometimes, they are observing at the animals, plants, and even going back to basics principle of how objects are formed. However, they are those who copy at an everyday object as if to amuse the people. Wong [2] claims that design is a visual development process. Forms is a visual language as the basis of design. The conceptual, visual, emotional and functional aspects are the common design considerations perceived by designers. It is from here that the design factors are developed into a conceptual and afterward a visual component in architectural design.

The category of relational elements determines how the forms are placed and interrelated. Others are to be perceived, such as direction and position; while space and gravity are to be deliberated to accelerate with the design principles. The quality and extension of a concept are based on the functional elements. They are representation of realistic, stylized, or near-abstract objects that enhanced the meaning and function [2]. In architecture, the function of the building is the ultimate objective of the design and no other design exist or built just for the fun or sculptural purposes without functional usage.

Since the dawn of the century, our forefathers built the shelter to fulfil the physiological needs of the occupants and protect them from natural occurrences. This is the beginning of architecture. Curtis [3], allowed us to look back at Laugier's print to realize the basic requirements of human for shelter. It is for the purpose of shelter that civilization developed and flourished. It has been quite sometimes that we tend to ask the question of "what makes architecture art of freedom until an architect can do whatever he/she wants just to express his/her ideas as well as making the object a functional art form". This research is trying to show the reader the crazy ideas designed by architects to express their work as a sculpture or functional objects. The methodology of this article is based on the exploratory survey of secondary data available from the literature and also longitudinal information from architectural history. In this paper, we attempt to show few examples of the work articulated from the platonic object until the everyday items converted into the structure for a human. It is produced by architects to illustrate the features of what is said to be trending architecture to the masses. The illustration of the façade and built form that comprises the mass, scale, style and location of the building explained the matrix of the distinctions. The matrix represents the criteria use in comparison to show the scale, massing, form and characteristic of the façade. It helps to explain the differences between the design approaches and constraint either the site or the budget. The result from these analyses will help us to understand the creation of an individual with a reference to a theme of gallivanting.

2. Literature: Objects and Form

In architecture, symbols and sign that are strictly related to human are the immediate objects that we see every day. The difference between an object and a form in the built environment covers a wide array of things. An object can be paraphernalia created for the daily purpose that we took for granted and a form is an object that believed to have a value of compositional unity out of intentional arts [9]. Education has a tremendous impact on future followers who inherit architecture discipline. While pursuing the course of architecture, the definition of form is not clearly elucidated in relation to the intention of the design subject. Instead, it is becoming a culture of architecture student to study the work of their masters and trying to discover their philosophy and translate according to their understanding.

Parker [5] stated that form is not only physical or aesthetic, but it composed of all elements that constitute any aspect of the character of a product, including a society to which the product belongs, and therefore its shape is a representation of the usability and desirability of a creation. Music without rhythm is noise, letters will remain invocative words without composition of good sentences and painting on canvas will just be a brushstroke [5]. Generally, all art is created through the art of processes and actions. Similarly, to the design of buildings, the end-product would not be relevant without achieving its intention. Thus, form in the area of design is an integral component that became an analytical tool for architecture.

The point, line, plane and volume are the primary elements of architectural design [6]. Conceptual components are not apparent, non-measurable, perhaps they are no longer conceptual if they do exist instantaneously. Thus, the visibility of conceptual elements implies shape, scale, colour, light and texture in the composition. Yilmaz [7] argued that they are all combined to add a spatial consistency for the purpose of good design. Architectural elements are seen differently from different angles and distances as a three-dimensional shapes or objects. Visual elements that make the form look different in different situations in various conditions especially when lighting were added, thus creating a visible colour or texture of the building. This will create a timeless architecture.

Adjacent to this idea few authors are trying to establish different forms of classification schemes. Ching [6], is a remarkable example. He makes a distinction between standard and unusual shapes and subsequently classifies architecture into three significant categories: platonic solids, additives and substrates (Figure 1.0).

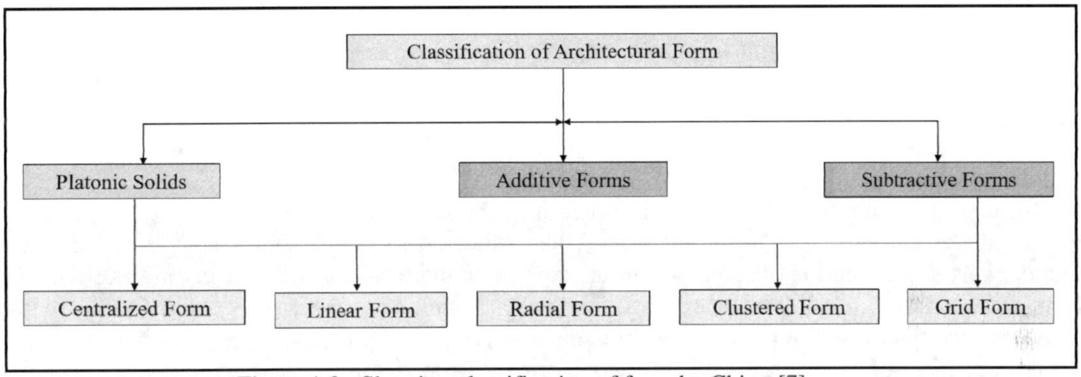

Figure 1.0: Showing classification of form by Ching [7].

The courses of self-formation of objects in the physical world, the fabrication of forms by a living organism expresses the human existence or the attitude of a human being towards the environment. The action and involvement of human in creating a place to get shelter are subject of the theory of form formation. On the other hand, the Bauhaus movement led by Walter Gropius, unite the arts and reimaging the material as a core objective in the expression of the building [3]. This movement has a massive impact on the progress of architecture education, especially in America. Even the students from the developing world are influenced by the changes and the progress. The movement continues until architecture spread to every corner of the world with the same processes and procedures.

While at school, we were taught of the design process before we could attempt to transfer the floor plan into a three-dimensional building form with a façade and massing. However, with the advent of the latest computer software and printing mechanism, the mass model-study technique became an array of several trial and error experiment in determining the structure and form. Apparently, the new digital technology for architecture student changes the way of design and revolutionized the styles of presentation in their works. This approach had a tremendous impact on the design process, and it gives the freedom of expression without limitation in design. The epitome of architecture also changed from the allude of Mies Van Der Rohe "Less is More" to a more challenging word of promises to the client [3]. Thus, architecture as visual arts are becoming the symbol of economics progress with modern technology refashioning from that philosophy into endeavouring to design harmonising the trend of innovative research for the university in a market-oriented culture. It shows how impactful these trends were and how they work in the real world, giving so many variations and approaches in designs. The western architects are taking these opportunities to experiment with their creative mind to design and build buildings in the developing world. The high economic progress through the value of petro-dollars in the Middle East and the new Pacific Rim Region became the new ground for these architectural styles to flourish. The building shows different approaches and amplifies the trends with the characteristics of the façade, form and massing. This is the ground that the architects are gallivanting with architecture.

3. Methodology

Klassen [9] identified the correlation of sign and symbols of architecture with ideology and design according to trends. However, the trends and the philosophy of the design is much stronger while he or she is capable of strategizing his or her works. In order to investigate the parameter of architectural process and procedures in producing a building design, this paper applied the exploratory methods together with the qualitative technique by extracting and analysing the existence of the design work published in the journals or any printed materials to evaluate the theme of gallivanting approach of the design concerning the scale, form, design elements and materials application of the building. Pictorial techniques of repertory grid will be combined in representing the analysis of the architecture [14] The work of early architects such as Etienne-Louis Boullée until Rem Koolhaas will be analysed according to the criteria as mentioned earlier (Table 2.0). The basic design principles will be used to utilize matrixes in evaluating the existence of all these elements. In architectural practice these are the variable of design elements and principles commonly used to articulate and defence the architecture works.

4. The Basic Form

Square or rectangle, round or circle and triangle are the three main basic shapes used in all design of built form. Out of this form, the architect turned it into platonic form and attempt to play around with the design by manipulating the perimeter of the boundary to create the space. Architecture as a functional art is the creation of space for human and not a mere functionless object in the environment. From a two-dimensional perimeter together with a height of the user conforming with the anthropometric and ergonomic, the liveable space is created. A central, linear, radial, clustered, or grid organization of form may lead to spatial organization. Forms can be regular or irregular. In general, regular forms are stable and symmetrical on one or more axes. According to Ching [6], even when transformed dimensionally or by deducting or adding components, forms will retain their regularity. A form can be changed, and its identity still retained by changing one or more of its dimensions. By removing part of their size, a shape can be changed. The type may maintain or be transformed into another type, depending on the degree of the subtractive process. By adding elements to its volume, a form may be transformed or altered. Whether an identity of an original form is changed or preserved, is resolute by the character of the additive or subtractive process occurred in the form itself. Irregular forms are typically asymmetric and more complex than regular forms. As the focus is both solid masses and spatial voids in architecture, regular forms can be part within irregular forms. Likewise, regular forms of irregular form can be integrated depending on the function of the design [5]. This process is known as transformation and it is regulated by the principles of architectural design.

4.1 The Plane

Le Corbusier, Mies Van Der Rohe and Frank Llyod Wright were the father of modern architects experimented with a different plane in design to achieve an aesthetic art in their projects. The recorded work of these architects became a breakthrough in our modern time. Their significant contributions readdress the pragmatic of the elements of design in their own manner. They were pursuing methods of expressing poetic reaction to the social and technological realities of their era. Departing from the formal horizontal and vertical lines of the modern masters, Rietveld explored the different sizes of planes from cubist perspectives by introducing abstract plane to his building façade to express the lightness of the mass of the building [3]. On the other hand, Peter Eisenman together with his New York Five colleagues subtracted the 'box' by defining a clear definition of planes from

IOP Conf. Series: Materials Science and Engineering 1101 (2021) 012025 doi:10.1088/1757-899X/1101/1/012025

cubist approach to show the drama of solid and void which make the building embody a piece of art. Conversely, these were the beginning of architectural transformation, which classified as form extraction [11].

4.2 The Platonic Form

Architecture develops through the evolving of ideas. The pure platonic form has been used in expressing architecture since the early 19th century. The closest to our discussion is the Einstein Tower by Erich Mendelsohn (Figure 2 in Table 1.0) and Cenotaph for Newton Monument by Etienne-Louis Boullée (Figure 3 Table 1.0). This building is expressively saluted delicately in a nice phrase as "*... the emblematic of the particular historical precipice and an artistic feat that foreshadowed the modern conception of architectural design*". On the other hand, CCTV Headquarters (Figure 4 *Table 1.0*) does not carry the continuity of architecture for the culture, especially for the city of Beijing, as commented by the President of China. The direct translation of platonic form in the designing of building signify lacks of ideas among architects. Nonetheless, such accidental design is becoming an iconic building. The Clam Building has an excessive unused space just for the sake of maintaining the circular form. On the other hand, the CCTV was criticized by the Beijing residents as ", ... *as sexual undertones of the structures*" - and Koolhass apologized and disputed with such claim [11].

The surface of urbanism in developing world is decorated with high-rise building to illustrate its skyline. The scale of the buildings is noticeable as each trend had a specific approach towards an exact gauge in judging the height of a building. The modernism and post-modernism era with epitome of 'less is more' and such, did not portray a wide difference in the building shape or height of similar scale towards the high rise. On contrary, a building in United Arab Emirates (UAE) that is Cayan Tower in Dubai Marina (Figure 5), considered as a destructive design conforming to Rem Koolhaas philosophy and language of architecture related buildings and technology. As compared to Emirates Towers (Figure 6) which is also an example of modernism of similar scale - the massing of buildings is similar to the latter but differs in elements. Cayan Tower is one large tower with no variation in levels or masses. However, the Emirates Towers consists of two towers of different size integrated with landscape and blending with the human scale on the ground and its surroundings [11]. This approach attempts to contain a sustainable city as a theme of UAE development. It has a diverse massing of form stretching from medium to low with a range of housing typology blending with the traditional and contemporary middle-eastern style showing different scale and height. Dubai attempts to suit the sustainable city with a futuristic type of architectural feats where the transition of the standard housing has changed to a more sustainable way of living. The sustainable city adapts to the most recent types architectural technology comprising the façade treatments using the latest technology with the most durable materials. The façade combines an expression of minimalism with its futuristic stance to retain the sustainable factors. The latest technology for the shading device from recycled materials and the application of the solar panel elevate the iconic status of this building. The innovative and responsive intervention of the architects to the environment will accentuate the quality of our living space.

Table 1.0 Table of Figures

Figure	Building	Name of Building/ Architect	Purpose
Figure 2		Einstein Tower by *Erich Mendelsohn (Source: https://www.archdaily.com)*	Observatory

Figure 3		Cenotaph for Newton *by Etienne-Louis Boullée* *(Source: https://www.archdaily.com)*	Monument
Figure 4		CCTV Beijing Headquarters by Rem Koolhas *(Source: https://www.dezeen.com/tag/cctv/)*	Office
Figure 5		The Cayan Tower by Skidmore, Owing and Merrill Architect *(Source:https://www.wowabouts.com/explore/post-cayan-tower)*	Residential
Figure 6		The Emirates Towers by Hazel Wong *(Source:http://www.skyscrapercenter.com/building/emirates-tower)*	Hotel

Table 2.0 Analysis of form

Images of building typology and its resemblance (F: Form, R: Resemblance, P: Purposes)			
 F: Curvilinear R: Animal P: Kindergarten	 F: Elongated R: Serpent P: Residence	 F: Criss-cross R: Auxiliary geometry P: Office	 F: Simple form R: Coin/Shell P: Office

IOP Conf. Series: Materials Science and Engineering 1101 (2021) 012025 doi:10.1088/1757-899X/1101/1/012025

 F: Sphere R: Ball/Marble P: Observatory	 F: Ordinary Object R: Picnic Basket P: Office	 F: Freedom R: Spiral pinnacle P: Church	 F: Triangular R: Origami P: Academy
 F: Elliptical R: Ball P: Office	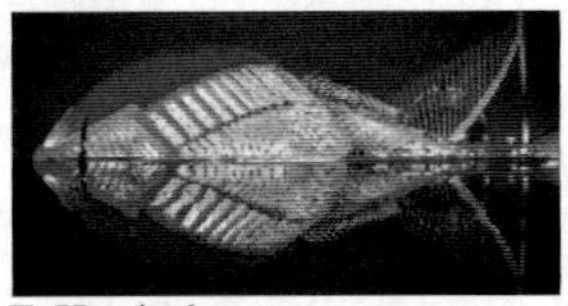 F: Hemisphere R: Skeleton P: Theme park	 F: Free form R: Blob P: Museum	 F: Cube R: Hexagon shape P: Housing
 F: Cylinder R: Pipe P: Office and Laboratory	 F: Curves R: Floral P: Museum		 F; Free form R: Music instrument P: Concert hall & exhibition

3. Discussion

For centuries architecture is a symbol of progress and civilization. On the other hand, the above examples summarize the built-form produced by architects to create the diversity of building types in the environment. It is a recollected statement that, the very idea of architecture is for human. Creation of the design is not for fun or introducing something against the norm of human habitation. Instead, the setting and stability of the building designed by architect to be different from ordinary form such as CCTV and Serpent Building seems against the will of the people due to its height and awkward shape. Typically, cylindrical and curvilinear form is conspicuous once it is built as compared to a straight linear form, unless the later resembles an object that is accustomed to our eyes. For example, the Kindergarten Cat Building (animal) and a Picnic Basket (everyday object) Building, attract the attention of the participants. The space that is created for our daily activities shaped the culture originate from our lifestyle [12]. Churchill [13] profound that "we shape the building and the building will shape us". Such a statement comprehended the creation by the architects manifestly to be different from the mainstream. As a result, all sorts of design started to flourish especially in the new economic progressive world of developing countries. This is the playground for ambitious architects to set his mark on new design ideas. He added that *"... the battles fought over class, ideology and language are represented most clearly in the explosion of new building types during the Century of Revolutions"*. The pure platonic of building may be gone, but not forever, the shape of the sphere, square or pyramid may be transformed into something else to suit the taste of the client. The platonic form takes into different volume, taking the unusual shape of the form either elongated or otherwise. With the aid of complex computer software, more and more sophisticated design will appear. The fun in designing a building to muse the participants may not be extensive, similarly the application of colourful design in architectural presentation sometimes temporary and deceitful. The beautiful "pictures" together with the powerful words of an architect to lock the design is dissimilar with the building when built. The ensemble of the building from the lines of form either horizontal, vertical, curvilinear or freeform carries different impression and interpretation to the viewers and participants: on the other hand, each lines produce by architect on his design will leave a foot prints and the building will be there for a long period of time. Time and again, discontinuity of communication and literary paradigm weaken the truth of architecture intentions. Thus, once built, we have to endure with it, either good or bad, positive or negative of the design to the environment and the passes-by.

4. Conclusion

The world is changing, economy is progressing and technology are also revolutionizing. The evolution of technology in other fields are remarkable but the progress of building technology is at snail pace. The production of architecture did not progress exponentially instead it revolutionized in its design focussing on form. The architect is still seeking for unique form to illustrate their design as discussed earlier in the analysis. The integration of form, spatial distribution of the interior, services and the economic aspects of the construction has to comply on the reality of the architectural discipline as taught during the school days. The image of building represented in Table 1.0 and 2.0 reflects a mere embellish façade, in fact, the interior of the building is still composed of linear straight line due to the construction technique as well as the nature of the materials used, for example the Cat Kindergarten Building and Coin/Shell building. The curvilinear space is difficult to match the furniture which are basically straight edges. The splendour of the building is appreciated from outside but the design elements comprising the ventilation, structure, services and technique of construction are far behind from other fields. Even the latest building design by prominent architects was customarily done and not by machine. The arrangement of bricks and the casting of concrete structure is similar to the one that has been practiced since the early stage of modern era. Apparently, the twisted Cayan Tower of Dubai takes eight years to complete. The assemblage of the building envelope and the integration of services to serve the occupants sometimes more complicated to install in a building – some examples are buildings designed by Moshe Safdie at Montreal Expo, Frank Gerry at Bilbao and Zaha Hadid at Cultural Centre at Baku. Similar experience found in the other building such as the CCTV in China and the Coin/shell in Abu Dhabi. In order to achieve such form, the cost of the building is escalating and the targeted period of completion is often miscalculated. Irrevocably, the essence of design with the idea of gallivanting with architecture remains among the practising architect. The architect knew that conventional design form for a building does not reflect iconic features. Perhaps, the capability of parametric design software will produce another version of architecture in the future.

5. References

[1] Antoniades A C 1998 *Architecture and Allied Design: An Environmental Design Perspectives* (USA: MSS Corp).
[2] Wong W 1993 *Principles of Form and Design* (New Jersey: John Wiley & Sons).
[3] Curtis W J R 2018 *The Modern Architecture Since 1900 3rd Edition* (Oxford: Phaidon Press).
[4] Allsopp B 1977 A Modern Theory of Architecture, Routledge, London
[5] Parker D W 2005 *Principles of Aesthetics* (USA: University of Michigan).
[6] Ching F D K 2007 *Form, Space and Order* (New Jersey: John Wiley & Sons).
[7] Yilmaz S 1999 *Evolution of the Architectural Form Based on the Geometrical Concepts* PhD Thesis (Izmir: Institute of Technology).
[8] Zheleva-Martins D 2006 Theory of Form Formation/Theory of Composition: In Search of Differentia Specifica, *Proc Engineering and Product Design Education Conference, Salzburg University of Applied Sciences, Salzburg, Austria.*
[9] Klassen W W 1990 *Architecture and Philosophy* (Philippines New Day, Publisher).
[10] Ching F D K, Onouye B S, Zuberbuhler D 2013 *Building Structures Illustrated: Patterns, Systems, and Design* (New Jersey: John Wiley & Sons).
[11] Wu J 2009 *China Netizens Slam 'obscene' CCTV HQ* Retrieved 26 March, 2019 from http://www.china.org.cn/culture/2009-09/10/content_18500150.htm.

[12] Rapoport A 2006 Vernacular Design as a Model System *Vernacular Architecture in the Twenty-First Century (Theory, Education, and Practice)* eds L. Asquith and M. Vellinga (London: Taylor and Francis) pp. 179 – 198.

[13] Markus T A 2006 *Building and Power: Freedom and Control in The Modern Building* (London: Routledge).

[14] http://www.emeraldgrouppublishing.com/archived/research/guides/methods/repertory grid htm, Emerald Publishing, retrieve 17th June 2020

IOP Conf. Series: Materials Science and Engineering 1101 (2021) 012026 doi:10.1088/1757-899X/1101/1/012026

The Wood Carvings of Two Traditional Timber Mosques in Kelantan and Terengganu

J Wahid[1], **A Abdullah**[2,] **B M Kamel Saleh**[3,] **J Awad**[3]

[1] Architecture Department, Faculty of Built Environment, University Malaysia Sarawak (UNIMAS), Malaysia
[2] Department of Architecture, Faculty of Architecture and Ekistics, University Malaysia Kelantan (UMK), Malaysia.
[3] Architecture Engineering Department, College of Architecture, Art and Design, Ajman University, U.A.E.
*Corresponding author: wjulaihi@unimas.my, azli.ab@umk.edu.my, basseemsaleh@hotmail.com, j.awad@ajman.ac.ae

Abstract. Wood carving has a long history in Malaysia. It is greatly influenced by the rich landscape of tropical forest in the living environment of the locals. The motifs of the carving are primarily derived from the images of flora and fauna. It mimics the intricate nature of leaves and flowers of the plants. The craftsmen initially use the bas-relief design consisting also of animal imageries on the boat's sides or sometimes on the house components and even the dagger handle. However, with Islam's arrival in the early 15th century, due to its teachings, contributed to the ban of animal images on art form. While most of the carvings are done on wood, there is no ample evidence of records available due to time and tides of the decaying process of the artwork. Likewise, there is no extensive research and documentation produced by academics to record the traditional wood carving of the Malays. Although few collections had documented the contemporary Malay houses produced by local scholars, sadly, there was no recount on the importance of housing elements and the technique of construction neither on wood carving. Hence, this paper attempts to examine the influence and pattern of wood carving on traditional mosques in Kelantan and Terengganu. The research conducted case studies, to seek the signs of the features and elements of wood carving on the traditional mosque in Kelantan and Terengganu, disclosing the influence of patterns and motifs on the selected images of its architecture. A measured drawing technique commonly used in conservation works guided the research operation which comprised of hand-drawn sketches to capture the details, measurement of building components, as well as transferring the images using tracing paper, while the other methods in gathering the information extended to direct interviews and photography. Then, the data collected were analyzed to determine the differences of each carving patterns and motifs in the two mosques in Kelantan and Terengganu. The result of the research established the future records on the variety of motifs and patterns of the Malay wood carving. Subsequently, it further elaborates the extensive techniques of wood carvings system and records the patterns found at Kelantan and Terengganu's mosques for future inventory and recorded history.

Keywords. *traditional timber mosques, wood carving motifs, Terengganu and Kelantan*

1. Introduction to Traditional Mosque

The arrival of Islam in Malaysia has dramatically changed the way of thinking and the way of life of the locals [1]. Naturally, the people respect the environment and responses to cultural practices which are exemplified in the built form. Likewise, the architecture of dwellings is also adapted to the guidance of Islam. A Malay-Islamic architecture such as mosques, suraus, traditional Malay royal palaces and residences including tombs, or better known as *pusara*; were part of the historical remnants of the Malay built form that is often used as a reference for the architecture of the Malays [2]. Furthermore, the mosque is the center of the community. Thus the agglomeration of houses around the mosque symbolizes the unity in Islamic teaching evidently through architecture as well. As a result of its historical development, the mosque architecture has undergone various changes and transformation paralleled to the events of Islamic practice, as reflected in different forms and patterns of Islamic religious buildings in the region.

The traditional mosque architecture in the Archipelago and the mosque architecture built during the early Islamic civilization shows distinct form and construction. The Mosque in the Malay Archipelago is much simpler and greatly influenced by the availability of the building materials and the technique of construction exclusively in timber. Consequently, timber as a material that is hardly durable results in the traditional mosques in the Archipelago as almost forgotten by historians [3] and yet the architecture of these mosques is one of the essential architectural relics of Islamic civilization. Additionally, the power struggle and civil war also destroyed the buildings amidst the scarcity of citizen's involvement in the further creation of arts and crafts. Thus, this condition led to the architecture of the traditional building only being built for survival and immediate usage with weak materials [4]. On the other aspect, the deterioration in timber building construction technology is due to the Hindu and Buddhist civilization, which introduced the custom of stone and masonry as building materials that was then widely used since it was more durable.

2. Research Objective

There exists a dialectic discussion on the conservation of Malay Architecture, primarily focusing on the Malay house. The interests itself has been engaging scholars since the early 1980s, but with a little continuation on the subject of the Malay Mosque's ornamentation and relics. Therefore, this paper aims to investigate these elements in the traditional timber mosques both in Kelantan and Terengganu. Both of these regions are the centres of wood carving as well as related arts and craftworks that is fundamental to the Malay world. Finally, a comparative research on the motif and its influence in the mosque buildings was evaluated visually.

3. Research Methodology

The documentation of the written aspects related to carvings and artworks were reviewed from books and various journals. The field-work was carried out in July 2018 until March 2019 at the site in Kelantan and Terengganu. Several visits were done to have a detailed inspection of the components of the buildings. The visual interpretation of the carving's motifs and patterns are gathered by using carbon paper that was traced and drawn on the site. Meanwhile, the stages of inquiry are divided into three parts. The first part is on the floor plan, and the surrounding area, including the tombs and ablution area - annexed to the main building. The second level of the examination was on the envelope of the building that includes the fenestration elements and vent light of the wall sections and, lastly, the roof structure, and the fascia boards. Here, the technique of measured drawing was highly applied. Sketching method and tracing on-

IOP Conf. Series: Materials Science and Engineering 1101 (2021) 012026 doi:10.1088/1757-899X/1101/1/012026

site and photography were used and labelled according to the components of the stages, as mentioned earlier. Meanwhile, two traditional mosques were selected in Kelantan and Terengganu due to its original state and high-quality carvings as highlighted by Farish Noor and Eddin Khoo [5]. Finally, after the fieldwork, the existing woodcarvers of the mosques were interviewed.

4. The Malay Mosque

The mosque in the Malay World has its own distinctiveness compared to other Islamic architecture especially to the Middle-Eastern styles [6]. In general, the building of a mosque varies according to the shape and form of the building plan. The most common shapes are either square or rectangular, this is to accommodate the maximum number of *jemaah* in the prayer line. Thus, the roof structure which is the most prominent structure to give shapes and coverage of the space below also create a dominant character of its own in the built form. The hip roof for eample, only allow to cover the elongated floor plan while the tiered roof differentiates the massive mass of the building as well as allowing lights and ventilation to the occupants inside. The style of architecture of these mosques are influenced by the architecture of Javanese, while some mosques also have elements of influence from the Middle-Eastern, Indian, Persian, and even European [7].

Indonesia, as a neighbour of Malaysia is a vast country comprises of thousand of islands and, therefore, it gives birth to different design of mosques according to their context. The typology of the building commonly corresponds to the context of the region. According to Wiyoso [4] the mosques design in the Archipelago is supported by the geographical situation of the Malay World and the transformation of culture. Evidently, the mosques in Indonesia are mostly built of tiered roof system befitting the application of timber as the main building materials. It is common to find that every district exhibit a different architectural language in design together with the Colonial influences and the context of the place where they live.

5. History Of Wood Carving

Since the beginning of the century, artisans who practice wood craftsmanship, in particular carvings, have enriched its historical development. Nonetheless, it is difficult to conclude who started the craft and where it originated from since there are no apparent records and documentation of its origin [8]. However, in the Malay Annals (*Sejarah Melayu),* Tun Seri Lanang described the beauty of Sultan Mansor Shah's palace (1459-1477). He recorded the beauty of its magnificent architecture that is decorated with a variety of attractive wood carvings on the walls, thresholds, doors, windows, staircases, and other components of the building. The palace was later known as the *'Istana Hawa Nafsu'* (Palace of Passion), and the *'Sejarah Melayu'* described that the palace was the most beautiful building of its time. *'Sejarah Melayu'* remains an early record of Malay wood carvings on traditional architecture. From there, it is vibrant with the recollection that the traditional Malay community was cautious about the intricate value of art and wood carvings in architecture [9]. Another recorded story in *'Hikayat Misa Melayu'* described the wood carvings at Istana Sultan Iskandar Zulkarnain (1756-1780). The palace was located on Pulau Indera Sakti, in Kuala Sungai Kinta, Perak. Unfortunately, the palace has disappeared and can not be traced due to development and modern progress. Conversely, the art of wood carving in Peninsula Malaya was heavily influenced by Hindu elements intricately devoted to the images of diversed Hindu deities.

Since the beginning of the 1st Century, it was contested that the Hindus influenced the distinctive socio-cultural formation of the Malay community. The elements of Hindu culture that had a profound influence on the socio-cultural aspects of the Malay society embraced literature, traditional medicine, and beliefs. It is alleged that handicrafts, especially wood carvings in the Malay world, have some traces of Hindu culture [10]. Additionally, the factors influencing Malay wood carvings were through the social structure of the feudal systems. Under this system, the community was divided into two groups. The first group was the aristocracy of the nobility and the palace, while the second group consists of the commoners living in coastal areas, hinterlands, and villages. The aristocrats' residence was decorated with high-quality carvings created by palace carpenters and carvers. The design and decorative elements symbolized the representation of administration and formal Royal organization [5].

The late Nik Rashidin Nik Hussein opined that the history of wood carving is only appreciated by the person who dedicated their works by doing the carving on the wood himself through the intricate finishing product. He further elaborated that, during such moments, the artisan will immerse himself in transmitting history through his experience and skills [5]. The authors added that the carving activities are very significant in three states, namely - Kelantan, Terengganu, and Patani. However, there were woodcarvers in the states of Pahang, Kedah, Perak, and Malacca too. Most of the time, though, the carvings are found at the palace, mainly to decorate the interior space of the hall *(balai rong).* The aesthetic of the designs are associated with the formal organization of the palace and indirectly has accorded recognition to the art of traditional Malay wood carving [9]. In the beginning, wood carving was created for a craft activity and as a medium for entertainment by ordinary people. It can be found in the handle of the Malay dagger – i.e., *keris* or walking sticks and boats. This situation changed when wood carving became one of the essential elements of palace decorations. Talented and highly-skilled craftsmen in the arts were usually designated as *'Seniman Di Raja'* (Royal Artists). As a result, traditional wood carving, which was conceptually based on symbols and culture of the Malay community, soon found a special place in the high society [11]. The Golden Age of traditional Malay wood carving began when the Sultan took over the art of wood carving as the patronage of the discipline. It is said that the art of wood carving and other traditional crafts such as traditional music, traditional textile, *keris*, traditional food and, etcetera flourished after it was under the authority of the Sultan. The age of brilliance of the art of Malay wood carving initiated when the national recognition lifted the status of the artisans as the palace artists [12]. The art of carving in the palace has its aesthetic value and its purpose and philosophy. In an esthetic sense, the art of carving customarily emphasized the artistic significance and symbols of the status of the palace, while at the same time respecting its purpose, and emphasis on the apparent and implied beauty of the wood carvings made. Furthermore, the motifs and patterns were specially designed for the palace. Thus, specifically, it has its significance, symbolism, and philosophy, which ultimately differentiates the ruling class from the commoners.

5.1 Wood Carving Philosophy

Generally, the beauty of the carvings found on the wall is aiming at giving treatment to the wood panels. Wood carving on the fascia board *(papan pemeleh)* added a series of embellishment to the eaves (Figure 3). The carving used on the side of the window enhanced the function of fenestration on the window's panel. Meanwhile, the perforated hole (*tebuk tembus)* in wood carving at the interior of the building can give a sense of beauty and add value to the art of the building. Although the figurine is laid on the panel, the concept and philosophy of carvings are still the same as the medium for beauty (Figure 1 & 2). It emphasizes more on the flora rather than the fauna.

Further, wood carving is an expression of love for beauty [8]. The beauty is also born from the nature of honesty to the art. Consequently, the art is regarded as the quality of patience and diligence, which is the philosophy of art and craftworks practised by most Malay craftsmen and carvers. Such intricacy of artistry is principally sourced from the environment and is often the inspiration for the carvers. The idea complements the ideal human life, to ensure that humans can live perfectly on earth and interact with nature while glorifying the Creator.

Figure 1. The concept of human relationship with Creator (Drawings by Hanafi, 2007)

Figure 2. The combination of human and Creator (Drawings by Hanafi, 2007)

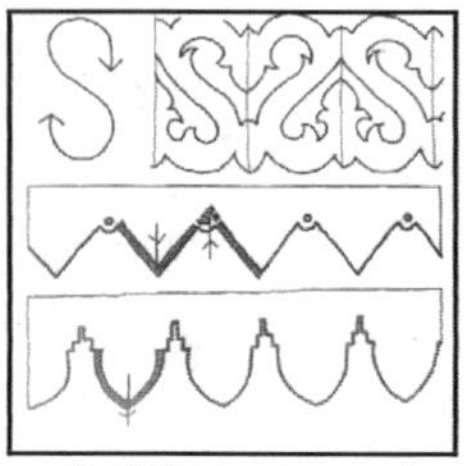

Figure 3. The concept of life (Drawings by Hanafi, 2007)

The abundance of plants in the surrounding is an essential element as a foundation of inspiration to the woodcarvers. Scholars have noted that the Malays have a close relationship with nature. Kling [13] profounded that the environment surrounding of the Malays comprises of swamps, hills, rivers, sea, and mangroves that enriched the creative practice of their daily lives (Table 1). This abundance of natural resources bound the close relationship of the Malays with the Creator. Islam encourages its people to create various works of art. However, the work must not contradict the tenets of Islamic teachings, the Quran and Hadith and does not prohibit its followers from appreciating music or visual arts. Notably, the beauty presented in music and visual arts must be of common interest to the people's acceptance within the scriptural teachings of Islam. Thus, Islam prohibits decorative elements that resemble animals and human figures. However, this prohibition does not impede the freedom to explore the work of art in Islam. There are also abstract motifs found in the art of pottery, metal, and the likes, which have proven to have high-quality skills and artistry.

Table 1. Evolution of Malay carvings motif. Source [10]

No	Influence of Hindu Cultural	Influence of Islam	Symbol and Meaning
1	Kala	Foundation of carved flower	Source of the carved flower
2	Makara	Sulur bayu	Tendrils
3	Mountain/ Cosmic Mountain	Head (Gunungan)	Anything triangle at the top
4	Bodhi Tree	Beringin Tree	Tree which is associated with Buddhism
5	Stupa	Stupa	The top of triangular shape
6	Padma/ Teratai	Tanjung, a fragrant flower	Beauty
7	Naga	Not used in the motif	Not applied
8	Geruda/ Jentayu	Petala Indera/ Gagak Sura	Mythical bird/ mitos

6. Case Study 1 : Kampung Laut Old Mosque, Nilam Puri, Kota Bharu, Kelantan (1730)

Figure 4. Kampung Laut Old Mosque, Nilam Puri, Kota Bharu, Kelantan.
(Image by : Wahid & Abdullah, 2019)

6.1 Background of Case Study 1

The Kampung Laut Old Mosque (KLOM) has a fascinating contested history regarding its origin. However, its historical value is priceless for architectural lovers, researchers, and students alike (Figure 4). The building was relocated to Nilam Puri, Kelantan, after the great flood in 1966. Its relocation was due to the efforts and initiatives of the Malaysian Historical Society. According to Mizan and Anuar [14], this mosque is three hundred to five hundred years old and the oldest mosque compared to the Great Mosque of Demak in Java. According to a local historian, Mr Nik Man bin Nik Mat, the existence of the mosque can be traced to his ancestors - Raja Imam, who founded it. He added that about 400 years ago, there was a Muslim Prince from Sri Wijaya - Syed Mahmud, also known as Raja Muda Laksamana. The Prince sailed from Java after a battle and power struggle over the throne. In his voyage, his ship was struck by a storm and stranded on South Siam Beach. Some of his followers remained at the coast while others explored the land and "opened" a small village called Kampung Laut. Syed Mahmud changed his name to Raja Iman and then built a mosque. He added that the mosque initially had an open pavilion. This story contested the findings of Al-Ahmadi [15], who collected oral history on the cases of Kampung Laut Mosque. He concluded that the mosque was 500 years old and was built by a Champa preacher. They came from Java and assigned to spread Islam and simultaneously constructed a mosque in the region. They established two mosques located in East Java Island, one of which is known as Masjid Agung Demak in Java. Meanwhile, the Kampung Laut Mosque was the first mosque established by them before they landed on Java Island. This traced record indicated that the Kampung Laut Mosque is the earliest mosque built before the Great Mosque of Demak and is a significant relic in the Malay Archipelago.

According to the third source though, Encik Abdullah bin Muhammad, who lives in Kampung Langgar Kota Bharu, claimed that the mosque was founded by the Wali Songo (*Nine Wali*) called Sunan Giri and Sunan Bonang. Both of them were Sufi members who came from Java Island and were students of Shaykh Mohammad Saman, a prominent religious figure of the time. It is said that Shaykh Mohamad Saman had instructed them to build the mosque. It was constructed with a four-pillar inclusive shape and supported by twelve exterior columns. The mosque was built to the Main Prayer Hall with a rectangular floor plan (Figure 5). Its roof consists of a three-tiered structure called *'Bumbung Limas bertingkat'* (Figure 6). Between the second and third layers of the roof, there are timber louvres functioned to provide natural ventilation and natural lighting at the Main Prayer Hall. At the topmost of the roof, there is a decorative element known as *'buah buton'* as the pinnacle pointing to the sky. The entire floor of the mosque is elevated about one meter from the ground. The mosque does not have a foyer, and the walls are decorated by luxurious wall panels called *'Janda berhias'* to the full height of the window opening.

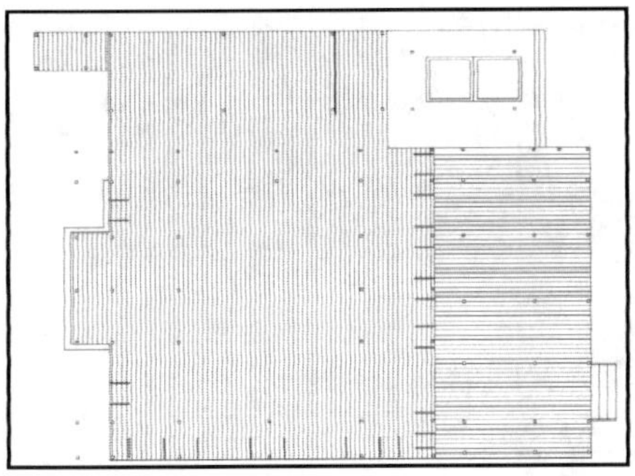

Figure 5. KLOM Floor Plan (Drawing by: Wahid & Abdullah, 2019

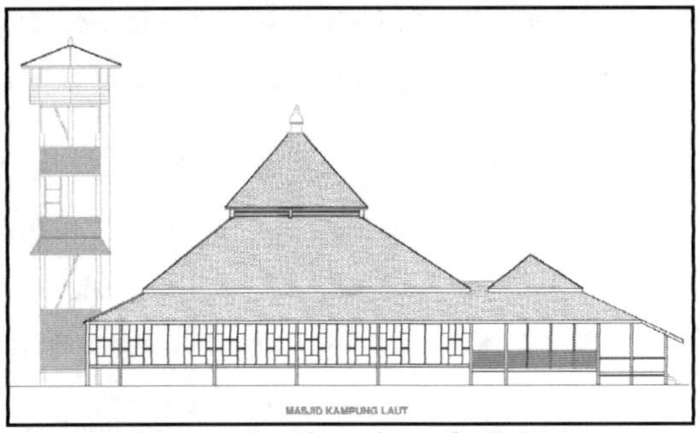

Figure 6. Elevation of KLOM (Drawing by: Wahid & Abdullah, 2019)

6.2 Carving Elements of Case Study 1

KLOM has eight bracket columns located on the front and back (Figure 7). Altogether, there are sixteen types of carvings found on the brackets: however, two of them are incomplete. The carvers may have died or moved to another area before he could finish the work. The motifs are entirely different in comparison to the motifs at the existing KLOM pulpit or that of Pasir Pekan Mosque. The panels are fabricated from *cengal* wood. This motif is known as the '*Daun Dewa*' or the Langkasuka motif. According to Mrs Rosnawati Othman, it is identified as the oldest motif in wood carving artistry. This motif is said to be found in a 7th Century old *Keris*. The Langkasuka motif came from the Langkasuka Civilisation that flourished in the Archipelago in the early century. If observed at its outline, it is a motif that arises from an abstract lotus flower. If the Siamese carvings are famous for their fire motifs, the Old Malay Empire is renowned for its lotus flower motifs, which illustrates the symbol of holiness and the living. The pattern does not have any source of seed, source of water, source of vases, or soil as customary in the philosophy of carving. However, its richness is in the leaves, which are twisted and repetitive to produce patterns and motifs achieving visual balance. While the works featured symmetrical design and method through a 45-degree axial angle, however, through scrutiny, there is an imbalance from the right and left. The truth in wood carving philosophy is practised by the carver to show that the wood carving design is a static visual connection. In the Malay wood carving, the source of the technique is seen through the 'S' movement that is presented (Figure 2). In the Langkasuka motifs, there is no 'S' element, and that is why it was called the early-motifs in Malay woodcarving. Later, the motif is transformed by other carvers into *'daun tua', 'kerak nasi'* leaves, *'tikam seladang'* leaves, and so on. This transformation was based on their manifestation, knowledge, skills, and philosophy with the natural context, surrounding, and evolved religious beliefs. The Langkasuka motif is, therefore, an older technique, which is uncommon now in our contemporary wood carving form. Nonetheless, the motif is still abundant in the traditional *Keris* making.

Figure 7. Lotus Flower Pattern at the Column (Image by: Wahid & Abdullah, 2019)

7. Case Study 2 : Old Mosque Langgar, Kota Bharu, Kelantan (1874)

Figure 8. The front view of the Old Mosque Langgar, Kota Bharu, Kelantan was built in 1874 (Image by: Wahid & Abdullah, 2019)

7.1 Background of Case Study 2

Old Langgar Mosque (OLM) describes the situation and the environment of the lifestyle of the Kelantan Sultanate and its kin (Figure 8). There is a cemetery in front of the mosque built by the Kelantan Sultan known as '*Makam Di Raja Langgar'* (Royal Tombs). In 1756, Sultan Long Yunus was a renowned personality who managed to unite the smaller kingdoms in Kelantan into a powerful government. He was laid to rest in Kampung Langgar, and later the area was designated as *Makam Di Raja Langgar*. From here, the settlement was established and opened to the public; at the same time, the mosque was built (1870). The OLM mosque has an elongated prayer hall (Figure 9), and the façade of the mosque has some influence from Siamese architecture (Figure 10). In 1886, a civil war broke out in Kelantan during the reign of the Long Senik Mulut Merah, and he was killed and buried in the *Makam Di Raja Langgar* too.

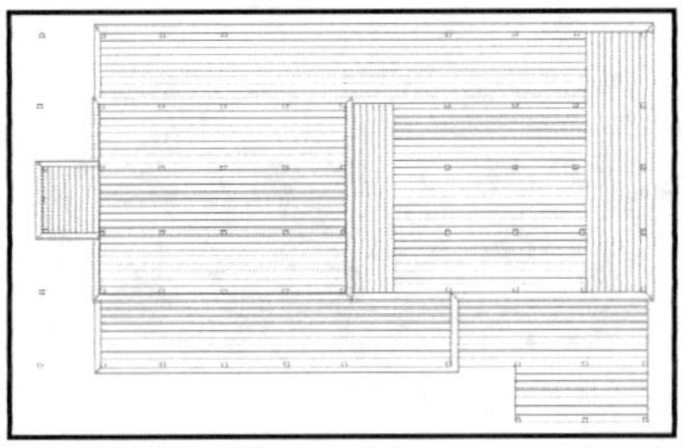

Figure 9. OML Floor Plan, which shows the original layout
(Drawing by: Wahid & Abdullah, 2019)

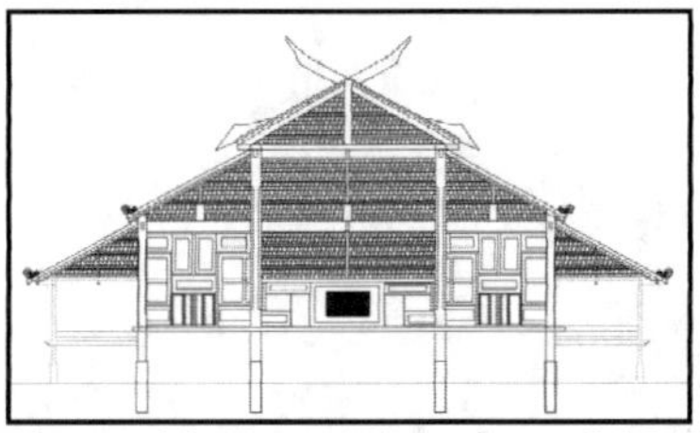

Figure 10. Front Elevation of OML
(Drawing by: Wahid & Abdullah, 2019)

7.2 Carving Elements of Case Study 2

The mosque has only one 2-D carving panel located at the pulpit. The motifs of the subject were from Chinese influence, showing the floral carving along with the panels. The panels were claimed to have been taken from other buildings and placed here because it omitted the peacock motifs and the bird's head (Figure 11). As explained earlier, the symbol of animals is against the tenets of Islam. The peacock's motifs are also found on the mosque's original pulpit, which is under the care of Academy Nik Rashidin in Kandis, Bachok, Kelantan, and is also its new location. The effort to trace its original pulpit was conducted by researchers from interviewing the Director of Academy Nik Rashidin and Mrs Rosnawati Othman.

According to Mrs Rosnawati, this pulpit was almost forgotten because it has been traded at antique stores and eventually salvaged by Nik Rashidin due to its historical significance. It is believed that this pulpit came from the mosque that was built in 1874. The pulpit seat is made from a carved wooden panel with an ornament of *'ketam guri'* and *'kerak nasi'* leaves. The panels are placed around the decorations with *'tebuk tembus bersilat'*. There are four pillars of carved lotus motif flower ascending its stairs. At the lower side of the carved panel of the column, elaborative writings in *khat* bestowing *"May Allah bless the family of Sultan Mahmud Shah Kelantan"* and, *"Ya Allah, show us taufik and hidayah over all our good deeds"*, is written. The *'gunungan'* carving in the flower pattern of '*ketam guri'* and '*kerak nasi'* with the characteristics of *'sulur bayu'* is located at the end of the pulpit.

Another artefact in Nik Rashidin's collection is the pulpit ceiling panel. It is made of *cengal* (hardwood) and measures 21.8 cm x 21.8 cm x 1.29 cm. It is carved with the Langkasuka motif similar to the Tomb of Che Ku Tuan Nawi, that signifies its heritage value. The whole design is from the pattern of lotus bud or stupa that is kept safely by the Academy Nik Rashidin.

Figure 11. The Figure shows the carving in the mosque panel. The lower panels shows the carvings from Chinese motifs with the hidden peacock outline. This carving has been identified for use in the mosque's original pulpit. (Image by: Wahid & Abdullah, 2019)

8. Case Study 3: The Kampung Tuan Mosque, Kemaman, Terengganu (1830)

Figure 12. The Kampung Tuan Mosque, Kemaman, Terengganu was built in 1830. (Image by: Wahid & Abdullah, 2019)

Figure 13. The Interior view of Kampung Tuan Mosque, Kemaman, Terengganu was built in 1830. (Image by:Wahid & Abdullah, 2019)

8.1 Background of Case Study 3

Kampung Tok Tuan Mosque (KTTM), located in Kemaman, is among the oldest mosques in Malaysia (Figure 12). This mosque was established in the early 19th century. It is located in the village of Tok Tuan, Bandar Chukai, in the district of Kemaman, Terengganu. Chukai, which means 'tax', is regarded as a regional centre in Kemaman. Due to its proximity to the river, it is the main route used as a water transportation system and traditional food sources. The mosque is located at the fishing village and sits facing the Limbong River, which is one of the gateways to the area. It is still used as a place of worship and the dissemination of Islamic knowledge and social affairs of the community.

The mosque was built in 1830, along with the opening of Kampung Tuan (Figure 13). The founder and design of the mosque were Syeikh Abdul Rahman Syed Abdul Samad Al-Palembangi. He is also known as Tuan Mandak, a teacher and educator of the Islamic religion for the local community. He is the son of Sheikh Abdul Samad Al-Palembangi, a renowned religious teacher in Malay World, who was also known as the Islamic warrior against the Siamese Kingdom a long time ago. Nevertheless, the mosque famous in Terengganu due to the spirit and dedication of his great-grandson to Sheikh Abdul Samad Al-Palembangi, also known as Tuan Wok. He is also a religious teacher and has contributed much to the Islamic knowledge for the local community.

This mosque is also an institution for the study of Sufism knowledge known as *Tariqat Syeikh Saman or Sheikh Waliullah. Zikir* (hymn) ceremonies are often performed at this mosque that fascinates the attention of the nearby communities. The congregants of the mosque believe that prayers enliven the spirit of the place as well as significantly upholds the spiritual welfare of Tok Guru Wok. The faithful believed that their prayers are received by the Al-Mighty conclusively if they pray at this mosque. Therefore, the mosque extraordinarily becomes the focus of the community from all over the region during that time.

The mosque is used for Friday prayers in Chukai (Figure 14 & 15). However, in 1920, there were a few new mosques built to accommodate the growing population. Consequently, the number of prayers to this mosque decreased over time due to the ever-increasing number of new mosques. Eventually, the golden age of the KTTM mosque vanished following the demise of Tok Guru Wok. Later, his family members moved to Beserah in Pahang after a dispute with the locals. However, this mosque is still functioning for daily prayers until now. And there are even dedicated followers of Sufism who perform prayers here and conduct religious classes with the locals and ceremonies at the mosque.

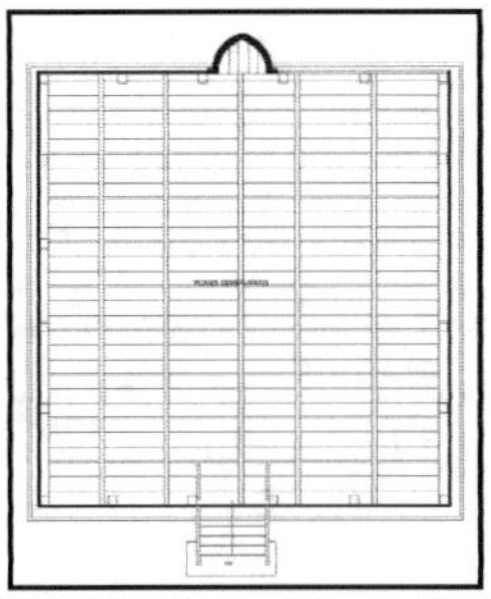

Figure 14. KTTM Floor Plan. This rectangular floor plan is designed without a column structure in the middle that supports the load from the roof structure. Looks like a curved-shaped mihrab on the forefront of the imam's use in congregation prayers.
(Drawing by: Wahid & Abdullah, 2019)

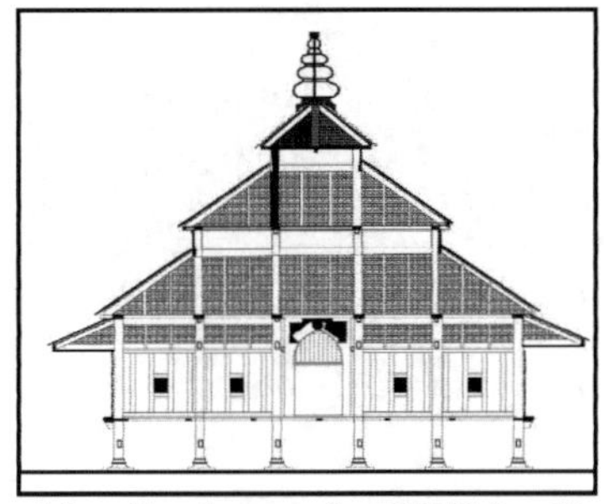

Figure 15. Section of KTTM, Kemaman, Terengganu.
(Drawing by: Wahid & Abdullah, 2019)

8.2 Carving Elements of Case Study 3

The carving motif on the mihrab of this mosque was decorated with lotus petals carved with *'tebuk timbul'* technique. There are not many decorative elements found on the KTTM. However, its pulpit has a different shape of wood carvings and the motifs of *'gunungan'* at the front panels. The walls of this mihrab are written with *'khufi'* calligraphy from the verses of Qur'an but are left incomplete (Figure 16).

According to on-site interviews, *'Khufi'* inscriptions are incomplete in the *'gunungan'* of mihrab panels, was deliberately left unfinished to interest readers to research the whole verse for comprehension further. However, according to Mrs Rosnawati Othman, there is a possibility that the carvers who did the *khat* panels had possibly moved elsewhere or had died before the carving was completed. There is also an urban legend that, if someone can practice and finish the verses of the Qur'anic inscription on the mihrab, a mystical power that can not be overwhelmed by any ordinary person engulfs him. Moreover, in carving, there is also a philosophy and principle saying that carving is deliberately prepared by carvers who are weak and imperfect, testifying the nature of human beings. Such recurrence is also the case in the brackets section of KLOM, Nilam Puri, Kelantan in which only two panels are completed out of the 16 sections of the bracket.

Figure 16. The Figure shows the carving of the khat at the mihrab. This symmetrical carving has a focal point in the center using lotus flower motif. On the right side of the panel is almost a perfect panel while on the left is unfinished works. (Image by: Wahid & Abdullah, 2019)

9. Case Study 4 : The Kampung Bukit Bayas Mosque, Kuala Terengganu (1830)

Figure 17. The Figure shows the KBBM, Kuala Terengganu from the outside.
(Image by: Wahid & Abdullah, 2019)

Figure 18. The Figure shows the KBBM, Kuala Terengganu from the outside.
(Image by: Wahid & Abdullah, 2019)

9.1 Background of Case Study 4

Kampung Bukit Bayas Mosque (KBBM) (Figure 17 & 18) was built earlier than Kampung Masjid Mosque with high-quality and refined artworks. According to Ayah Ngah Ustaz Ahmad Awang; Kampung Masjid was handed to Hj. Kadir in 1870. At that time Sheikh Kadir Al-Bukit Bayas was only a child. Sheikh of Bukit Bayas died in 1867 when Sultan Umar's timber palace was handed over to Hj. Kadir Kampung Mosque.

According to historians, Sheikh Haji Wan Abdul Kadir Bukit Bayas, at the age of 30 years old, came to Terengganu from Patani in 1824. Although his background in Patani was not announced, it is believed that he came from a devoutly religious family. Equally related to the story, he moved to Terengganu from the Siamese attack in Patani in 1823. He was accepted as a cleric by the people of Terengganu, the similar role he played while in Patani. His success in Terengganu spread, which led many Patani fighters to migrate to Terengganu in stages. Most of them came by sea and opened new settlements along the coast of Terengganu, namely; Besut, Batu Batu, Rusila, Dungun, and so on. Terengganu became their choice because of its stable political situation protected by the Kelantanese Kingdom, who was also the enemy of the Siamese. Coincidentally, the migration to Terengganu coincided with the era of the Dutch handing over Malaya to the British. The Patani population moved to the interior once they received community recognition. The age of Shaykh Haji Kadir Bukit Bayas was the period of strengthening Islamic knowledge in the Islamic society of Terengganu. During this time, the Patani Islamic fighter, such as Sheikh Haji Abdul Kadir Bukit Bayas, built the Islamic school in 1832 at Bukit Bayas. He was entrusted as a teacher to Sultan Daud (1830 - 1831), Sultan Mansur (1831 - 1837), Sultan Muhammad (1837 -1839) until Sultan Umar, who reigned from the year 1839 to 1877. Among his students was Tok Ku Paloh and Haji Abdul Rahman Limbung. According to another annotation, Sheikh Wan Abdul Kadir Bukit Bayas migrated with his friend, Tok Duyung (Haji Wan Abdullah bin Wan Mohammad Amin), to open an Islamic school in Duyung Kechil in the 1840s.

9.2 Carving Elements of Case Study 4

There are some exciting elements and motifs in this mosque. Different motifs can be identified on every panel with its function. There are also parts of carving that had been removed and installed in other places. The carved motifs on the walls of the mosque consist of flora, *khat,* and geometrical patterns. Most of the interior wall is decorated with sunflowers motif with *'tebuk tembus bersilat'* technique, while some parts are decorated with calligraphy and geometrical designs. The combination of sunflower and calligraphy motifs represent unique decorative elements. After scrutinizing the pattern, the motifs of the sunflower are not repetitive and not even the same between one panel to another (Figure 19). However, the source of the concept in the technique is similar, to 'the seed behind the stem". The shape of the symmetry is also identical, although the carvings on the left and right are different. All of these carving elements are Malay motifs that originated from the philosophy of the 'S' shaped then developed into

IOP Conf. Series: Materials Science and Engineering 1101 (2021) 012026 doi:10.1088/1757-899X/1101/1/012026

flower signifying a return to the source (Figure 2). These panels are surrounded by frames to accentuate its intricate carving.

Figure 19. The Figure shows an origin of a symmetrical carved panel with sunflower motifs designed using *'tebuk tembus bersilat'* technique. The concept of 'the seed behind the stem' circulates to form the 'kudup', flower and subsequent to the origin. An 'S' shaped is one of the essence of traditional Malay carving technique where solid and void silhoutte quite low but still maintains the balance to represent the entire panel of decorative elements. There is a frame bordering decoration that shows its scale and minute detail (Image by: Wahid & Abdullah, 2019)

Table 2. Table of 2D carving panel element at selected case studies Traditonal Timber Mosques. (Source: Wahid & Abdullah, 2019)

No	Item	Mosques			
		Kampung Laut Old Mosque, Nilam Puri Kelantan (1730	**Old Mosque Langgar, Kota Bharu (1874)**	**Kg. Tok Tuan Mosque, Kemaman, Terengganu (1830)**	**Kg. Bukit Bayas Mosque, Terengganu (1840)**
i	**Carving panel on wall (2D carving)**		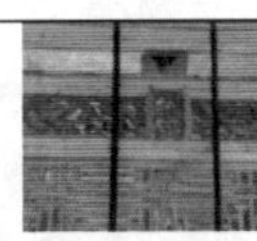		
ii	**Carving Style**		'Tebuk tembus bersilat' (symmetrical design)		'Tebuk tembus bersilat', and 'tebuk tembus' (symmetrical design)
iii	**Carving Motif**		'Bunga ketam guri', 'daun kerak nasi', Chinese Motif, with callighraphy		'Bunga ketam guri', 'kekacang', 'sesayap'
Total		**0**	1	**0**	2

The table above (Table 2.0) shows that there are only two mosques with 2D carved panels. For the Old Mosque at Langgar, Kota Bharu, there is only one panel using *'tebuk tembus bersilat'* technique and symmetry. This panel is located on the walls of the mosque. Among the motifs used are *'ketam guri', 'kerak nasi'*, a resemblance of Chinese motif and calligraphy.

For the mosque at Kampung Bukit Bayas, Terengganu, three panels use *'tebuk tembus bersilat'* technique, *'tebuk tembus,'* and applied symmetrical principles in panel design. Most of the motifs available are *'ketam guri,' 'kekacang,'* and *'sesayap'*. While for the other two mosques, there are no traditional carved panels. Most of the motifs used in the carving panel at this mosque are traditional Malay carvings, where the patterns, motifs, and designs are typical of its style.

IOP Conf. Series: Materials Science and Engineering 1101 (2021) 012026 doi:10.1088/1757-899X/1101/1/012026

Table 3. Schedule of comparison for 2D carvings element at selected Case Studies Traditonal Timber Mosques. (Table by : Wahid & Abdullah, 2019)

No	Item	Mosques and Location				Note
i	**Carving panel (2D panel)**	**Kampug Laut Old Mosque, Nilam Puri Kelantan (1730) = A**	**Old Mosque Langgar, Kota Bharu (1874) = B**	**Kg. Tok Tuan Mosque, Kemaman, Terenggau (1830) = C**	**Kg. Bukit Bayas Mosque, Terenggau (1840) = D**	
		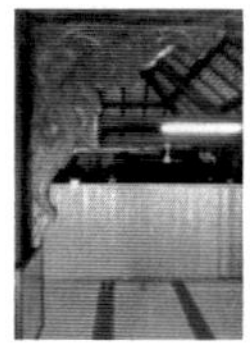			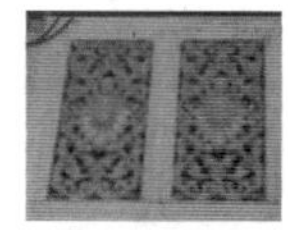  	
ii	**Overall Design**					
	Symmetrical	**O**	**O**	**O**	**O**	**A/B/C/D**
	Assymetrical	**X**	**X**	**X**	**X**	**-**
iii	**Overall Motifs:**					
	Daun sesayap	**X**	**X**	**O**	**O**	**C/D**
	Ketam guri	**X**	**O**	**X**	**O**	**B/D**
	Kekacang	**X**	**X**	**X**	**O**	**D**
	Matahari	**X**	**X**	**X**	**O**	**D**
	Langkasuka	**O**	**X**	**X**	**X**	**A**
	Teratai	**X**	**X**	**O**	**X**	**C**
	Chinese	**X**	**O**	**X**	**X**	**B**
	Kerak nasi	**X**	**O**	**X**	**X**	**B**
	Daun Tua	**X**	**X**	**X**	**X**	**-**
iv	**Overall Technique :**					
	Freeform	**X**	**X**	**X**	**X**	**-**
	Framing	**O**	**O**	**O**	**O**	**A/B/C/D**
	Tebuk tembus	**X**	**X**	**X**	**X**	**-**
	Tebuk tembus Bersilat	**X**	**O**	**X**	**O**	**B/D**
v	**Location :**					
	On Wall	**X**	**O**	**O**	**O**	**B/C/D**
	Above window	**X**	**X**	**X**	**O**	**D**

	Above door	X	O	X	X	B
	Mihrab	X	X	O	O	C/D
	Post	O	X	X	X	A
	Ceiling surface	X	X	X	O	D
	Total	4	8	7	11	

Concerning the above analysis, the entire A, B, C, and D mosques had applied symmetrical principles in the forming of carving panels. Whereas in the carved motifs; the leaves motifs are found in mosque C and D. The mosque B and D use the *'ketam guri'* and mosque C using *'sesayap'* motif. Most elements of these motifs are combined with other motifs. The mosque D panels introduced the sunflower motifs. Meanwhile, mosque A also uses bracket with the Langkasuka motif. Mosque C uses the motifs of the lotus flowers, which are among the earliest motifs in traditional artistic sculpture. At mosque B, there are many applications of Chinese-style panels and *'kerak nasi'*. Most of the panel uses framing forms in all the mosques. Mosques B and D use the *'tebuk tembus bersilat'* while mosque A and C decoratively use a panel of *'tebuk timbul'* technique.

The location of most of these carving panels for mosques B, C, and D are on the walls (Table 3.0). Meanwhile, some carvings at mosque D, which are evidently on the components of the windows. Only one panel of the carvings are on the top of the door that is at mosque B. Besides that, the mihrab section of mosque C is heavily decorated with plant motifs carving. While, at mosque A, there are brackets located at the mosque. As for mosque D, the carved elements on the ceiling is visible.

10. Conclusion

In summary, the study found that the Malay wood carving is divided into several facets where a typical traditional Malay carving can be identified from its character. Through the interview with Mrs Rosnawati Othman, she confirmed that the motifs and elements of the decorative subjects could be distinguished from its composition. The most obvious indication of this statement is the frame or border of each work signifying prominent aspects. Nevertheless, there are carvings with no frames, especially those designed in free-form such *'gunungan'* or continuous piece.

The second finding shows that there are symmetrical elements in the carvings. Furthermore, most carvings are repetitive of the same motifs, which gave a balance in visual impression to the naked eye. It is a justification for the concept of balance in life. The third is that it demonstrates a significant factor in the design of traditional Malay carvings, where there is a different pattern of carvings between one side and the other. Conspicuously, when viewed at a close range, most of the panels in traditional carving have a slight difference between the left and right. The imperfect symmetry is deliberately to highlight the human weakness in poetic comparison towards the nature of design by the Creator. Furthermore, it justifies a symbolic representation of the natural element, which is not equal to each other. The philosophy of this carving makes it visually balanced.

The fourth is identified through the source of the conception of the wood carving motif. Most of the traditional Malay carvings analogy resembles the seed in the pot, then the soil, later the water source, and in the end the flower or the stem. The interpretation of lines started from simple circular line-form of the letter 'S' and connected with leaves from the stem (*kudup*), added with flowers that show the complete carving results. The artistry illustrates the philosophy of Malay art, which is gentle, minute-detailed, and intricate-implication to the environment, society, and obedience to the Creator. Accordingly, the philosophy of carving and nature in the craftsman's identity makes the carvings seemingly fluid. As Noor and Khoo [5] profounded that *"Asal itu punca, punca itu rahsia, tajam tidak menikam lawan, tetapi berlegar mesra" (literally translated – the beginning has a source, the source is secret, sharp as fighting an opponent but it matches harmoniously).*

All of these elements are the arch characteristic of traditional Malay carvings. The case study established that the motifs of woodcarving in all of the four mosques are somewhat similar in their

IOP Conf. Series: Materials Science and Engineering 1101 (2021) 012026 doi:10.1088/1757-899X/1101/1/012026

motifs' appearance. It may be due to environmental factors, skills, knowledge, craftsmanship, and culture. In Kelantan, for example, most of the motifs found are the motif of the *'ketam guri'*, the *'ketumbit'* and a mixture of other motifs. This repetition is because, most carvers in Kelantan decline to mark carvings in surrealist concept and negated to imitate the nature directly, replicating the same design. However, the final design is in abstract form, which unrivalled the Creator's creation. Most carvings are simple but still retain their characteristics as a traditional style. Also, the application of solid and void is to balance the carving panel. Therefore, most carvings in Kelantan are simple in terms of design, motifs, and intricateness.

In contrast to the carvings in Terengganu, most of the motifs used are from the *'ketam guri'*, *'kekacang,'* and *'sesayap'*. According to Mrs Rosnawati Othman, sculptors in Terengganu considered carving as art compared to carving in Kelantan which received the influence of wood carving from Patani. Therefore, the complexity of Terengganu's carving style makes it rather prominent and refined in terms of quality and results. Although most of the designs of carving patterns are taken from carving designs in Kelantan and Patani, yet, they further refined the work. Consequently, Terengganu carvers focused their skills on the fineness and quality of the carving. The discovery of carvings at the Kampung Baruas Mosque in Kuala Terengganu confirms this argument. Most notably since the carvings here are elaborate, and the intricateness of solid and void elements is evident. Nevertheless, it is also likely that there is a personal touch of Patani sculptors that had immigrated to Terengganu, had also contributed their expertise in the carving artistry.

11. References

[1] Abdullah, A. R. 1998 *Pemikiran Islam di Malaysia*. Kuala Lumpur: Dewan Bahasa dan Pustaka.

[2] Yatim, O. M. 2000 *Seni Bina Melayu Islam: Pengaruh Dalam Alam Bina Dalam Tamadun Islam di Malaysia*, Kuala Lumpur : Dewan Bahasa dan Pustaka.

[3] Surat, M. 2008 *Pembangunan Tipologi Masjid Tradisional Melayu Nusantara*, Thesis Sarjana (unpublished), Bangi. Universiti Kebangsaan Malaysia.

[4] Yudoseputro, W. 1986 *Pengantar Seni Rupa Islam di Indonesia,* Bandung : Penerbit Angkasa.

[5] Noor, F. A. & Khoo, E. 2003 *Spirit of Wood: The Art of Malay Woodcarving*. Singapore. Periplus.

[6] Ahmad, A.A., Ali. A. & Arbi, E. 2013 Typology of Interior Architecture for Traditional Masjids in. *Jurnal Al-Tamaddun* **8**(2) pp 89–100.

[7] *Ekslopedia Islam* 2006 Kuala Lumpur: Dewan Bahasa dan Pustaka.

[8] Hanafi, Z. 2007, *Reka Bentuk Pola Hiasan Dalam Bangunan Tradisional Melayu*. Kulim, Kedah: Amber-Solara Publication.

[9] Nasir, A.H. 1987 *Traditional Malay Wood Carving*. Kuala Lumpur: Dewan Bahasa dan Pustaka.

[10] Hussin, H., Baba, Z. Hassan, A. & Mohamed, A. E. 2012 The Philosophy in the Creation of Traditional Malay Carving Motifs in Peninsula Malaysia, *Malaysia Journal of Society and Space* **8** (7) pp 88 - 95

[11] Perbadanan Kemajuan Kraftangan Malaysia 2009 *Ukiran Kayu*, Kuala Lumpur: Perbadanan Kemajuan Kraf Tangan Malaysia.

[12] Hassan, M. A. 1992 *Pendidikan Estetik Daripada Pendidikan Tauhid*. Kuala Lumpur : Dewan Bahasa dan Pustaka.

[13] Kling, Z. 1989 'Kebudayaan Melayu Sarawak', Special Issue no.4 part II, Sarawak Museum Journal, Kuching, Sarawak.

[14] Mizan, D. & Anuar, T. 2005 The Malaysian Mosque: Evolution, element and meaning. *Journal of Built Environment*, UiTM **2** (1) pp 1-10.

[15] Al-Ahmadi, A.R. 1978 *Sejarah Masjid Kampung Laut*. Kuala Lumpur: Kementerian Belia dan Sukan.

Eclecticism in Architecture of *Masjid Bandaraya Kuching*, Malaysia

B Karsono[1], D R Koesmeri[1], J Wahid[1], B M Saleh[2]

[1] Architecture Department, Faculty of Built Environment, University Malaysia Sarawak (UNIMAS), Malaysia
[2] Architecture Engineering Department, College of Architecture, Art and Design, Ajman University, United Arab Emirates (U.A.E.)
Corresponding author: kbambang@unimas.my

Abstract. The eclecticism approach is unconsciously applied in the design process by selecting elements considered as the best elements from different architectural styles or eras to produce a new distinctive building model commonly applied by architects in design. Nowadays, the combination of various architectural elements from different styles into one building is often created, especially in Malaysia. In this research, the author focuses on the design of mosques in Kuching. Throughout history, the *Masjid Bandaraya Kuching* (known as *Masjid Bahagian Kuching* by the locals) has undergone four (4) stages of transformation process since it was first built in 1840 A.D until the latest changes that took place in 1968 A.D. This study aims to reveal the eclecticism approach applied in every stage. Qualitative method is applied in this research by examining the photographs from the archives and analysing the longitudinal timeline and cases through the history of Kuching, literature and observations concerning the *Masjid Bandaraya Kuching* to understand the initial forms and changes that have occurred from the past to the present. The analysis takes into account both comparative and contextual analytical techniques to represent a close interpretation. Findings indicate that the building of *Masjid Bandaraya Kuching* has been transformed through the traditional eclecticism as well as radical eclecticism approach.

Keywords. *traditional eclecticism, radical eclecticism, transformation, masjid*

1. Introduction

Undoubtedly, that before the colonialization, the architecture identity of mosques (*masjid*) in Southeast Asian region was very distinctive and reflected the origins of the indigenous characters of its people conforming to their environmental and social interests. Nowadays, these identities are slowly diminishing because of the influence of the foreign perceptions and thoughts that interfere with the thinking of the architects especially from Western countries as well as the influence of the Islamic conceptualization of images of domes and arches from the Middle East and Southern Asia.

The beginning of mosque architecture development in Southeast Asia was closely related to the process of Islamization of the locals in this region. Scholars argued that the arrival of Islam in Southeast Asia was brought by Muslim traders from the Arabian Peninsula since the 7th century, but there is no sufficient evidence to support such statements. Based on historical records, Islamization began in the late 12th century, when *ulama* from various places in the Arabian Peninsula wanders in Southeast Asia [1]. The *ulama* from Demak (Central Java-Indonesia) in 1498 spread Islam in South Kalimantan, formerly known as Banjarmasin [1]. At the end of the 15th century, the Sultanate of Brunei accepted Islam and then spread it throughout their territory, including Sarawak [1]. In the mid-19th century, the

IOP Conf. Series: Materials Science and Engineering 1101 (2021) 012027 doi:10.1088/1757-899X/1101/1/012027

development and appreciation of Islam became more widespread and stronger in Kalimantan and Borneo. However, towards the end of the 19th century, the blooming of Islam in Sarawak showed a slight decline due to the arrival of Brookes and the Christian missionaries that began in 1839 or earlier. The spread of Christianity brought by Western Missionaries and explorers endured a strong based by building Churches and Mission schools in the state of Sarawak. However, a year before Brooke's family established their stronghold, the first typology of the mosque known as *Masjid Bandaraya Kuching* was built, today known as Masjid *Bahagian* among the locals living in Kuching.

Throughout history, the *Masjid Bandaraya Kuching* has experienced several gradual changes, specifically four stages due to the renovation and maintenance process. The first stage was in 1840 initiated by one of the Malay community leaders, Datuk Patinggi Ali. The shape and form of the mosque were very simple, built on stilt with pyramid-shaped roof and constructed in timber. The second stage was in 1880 when the first renovation took place by lowering the floor on to the ground and strengthening the structure of the main building by introducing concrete cast-in-situ columns, concrete floors and brick walls, however, the building form still reflected both the traditional Demak Mosque in the island of Java, Indonesia and Masjid Kampung Laut, Kelantan, Malaysia. The form and configuration of the floor plan and the three-storey pyramid-shaped roof show an integrated image within the surrounding. The third stage took place in 1929 when the second renovation was carried out and the major changes were made with the top level of the pyramid-shaped roof replaced by an Egyptian Mamluk-style dome.

The first Prime Minister of Malaysia visited Sarawak in February 1966 and he commented that the State Mosque of Sarawak was not suitable for the growing population of the state and ordered the Sarawak State Mosque to be replaced with a new representative building. A year later the old mosque was demolished and the process of constructing a new mosque began. The design for a new mosque was new, emulated the Moghul-style building with a main onion-shaped dome in the centre and smaller domes around it. When the construction was completed in 1968 it became a distinct building for the locals and it remains until today.

Thus, this research attempts to focus on the transformation and physical changes in the architecture of *Masjid Bandaraya Kuching* which is expected to correlate with an eclectic approach in the design process. The purpose of this study is to reveal the eclectic approach at every stage of the physical changes in the *Masjid Bandaraya Kuching* and its influence on the local built form.

2. Methodology

This research applies a qualitative method with an analytical interpretation by taking *Masjid Bandaraya Kuching* as the case study. The initial technique of the research is to gather the information and data for accurate deliberation of the process. The photographs from archives and longitudinal timeline from the history of Kuching as well as an observation on the *Masjid Bandaraya Kuching* to comprehend the changes that have taken place from the past to the present [2]. The analysis takes into account both comparative and contextual investigative techniques to represent a close interpretation of how eclecticism occurs. The case study helps to provide a comprehensive overview of the study to draw specific conclusions on the design approach of the mosque [2]. The exploratory technique also applied by studying the works of other authors and various historical literature concerning the *Masjid Bandaraya Kuching*. The field observations and documentation related to the current building form, pattern and details help to transmit closely to the objective of the study in detail.

3. Literature: Eclecticism in Architecture

The word eclectic comes from the Greek word *eklectico*s which means to choose the best. The eclectic style of architecture became popular in Europe and America in the 19th and 20th century during the revival movement series. The eclectic style refers to each design incorporating elements of previous traditional motifs, decorative aesthetics, structures that come from other cultures or past architectural era. This style combines such elements from the past to create something new and original. In architecture, these elements include the structure and form of buildings, ornaments, decorations taken

IOP Conf. Series: Materials Science and Engineering 1101 (2021) 012027 doi:10.1088/1757-899X/1101/1/012027

from local-traditional culture or other countries [3]. An eclectic approach in architectural design means the process of choosing the best-selected elements derived from an array of different architectural styles. Eclecticism as a method of design approach can be interpreted as a process of selecting parts of thinking, principles and architectural elements from the past which are then altered in such a way that parts of thought, principles and elements are composed to create a new theory, new principles and new styles despite having the same name.

Jencks [4] clearly states that eclecticism has a wide range of exploration. He added that [4], eclecticism is a kind of weak compromise, a mishmash in which second-class thinkers can take refuge in the confusing complexity of antinomy; they combine contradictory material with the hope of avoiding different choices or looking through problems to creative conclusions. Eclecticism is a natural evolution of culture with choices and can be categorized into traditional eclectic and radical eclectic. Traditional eclecticism is a process to choose the right style or sub-system in the right place, while radical eclecticism mixes these elements in one building [4]. An architect can glean various architectural replicas or artificial components that might be a part of a roof, wall, column, entrance, courtyard, building façade, arcade, arches and others. Afterwards, assemble them in one building according to his design concept or the owner's wish for the building [5] [6] [7] [8].

Throughout history, eclecticism has played an important role in the creative process of combining such elements from various styles leading to new design models. The eclectic approach was used to clench elements of such style and assemble it into one building, generally, the result will show strange new features and consequently considered as abusing the concept of notable buildings [9]. This situation will fabricate the local understanding of the architectural features. Sometimes, such practices also show simplicity in designing buildings of similar importance such as masjid, however, the eclecticism must adhere to the function of the potential buildings to avoid wastage of space and developing a new vocabulary of vernacular elements of architecture [10].

4. Findings and Discussion

4.1. Features of the first built

There are only a few historical records that explain the physical condition and form of the first *Masjid Bandaraya Kuching*. Limited notes and photographs from the past were obtained and used for analysis. From the written record the original building was built in 1840 initiated by Datuk Patinggi Ali, a community leader in Kuching. The mosque is located on the bank of the Sarawak River and its location is still maintained until today. A special feature of the building is on stilt with a pyramid roof where most of the material is dominated by local timber. Construction work was carried out by the locals with techniques that are inherited from generation to generation. This gives the impression that this first building features have a strong local vernacular character.

Figure 1. Site of *Masjid Bandaraya Kuching*, never moved to another location until now.

Figure 2. Image of first built in 1840 A.D [11]

4.2. Features after the first renovation

In 1880, along with population growth in the area, the existing mosque could no longer accommodate the growing congregation, so the first renovation was carried out. This renovation took place by the local community with assistance from Brooke's government. The usage of new materials such as brick walls, cast-in situ-columns and concrete floors was implemented. The floor level of the building which was previously built on stilts was lowered to the ground and elevated higher than the level of the street to prevent flooding from Sarawak River (figure 3). The influence of Masjid Demak, Java-Indonesia (figure. 4) and Masjid Kampung Laut, Kelantan-Malaysia (figure.5) are clearly shown through the roof shape with a three-tiered pyramid roof. This influence is most likely brought by *ulama* who came from Java Island or the Malay peninsula who teach Islamic knowledge to the locals in Sarawak. This phenomenon indicates that the traditional eclectic approach was carried out by the community to choose the right style that is suitable for their culture and their understanding. This approach is tallied with Jenks [4] idea of choosing the right style or sub-system in the right place which is claimed traditional eclecticism process in architecture.

Figure 3. Image of the first renovation in 1880 A.D

Figure 4. Image of Masjid Demak, Java-Indonesia, 1479 A.D.

Figure 5. Image of Masjid Kampung Laut, Kelantan-Malaysia, 18th century

4.3. Features after the second renovation

In 1929, initiated by Islamic leaders and the Muslim community with the assistance of Brooke's government, the second renovation was carried out to the *masjid*. The main concern of the renovation was to add a dome structure on the top level of the pyramid-shaped roof which was replaced by an Egyptian Mamluk-style dome (figure 6). The renovation also replaced the doors of the masjid with an entrance doorway and large windows of typical European buildings. The entire renovation was completed in 1930.

There's no historical explanation found why Egyptian dome style was selected to replace the top pyramid-shaped roof. Nevertheless, this phenomenon can be judged through a famous trend during that time. This was associated with the upsurge of sea expeditions by European explorer to Southern Asia and Southeast Asia region. The earlier great sea expedition at the end of 15th century pioneered by the

Portuguese through the Cape of Hope (South-Africa) and the Indian Ocean and later followed by the Dutch, British and French left a cross-cultural traces of social and physical influence. Even the second expedition around the colonization of the Philippines by the Americans and taken over by the Spaniards [11]. The discovery of the new world and expeditions by the Western missionaries also change the worldviews of the locals due to the information and new technology in building techniques that slowly change the way the local behaves.

Figure 6. Second renovation, top pyramid-shaped roof replaced with a dome, 1929 – 1930 A.D

The implemented dome was influenced by the building elements during the Mamluk dynasty in Egypt around the 15th century. The dome at the Complex of Sultan al-Ashraf Qaytbay, the City of Dead (the Northern Cemetery), Cairo, Egypt which was built in 1470-1474 AD can be used as a reference as in Figure 7. The basic shape of the Complex of Sultan al-Ashraf Qaytbay dome was copied and implemented in the second renovation of Masjid Bandaraya Kuching, proportions and scale between the dome and its base were maintained. The modification only appeared in the usage of materials and construction techniques. Mud bricks were implemented at the dome in the Complex of Sultan al-Ashraf Qaytbay, a favourite material that was widely used during the Mamluk and Abbasid Dynasty in 14th-15th centuries [12]. While, for Masjid Bandaraya Kuching, the materials were a combination of a cast in situ concrete for the dome base, brick walls and steel for the dome structure.

To the locals, the combination of the pyramid roof and the dome became a new sight that they had never seen before. The dome is believed to be the first dome structure built in the state of Sarawak. The phenomenon indicates that the radical eclecticism approach was implemented by introducing Egyptian-dome as an element to be added to the existing building to fulfil the desire of the influential figures in the society during that time. Adversely, this situation forced the local people to replenish their memory with a new image, gradually they will accept it as a part of their daily life and culture.

Figure 7. Dome similarities between Complex of Sultan al-Ashraf Qaytbay (built-in 1470-1474 A.D) and second renovation of *Masjid Bandaraya Kuching* (1929-1930 A.D)

4.4. Features after the new design in 1968

In February 1966, His Excellency Tunku Abdul Rahman Putra Al-Haj, Malaysia's first Prime Minister considered that the existing building of Masjid Bandaraya Kuching was no longer suitable to serve the Muslim community in Kuching. Therefore, he proposed to replace the mosque with a new and more representative building to commemorate Sarawak independence in Malaysia. By the time of the new design started to emerge in 1967, an existing Masjid Bandaraya Kuching was demolished and the process to construct a new building began (Figure 8). However, the new design has a close influence of Mughal architecture (1494-1707 A.D) from North India [11], that appear on the new *Masjid Bandaraya Kuching* as in Figure 9. In Figure 10 the Badshahi Mosque in Lahore, Pakistan built by Mughal emperor (1672-1674 AD) also showed a close reference for the new mosque too.

Figure 8. Demolition process in 1967 due to constructing the new design

Figure 9. Image of the new design, finished in 1968

Figure 10. Masjid Badshahi in Lahore, Pakistan, built by Mughal emperor (1672-1674 A.D)

The characteristics of Mughal buildings generally dominated by specific elements on the roof, such as grand dome located in the centre of rooftop; followed by smaller domes beside the grand dome with four minarets in each corner of the building. The Mughal style also showed a similar example with a small dome of the same height plus a grand dome in the centre as well as two smaller pinnacles with small domes above the main entrance to the prayer hall. The wall fenestration as the elements of building facade was decorated with a repetition of arches arranged geometrically and a larger arch as the main entrance with a specific Mughal-ornaments covered the external wall with attractive geometrical compositions. The arches were functioned as access to the praying hall so that the permissively high entrance gave a sense of welcoming and devotion to the Almighty.

It is obvious that the arrangement of Mughal's elements was duplicated at the *Masjid Bandaraya Kuching*, the architect had modified the shapes and scales of such elements to adapt to local circumstances. The roof as the main elements of grandeur building especially the religious building as a mosque is covered with an onion-shaped dome together with four minarets on each corner of the building and two small minarets on the main entrance were implemented similarly with the Mughal design. The wall fenestration is implemented repetitively with concrete arches to frame the windows surrounded the perimeter of the mosque. Unlike the Mughal mosque, there was no ornament applied on the external wall, consequently, the building façade became plain and clean.

This building is considered an eclectic example of Mughal style applied to the new building as a new architectural feature in the Southeast Asia region and it was believed to be the first Mughal-style building in Sarawak. According to Jenks [4], this process can be categorized as a radical eclecticism due to the conscious interpretation of design element by the architect concerning Mughal's elements and arrangement principles to be implemented into one building. On the other hand, the previous mosque that was demolished did not carry any resemblance of design elements associated with the new mosque to commemorate the essence of the old *Masjid Bandaraya Kuching* which was never seen before by the locals.

5. Conclusion

Table 1 summarizes the transformation that was carried out at the *Masjid Bandaraya Kuching*. The first building typology (1840 A.D) correspond to the origin in the local and indigenous technology and culture. However, forty years later, the first renovation took place in 1880 A.D and the traditional eclecticism elements borrowed from the Nusantara region (Java Island and Malay Peninsular) emerged. Paradoxically, during the second renovation in 1930 a radical eclecticism transformation occurred and in 1968 with a new design influenced by Egyptian-dome and Mughal-style elements implemented in the building. Each stage of renovation, radical changes visibly detached the indigenous style with alien features appear to the mosque.

Table 1. Transformation of *Masjid Bandaraya Kuching*

Stage	Date	Building appearance and data	Architecture Style
1. First Built	1840 A.D	- On stilts - Pyramid-shaped roof - Strong vernacular feature - Local timber used as main materials - Built by local people	Vernacular Architecture
2. First Renovation	1880 A.D	- Lowered on the ground level by taking soil to elevate floor level. - Three-tiered pyramid-shaped roof, similar features with Masjid Demak, Indonesia (1479 A.D) and Masjid Kampung Laut, Malaysia (18th century) - Brick walls, cast-columns, cement floors and timber for roof frame were used for construction technique and materials - Built by local people with assistance from Brooke's government.	Traditional eclecticism [4], similar style with: - Masjid Demak in Java, Indonesia (1479 A.D) - Masjid Kampung Laut in Kelantan, Malaysia (18th century)
3. Second Renovation	1930 A.D	- Changes on the highest pyramid-shaped roof, replaced with Egyptian-dome (Mamluk era from 15th century). - Strange/new appearance in Sarawak, a combination between pyramid-shaped roof with a dome on the top, was considered as the first dome structure built in Sarawak. - Brick and cast-cement were still maintained, walls and columns, timber for roof frame, steel for dome structure. - Built by local people with assistance from Brooke's government.	Radical eclecticism [4]: - Awkward appearance never existed before.
4. Old building demolished, replaced with new design	1968 A.D – nowadays	- Double volume interior with mezzanine floor. - Mughal influences with minor modification, the onion-shaped dome as the main dome. - Minarets were inspired from Mughal style and implement with some modification on the dome. - Repetition of arches which was originated from Persia. - New appearance for the locals which had never existed before.	Radical eclecticism [4]: - The usage of Mughal style elements on the minarets and centre dome popped up.

This study found out that eclecticism approach stated in this research was mainly showing the selection of the best elements to be integrated into the existing building to produce a distinctive design. Perhaps, the main purpose of this eclectic approach is to provide historical and technical value to the building. On the contrary, this approach came in the incorrect direction due to the elements of the past style were resurfaced with factors and circumstances different from their original conditions. For example, the addition of Egyptian-dome during the second renovation and duplication of Mughal-Indian style on the new building eventually became a shallow appearance to the locals. Elements which were taken from architectural works from past styles, then applied or implemented is just an embellishment to buildings that did not represent the era from different regions and cultures that may mislead local cultural heritage and understanding. These buildings will become a collection of incompatible models that do not express the continuity of local culture. Eclecticism that disseminates in the public via misrepresentation of protruding excessive pride will create confusion for the locals. The younger generation may blindly copy this element to represent their architectural identity. We should not overlook the regional cultural treasures in Southeast Asia and the *Nusantara* regions whose elements are more suitable to be used and implemented in our architecture.

6. References

[1] Rahmawati 2014 Islam di Asia Tenggara, *Jurnal Rihlah,* Vol. **II**, **1** pp. 104-113.

[2] Groat L, Wang D 2013 *Architectural Research Methods* (New Jersey: John Wiley & Sons, Inc.).

[3] Burden E 2002 *Illustrated Dictionary of Architecture* (McGraw-Hill Companies, Inc.).

[4] Jencks C 2002 *The New Paradigm in Architecture: The Language of Post-modernism* (Yale: Yale University Press).

[5] Kaji-O'Grady S 2012 Authentic Japanese architecture after Bruno Taut: the problem of eclecticism, *Fabr. J. Soc. Archit. Hist. Aust. New Zeal.,* vol. **11**, **2** pp. 1–12.

[6] Carpo M 2005 Architecture: Theory, Interdisciplinarity, and Methodological Eclecticism, *J. Soc. Archit. Hist.,* vol. **64**, **4** pp. 425–427.

[7] Pachta V, Papayianni I 2017 The Study of the Historic Buildings of Eclecticism in Thessaloniki Under the Prism of Sustainability, *Procedia Environ. Sci.,* vol. **38** pp. 283–289.

[8] Haddad E 2009 Charles Jencks and the historiography of Post-Modernism, *The Journal of Architecture,* **14:4** pp. 493-510.

[9] Sami I 1967 *Theories of Architecture, First Year Course for Architecture Students* (United Color Printing Press).

[10] Karsono B. *et al* 2018 The Usefulness of Main Prayer Hall of Masjid Agung Islamic Center of Lhokseumawe, Aceh-Indonesia, *IOP Conf. Ser.: Mater. Sci. Eng.* **401** 012012.

[11] Wahid J, Karsono B, 2011 *Desain dan Konsep Arsitektur Lansekap dari Zaman ke Zaman* (Yogyakarta: Graha Ilmu).

[12] Khali M A O 1997 *Reflections of the cultures of architecture and architecture in Egypt with a special mention for the city of Cairo (a suburb of Maadi)* (University of Cairo).

IOP Conf. Series: Materials Science and Engineering 1101 (2021) 012028 doi:10.1088/1757-899X/1101/1/012028

Place Attachment in the Riverfront Public Space Case of Malaysia

B Karsono[1], **B M Saleh**[2], **C O Chung**[2] **and J Wahid**[1]

[1] Architecture Department, Faculty of Built Environment, University Malaysia Sarawak (UNIMAS), Malaysia
[2] Architecture Engineering Department, College of Architecture, Art and Design, Ajman University, United Arab Emirates (U.A.E.)
Corresponding author: kbambang@unimas.my

Abstract. Many cities in Malaysia established their first development on the banks of the river. Urbanization phenomenon and technological advances then cause these riverfront areas developed and changed gradually. Progressively, inappropriate physical changes at the riverfront environment can reduce the usage and meaning of place together with the level of place attachment. The role of physical appearances related to the activities become famous issues in urban design discourses as an effort to establish the identity to the place; nevertheless, the research to measure the attachment level in public space is rarely done. This research is to identify the place attachment and its contributed components focus on pedestrian space in the riverfront area, public spaces which are favoured in Malaysia. Mixed-method approaches are implemented in this study by developing a framework to indicate the elements that affect the place attachment by the user. Surveys and interviews carried out in two riverfront promenade in Kuching and Malacca, there are Sarawak and Malacca riverfront promenade. 330 respondents involved in the questionnaire while 26 people were interviewed. Triangulation method is used to analyse the data and the results are achieved based on the objective of the research. The results found that some factors affected the level user's attachment, namely: their familiarity to place, length of engagement, personal background and user's role.

Keywords. *place attachment, familiarity to place, length of engagement, place dependence*

1. Introduction

Place attachment is related to the quality of user's experiences within the urban space, concerning the importance and function of a place who use and occupy the space. It is important to recognize precisely the specific quality embedded in a place resulted from the user's perception in a correct interpretation. Therefore, to investigate the quality of place attachment and its elements become an important effort to understand the identity of the place. Place attachment is an important component to generate and giving meaning to the place in supporting urban life and sustaining the activities within the city.

This study attempts to evaluate the physical elements in urban centres in Malaysia by identifying the characters and identities. Besides, external influences have led the cities to emerged in global features, which have reflected on the homogeneity of buildings shape and appearance, instead of loss of local cultural values, loss of urban attractiveness, the role of public space for cultural as well as social interaction. This transformation will influence the way users perceive and experience the places and will giving effects on their existing psychological meanings. Loss of psychological meanings can affect the

IOP Conf. Series: Materials Science and Engineering 1101 (2021) 012028 doi:10.1088/1757-899X/1101/1/012028

identity of place and interfere with the level of emotional attachment to the place.

This study aimed to examine the place attachment and its elements that influence the perceptions of the user in the riverfront promenade in Malaysia. Further to this, the research will explore the causal relationship between place attachment, place meaning, attributes as well as characteristics of a place that strongly influence the level of attachment.

Two riverfront open spaces that pioneered riverfront urban renewal as public spaces in Malaysia were selected as the case study, i.e. the Sarawak riverfront promenade (S-RP) in Kuching and the Melaka riverfront promenade (M-RP) in Malacca. The two case studies are not analysed, by comparison, however, the context of the discussion will have a bearing on the objective of the study. The picture from these two places will enlighten the discussion further. Both of the study areas was dominated by important historical buildings, sites and streets which were constantly changing according to the time of day. Concerning this place, the historical value is one of the reasons for visitors to stopover at the sites along with the existent of modern physical design features available at the site which become alternate destination and attraction for the visitors.

From the above background, this study identifies the element which covers the familiarity related to the place, length of engagement, the background of the user and user's role as well as the strong characteristics that influence the place attachment. This study will assist in developing parameters for the development of local urban areas more identifiable and meaningful to users.

2. Literature Review

There are three elements needed to figure out the quality of a place i.e. physical features, functional concern, and meanings or symbols [1]. According to Relph [1], a place is made up of physical forms, supported by activities and developed meanings. Canter [2] has a similar view in describing a place – he profound that it is three-dimensional generated by its activities, physical environment as well as the concept of place. In light of this, this research aims to find out the level of attachment in the riverfront promenade within the city in Malaysia. This research argues that the attachment means the relatedness of physical elements and its function or relatedness of the user to a specific place. It can be understood that the level of attachment will influence the meaning of the place.

Definitions for place attachment is known as a good emotional attachment among individuals within their groups to a certain environment [3]. For the last three decades, attachment of places and concepts related to the sense of place has become a debated theory among scholars. [1], [3], [4]. Various efforts through empirical study have been conducted to explore these concepts for the last 15 years, and this study becomes part and partial of the contribution to the study focusing on measuring the quality of attachment.

Majority of scholars agree that the level and form of attachment are affected by several elements, i.e., by characteristics of socio-demographic [5], [6], activity and management relationships [7], [8] and landscape type [9]. Lobo [4] argue that the sense of place in North America's community emerges from the length of engagement and also concerning all the values of its emotional and physical and perceptual as well as attributes that attached to the place.

The physical appearances and its characteristics contribute to the formation of a sense of place and followed by intense attachment and place satisfaction [10]. Researchers sometimes failed to describe the degree of influence and the causes of attachment formation [10]. Corresponding to this opinion, the research is composed to examine the degree of the characteristics and attributes of riverfront promenade affect the place attachment.

Research in place attachment usually encompasses the dimensions of physical appearances, perceptual and psychological as well as socio-cultural aspects to the place. Urban design has accentuated the relatedness of these dimensions in research into the perception and the quality of place within the city. Place attachment delineates the bonding of people with their place and interdependence among them which create the characteristics and attributes within the place. Assessment of the attachment level and recognition of place differences may clarify the relationship between perception and meaning through the attributes that they have on the place. By taking the riverfront promenade as the case study

IOP Conf. Series: Materials Science and Engineering 1101 (2021) 012028 doi:10.1088/1757-899X/1101/1/012028

it is believed that has a strong influence on the nature of the city centre and acts as a contributor to the lives of urban users. In this context, place attachment research is important to preserve the common sense of the city as well as enhancing the well-being of users who have an affinity for the place.

3. Methodology

Many researchers and scholars conduct quantitative and qualitative methods in evaluating place attachment [6], [11], [12], [13]. They had established two methods of measuring environmental perceptions and attachment either direct or indirect [14], [15]. Indirect methods are self-assessment reports (including questionnaire, interviews, checklists and discussion) and time-sampling (reporting at several time intervals on what to see), whereas the direct method implicates a phenomenological approach to examine people's activities and events, as they occur spontaneously in daily life [16]. This technique can be produced by a critical description concerning the phenomenon and discarded all preconceived notions about the object [17]. The purpose is to conceive the qualitative uniqueness concerning the place's holistic meaning. This method has been broadly used by scholars and researchers [6], [13], [18], [19], [20] in urban studies.

This research implements both methods, direct and indirect, to examine place attachment and place distinctions. Indirect methods (questionnaires, interviews, checklists and discussions) explain common and specific patterns of human contact with places, while direct methods (field observations) provide researchers with an opportunity to examine the reality of a place and the activities involved in influencing attachment. By implementing these methods, the holistic patterns and meanings of attachment to the place can be uncovered.

4. Findings and Discussion

4.1. Length of engagement

Length of engagement affects the level of attachment in the specific place variously. Conforming to the existing research by many scholars, place attachment is affected by the length of engagement, frequent visits also are often generated by the length of stay in a particular area [20]. Evidence gathered from surveys and interviews has shown that continuous and frequent engagement contributes to familiarity with places and long-term involvement creates a sense of belonging.

Respondents linked their awareness and familiarity into the place to the frequency of visits and the duration of engagement indicated in such reactions:

> *"I'm familiar with this place. I've been running my family's souvenir shop for the past 20 years. It's a special place because of the old building" (Respondent 05: Store Owner, 20 years of involvement). M-RP*
>
> *"I came here because I've been shopping here for the past 10-25 years. The facilities here are nice and convenient for walking. I come here if I'm looking for certain foods. It's very special because it's been there a long time ago. " (Respondent 6: buyer, 20 years involvement). S-RP*

The results in Table 1 shows the respondents frequent the area regularly. Daily activities mostly implicate static users whereas mobile users (visitors and buyers) have engaged more weekly and monthly. S-RPs show the highest percentage of daily engagement while M-RPs show higher frequency for monthly engagement.

IOP Conf. Series: Materials Science and Engineering 1101 (2021) 012028 doi:10.1088/1757-899X/1101/1/012028

Table 1. Engagement Frequency (N=330)

Place	S-RP	M-RP
FREQUENCY	%	
Daily	58.2	47.3
Once a week	9.1	10.0
2 – 3 times per month	9.1	14.5
Once a month	23.6	26.4

Table 2. Engagement period (N = 330)

Place	S-RP	M-RP
LENGTH OF ENGAGEMENT	%	
1-4 hours	35.45	42.7
5-8 hours	10.91	15.5
9-12 hours	50.00	29.1
More than 12 hours	3.64	10.9

The presence of significant numbers of office workers as well as those who do business and work in the buildings next to S-RP and M-RP has contributed to the value of urban usage of space. This situation is proved by the period of engagement shown in the results of the study (refer to Table 2) indicate that in S-RP 50% of respondents engage 9-12 hours in the place, whereas in M-RP only 29%. The highest number of respondents engage 1-4 hours in both S-RP and M-RP for leisure and shopping purposes. Resultant from the frequent visits increases the familiarity level of people within the place.

Table 3. Last visited (N = 165)

Place	S-RP	M-RP
LAST VISIT To Place	%	
Last week	25.5	21.82
Last month	15.5	22.73
A few months ago	6.3	5.45
Last year	2.7	0.0

Table 4. Do you visit this promenade every time you are in this city? (N = 165)

Place	S-RP	M-RP
FREQUENCY	%	
Must	25.5	21.82
Always	15.5	22.73
Sometimes	6.3	5.45

Patterns of engagement are also supported by research findings showing the frequency of visits indicated by respondents' during their last visits. Table 3 shows that the majority of respondents' of their last visits were 'last week' and 'last month' and the most of them also came to see this places all the time or most of the time when they came to Kuching or Malacca (refer to Table 4).

Figure 1 indicates higher levels of an emotional attachment appeared from respondents who engaged in the promenade daily or who have been implicated in promenade for more than five years. Both user's groups (mobile and static) in S-RP and M-RP were identified as having strong feelings about attachment, meaning, positive affect, satisfaction and enjoyment. Although the study also shows that less positive stances about safety and security were mentioned.

The differences in reaction between groups are significant in S-RP and M-RP. Based on our depth interviews, the pattern may be related to the strength of the place's identity as reflected in the historical open space, the sense of belonging due to the period of engagement and feelings towards changes and physical improvement in the area.

Although there are differences in the types of engagements, for those who have been involved less than five years in S-RP and M-RP, the values of 2.88 and 2.43 in Table 5 indicates the different degree of attachment to place. This can be related to the issues of insecurity that appeared during the interview. This finding suggests that riverfront promenade as a place of attraction also reflects diversity in attachment

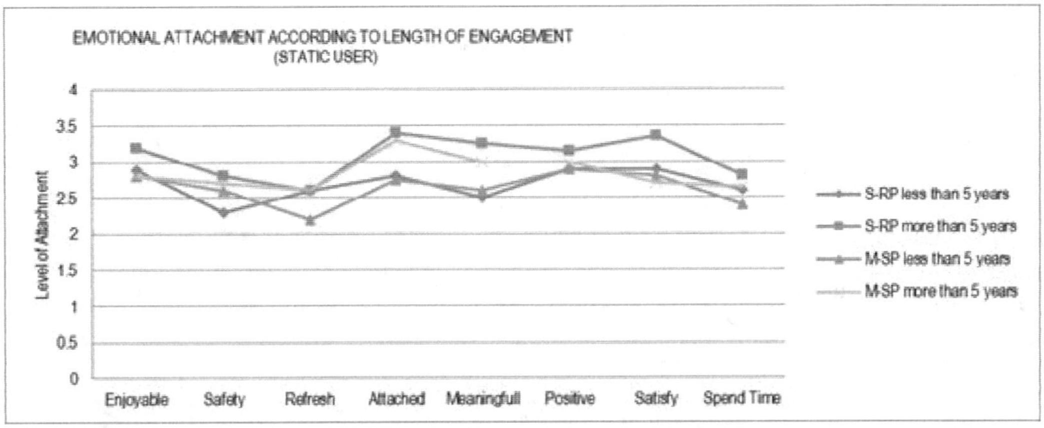

Figure 1. Emotional attachment levels referring to its length of engagement for static respondents in S-RP and M-RP.

Table 5. Emotional attachment levels referring to its length of engagement for static respondents in S-RP and M-RP

ENGAGEMENT	LEVEL OF EMOTIONAL ATTACHMENT			
	S-RP		M-RP	
Duration of stay	< 5 years (N=29)	≥ 5 years (N=28)	< 5 years (N=26)	≥ 5 years (N=28)
Average	2.88	3.25	2.43	2.78

The results of the survey indicate that the functional attachment is varied according to its length of engagement. Figure 2 shows that static respondents who have been engaged in promenade for more than five years have higher levels of functional connectivity than those who have been there for less than five years. Static respondents from S-RP who have more than five years of experience know this place have a strong attachment with promenade as the best place to do whatever they need to do, feel comfortable and that place is important for them. However, positive reactions have led to the process of improving physical quality to the promenade.

Table 6 indicates the functional attachment level for static respondent associated with its length of engagement. Respondents who were involved with open space less than five years of 2.78 (S-RP) and 2.57 (M-RP) accordingly indicated that the attachment level to promenade open space was above the average value. This scenario is corresponding to its long term influence of the engagement on open space in the sense of supporting pride, belonging and attachment. Many static respondents were those implicated in economic transactions and thus their continued involvement in open space created a sense of security in their activities and dependencies.

Table 6. Functional attachment levels in S-RP and M-RP referring to its length of engagement for static respondents

ENGAGEMENT	LEVEL OF FUNCTIONAL ATTACHMENT			
	S-RP		M-RP	
Duration of stay	< 5 years (N=29)	≥ 5 years (N=28)	< 5 years (N=26)	≥ 5 years (N=28)
Average	2.78	2.96	2.57	2.86

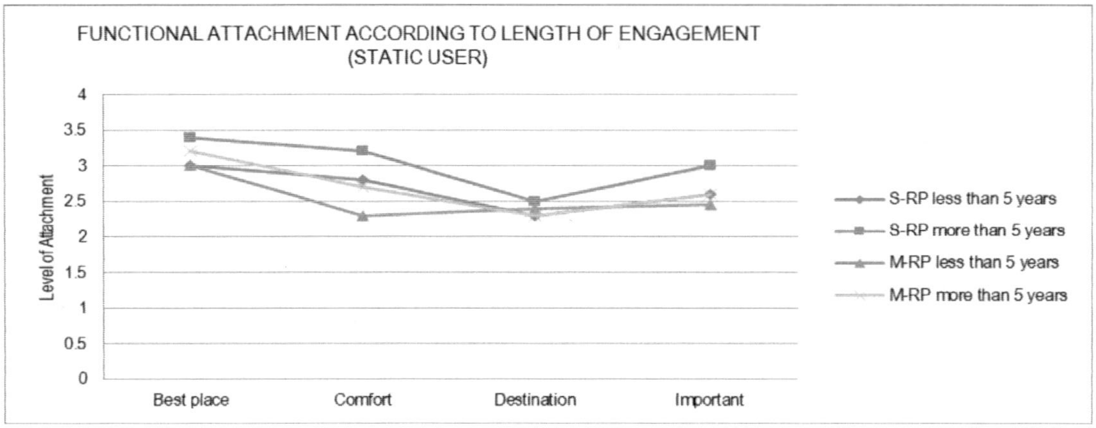

Figure 2. Functional attachment levels in S-RP and M-RP referring to its length of engagement for static respondents

4.2. Length of familiarity

Table 7. Familiarity Level to places (N=330)

FAMILIARITY	S-RP	M-RP
	%	
Very familiar	47.3	39.6
Familiar	52.7	60.4

Table 8. Relationship between familiarity and emotional attachment

FAMILIARITY (N=330)			
Attachment		Very familiar	Familiar
Place	S-RP	89.0 %	55.7 %
	M-RP	81.3 %	58.4 %

Table 9. Familiar places in S-RP and M-RP

S-RP	M-RP
Square Tower	Boat Museum
Floating mosque	Open space
Food and drinks vendor	Tan Kim Seng Bridge
Shaded seat	Clock tower
Boat jetty	Jonker street
Shophouses	Kampung Jawa
Pedestrian bridge	Rempah Garden
DUN building	Kampung Morten
River cruise boat	Hang Tuah & Chan Koo Chen Bridge

The respondents were asked about their level of familiarity in S-RP and M-RP about their continued engagement, knowledge and familiarization. Table 7 indicates the high level of familiarity with the places founded from most respondents. This is endorsed by static users involved daily and has a close relationship with other people who engage in the places.

Based on interviews, familiarity is the highly frequent mentioned aspect from respondents while asked about their reason why they attached to a place. One such response was from a street vendor in S-RP:

"I chose to vend here because here is where I started doing business for many years and got used to this place, already know many people here and know them well ..." (Respondent 03: hawker, 12 years of involvement) S-RP

IOP Conf. Series: Materials Science and Engineering 1101 (2021) 012028 doi:10.1088/1757-899X/1101/1/012028

Length of engagement and frequent visits become reasons for most of the respondents to mention that they recognize the place very well. The familiarity is also shown from respondents' understanding about the physical appearances and changes in the place.

"Uncle is very familiar with this area. Names of the street have changed a lot. The old market used to be next to the dockyard but the market was demolished and it became a field and parking lot... "(Respondent 04: pedestrian, 20 years of involvement) S-RP.

The following illustrates the respondent's familiarity with other people:

"No, I will not leave this place. Because I have many regular customers and I know other traders here ..." (Respondent 05: street vendor, 15 years of involvement) M-RP.

It is found that the respondents' familiarity level is related to the frequent visit to special physical appearances, shops kiosk and open spaces. Table 9 indicates the most commonplace is the place that represents the main function of the place. It is noteworthy that respondents also described the open space as a whole for their familiarity while others noted the existence of a well-known and popular building that attracted visitors.

Referring to the influence of familiarity, the findings are in line with Manzo [21] which argue that the most commonplace is the space that users often use or visit. An intense familiarity with places can support the cultivation of a sense of belonging to a particular place.

5. Conclusion

Level of attachment to S-RP and M-RP is strongly influenced by the role of the user and also by the length of engagement within the place. It can be concluded that the stronger attachment can be developed if more engagement frequency emerges within the place. Length of engagement as well as familiarity is the elements that link to the level of attachment. Studies agree with the idea that in some places people can become more attached to such places because of a high level of experience, which is usually shaped by their long-term settlements in specific locations, witnessing important events and frequent visits to the place [6], [20]. Experiences that related to long term involvement in an economic transaction, regular business and socio-cultural activities, can be associated with increasing levels of engagement with the public and activities in open space.

The results of the study clearly show that the length of engagement contributes to the familiarity to the place, thus affecting the level of attachment. These findings bear out that the level of attachment is affected by the length of engagement. Whereas the cultural characteristics of the place and the ethnic background of the consumer influence the level of attachment which was reflected on the strength of the sense of belonging to the place. Level of attachment was reflected by the memory, knowledge about the place used by the high level of familiarity and long-term integration within the place.

The study indicates that there are slight differences in emotional and functional attachment amongst static and mobile users. The static user expresses a stronger sense of identity and dependence to the place, while the mobile user is closer to the functional aspects of the riverfront promenade. The study results support the assumption that the attachment level of a place differs depends on the role of the user in that place.

6. References

[1] Relph E 1976 *Place and placelessness* (London: Pion Publication).
[2] Canter D 1977 *The psychology of place* (London: The Architectural Press Ltd).
[3] Low S 1992 Symbolic ties that bind: Place attachment in the plaza. In I. Altman & S. Low (Eds.), *Place Attachment* (New York: Plenum Press).
[4] Lobo C 2004 *The role of environmental perceptions in a sense of place: A case study of neighbourhoods in Phoenix, Arizona*. Unpublished PhD Thesis, Arizona State University, UMI Proquest Digital Dissertation,
[5] Altman I, Low S, (Eds.) 1992 *Place attachment* (New York: Plenum Press).
[6] Gustafson P 2001 Meanings of place: Everyday experience and theoretical conceptualizations, *Journal of Environmental Psychology* **21** pp. 5-16.

[7] Kyle G, Graefe A, Manning R, Bacon J 2004 Effects of place attachment on user's perceptions of social and environmental condition in natural setting, *Journal of Environmental Psychology* **24** pp. 213-225.
[8] Davenport M A, Anderson D. H 2005 Getting from sense of place to place-based management: An interpretative investigation of place meanings and perceptions of landscape change *Journal of Society and Natural Resources* **18** pp. 625-641.
[9] Williams D R, Anderson B S, McDonald C D, Patterson M. E 1995 Measuring place attachment: More preliminary results. *Paper presented at the Leisure Research Symposium, NRPA Congress, San Antonio.*
[10] Stedman C R 2003 Is it just a social construction? The contribution of the physical environment to sense of place, *Journal of Society and Natural Resources* **16** pp. 671-685.
[11] Gieryn, T F 2000 A Space for a place in sociology. *Journal of Annu. Rev. Social* **26** pp 463-496.
[12] Karsono B, Indira S S, Deni 2015 The significance of uniqueness, comfort, security and safety to place attachment. *Jurnal Teknologi* **78** pp. 179-183.
[13] Karsono B, Deni, Fithri C.A 2016 Assessment of Functional and Emotional Attachment in Malacca Riverfront Promenade, *Jurnal Teknologi (Sciences & Engineering)* **78** pp. 153-157.
[14] Proshansky H M, Ittelson W H, Rivlin L G 1976 *Environmental psychology: People and their settings* (New York: Holt, Rinehart and Winston).
[15] Gifford R 1987 *Environmental psychology: Principles and Practice* (Boston: Allyn and Bacon, Inc).
[16] von Eckartsberg R 1998 Introducing existential-phenomenological psychology. In R. Valle (Ed.), *Phenomenological inquiry in psychology* (pp. 3-20), (New York: Plenum).
[17] Seamon D 2003 *A geography of the lifeworld,* (New York: Routledge Keegan Paul).
[18] Jacobs J 1984 *The death and life of great American cities: The failure of modern town planning*, (London: Peregrine Books).
[19] Shuhana S 2004, *Kriteria kejayaan jalan membeli-belah tradisi di Malaysia: Kajian kes Kuala Lumpur*, Universiti Teknologi Malaysia.
[20] Karsono B, Wahid J 2015Attributes and Characteristics of Place Attachment, *Applied Mechanics and Materials* **747** pp. 132-135.
[21] Manzo L C 2009 A search for social connection in America's town square: Times square and urban public life, *Southern Communication Journal* **69** pp. 1-21.

IOP Conf. Series: Materials Science and Engineering 1101 (2021) 012029 doi:10.1088/1757-899X/1101/1/012029

Barriers for green building implementation in Malaysian construction industry

S Y Wong[1], W W Low[2], K S Wong[2], Y H Tai[3]

[1]School of Built Environment, University College of Built Environment, Persiaran Brooke, 96000 Sibu, Sarawak, Malaysia.
[2]Department of Civil and Construction Engineering, Curtin University Malaysia, CDT250, 98000 Miri, Sarawak, Malaysia.
[3]DGYRE Pte Ltd, 55 Kim Chuan Drive #06-07 Precise Tree, 537098 Singapore.
Corresponding author: wongshiyee@ucts.com.my

Abstract. The high amount of greenhouse gases emissions in construction industry has alerted the experts in construction industry to look into the opportunity of green building. Green building claimed to have positive impact on the environment by reducing the carbon emissions, enhance workers productivity and provide long-term economic benefits. However, the adoption rate of green building in Malaysian construction industry is not that high compared to other developed countries. This paper aims to identify the key challenges that faced by the construction industry supply chain professions, such as engineers, developers, architects and contractors. Questionnaire survey was used to collect data in this paper. Thirty-six valid responses were received and used for analysis. The results shown that the most important barrier was the lack of market demand on green building. This seems to imply that the supply chain agents are rely on the demand of clients to decide on their level of involvement in green building. The results of this paper could contribute in raising the awareness among the supply chain agents in taking proactive steps in investing in green building by stressing its benefits to their clients.

1. Introduction

As of 2019, the building works in construction industry were accountable for 30% of greenhouse gases (GHG) emissions [3]. This significantly high amount of GHG emissions together with inefficient energy consumption from buildings could contribute to climate change, such as global warming and affects the sustainability of natural habitats [5]. To address these issues, green building is regarded as one of the potential solutions, by most of the developed countries, such as Australia, China, Italy, and the United Kingdom (UK) [8, 9].

Green building could be defined as a building that help to minimize negative impacts to the environment, while enhancing the occupants' comfort and financial aspect [3]. Various research had proven the benefits of implementing green building. Such benefits include the reduction of GHG emissions [11], and enhancement of occupants' health and productivity [3, 8]. However, some people may regard green building as high construction cost and hence reluctant to invest in green building. Studies had been conducted by various researchers on the viability of green building and identified it to be profitable to the occupants throughout the lifecycle of the building [12]. The results of cost premium analysis for the green building had proved on its feasibility [13].

Rating tools to access green building features, such as BREEAM, green star and LEED, had been developed to assess the building performance, which generally based on the efficiency of energy and water; sustainable management and planning; innovation; and quality of indoor environment [14-16]. With the reference to the developed rating tools, Malaysia had developed its own rating tools, namely Green Building Index (GBI), that specifically suitable for its tropical weather and community needs, to nurture the green building development in Malaysia [17].

Despite the benefits and rating tools existed for the green building in construction industry, the adoption of green building in Malaysia could consider at the beginning stage if compared to other developed countries [18]. There are only a few examples of building that embrace green building concept, such as Securities Commission Mesiniaga Tower, Perdana Putra, University College of Technology Sarawak and Diamond Building [19-21]. These buildings are mostly commercial building, but little residential property be classified as fully adopting the green building practices [22].

To provide a better overview and strategy for the adoption of green building, this research aims to examine the barriers to green building application in Malaysian construction industry. This research focuses on the opinions of supply chain agents to provide an overview of such professions on the adoption of green building technologies and features in Malaysia.

2. Identification of potential barriers

Different studies had been conducted to examine the challenges of green building implementation. For examples, [10] utilised questionnaire survey to collect the opinion of project managers in developers firms in Kuala Lumpur and Selangor, [2] examined the opinions of developers in Indonesia with regards to the challenges in adopting green building, [6] looked into this area from the building experts' perspectives in the United States, and [1] conducted a study in Ghana, specifically discussed on the barriers related to green building technology adoption. However, most of the studies were conducted outside Malaysia (such as Indonesia) and did not combine the perspectives of a wider group of professions (such as engineers, designers, quantity surveyors). The barriers extracted from literature review were tabulated in Table 1.

Table 1. List of barriers

Barriers	1	2	3	4	5	6
Lack of green building regulations and codes	√		√	√	√	√
Lack of subsidies on green technology						√
Lack of training related to green building technologies	√	√	√	√	√	√
Lack of awareness of existing incentives	√	√	√	√		√
Lack of awareness on the benefits of green building technologies	√			√	√	
Lack of environmental awareness among developer and consultant					√	√
Lack of information and databases on the green technologies	√	√	√	√	√	
Low market demand by clients	√	√	√	√	√	√
Lack of research in local context on green building technologies	√	√	√	√		√
Limited time for green practices implementation	√			√		
Resistance to change by the supply chain agents	√	√		√		√
High cost	√	√	√	√	√	√
Limited finance support for the up-front cost					√	√
Clients' preference on the instant payback benefits					√	√
Risk and uncertainties for new green technologies	√	√		√	√	√

Note: 1 = [1]; 2 = [2]; 3 = [4]; 4 = [6]; 5 = [7]; 6 = [10]

3. Questionnaire development

The barriers that identified from the literature review were further classified into four categories, which are barriers related to government and policy, barriers related to knowledge and awareness, barriers related to local professionals, and barriers related to economic and finance (see Table 2). Codes had been assigned to each barrier for the ease of data presentation and discussion.

Table 2. Barriers and codes

Code	Barriers
G	**Barriers related to government and policy**
G1	Lack of green building regulations and codes
G2	Lack of subsidies on green technology
G3	Lack of training related to green building technologies
K	**Barriers related to knowledge and awareness**
K1	Lack of awareness of existing incentives
K2	Lack of awareness on the benefits of green building technologies
K3	Lack of environmental awareness among developer and consultant
K4	Lack of information and databases on the green technologies
K5	Low market demand by clients
P	**Barriers related to local professionals**
P1	Lack of research in local context on green building technologies
P2	Limited time for green practices implementation
P3	Resistance to change by the supply chain agents
E	**Barriers related to economic and finance**
E1	High cost
E2	Limited finance support for the up-front cost
E3	Clients' preference on the instant payback benefits
E4	Risk and uncertainties for new green technologies

The targeted respondents were developers, builder, architect, engineer, contractor, designer, consultant and project manager who works in the Malaysian construction industry (both East and West Malaysia). The targeted respondents were randomly selected from the membership list in Board of Engineers, Construction Industry Development Board (CIDB) and the web search. Two hundred questionnaire surveys were distributed through online and/or face-to-face methods. The data analysis method for this paper was descriptive statistics.

4. Results and discussion

This section discussed on the respondents' details including their demographic details, and the barriers that affecting the implementation of green building.

4.1. Respondents' details

Thirty-six valid responses were received, which is 18% of the response rate. Out of these 36 responses, 50% of the respondents were engineers (18), 25% of them were consultant (9), followed by contractors (14%, 5), architect (5%, 2), builder (3%, 1) and developer (3%, 1). Therefore, the majority of the respondents have engineering backgrounds. Sixty-seven percent of the respondents had more than five years working experience, while 28% of the respondents had more than 21 years of working experience. This seems to imply that the respondents who have varies level of experience, could provide a better overview of their opinions from different stages of working experiences.

In terms of the participants' involvements in green building, 64% of the respondents had experienced green building technologies in their project, but out of these number of respondents, 83% of them identified their involvement in green project was not more than five projects related to green building. Sixty-five percent of the respondents involved project related to green building in West Malaysia, 30% of them have involved in green building project in East Malaysia, and 4% of the respondents involved green project in both East and West Malaysia.

4.2. Barriers of green building implementation

The respondents were asked to rank the level of importance of each barrier, with 1 being not important and 5 being extremely important. Table 3 showed the descriptive statistics of the barriers.

Table 3. Means, standard deviation and ranking of barriers

Code	Mean	Standard Deviation	Rank
G1	3.53	0.97	7
G2	3.44	1.34	12
G3	3.44	1.00	11
K1	3.89	1.06	2
K2	3.81	0.89	3
K3	3.56	0.87	6
K4	3.61	0.87	5
K5	4.25	0.81	1
P1	3.39	1.05	13
P2	3.25	1.00	14
P3	3.53	1.18	8
E1	3.75	1.16	4
E2	3.44	0.98	10
E3	2.53	1.00	15
E4	3.47	1.27	9

The results showed that "K5 - Low market demand on green building" was ranked as the most important barrier. This seems to imply that the respondents who are the supply chain agent relied on the client demand to provide green building. As supported by [23], this scenario could be classified as vicious circle of blame, as the supply chain agent and demand-side stakeholders are blaming each other for not providing/investing in green building. There is possibility that the clients are unaware about the benefits of green building and hence leads to the low demand, and low implementation of green building.

Lack of awareness on the existing incentives (K1) was ranked as the second most important barriers. This was supported by the literature in [1, 2, 4]. It seems to imply that the respondents may not be fully aware on the supports and/or incentives that provided by the local government for investing in buildings that have green features. It is interesting to note from Table 3 that the three most important barriers that selected by the respondents were all classified under the category of knowledge and awareness barrier. The other barriers that classified under policy and economics seems to be least important if compared to the awareness barriers.

The fourth most important barrier was "E1 – High cost". The respondents could opt that building which consists of green features is having high capital cost or high maintenance cost. This may create an impression to the developers that selling properties with green features could be burden, as most of the buyers are looking for affordable property price.

The fifth most important barrier was the limitation of information and databases (K4). Different professions and/or stakeholders are involving in the construction industry, and hence created information at different project stages [24]. There is no proper system which could collate the information into a single database in Malaysia, and hence leads to the loss of information at different project stages [25].

The following barrier was "K3 – Lack of awareness on green technologies benefits". This seems to contradict with a studies conducted in US, as the US green building experts were reported that lack of awareness as the second most important barriers [8]. However, the differences in geographical location and cultures of different countries, may lead to the discrepancies on the barriers' rankings.

In terms of the lack of regulations and codes in relation to the green building (G1), the respondents may perceive that the current codes and rating tools in Malaysia, such as Green Building Index (GBI), is not being fully understand by all the stakeholders. This barrier is consistent with the findings in [8, 9].

Two barriers under the category of economic and finance, which are Risk and uncertainties for new green technologies (E4), and Limited finance support for the up-front cost (E2) were ranked as the ninth and tenth most important barriers. These seems to imply that the stakeholders are looking for the financial support or more affordable price of the green technology.

The five least important barriers classified by the respondents were "G3 - Lack of training related to green building technologies", "G2 - Lack of subsidies on green technology", "P1 - Lack of research in local context on green building technologies", "P2 - Limited time for green practices implementation" and "E3 - Clients' preference on the instant payback benefits". All of these barriers were from the categories that related to government policy, local professionals and economics. This may imply that the respondents reckoned they have sufficient awareness on the green building, but there are hindrances from the external sources.

5. Conclusion

This paper provided an overview on the factors that affect the implementation of green building in Malaysian construction industry from the perspective of different professionals, which mostly consist of engineers. The main barriers identified from the questionnaire was lack of client demand on building that have green features. This seems to imply that the demand side stakeholder does not have the interest to invest in green building. The tropical weather in Malaysia may lead to the situation that the residents are not aware on the importance of green building if compared to countries that have four seasons. High cost was ranked as the fourth most important barriers. It is undeniable that green building could help with mitigating the climate change but its adoption in Malaysian construction industry still require a considerable good amount of time. As the respondents in this paper was mostly engineers' opinions, it would be good to conduct a future studies related to clients or developers' opinions in relation to green building. Future research could look into the barriers related to the knowledge and awareness, as there are limited studies carried out in this area, specifically related to the storage of green building information and providing a standardised database.

References

[1] Chan, A.P.C., et al. 2018 Critical barriers to green building technologies adoption in developing countries: The case of Ghana, *Journal of cleaner production* **172** p. 1067-1079.

[2] Wimala, M., E. Akmalah, and M.R. Sururi 2016 Breaking through the barriers to green building movement in Indonesia: Insights from building occupants, *Energy Procedia* **100**(100) p. 469-474.

[3] World Green Building Council, *New report: the building and construction sector can reach net zero carbon emissions by 2050.* 2019.

[4] Deng, W., et al. 2018 Barriers and policy recommendations for developing green buildings from local government perspective: a case study of Ningbo China, *Intelligent Buildings International* **10**(2) p. 61-77.

[5] Sinha, A., R. Gupta, and A. Kutnar 2013 Sustainable development and green buildings, *Drvna industrija* **64**(1).

[6] Darko, A., et al. 2017 Examining issues influencing green building technologies adoption: The United States green building experts' perspectives, *Energy and Buildings* **144** p. 320-332.

[7] Häkkinen, T. and K. Belloni 2011 Barriers and drivers for sustainable building, *Building Research & Information* **39**(3) p. 239-255.

[8] Darko, A. and A.P. Chan 2016 Critical analysis of green building research trend in construction journals, *Habitat International* **57** p. 53-63.

[9] Li, Y., et al. 2014 Green building in China: Needs great promotion, *Sustainable Cities and Society* **11** p. 1-6.

[10] Abidin, N.Z., N.A. Yusof, and A.A. Othman 2013 Enablers and challenges of a sustainable housing industry in Malaysia, *Construction Innovation* **13**(1) p. 10-25.

[11] Balaban, O. and J.A.P. de Oliveira 2017 Sustainable buildings for healthier cities: assessing the co-benefits of green buildings in Japan, *Journal of Cleaner Production* **163** p. S68-S78.

[12] Zhang, L., J. Wu, and H. Liu 2018 Turning green into gold: A review on the economics of green buildings, *Journal of cleaner production* **172** p. 2234-2245.

[13] Dwaikat, L.N. and K.N. Ali 2016 Green buildings cost premium: A review of empirical evidence, *Energy and Buildings* **110** p. 396-403.

[14] BREEAM 2020 *How BREEAM Certification Works*. Retrieved on June 18, 2020 from: https://www.breeam.com/discover/how-breeam-certification-works/.

[15] LEED 2020 *LEED rating system.* Retrieved on June 18, 2020 from: https://www.usgbc.org/leed.

[16] Wong, S.Y., et al., *A comparison of international and australian rating tools for sustainability elements of residential property*, in *RICS COBRA 2015.* 2015: Sydney, Australia.

[17] GBI Sdn Bhd 2020 *GBI organisation.* Retrieved on June 18, 2020 from: https://www.greenbuildingindex.org/how-gbi-works/gbi-organisation/.

[18] Samari, M., et al. 2013 The investigation of the barriers in developing green building in Malaysia, *Modern Applied Science* **7**(2) p. 1.

[19] Clean Malaysia 2015 *The rise of green building in Malaysia.* Retrieved on June 18, 2020 from: https://cleanmalaysia.com/2015/08/28/the-rise-of-green-building-in-malaysia/.

[20] UCTS 2020 *Green Building Index (GBI).* Retrieved on June 18, 2020 from: https://www.ucts.edu.my/home/green-building-index-gbi/.

[21] Suruhanjaya Tenaga 2012 *Diamond Building.* Retrieved on June 18, 2020 from: http://www.mgbc.org.my/Downloads/IGEM2012/IGEM2012%20-%20MGBC%20ST%20Diamond%20Building.pdf.

[22] Shan, M. and B. Hwang 2018 Green building rating systems: Global reviews of practices and research efforts, *Sustainable Cities and Society* **39** p. 172-180.

[23] Cadman, D. 2000 The vicious circle of blame. Cited in M. Keeping. 2000, *What about demand? Do investors want "sustainable buildings*.

[24] Miller, W., et al., *Strategies and solutions for housing sustainability: building information files and performance certificates*. 2014, Queensland University of Technology.

[25] Wong, S.Y., et al. 2018 Improving information gathering and distribution on sustainability features in the Australian residential property market, *Journal of Cleaner Production* **184** p. 342-352.

IOP Conf. Series: Materials Science and Engineering 1101 (2021) 012030 doi:10.1088/1757-899X/1101/1/012030

Extension of Time (EoT) Considerations in Construction Duration Estimate for Public Construction Projects

S N Ting[1], V C Darrell[1], A B H Kueh[1], Y Y Lee[1] and C K Ng[1]

[1] Department of Civil Engineering, Faculty of Engineering, Universiti Malaysia Sarawak, 94300 Kota Samarahan, Sarawak
Corresponding author: snting@unimas.my

Abstract. Construction duration is defined as the time frame given by the Client of a project use to complete the project under normal working conditions, practice of construction. However, often times, projects faced time overruns and hence, needing to activate contractual remedies like Extension of Time (EoT). This paper aims to investigate the factors related to award the Extension of Time (EoT) for construction projects in especially in the public sector in Sarawak. A structured checklist was created as an instrument and employed to accomplish the objective of this research. A total of 111 previous public projects which were granted EoT were studied. The findings revealed several detailed factors that had typically led to project delays in the State. These findings are expected to offer the construction industry stakeholders in Sarawak a focal point on those delay that could be included in construction project duration estimates and management of construction performances, followed by the appropriate corrective and improvement measures for time management for construction projects.

1. Introduction

The construction sector is among the industrial sectors contributing significantly to development and the economic system in the State [1]. However, most construction projects nowadays are much more complex [2] involving many specialised material, equipment, labour and also complicated project delivery and contractual relationship. This has led to more problems and difficulties with construction duration estimation, bringing on board another issue frequently extensive project schedule and timeline [3]. Delays occurrence are becoming more rampant and this may not be cause by mismanagement of the projects but poorer time planning in the first place.

Contractual remedy laid out that if a project delay occurs, contractors are penalised and requird to pay the sum of liquidated damages (LAD) as stated in the contract. And to prevent such loss and to gain additional time for completion of the projects, contractors often times will find ways to apply for extension of time (EoT). The definition for EoT by The Society of Construction Law (SCL) as contained in the Delay and Disruption Protocol (SCL 2002), is "*the additional time granted to the contractor to provide an extended contractual time period or date by which work is to be, or should be completed and to relieve it from liability for damages for delay (usually liquidated damages)*".

It is not unusual for delays to take place in construction projects, and in other projects which the schedule is used for planning of activities. Delay dictates whether a project or some other date is finished late, such as a milestone. A number of unforeseen incidents during construction can result in a delay in completing the project, or increase the amount of work to be done within a given time frame.

Ibironke, Oladinrin, Adeniyi and Eboreime [4] believes that delays in construction may be caused by the client, the contractor, the consultants, acts of God, or a third party and they may occur early or late in the job.

2. Problem Statement

The construction sector in Sarawak is no exception to the worldwide delay trend whereby major projects are carried out beyond the initial project time set out in the contract agreements. According to most researches conducted during the last two decades, the important factors that must be taken into account during construction duration estimation and causes of delays related to construction projects summarised as below. The following are factors which have been studied by the researchers and detailed in Table 1 as given below:

Table 1. Factors taken into construction duration estimation and causes of delays by past researchers.

Researchers	Country	Findings
[5]	Australia	Ground floor area, number of floors, facility type
[6]	UK	Facility type, type of client, contractor selection method, procurement method
[7]	Turkey	Weather conditions, construction methods, labor productivity, site condition, project cost, facility type, number of floors, floor area, project complexity, design comprehensiveness, experience of consultant, experience of owner, experience of contractor, procurement method
[8]	US	Precipitation, wind velocity
[9]	US	Number of floors, floor area
[1]	Australia	Shortage of labor, skill shortage, site condition, design comprehensiveness
[10]	Singapore	Equipment availability and failure, construction methods, shortage of labor, availability of staff, availability of material, site condition, experience of consultant, experience of owner, project duration set, experience of contractor
[11]	Vietnam	Site condition, design comprehensiveness, subcontractor skills
[12]	Palestine	Terrain condition, soil and rock suitability, soil and rock drill ability, project size, road length, road width,
[13]	Malaysia	Late equipment delivery, insufficient number of equipment, equipment availability and failure, construction methods, shortage of labor, shill shortage, labor productivity, availability of material, late material delivery
[2]	India	Weather conditions, air quality, equipment availability and failure, shortage of labor, availability of material, late material delivery, site preparation time, site condition, project cost, complexity, reworks and repair
[14]	Egypt	Project location, facility type, floor area, total height, finishing quality, design comprehensiveness, type of client. Contractor selection method, procurement method
[15]	Korea	Site condition, bedrock, number of floors, floor area, lot area, number of households, structural system, foundation

		system, roof type,
[16]	Kuwait	Project cost, number of floors, floor area
[17]	Slovak Republic	Project cost, number of floors, floor area
[18]	Serbia	Earthworks, amount of crushed stone, number of curbs, amount of asphalt base layer, number of asphalt surface later, amount of concrete prefabricated elements, drainage works, traffic signalisation works
[19]	Korea	Number of floors, number of buildings,
[20]	Korea	Construction commencement month, facility type, number of floors, floor area, number of building, lot area, structural system
[21]	Iraq	Site accessibility, facility type, number of floors, floor area, total height, structural system, foundation system,

Through the understanding of delay factors and reasons for delays in construction projects, better time management measures can be executed. It is important to identify the reason of delays, so that stakeholders can take appropriate steps in order to prevent such scenarios once these causes become noticeable. More importantly, from these factors and reasons for delays, those that are eligible for EoT will be further studied and considered so that initial planning for schedules can also be inclusive of these factors. Besides avoidance and mitigating them when they occurred, this research advocate incorporating them especially in time planning especially in construction duration estimation

2.1 Research objectives

The main objective of this research is to identify the factors of EoT that has been awarded to the contractors for the Sarawak construction projects. This will enable the construction industry stakeholders in Sarawak not only to understand factors that most delays and affecting project time performances but also to include them in planning the project schedule.

3. Research methodology

From the literature review, these 61 factors of delay in construction projects were chosen and were divided into 10 groups. Characterisation of delay factors are as follows: contractors; consultant; client; environmental; equipment; labor; material; project characteristics; site; and others.

A structured checklist was used as the main instrument in the data collection and is found to be useful to gather the data to be used in the analysis. Three (3) checklists was prepared based on project type (buildings, roadwork/bridges and others). Each of the checklist was separated into two sections: section I and section II. Section I comprises of inquiries associated with the project's title, tender number, year of project, project duration (predicted and actual) and project budget. The total days of EOT was measured by substituting the actual and predicted project duration. Section II comprises of the category of delay factors and its detailed factors (gathered from relevant materials through a detailed literature review of previous related researches), each with a checkbox and a remark section for any additional or important information, e.g., total days of delay.

Data collection was conducted at Quantity Survey (QS) Branch, Jabatan Kerja Raya Head Quarters (JKR HQ) Sarawak and it took place in February 2020 for a period of 10 days. A total of one hundred and eleven (111) projects data with EoT were managed to be collected. The raw data obtained from the checklists were compiled, organised and analysed with the aid of Microsoft Excel. A quantitative interpretation was found to be the most appropriate analysis to analyse the data.

4. Results and discussions

The analysis from structured checklists led to development of several reasons behind these construction projects' EOTs. There are in fact many factors which influence the actual duration of project, these factors include: inclement weather; site possession; low water level; change in design; contractor selection method; delay by SESCO; SO instruction; authority inspection and approval; unforeseen reasons; suspension of work; existing utilities; site condition; client late decision making; land blockage; 3rd party obstruction; air quality; flooding; and other reasons.

The results showed that some of the factors gathered from the literature reviews from related papers, does happen to be the reason for project's EOT. However, these factors have been lumped into a few major categories based on the nature of the factor itself.

4.1. Inclement weather

Inclement weather level is one of the factors that will be considered in granting the EoT to contractors. Based on the data of past construction project's EoT collected, the factor of inclement weather has affected most of these projects, 92 projects out of 111 projects. According to the EoT approval form, this factor will be taken into consideration if the rainfall intensity is more than 10mm in a day. The contractor will apply for the EoT due to exceptionally inclement weather based on the rainfall data in the Malaysia Hydrological Yearbook by Malaysian Meteorological Department (MMD).

As mentioned earlier, 82% of the 111 construction projects in Sarawak with EoT was affected by the exceptionally inclement weather. These projects include RC pontoon, jetty, roadwork, building, waterworks, airport, bridge, earthwork, agriculture sub-station, playground, gate and electrical system projects. However, based on the data, only 1 project was seriously affected by this factor, which is a jetty project, having an EOT of 397 days due to inclement weather. Other than that, only twenty-three (23) projects were least affected by inclement weather, having EoT less than 10 days.

Besides that, it is also noticeable that all of 92 projects affected by exceptionally inclement weather are located all around Sarawak. This suggests that the EoT due to this factor is not depending on its location, but it is based on the rainfall intensity distribution data provided by the MMD, which will be updated annually.

Based on the data analysed, the predicted project cost of the 92 projects is between RM488.6k to RM302.6mil. However, it is noticeable that the percentage of project delay due to inclement weather is not highly affected by the cost of project. As for instance, the jetty project with a cost of RM187.62mil was having a delay of 27.94% project delay due to inclement weather as compared to a project of RM302.6 with 100% delay. There is no pattern of any cost that will result in a high or low EoT due to this factor.

4.2. Internal issues

Internal issues area another main category that will be considered in granting the EoT to contractors. Based on the data of past construction project's EoT collected, the factors under internal issues has affected 42 projects out of 111 projects. A few factors that has been lumped into internal issues are including site possession, change in design, contractor selection method, variation order, etc.

As indicated earlier, 42 out of 111 construction projects in Sarawak with EOT was affected by the internal issues. These projects include building, roadworks, waterworks, bridges, airport, jetty and seashore projects. However, based on the data, only 13 projects were affected 100% by this factor, which most of them are building projects, with a maximum EoT of 162 days. Other than that, only 2 building projects having less than 20% project delay due to internal issues with maximum EoT of 30 days.

It is notable that all of 42 projects affected by internal issues have different locations, which included Asajaya, Bau, Betong, Bintulu, Julau, Kapit, Kuching, Lawas, Limbang, Lundu, Marudi, Miri, Mukah, Pusa, Sarikei, Serian, Sri Aman. It is noticeable that these projects are affected by its location, especially projects with 100% delay due to this factor. As for instance, a building project located in Miri with no access available have the highest EoT of 162 days, solely due to internal issues, but another building project in Miri which is considered accessible, have the lowest EoT which is 10 days due to internal issue alone.

Based on the data analysed, the predicted project cost of the 42 projects affected by internal issues is between RM599.5k to RM187.6mil. However, it is noticeable that the EoT is not highly affected by the cost of project. As for instance, the jetty project with a cost of RM187.6mil has 49 days EoT, which is almost similar EoT to a project cost of RM888k with 48 days EoT. This means that there is no pattern of any cost that will result in a high or low EoT due to this factor.

4.3. Environmental and unpredictable reasons

Environmental and unpredictable causes are one of the main categories that will be considered in granting the EoT to contractors. Based on the data of past construction project's EoT collected, the factors under environmental and unpredictable reasons has affected 27 projects out of 111 projects. A few factors that has been lumped into environmental and unpredictable reasons includes unforeseen reasons, air quality, flooding event, etc.

As mentioned earlier, 27 out of 111 construction projects in Sarawak with EOT was affected by the environmental and unpredictable reasons. These projects include building, roadworks, bridge and waterworks projects. However, based on the data, only 3 projects were affected 100% by this factor, which all of them are building projects. The project with a most EoT is a roadwork project with 238 days of EoT while, only 1 roadwork project having less than 10% project delay due to environmental and unpredictable reasons with EoT of 28 days.

It is notable that all of 42 projects affected by environmental and unpredictable reasons have different locations, which included Asajaya, Bau, Belaga, Betong, Bintulu, Dalat, Julau, Kabong, Kanowit, Kapit, Kuching, Miri, Samarahan, Sarikei, Simunjan, Song and Sri Aman. It is noticeable that these projects are affected by its location. As for instance, a building project located in Kuching of less than 25km from nearest city, have a lower EoT of 5 days, compared to a roadwork project in Dalat, of 25 to 50km from nearest city, have a higher EoT which is 238 days due to environmental and unpredictable reasons.

Based on the data analysed, the predicted project cost of the 27 projects affected by environmental and unpredictable reasons is between RM677.7k to RM88.4mil. However, it is noticeable that the EoT do not affected by the project cost. As for instance, the building project with a cost of RM677.7k has 5 days EoT, which is almost similar EoT to a roadwork project cost of RM79.1mil with 3 days EoT only due to environmental and unpredictable reasons. This means that there is no pattern of any cost that will result in a high or low EoT due to this factor.

4.4. Third party obstruction

Third-party obstruction is one of the main categories that will be considered in granting the EoT to contractors. Based on the data of previous construction project's EoT collected, the factors under third-party obstruction has affected 21 projects out of 111 projects with EoT. A few factors that has been lumped into third-party obstruction includes delay by SESCO, authority inspection and approval, existing utilities, land blockage, etc.

As stated earlier, 21 construction projects in Sarawak with EOT was affected by the third-party obstruction. These projects include building, roadwork, riverine terminal, earthwork, bridge, waterwork and agriculture sub-station projects. However, based on the data, only 3 projects were affected 100% by this factor, which the waterworks project having a highest EoT of 135 days. Other than that, only 1 roadwork project having less than 25% project delay due to third-party obstruction EoT of 54 days.

All 21 projects affected by changes in design have different locations, which are Bau, Dalat, Kapit, Kuching, Lawas, Limbang, Lubok Antu, Miri, Pusa, Samarahan, Sarikei and Simunjan. It is noticeable that these projects do not affected by the construction project's location. As for instance, an earthwork project located in Kuching of less than 25km from nearest city, have the lowest EoT of 20 days, but another building project in Kuching of similar distance from nearest city, have a higher EoT which is 181 days due to third-party obstruction.

Based on the data analysed, the predicted project cost of the 21 projects affected by third-party obstruction is between RM738.1k to RM85.5mil. However, it is notable that the EoT is not affected by the project sum. As for instance, the roadwork project with a cost of RM738.1k has 103 days EoT, which has an almost equivalent EoT to a project cost of RM85.5mil with 124 days EoT due to third-party obstruction. This signifies that there is no pattern of any cost that will result in a high or low EOT due to the third-party obstruction.

5. Conclusion

The general factors that affect a schedule of a construction projects has been identify this research. 61 factors of construction project delay were divided into 10 groups as follows: contractors; consultant; client; environmental; equipment; labour; material; project characteristics; site; and others. From the factors of delays identified, a checklist was put together and was used an instrument to collect the practical data related to delays in 111 public construction projects of various locality, cost, natures and types that had been carried out in the state of Sarawak for the past 5 years. Based on the analysis, these project characteristics such as type of projects, project locality and project cost that may been affect construction projects timeline were analysed through the quantitative interpretation.

From the delay causes identified, those that were awarded with EoT were further studied. Based on the results, construction projects in Sarawak has been awarded EoT for a few reasons, which includes inclement weather, site possession, low water level, change in design, contractor selection method, delay by SESCO, SO instruction, authority inspection and approval, unforeseen reasons, suspension of work, existing utilities, site condition, client late decision making, land blockage, third-party obstruction, air quality, flooding and other reasons. These reasons were further classified into category with the top categories of delays granted with EoTs being Inclement Weather, Third Party Obstruction, Internal Issues and Unpredictable especially environmental factors. It is believed that these factors could have and should be incorporated into initial schedule planning and construction duration estimation.

This research advocates that a good schedule need to be planned properly and then managed rigorously. By understanding the reasons for delays and why EoT are granted, it is much possible to optimised the schedules. Factors that can be incorporated accurately in the schedule can further reduce potential of EoT application. In other words, comprehensive schedules can then be better understood, controlled and executed. Delay reductions efforts should be in both planning and mitigation through a comprehensive and concise time duration estimation effort.

6. References

[1] K Wong and V Vimonsatit 2012 A study of the factors affecting construction time in Western Australia. *Scientific Research and Essays*, **7**(40), 3390–3398.

[2] N J Babu 2015 Factors Affecting Success of Construction Project. *IOSR Journal of Mechanical and Civil Engineering Ver. V*, **12**(2), 2320–2334.

[3] J A Alsuliman 2019 Causes of delay in Saudi public construction projects. *Alexandria Engineering Journal*, **58**(2), 801–808.

[4] O T Ibironke, T O Oladinrin, O Adeniyi and I V Eboreime 2013 Analysis of non-excusable delay factors influencing contractors' performance in Lagos State, Nigeria. *Journal of Construction in Developing Countries*, **18**(1), 53–72.

[5] P E D Love, R Y C Tse and D J Edwards 2005 Time-cost relationships in Australian building construction projects. *Journal of Construction Engineering and Management*, **131**(2), 187–194.

[6] T K Burrows, I Pegg and J L N Martin 2006 Predicting construction duration of building projects. *AACE International Transactions*, *2006*, 1–13.

[7] E Odabaşi 2009 *Models for Estimating Construction Duration: An Application for Selected*

Buildings on the Metu Campus

[8] S Apipattanavis, K Sabol, K M Molenaar, B Rajagopalan., Y Xi, B Blackard and S Patil 2010 Integrated framework for quantifying and predicting weather-related highway construction delays. *Journal of Construction Engineering and Management*, **136**(11), 1160–1168.

[9] I Choudhury,W Bank and F Scholar 2012 AC 2012-3382 : Effects of Construction Cost and Volume on A Study of the Factors of Construction Time for Educational Projects in Texas.

[10] B G Hwang, X Zhao and S. Y Ng 2013 Identifying the critical factors affecting schedule performance of public housing projects. *Habitat International*, **38**, 214–221.

[11] L Le-Hoai, Y D Lee,and A T Nguyen 2013 Estimating time performance for building construction projects in Vietnam. *KSCE Journal of Civil Engineering*, **17**(1), 1–8.

[12] I Mahamid 2013 Effects of project's physical characteristics on cost deviation in road construction. *Journal of King Saud University - Engineering Sciences*, **25**(1), 81–88.

[13] A H Memon, I A Rahman, A A A Aziz and N H.Abdullah 2013 Using Structural Equation Modelling to Assess Effects of Construction Resource Related Factors on Cost Overrun Faculty of Civil and Environmental Engineering , Faculty of Technology Management , Business and Entrepreneurship ,. *World Applied Sciences Journal*, **21**, 6–15.

[14] A A Gab-Allah, A H Ibrahim and O A Hagra 2015 Predicting the construction duration of building projects using artificial neural networks. *International Journal of Applied Management Science*, **7**(2), 123–141.

[15] R Jin, S Han, C Hyun and Y Cha 2016 Application of case-based reasoning for estimating preliminary duration of building projects. *Journal of Construction Engineering and Management*, **142**(2), 1–8.

[16] A M Jarkas 2016 Predicting Contract Duration for Building Construction: Is Bromilow's Time-Cost Model a Panacea? *Journal of Management in Engineering*, **32**(1), 1–8.

[17] D Mackova, M Kozlovska, R Baskova, M Spisakova and KKrajnikova 2017 Construction-duration prediction model for residential buildings in Slovak republic based on computer simulation. *International Journal of Applied Engineering Research*, **12**(13), 3590–3599.

[18] I Peško, VmMučenski, M Šešlija, N Radović , A Vujkov, D Bibić and M Krklješ 2017 Estimation of costs and durations of construction of urban roads using ANN and SVM. *Complexity*, *2017*.

[19] J H Lim, D Y Kim, D Kim, S C Jeong, D K Seol and Y K Huh 2018 Developing a construction duration estimation model to ensure the safety in apartment housing construction sites. *KSCE Journal of Civil Engineering*, **22**(7), 2195–2205.

[20] D J Yeom, H M Seo, Y J Kim, C S Cho and Y Kim 2018 Development of an approximate construction duration prediction model during the project planning phase for general office buildings. *Journal of Civil Engineering and Management*, **24**(3), 238–253.

[21] E D A Al-Zubaidi, A H Yas and H F Abbas 2019 Guess the time of implementation of residential construction projects using neural networks ANN, **7**(3), 1218–1227.

Acknowledgments

This research will like to acknowledge Universiti Malaysia Sarawak and Jabatan Kerja Raya Sarawak for a research grant with number GL/F02/JKRS/2020 for this joint research collaboration.

Comparative study of adsorptive removal of oxyanion ($Cr_2O_4^{-2}$) from aqueous media over unmodified and magnetically modified *Melia azedarach*

M. Nosheen[1], A. A. Khan, M. Salman[1], S. S. Shafqat[3]*, M. Sajid[4], M. Rizwan[5], M. Aslam[3] and F. Ijaz[1]

[1]School of Chemistry, University of the Punjab, Lahore, Pakistan
[2] Department of Mechanical and Manufacturing Engineering, Universiti Malaysia Sarawak, Kota Samrahan, Malaysia
[3]Department of Chemistry, University of Education, Lahore, Pakistan
[4]Department of Materials and Chemical Engineering, Yibin University, Yibin 644000, Sichuan China
[5]Department of Chemistry, The University of Lahore, Lahore, Pakistan
*Corresponding author: salman.shafqat@ue.edu.pk

Abstract. In the present study, unmodified *Melia azedarach* (UMA) and magnetically modified *Melia azedarach* (MMA) were used as a low cost biosorbent for the adsorptive removal of oxyanion (chromate ion $Cr_2O_4^{2-}$) from aqueous solution. The UMA and MMA were characterized by Fourier Transform Infrared Spectroscopy (FTIR) and Scanning electron microscopy (SEM). Batch biosorption process was used to check the effect of well-known process parameters including biosorbent dose, contact time and pH of solution. Experimental results revealed that MMA has more adsorption capacity as compared to UMA. Maximum % age removal (60.49%) was observed under acidic condition (pH=2) using MMA. The obtained experimental data was evaluated using Langmuir and Freundlich isotherms as well as pseudo 1st and ^{2nd} order reaction kinetic models. Results showed that equilibrium mechanism can satisfactorily be explained by Langmuir model as depicted by the correlation factor (close to unity) and follows 2nd order reaction kinetic models.

1. Introduction

In present era, the largest component of the life affected is the water contaminated due to anthropogenic activities. As well observed, there is always increase in water demand in the agricultural and domestic sectors, hence it is of utmost importance to properly treat the industrial effluents and to remove heavy metals and other toxic chemicals before disposing into the water streams. In Pakistan, a large industrial sector consists of textile, leather, paper and pharmaceutical industry, hence releasing a large amount of heavy metals in the environment [1]. Heavy metals are major concern from the previous years because of their persistence in environment, non-biodegradable nature and toxicity. Hexavalent chromium (Cr^{+6}) a heavy metals is highly toxic and also categorized as a carcinogenic metals by US-EPA and many other agencies. From the various sources of Cr in the environment, tanning industries are the major source of disposal of chromium-contaminated sludge [2]. On the other hand trivalent chromium (Cr^{+3}) is needed by the plants and animals in trace quantity and act as an essential heavy metal [3,4]. The hexavalent chromium (Cr^{+6}) is a cause of sensitization and allergic reactions on the skin. Moreover the inhalation of Cr^{+6} is responsible for the absorption of chromium in the blood and serum and thus reached into the kidneys whereby producing chronic problems. It can also be absorbed by the gut and dangerous for the respiratory tract [5].

A number of physical, chemical, biological as well as physiochemical approaches have been developed to remove toxic heavy metals from industrial effluents [6]. All of these approaches have their own demerits and setbacks [7,8]. The very accurate method for removal of such heavy metals is still not known because of complexation. However, biosorption is one of the effective and efficient physical methods for the removal of such metals from the contaminated water. Different types of biosorbents have been used for the removal of chromium from water including coconut coir pith, mucor, duneliella, composite chitosan, green algae, spirogyra and so on [9-12]. It has been found from the literature that chemical or physical modification of adsorbents can enhance their adsorption capacity due to increased surface area due to the exposure of more active groups on the surface [13-15].

Thus the development of modified, more efficient, inexpensive biosorbent for waste water treatment is the subject of intensive studies. One of the biosorbent is *Melia azedarach*. The adsorption capacity of *Melia azedarach* can be enhanced by the magnetic modification because it causes ring opening of cellulose subunits that leads to the exposure and availability of more hydroxyl (-OH) groups onto the surface.

In this study unmodified and magnetically modified *Melia azedarach* were used for removal of oxyanion ($Cr_2O_4^{2-}$) from aqueous solution. Various well-established parameters including biosorbent dosage, *p*H, contact time and temperature were studied to explore optimum conditions. Langmuir isotherm and Freundlich isotherm were used to explain the adsorption phenomenon on the surface of adsorbate (heterogeneous surface). Date obtained from the experiments was also verified to investigate rate of adsorption and capacity by applying pseudo 1st and 2nd order kinetic model.

2. Materials and Method

2.1 Chemicals

Chemicals used in the current study were of analytical grade. Potassium dichromate ($K_2Cr_2O_7$), ferric chloride ($FeCl_3.6H_2O$), sodium hydroxide (NaOH), hydrochloric aicd (HCl), sulphuric acid (H_2SO_4) were purchased from Fluka and Sigma Aldrich. Distilled and deionized water was used to prepare the solutions.

2.2 Method

2.2.1 Preparation of Adsorbent. The branches of *Melia azedarach* were collected from the vicinities of University of the Punjab and were grinded by using pin mill to obtain particle size upto 60 mesh. The particles were washed and subjected to boiling for the ultimate removal of lignin and then these particles were allowed to filter. The continuous washing was continued until a clear filtrate was obtained. The residue was then dried below 80°C.

2.2.2 Modification of Adsorbent. 25g of raw adsorbent was added in 50mL of concentrated sulphuric acid and was left for 24 hours at room temperature so that charring may proceed. Next day the charred material was washed with distilled water till the neutral *p*H. The material was dried below 80°C. In this way the structure of adsorbent was modified by the ring opening of sugar molecules and thus increases the surface area (active sites) of adsorbent.

This material was then named as charred *Melia azedarach*. Charred *Melia azedarach* was poured into 0.2M solution of hydrated ferric chloride and was allowed to stir for 4 hours at 150rpm. Afterwards the mixture was decanted off and the modified adsorbent was washed with distilled water.

IOP Conf. Series: Materials Science and Engineering 1101 (2021) 012031 doi:10.1088/1757-899X/1101/1/012031

The adsorbent was allowed to dry and was labelled as Fe^{+3} loaded charred *Melia azedarach* or modified *Melia azedarach* (MMA).

2.3 Characterization

The prepared as well as processed samples were characterized by Infrared Spectroscopy (IR), Scanning Electron Microscopy (SEM) and by Absorption spectrometer.

IR spectra were recorded on Perkins Elmer FTIR (spectrum RXI) spectrometer. IR spectrographs helped to study the changes in the functional groups of unmodified and magnetically modified products.

Morphology and surface of unmodified as well as modified *Melia azedarach* was studied before and after modification using SEM Jeol (JSM-6390LA). The surfaces of the samples were coated with gold before the measurements.

Absorption spectrometer was utilized to investigate the adsorption of chromate at various established process parameters.

Orbital Shaker (Vortex Modal-OSM-747), Vacuum Filtration Assembly, pH Meter (Jenway Modal 3320) was also used during the investigation of process parameters.

2.4 Adsorption study of $Cr_2O_4^{-2}$ by UMA & MMA Melia azedarach

For adsorption studies, 0.2 g of UMA and MMA *Melia azedarach* were added in 50ppm of chromate at *p*H 5.0 and the beakers were then stirred on a magnetic stirrer for 30 minutes at 150rpm, separately. Afterwards, the flasks were taken out from shaker and the contents were filtered for the removal of adsorbents. The filtrates were then subjected to atomic absorption spectrometric analysis for the concentration determination of Cr^{+6} at wavelength 357.9nm. A decrease in adsorption capacity (q) was observed with the increasing amount of adsorbent (per gram). This observation is in accordance with the equation below

$$q = (C_o - C_e)V/m \quad (1)$$

V is volume of sample, Co is initial concentration, Ce is concentration at equilibrium and m is mass of adsorbent

To investigate the UMA as well as MMA *Melia azedarach* dose, the experiments were performed for 0.2, 0.3, 0.4, 0.5, 0.6, 0.7, and 0.9g of adsorbents at temperature 25°C for 30 min at stirring speed of 150rmp. The effect of contact time was studied by varying the time at 5, 10, 15, 20, 25, 30, 35 and 40 minutes. To optimize the effect of initial pH on the adsorption of chromate onto UMA and MMA *Melia Azedarach*, the experiments were performed at initial pH of 1.0, 3.0, 5.0, 7.0, 9.0 and 10.0 using 0.5 g of UMA and 0.3g of MMA adsorbent at temperature 25◦C and stirred for 15 min.

3. Results and Discussion

3.1 FTIR Studies

To investigate the magnetic modification of *Melia azedarach* FTIR studies of biosorbent was conducted before and after the modification. It can be concluded from FTIR plot that the modification of adsorbent has been done. By comparing the spectra shown in figure 1, it can be observed that there is a broad spectrum in the region of 3700-3000 cm^1 for modified adsorbent (MMA). This shows that the modification of adsorbent has been done by the use of $FeCl_3.6H_2O$ and H_2SO_4. By modification the binding capacity of adsorbent has been enhanced by introduction of Carboxylate, carbonyl and amino groups.

3.2 SEM Analysis

Scanning electron microscopy (SEM) was used to investigate morphology of unmodified and modified adsorbents. The operating parameters were 100mm, 8.0 kV and 200–300x magnification power. Figure 2(A) showed the SEM of unmodified *Melia azedarach* and figure 2(B) showed modified *Melia azedarach* biosorbent and reflected that the biomass consisted of heterogeneous pores, roughness and spongy surface. The adsorption capacity of biosorbent was dependent upon small pores present on heterogeneous surface. The modified *Melia azedarach* developed more porosity, roughness and spongy surface that lead to better adsorption.

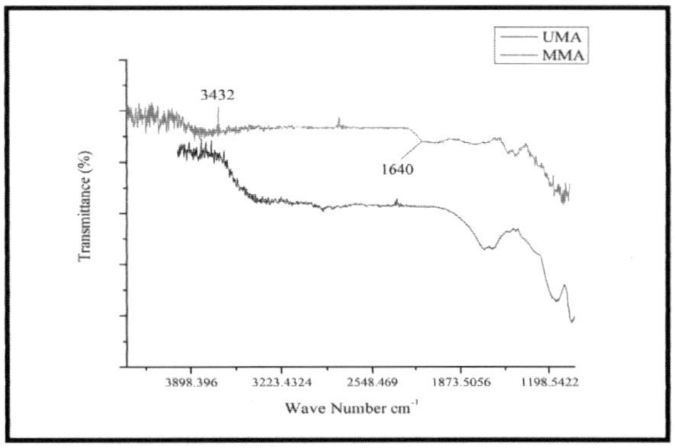

Figure 1. FTIR of UMA and MMA *Melia azedarach*

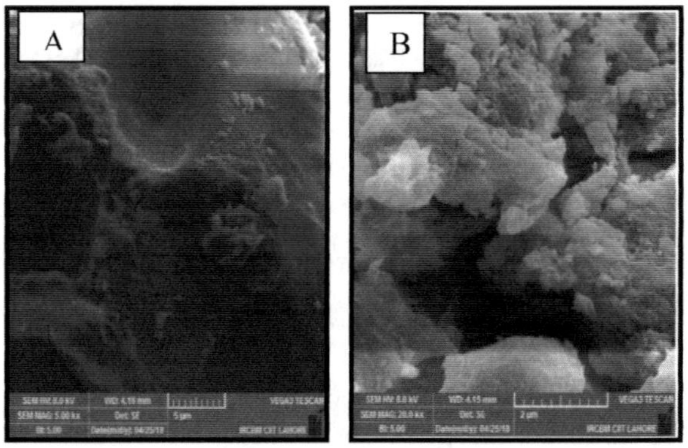

Figure 2. SEM of UMA (A), MMA (B) *Melia azedarach*

3.3 Effect of Adsorbent dose

Figure 3A demonstrates the effect of adsorbent dose on the removal of chromate ion ($Cr_2O_4^{-2}$) using UMA and MMA respectively. The adsorption parameter was studied by using initial concentration of 50ppm.

Decrease in adsorption capacity was also found that at constant concentration of metal ion. This observation is related to the fact that with the increases in the adsorbent dose the number of binding sites is increased, in this way the chromate ions will spread on the adsorbent. Hence there is a decrease in the adsorption capacity per gram of dose. The percentage adsorption ranges from minimum percentage (5.5%) to maximum percentage (24.79%). Under these conditions, MMA showed greater q values as compared to UMA. For instance by having similar conditions (0.3g/50mL) the adsorption capacity noted to be 0.77mg/g for UMA while MMA showed 3.31mg/g for chromate ions. Hence it is proved that the modified *Melia c* is more favorable.

3.4 Effect of pH

The maximum percentage adsorption (60.49%) was observed at acidic (pH=1). The working pH range for the removal of chromate ions ($Cr_2O_4^{-2}$) is 2-3 due to increase in the H^+ ions. At higher pH values, an increase in concentration of OH^-(negative ions) was observed and meanwhile there was decrease in the H^+ ions (positive ions). The chromate ions removal would be favorable in the presence of positive species (H^+ ions) at low *p*H because of the attraction of negative ions ($Cr_2O_4^{-2}$) towards positive ions (H^+).Thus the working *p*H for the removal of dichromate ions is acidic as shown in figure 3B.

By applying similar conditions only with the change in adsorbent dose MMA showed greater q values as compared to UMA. For instance by similar conditions at *p*H=3 (0.5g/50mL for UMA, 0.3g/50mL for MMA) the adsorption capacity noted to be 3.13mg/g for UMA while MMA showed 6.96mg/g for dichromate ions. Hence it is proved as demonstrated in figure 3B that the modified *Melia azedarach* is more favorable for the removal of dichromate ions as compared to unmodified *Melia azedarach*.

3.5 Effeect of Contact Time

An increase in the time results in increased adsorption capacity has been presented in figure 3C using UMA and MMA respectively. This is due the fact initially chromate ions move rapidly towards adsorbent because at this stage the biosorbent is starved for adsorbent. With the passage of time the adsorption capacity becomes constant, because at that point the amount of chromate ions being adsorbed by the adsorbent establishes a dynamic equilibrium with the amount of chromate ion desorbed from adsorbent.

By applying similar conditions, MMA showed greater q values as compared to UMA. For instance at contact time of 10 minutes (0.5g/50mL for UMA, 0.3g/50mL for MMA) the adsorption capacity noted to be 2.63mg/g for UMA while MMA showed 4.32 mg/g for chromate ions. Hence it is proved that the modified *Melia azedarach* is more favorable for the removal of chromate ions as compared to unmodified *Melia azedarach.*

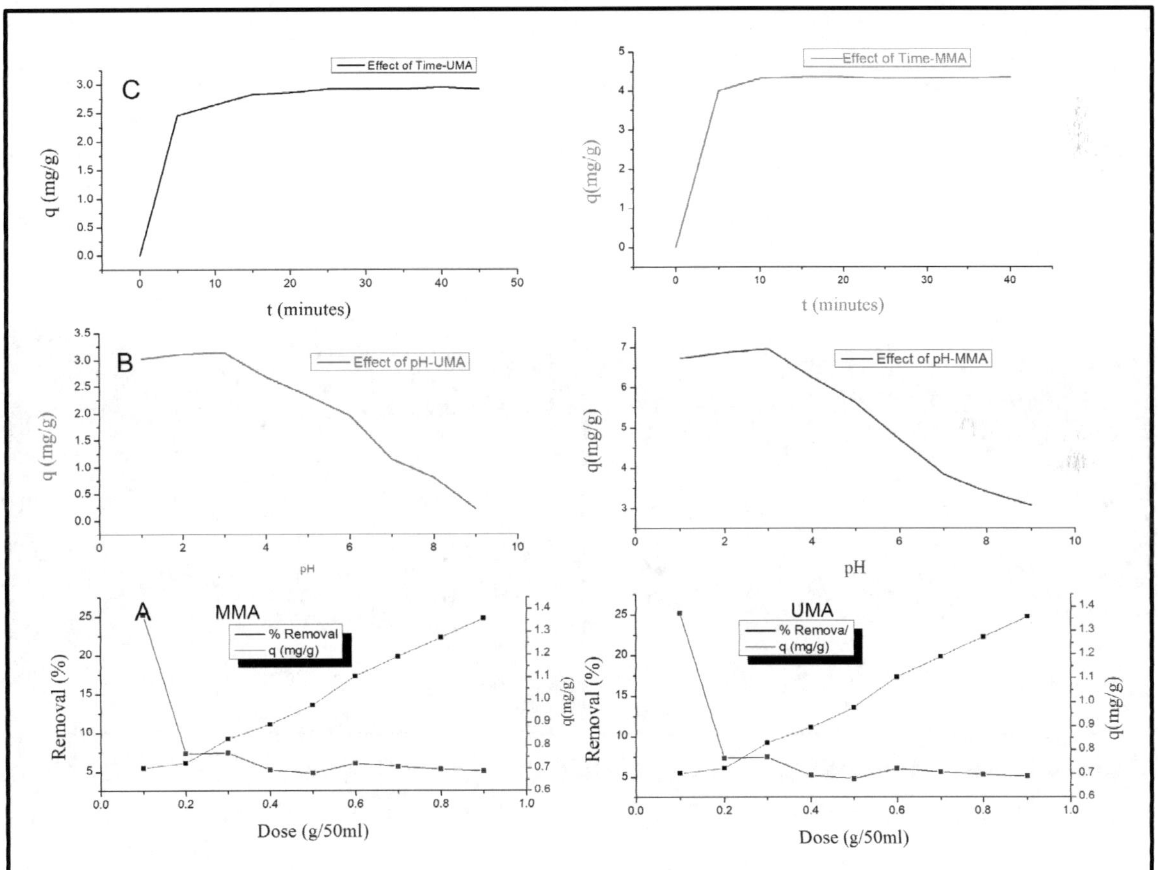

Figure 3. A= Effect of Adsorbent dose, B= Contact time, C=*p*H for ($Cr_2O_7^{-2}$) removal

3.6 Adsorbent Isotherms

Data obtained during the experimentation was analyzed by applying well established isotherm viz. Langmuir and Freundlich isotherm models.

3.6.1. Langmuir Isotherm. Langmuir isotherm is plotted according to the equation:

$$1/q_e=1/bq_{max}.1/C_e+1/q_{max} \quad (2)$$

IOP Conf. Series: Materials Science and Engineering 1101 (2021) 012031 doi:10.1088/1757-899X/1101/1/012031

For Langmuir isotherm model the correlation coefficient values were greater and fitting best to the adsorption process (R^2 value for dichromate ion is 0.97 using UMA and 0.98 for MMA) thus showing that Langmuir model figure 4 is better for understanding adsorption behavior of dichromate ion than Freundlich model. Table 1 represents the slope and intercept for the derivation qmax mg/g and constant relevant to free energy from the Langmuir plot.

MMA showed enhanced qmax value as compared to UMA. For instance the qmax value for unmodified adsorbent (UMA) is 5.95mg/g and qmax value for modified adsorbent (MMA) is 16.86mg/g.

3.6.2 Freundlich Isotherm. Freundlich isotherm is an empirical relationship obtained by plotting graph between lnCe on x-axis and lnqe on y-axis.

This model assumes that adsorption of dye from liquid to liquid surfaces takes place on a heterogeneous surface which involves different sites with several adsorption energies. Freundlich isotherm is plotted according to the equation iii:

$$\log q = \log K_f + 1/n \log C_e \quad (3)$$

The value of n in the range of less than 1 showed average adsorption, 1-2 showed good and 2-10 showed better adsorption characteristics. The correlation coefficient value of Freundlich isotherm (Figure 5) showed that it did not fit well to the adsorption data. These values are 0.97 for UMA and 0.95 for MMA.

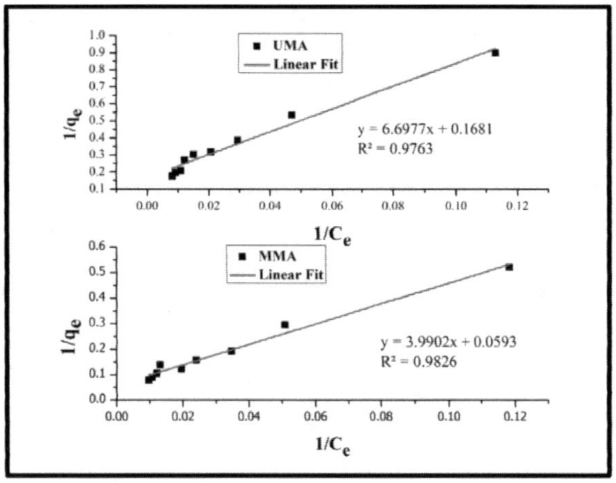

Figure 4. Langmuir isotherm UMA (A), MMA (B)

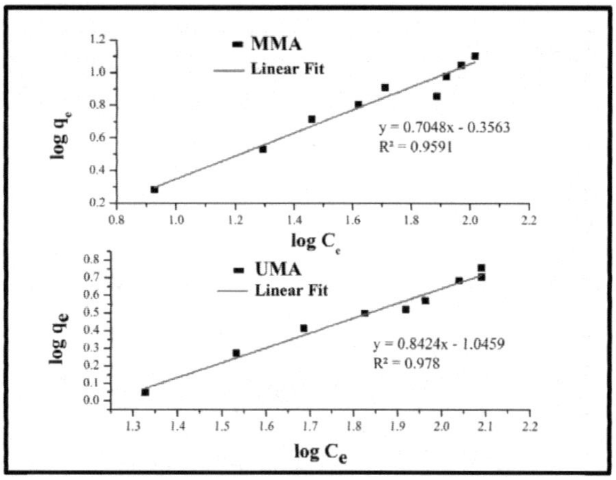

Figure 5. Freundlich Isotherm UMA (A), MAA (B)

Table 1. Langmuir and Freundlich parameters for $Cr_2O_4^{-2}$ Removal

Langmuir parameters					Freundlich parameters			
Adsorbent	Slope	Intercept	q_{max}(mg/g)	b(dm^3/g)	Slope	Intercept	n	K_f
UMA	6.69	0.16	5.948	0.02	0.84	-1.045	1.19	0.09
MMA	3.99	0.059	16.86	0.01	0.704	-0.3563	1.4204	0.44

3.7 Kinetic Modelling

The pseudo 1st and 2nd order kinetic model were applied to study the rate of adsorption and capacity.

IOP Conf. Series: Materials Science and Engineering 1101 (2021) 012031 doi:10.1088/1757-899X/1101/1/012031

3.7.1 Pseudo 1st Order Kinetic Model. First order reaction model is plotted according to the equation:

$$\ln (q_e-q_t)=\ln q_e-k_1.t \tag{4}$$

A graphical representation of Pseudo 1st order kinetic model for ($Cr_2O_7^{-2}$) removal using MMA is shown in figure 6.

The value of linear correlation coefficient R^2 for pseudo 1st order is less as compared to pseudo 2nd order model. For the prediction of mechanism involved Lagergren's pseudo 1st order kinetic model is usually applied.

By using *Melia azedarach* for chromate ion adsorption a pseudo 1st order kinetic model was obtained by plotting a graph between time and $\ln(q_e-q_t)$.

The R^2 value for UMA was 0.97 and that for MMA was 0.94 as shown in table 2 and didn't explain well to the adsorption kinetic.

3.7.2 Pseudo 2nd Order Kinetic Model. Second order reaction model is plotted (Figure 7) according to the equation:

$$t/q_t=1/k_2.q_e^2+t/q_e \tag{5}$$

By using UMA and MMA for the adsorption of chromate ion, the pseudo 2nd order model is represented in Fig. 7 and parameters for pseudo 2nd order are demonstrated in table 2. The value of correlation factor (R^2=0.99) value for UMA and MMA depicted that the pseudo 2nd order model was fitted better to the data. In a comparative analysis to the pseudo 1st order kinetic model, the q_e value derived for 2nd order kinetic model very appreciable related to the experimental value for UMA and MMA showing that pseudo 2nd order kinetic mechanism is best for the adsorptive removal of chromate ions by *Melia azedarach*.

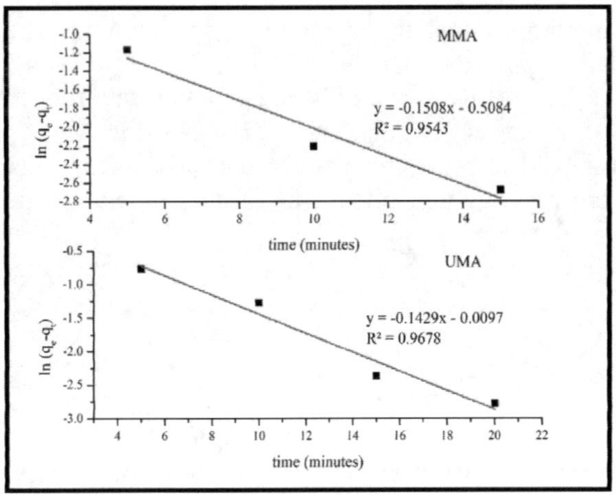

Figure 6. Pseudo 1st order kinetic model

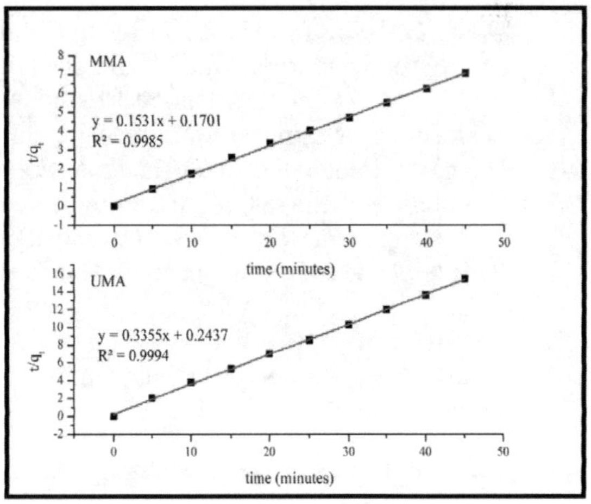

Figure 7. Pseudo second order model

Table 2. Pseudo 1st and Pseudo 2nd order order kinetic parameters for $Cr_2O_7^{-2}$ Removal

Pseudo 1st order					Pseudo 2nd order			
Adsorbent	k_1	R^2	q_e	q_e(experimental)	k_2	R^2	q_e	q_e(experimental)
For UMA	0.142	0.97	0.99	2.91	2.98	0.99	4.103	2.91
For MMA	0.150	0.94	0.60	4.32	6.53	0.99	5.88	6.37

4. Conclusion

In the present study cost effective and readily available *Melia azedarach* was applied as a adsorbents for the removal of dichromate ions from water bodies. *Melia azedarach* was magnetically modified to increase the efficiency for oxyanions uptake. FTIR characterization revealed that various functional groups are observed on the surface of UMA that can be responsible for chromate ions removal. Modification improves the surface morphology by ring opening of cellulose and producing free hydroxyl group (-OH) in MMA which is shown by the high uptake capacity of MMA as compared to UMA. Optimum pH for the maximum percentage adsorption removal (60.49%) of chromate ions were found in acidic medium (pH=2-3). Batch biosorption experiments showed that removal efficiency of chromate ions using MMA was better as compared with UMA. Equilibrium data was found better fit to Langmuir equilibrium model. The maximum Adsorption was observed i.e. 2.63mg/g for UMA while MMA showed 4.32 mg/g for chromate ions. Kinetic data was observed to follow pseudo 2nd order mechanism. Hence it can be concluded that MMA can be used for treatment of wastewater in terms of oxyanions removal in an effective and economical way.

5. References

[1] Mahmood Q, Shaheen S, Bilal M, Tariq M, Zeb B S, Ullah Z, and Ali A 2019 Chemical pollutants from an industrial estate in Pakistan: a threat to environmental sustainability *Appl. Wat. Sci.* **9** 47.

[2] Homa D, Haile E and Washe A 2016 Determination of spatial chromium contamination of the environment around industrial zones *International J. Anal. Chem.* 1-7.

[3] Boddu V M, Abburi K, Talbott J L and Smith E D 2003 Removal of hexavalent chromium from wastewater using a new composite chitosan biosorbent *Env. Sci. Tech.* **37** 4449-4456.

[4] Huang J, Cao Y, Shao Q, Peng X and Guo Z 2017 Magnetic nanocarbon adsorbents with enhanced hexavalent chromium removal: morphology dependence of fibrillar vs particulate structures *Ind. Eng. Chem. Res.* **56** 10689-10701.

[5] Park R M, Bena J F, Stayner L T, Smith R J, Gibb H J and Lees P S 2004 Hexavalent chromium and lung cancer in the chromate industry: a quantitative risk assessment *Ris. Ana. An Int. J.* **24** 1099-1108.

[6] Shah P 2018 Azo dye removal technologies *Adv. Res. Appl.* 169.

[7] Singh P K and Singh R L 2017 Bio-removal of azo dyes: a review *Int. J. App. Sci. Biotech.* **5** 108-126.

[8] Habte H, Zewge F, Redi M 2014 Removal of azo dye from water by absorption using waste residue from alum manufacturing process *Afr. J. Env. Sci. Tech.* **8** 482-490.

[9] Volesky B, and Naja G 2005 Biosorption: application strategies. *16th Int. Biohydromet.Symp.* 25-29.

[10] Burakov A E, Galunin E V, Burakova I V, Kucherova A E, Agarwal S, Tkachev A G and Gupta V K 2018 Adsorption of heavy metals on conventional and nanostructured materials for wastewater treatment purposes: A review *Ecotox. Env. Saf.* **148** 702-712.

[11] Sharma P K, Ayub S and Tripathi C N 2013 Agro and horticultural wastes as low cost adsorbents for removal of heavy metals from wastewater: A review. *Int. Ref. J. Eng. Sci.* **2** 18-27.

[12] Liu X, Hu Q, Fang Z, Zhang X Zhang B 2008 Magnetic chitosan nanocomposites: a usefulrecyclable tool for heavy metal ion removal *Lang.* **25** 3-8.
[13] Kousha M, Daneshvar E, Sohrabi M S, Jokar M and Bhatnagar A 2012 Adsorption of acid orange II dye by raw and chemically modified brown macroalga Stoechospermum marginatum *Chem. Eng. J.* 67-76.
[14] Liu, X., Hu, Q., Fang, Z., Zhang, X., & Zhang, B. (2008). Magnetic chitosan nanocomposites: a useful recyclable tool for heavy metal ion removal. *Langmuir*, **25**(1) 3-8.
[15] Kousha, M., Daneshvar, E., Sohrabi, M. S., Jokar, M., & Bhatnagar, A. (2012). Adsorption of acid orange II dye by raw and chemically modified brown macroalga Stoechospermum marginatum. *Chemical Engineering Journal*, **192** 67-76.

IOP Conf. Series: Materials Science and Engineering 1101 (2021) 012032 doi:10.1088/1757-899X/1101/1/012032

Design of EWMA-type control chart

H W You[1], M A Shahrin[2] and Z Mustafa[2]

[1] Pusat GENIUS@Pintar Negara, Universiti Kebangsaan Malaysia, 43600 Bangi, Selangor, Malaysia
[2] Pusat Pengajian Sains Matematik, Fakulti Sains dan Teknologi, Universiti Kebangsaan Malaysia, 43600 Bangi, Selangor, Malaysia
Corresponding author: hwyou@ukm.edu.my

Abstract. A control chart is typically examined based on the overall performance. This is crucial, as it will influence the final decision on the selection of a control chart. For this reason, the application of average run length (ARL) has been identified as the standard performance metric. In order to calculate ARL, practitioners must determine the size of the process shift in advance. Nevertheless, in practice, the process shift size cannot be predetermined, as practitioners may not have the requisite information associated with the process. Hence, in this work, the expected average run length (EARL) is employed as a performance metric for the unknown shift size in a process. This work evaluates the performance of a memory-type control chart, i.e. exponentially weighted moving average median (EWMA median) chart, with respect to ARL and EARL. The results indicate that, as long as the deterministic shift size is within the range of the process shift size, the chart's performance is nearly the same, regardless of the use of the optimal pair based on minimizing ARL or EARL.

1. Introduction

In the business world, the primary objective is to produce high quality products or services. In view of this, a collection of statistical tools with the objective of reducing the variability in the process has been suggested, namely, Statistical Process Control (SPC). A control chart is a useful tool in SPC [1], and has been widely adopted in business settings. This has motivated researchers to develop novel control charts that are more efficient with respect to the detection ability. For example, memory-type control charts have been introduced for better detection ability in a process. Some recent works by [2]-[4], to name a few, investigated memory-type control charts extensively.

The exponentially weighted moving average (EWMA) mean chart is the first EWMA chart introduced by [5]. The EWMA mean chart performs well in detecting small-to-moderate shifts in a process, under normality assumption. However, [6] showed that when the process contains contaminated normal data, the EWMA mean chart is insensitive to detect process shifts. In view of this, the EWMA median chart was introduced by [2] under such settings. The key benefit of the EWMA median chart is that it is robust against outliers.

After evaluation of the performance of a control chart, average run length (ARL) was considered as a standardized performance metric. The ARL is defined as the expected number of samples taken before the process is signalled as out-of-control [7]. The employment of ARL requires the practitioner to specify the magnitude of a shift. In real-world scenarios, the practitioners have limited knowledge to predefine the process shift size [8], [9]. This eventually leads to the introduction of the expected average run length (EARL), which is used to investigate the performance of the EWMA median chart. This is because the computation of EARL does not demand practitioners to define the process shift size. Moreover, research dealing with EARL as a performance measure in EWMA median charts is not yet available in the related literature.

This work aims to investigate the overall performance evaluation of the EWMA median chart using ARL and EARL. The run length profiles of the EWMA median chart are demonstrated in Section 2. Section 3 presents the comparison performance measure using ARL and EARL. Finally, some conclusions are drawn in the final section.

2. EWMA median chart

The EWMA median chart was pioneered by [2]. In this particular median chart, the sample median is used to monitor the process, as follows:

$$\tilde{T}_i = \begin{cases} T_{i,((n+1)/2)} & \text{if } n \text{ is odd} \\ \dfrac{T_{i,(n/2)} + T_{i,(n/2+1)}}{2} & \text{if } n \text{ is even} \end{cases}, \tag{1}$$

where n denotes the sample size, and we heuristically make the assumption that n is an odd number. With such an assumption, the sample median $\tilde{T}_i$ can be determined more quickly and easily. Work by [6] shows that the sample median is an ideal alternative to monitor if the process contains contaminated normal data.

Assuming that the process is distributed independently and identically (i.i.d) and has a normal distribution with the in-control mean and in-control standard deviation, i.e. μ_0 and σ_0, respectively, the statistic of the EWMA median chart is defined as follows:

$$U_i = (1-a)U_{i-1} + a\tilde{T}_i, \tag{2}$$

for $i \in \{1, 2, \ldots\}$. Note that $U_0 = \mu_0$ and a represent a smoothing constant, i.e. $a \in (0,1]$. The control limits, i.e. *UCL* and *LCL*, which represents the upper control limit and lower control limit of the EWMA median control chart, respectively, are defined as follows:

$$UCL / LCL = \mu_0 \pm b\sigma_0 \tag{3}$$

where b is the design constant.

Here, the Markov chain method is applied to obtain the run length profiles of the EWMA median chart. The transition probability matrix (tpm) for the transient states is denoted as **P**, as follows:

$$\mathbf{P} = \left[P_{i,j}\right]_{l \times l}, \tag{4}$$

where $i, j = 1, 2, \ldots l$. The interval between the control limits is divided into subintervals, $l = 2g+1$, each with a width of $2w$. Let K_j be the midpoint of the jth subinterval, for $j = 1, 2, \ldots, 2g+1$. The $P_{i,j}$ corresponding to the matrix **P** are as follows:

$$P_{i,j} = F_{\tilde{T}}\left(\frac{K_j + w - (1-a)K_i}{a} \middle| n\right) - F_{\tilde{T}}\left(\frac{K_j - w - (1-a)K_i}{a} \middle| n\right) \tag{5}$$

for $i, j = 1, 2, \ldots, 2g+1$. Here, $F_{\tilde{T}}\left(\ldots\middle|n\right)$ is the cumulative distribution function (cdf) of $\tilde{T}_i$, with $i = 1, 2, \ldots$.

Finally, the ARL is computed as follows:

$$\text{ARL} = v_1, \tag{6}$$

where

$$v_1 = \mathbf{p}^T(\mathbf{I}-\mathbf{P})^{-1}\mathbf{1}. \tag{7}$$

Note that **p** is the vector of initial probabilities and **I** is the identity matrix. The computation of the ARL requires the practitioners to predetermine the size of the process shift. In practical application, the practitioners may not know the precise size of the process shift in advance, as they do not have past knowledge of this process. Accordingly, EARL criterion is used as the EWMA median chart's performance measure, and is defined as follows:

$$\text{EARL} = \int_{\delta_{\min}}^{\delta_{\max}} f_\delta(\delta)\text{ARL}\, d\delta. \tag{8}$$

Note that $f_\delta(\delta)$ is the probability density function of the magnitude of the process shift, i.e. δ.

3. Performance of the EWMA median chart with unknown shift sizes

The efficiency of a control chart is generally assessed by the ARL. Here, in-control and out-of-control ARL are denoted as ARL_0 and ARL_1, respectively. The ARL_1 for various combinations of sample sizes, $n \in \{3, 5, 7, 9\}$ and the process shift sizes, $\delta \in \{0.25, 0.5, 0.75, 0.95, 1.25, 1.5, 1.75, 2.0\}$ are presented in Table 1.

Table 1. ARL_1s for different combinations of (n, δ), based on the optimal pair (a, b).

n	δ	(a, b)	ARL_1	n	δ	(a, b)	ARL_1
3	0.25	(0.1000, 0.4155)	46.53	5	0.25	(0.1000,0.3322)	31.78
	0.50	(0.1000, 0.4155)	14.87		0.50	(0.1265,0.3847)	10.64
	0.75	(0.1648, 0.5669)	8.09		0.75	(0.2273,0.5538)	5.76
	0.95	(0.2311, 0.7000)	5.66		0.95	(0.3163,0.6825)	4.03
	1.25	(0.3393, 0.8932)	3.74		1.25	(0.4768,0.8952)	2.67
	1.50	(0.4457, 1.0696)	2.84		1.50	(0.6294,1.0918)	2.01
	1.75	(0.5684, 1.2675)	2.24		1.75	(0.7571,1.2613)	1.57
	2.00	(0.6834, 1.4546)	1.81		2.00	(0.8521,1.3940)	1.29
7	0.25	(0.1000, 0.2845)	24.51	9	0.25	(0.1000,0.2528)	20.20
	0.50	(0.1589, 0.3795)	8.43		0.50	(0.1889,0.3753)	7.04
	0.75	(0.2830, 0.5444)	4.56		0.75	(0.3348,0.5383)	3.81
	0.95	(0.3968, 0.6772)	3.19		0.95	(0.4788,0.6823)	2.67
	1.25	(0.6072, 0.9098)	2.09		1.25	(0.7089,0.9088)	1.73
	1.50	(0.7588, 1.0813)	1.56		1.50	(0.8443,1.0503)	1.31
	1.75	(0.8663, 1.2103)	1.25		1.75	(0.9300,1.1461)	1.11
	2.00	(0.9362, 1.2995)	1.10		2.00	(0.9761,1.2005)	1.03

The optimal pair (a, b) is obtained in order to attain the intended $\text{ARL}_0 = 370$ for the corresponding n and δ. For example, in the case $n = 5$ and $\delta = 0.95$, the optimal pair, $(a, b) = (0.3163, 0.6825)$, yielding the lowest ARL_1 value, i.e. $\text{ARL}_1 = 4.03$. Meanwhile, these optimal pairs achieve the desired $\text{ARL}_0 = 370$.

According to Table 1, for a fixed *n*, the ARL_1 value decreases as δ increases. For instance, for $n = 3$, when $\delta = 0.25$, the corresponding $\text{ARL}_1 = 46.53$ with the optimal pair, $(a, b) = (0.1000, 0.4155)$, as opposed to when $\delta = 1.50$, the $\text{ARL}_1 = 2.84$ is attained using the optimal pair, $(a, b) = (0.4457, 1.0696)$.

Nevertheless, the calculation of the ARL required the practitioners to define the process shift. In real-world application, the practitioners may not have the requisite judgement on the process. Therefore, if a practitioner considers a particular shift size, δ, and uses the corresponding optimal pair, the performance of the EWMA median chart is affected if a different shift size is detected in the process. In light of this, the EARL is proposed to monitor the process when the size of the shift is unknown. Here, two EARLs are usually of interest, namely, the in-control EARL, EARL_0 and the out-of-control EARL, EARL_1.

Table 2 reveals the optimal pair (a, b) that is obtained by minimizing the EARL_1 of the EWMA median chart, based on the corresponding $(\delta_{\min}, \delta_{\max}) = (0.25, 2.0)$. Note that the EARL_0 is set at 370. For comparison purposes, the similar *n* value is shown in Table 2.

Table 2. EARL_1 and ARL_1 values for the EWMA median chart with the corresponding parameters when $n = \{3, 5, 7, 9\}$, $\text{EARL}_0 = 370$, $\delta_{\min} = 0.25$ and $\delta_{\max} = 2.0$

n	3	5	7	9
(a, b)	(0.1033,0.4240)	(0.1366,0.4035)	(0.1677,0.3925)	(0.2032,0.3928)
$EARL_1$	8.47	6.15	4.94	4.18
δ				
0.25	47.02	34.94	28.38	24.52
0.50	14.89	10.65	8.43	7.05
0.75	8.35	6.00	4.77	3.98
0.95	6.18	4.47	3.57	3.00
1.25	4.48	3.28	2.65	2.25
1.50	3.68	2.72	2.23	1.93
1.75	3.14	2.34	1.98	1.68
2.00	2.75	2.10	1.78	1.44

According to Table 2, when $n = 3$, $\delta_{\min} = 0.25$ and $\delta_{\max} = 2.0$, the optimal pair is $(a, b) = (0.1033, 0.4240)$, yielding the smallest EARL_1, i.e. $\text{EARL}_1 = 8.47$ while attaining the $\text{EARL}_0 = 370$. By considering $\delta = 0.5$ $(\text{i.e. } \delta \in (\delta_{\min}, \delta_{\max}))$ for the same *n* value, the $\text{ARL}_1 = 14.89$ is obtained using the optimal pair $(a, b) = (0.1033, 0.4240)$ from Table 2. Meanwhile, for the same combination (n, δ), the ARL_1 is 14.87 using the optimal pair $(a, b) = (0.1000, 0.4155)$. This demonstrates that the performance of the EWMA median chart is nearly similar in the case of using the optimal pair based on minimizing ARL_1 and EARL_1. In view of this, when the practitioner is unable to determine the precise process shift size, the optimal pair evaluated by minimizing the EARL_1 can be employed.

4. Conclusion

This paper explicitly demonstrated that the EARL can be applied as an alternative performance metric for the EWMA median chart in the case the process shift size is unknown in advance. Moreover, the proposed optimal pair (a, b) based on minimizing the EARL_1 can be implemented, as long as the considered shift size is within $(\delta_{\min}, \delta_{\max})$. This can overcome the inaccuracy of the process when monitoring using EWMA median chart, in the case a different shift size is detected. This study is based on the process parameters, i.e. the in-control mean and standard deviation are known. In real-world scenarios, the process parameters may not be known. Therefore, further research to examine the performance of the EWMA median chart with exact and random process shifts, when process parameters are unknown, is needed. The findings of this research are to be discussed in our future work.

5. References

[1] Montgomery DC 2019 Introduction to Statistical Quality Control. New York: John Wiley & Sons, Inc

[2] Castagliola P 2001 An $\overline{X}/R$–EWMA control chart for monitoring the process sample median *Int. J. Reliab. Qual. Saf. Eng.* **8** 123-135.

[3] Raza SMM and Siddiqi AF 2017 EWMA and DEWAM control charts for Poisson-Exponential Distribution: Conditional median approach for censored data *Qual. Reliab. Eng. Int.* **33** 387-399.

[4] Mitra A, Kok LK and Chakraborti S 2019 An adaptive exponentially weighted moving average-type control chart to monitor the process mean *Eur. J. Oper. Res.* **279** 902-911.

[5] Roberts SW 1959 Control chart tests based on geometric moving averages *Technometrics* **1** 239-250.

[6] Human SW, Kritzinger PE and Chakraborti S 2011 On the robustness of the EWMA chart *J. Appl. Stat.* **38** 2071-2087.

[7] Tang A, Castagliola P, Sun J and Hu X 2019 Optimal design of the adaptive EWMA chart for the mean based on median run length and expected median run length *Qual. Technol. Quant. Manag.* **16** 439-458.

[8] You HW, Khoo MBC, Castagliola P and Qu L 2016 Optimal exponentially weighted moving average charts with estimated parameters based on median run length and expected median run length *Int. J. Prod. Res.* **54** 5073-5094.

[9] Chong ZL, Khoo MBC, Teoh WL, You HW and Castagliola P 2019 Optimal design of the side-sensitive modified group runs (SSMGR) chart when process parameters are estimated *Qual. Reliab. Eng. Int.* DOI: 10.1002/qre.2395.

Acknowledgement

This research is supported by the Universiti Kebangsaan Malaysia, Geran Galakan Penyelidik Muda, GGPM-2017-062.

Reliable Smart Pet Feeding Machine Using Arduino Uno Starter Kit

M S Z M Suffian[1], A N H A Zaini[1], A Jamali[1], S Mohamaddan[1], and M F Ashari[1]

[1] Department of Mechanical and Manufacturing Engineering, University Malaysia Sarawak (UNIMAS), Kota Samarahan, Malaysia.
Corresponding author: msmsyazwan@unimas.my

Abstract. Pet feeding can be done manually by pet owners, or automatically with the help of the pet feeders. The main concerns here are to solve the issue of unavailability of pet owners and the hygiene aspect in pet feeding. The developed automated system enables pet owners to feed their pets while they are away. The main objective of operating this research is to develop an automated pet feeding device focusing on in house pets by utilizing Arduino Uno as its Central Processing Unit (CPU) or microcontroller which function to control a weight sensor (load cells) and a servo motor. Smart Pet Feeding Machine is an automated pet feeding device designed to run based on weight mechanism. It will automatically dispense pet foods when minimum amount of pet food weight is triggered and will automatically stop dispensing when it reaches maximum amount of weight. Its architectural design includes a vertical storage compartment to store pet foods, a square opening at the bottom of storage compartment as opening for dispensing, a motorized dispenser and an aluminium bowl to place dispensed pet foods. The motion of dispenser is operated by a servo motor attached to the dispenser.

1. Introduction

Nowadays, owning pet at home is no longer a trend. Taking care and feeding pet is no longer a burdensome task since there are huge numbers of pet feeder in the market. However, there are not much pet feeding machines that serve meal for pet freshly. Contaminated food will cause health issues to pets. One of the illnesses is called foodborne. This disease exists due to bacteria carried in contaminated foods.

As an overview, an automated pet feeding machine consists of a single-entry feeding zone, a feeding bowl, a food storage container, a sensor, a memory storage and at least one processor [1]. The processor selected to be integrated with Smart Pet Feeding Machine is Arduino Uno microcontroller, which is programmed using Arduino Software. Arduino Uno Starter Kit is a low-cost robotic kit. As the name implies, it suits starter. One driving concern of this work is the affordability and accessibility of educational tools to students and schools that do not possess enough financial support [2]. This device operates using weight mechanism and therefore a weight sensor is integrated to the system. Load cells are embedded to the device to sense the weight of pet food remaining. There are few reasons and advantages of using strain gauge type load cells as the weight sensor. One of them is its simple mechanism. A sensor probe has a first end disposed within the interior space. The sensor probe also includes a second end opposite the first end outside the interior space. At least one load cell is disposed within the interior space and is operatively engaged with the first end of the sensor probe [3,4].

2. Problems encountered by pet owners

The most primary problem is the unavailability of pet owners to feed their pets. Unavailability of pet owners due to work and other commitments are difficult to be solved without any helping device. Hence, here comes the usage of pet feeding machine. However, the operating mechanism of available of pet feeder has several rooms for improvement where the main concern is the cleanliness of pet food [5]. Besides, the design pet feeder available in the market does not suit pet owners who own more than one pet at home. This problem leads to issue where pets are not being feed accordingly [6]. Next problem is hygiene aspect in pet feeding. Improper pet feeding leads to low cleanliness level and might cause pet food contamination. Thus, this issue can cause health problems to pets.

3. Methodology

3.1. Designing process

Few sketches are made to ensure the design capable to achieve requirements needed to build the Smart Pet Feeding machine. Every design is reviewed and analysed to sort out advantages and disadvantages. From this stage, flaws in design can be detected. Hence, better modification can be done to make the design better. Based on review on designs, the best design is selected to be drawn via Autodesk Inventor 2015. The dimension of every part is reconsidered again in order to make the product suits the requirements and needs. During this stage, there might be several changes made due unsuccessful mechanism of design. Throughout designing stage, two designs are drawn with different specifications and dimensions. In Smart Pet Feeding machine, it can be classified into three main sections. The main sections are housing, hopper (dry pet food storage) and control gate dispenser. Other related parts are storage opening and pet food bowl. A preliminary study is conducted to determine the amount or weight of pet food to be dispensed. This study covers the average amount of pet food to be consumed by pet at once. Based on this study, 40 grams of pet food is set to be poured at single dispense. Hence, this will ensure the amount of pet food in the pet bowl is around 50 grams every dispense session, where initially it is left with 10 grams of pet food.

3.2. Programming stage

Arduino Software is utilised to develop the program. The programming is done on Arduino Uno via Arduino Software which programmed to control gate dispenser using weight sensor [7]. The weight sensor is used to detect the weight of remaining dry pet foods. Once the weight sensor detects a reading where it is set to detect weights which are less than 50 grams, the circuit will be completed. This means that the control gate dispenser will open and dispense the dry pet foods into pet food bowl. The mechanism of dispensing the pet foods is designed to be simple and minimalistic. By only dispensing pet foods when it is less in amount can ensure in providing clean meal to pets. This is to ensure the food dispensed earlier is eaten first by the pets before serving another meal.

3.3. Fabrication

The Smart Pet Feeding Machine model is fabricated using reliable materials. The materials can be classified into two categories which are software and hardware. The materials selected for the housing and base are mostly scrap materials. Hence, this gives more room on budgeting for software purchases [8]. The materials used for Smart Pet Feeding Machine's body are plywood, aluminium tin and Perspex. Special part such as dispenser is fabricated using 3-Dimensional printer [9]. Then, these components and parts are assembled together. Then, the finished product is tested and the result is analysed.

3.4. Experimental process

Analysis conducted is limited to weight sensor performance and servo motor functionality [10]. The experimental setups are shown below:

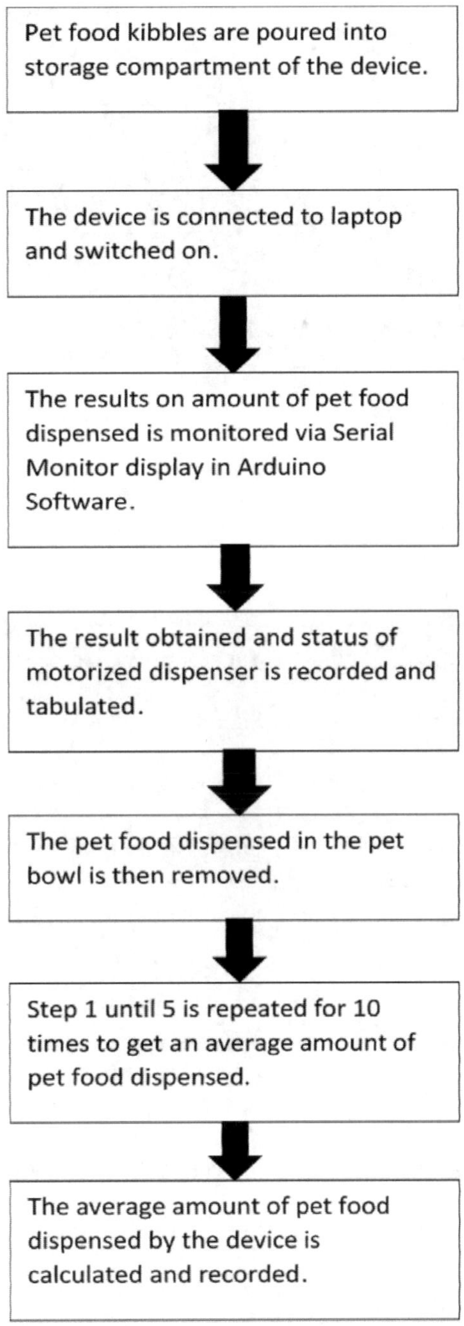

Figure 1. Experimental flow.

4. Result and analysis

Figure 2. Isometric View of the Smart Pet Feeding Machine (Zaini, 2018).

Table 1. Results of Testing on Smart Pet Feeding Machine Dispensing Accuracy.

Testing	Weight of Pet Food Dispensed (g)	Status of Dispenser	Percentage of Error (%)
1	42.7	Successfully dispensed	6.75
2	42.4	Successfully dispensed	6.00
3	41.8	Successfully dispensed	4.50
4	41.5	Successfully dispensed	3.75
5	41.1	Successfully dispensed	2.75
6	40.7	Successfully dispensed	1.75
7	40.3	Successfully dispensed	0.75
8	40.2	Successfully dispensed	0.50
9	40.1	Successfully dispensed	0.25
10	40.2	Successfully dispensed	0.50

4.1 Average weight of pet food dispensed

The average weight of pet food dispensed is obtained by dividing the sum of dispensed pet food weights with the number of testing of experiment.

$$\text{Average weight of pet food} = \frac{42.7+42.4+41.8+41.5+41.3+40.7+40.3+40.2+40.1+40.2}{10}$$

$$= \frac{411}{10} = 41.1$$

Based on the results tabulated in Table 1, it can be calculated that the mean weight of pet foods dispensed by Smart Pet Feeding Machine is 41.1 grams. Hence, from this data, it can be stated that the coding works on this project successfully.

4.2 Percentage of error

$$\text{Percentage of error} = \frac{Weight_{experimental} - Weight_{theoretical}}{Weight_{theoretical}} \times 100 \quad (1)$$

Example on calculation for percentage of error during individual test. Using result obtained in Test 1.

$$\text{Percentage of error} = \frac{42.7-40.0}{40.0} \times 100 = 6.75\%$$

Calculation for average of percentage of error by Smart Pet Feeding Machine. This approach is done to verify the efficiency of this design and hence obtaining the overall performance of device.

$$\text{Percentage of error} = \frac{41.1-40.40}{40.0} \times 100 = 2.75\%$$

4.3 Analysis of results

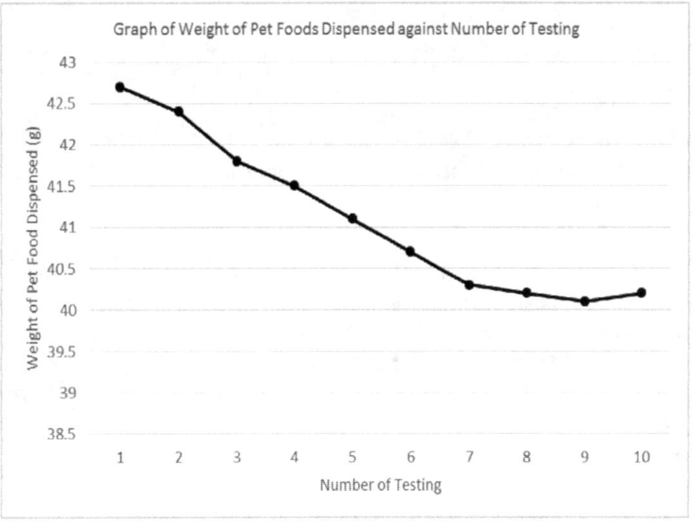

Figure 3. Graph of Weight of Pet Food Dispensed against Number of Testing.

IOP Conf. Series: Materials Science and Engineering 1101 (2021) 012033 doi:10.1088/1757-899X/1101/1/012033

Based on graph plotted above, Figure 2 shows the weights of pet foods dispensed in the total of ten tests. At Test 1, the weight of pet foods dispensed is 42.5 g, which exceeds 2.5 g from coded program where according to coding it should dispense 40 g of pet foods. Whereas the weight of pet food dispensed gradually decrease until Test 6. This situation occurred due to few possible reasons. Firstly, at Test 1, it dispenses the most due to weight of pet foods inside the storage compartment is at the highest point. Hence, the force applied might be slightly higher at initial tests. Next possible reason is due to delay in the control system. When the device is switched on, delay in time to close the dispenser might cause the pet food to be dispensed more than programmed. However, these errors do not affect the functionality of the Smart Pet Feeding Machine since the pet bowl can store more amount of pet foods. Furthermore, the amount of pet foods left in the pet bowl is not fixed. Hence, if the initial weight is lesser, the motorized dispenser will dispense more. Following tests which are Test 7 until 10 shows steady rate of dispensing pet foods. In these tests, it can be denoted that the control system is stable and ready to operate steadily as programmed. The initial error happened in earlier test is acceptable since it does not affect the functionality of the product.

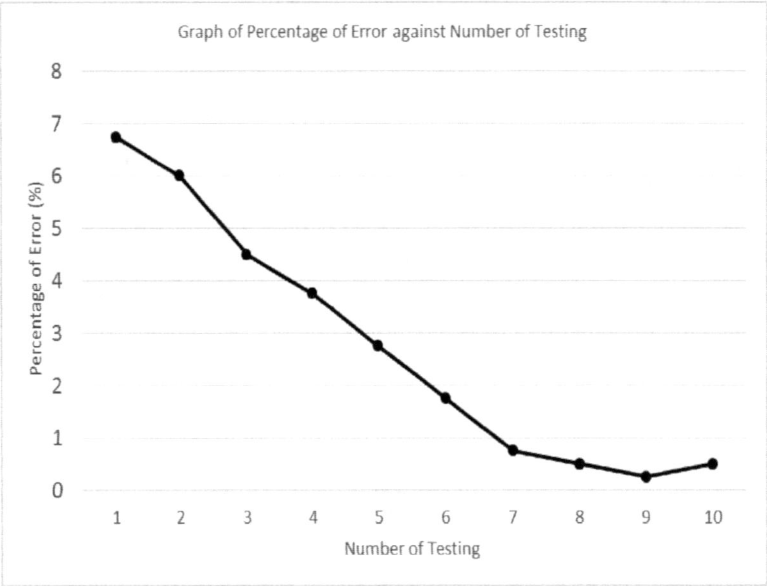

Figure 4. Graph of Percentage of Error against Number of Testing.

Figure 3 above shows the trend of percentage error of Smart Pet Feeding Machine performance. Based on this graph, it can be seen that the percentage of error is decreasing in every test and approaching zero. Hence, this proves that the device can operate smoothly as programmed.

5. Conclusion

This project had achieved the target and solved the issues stated in problem statements. The first issue that been solved is the cleanliness aspect during feeding pets. Whereas the automated pet feeder normally dispenses pet food at several times and this design is unsuitable for pet owners who owns more than one pet in house. Furthermore, the prices for automated pet feeding machines available in the market are very high. Hence, the enhanced controller system is developed for Smart Pet Feeding Machine where it can solve issues in cleanliness aspect, price issue and versatility of its function. Simple operating control system which implements Arduino Uno as the microcontroller and the easy to find materials for the housing allow Smart Pet Feeding Machine to be built at affordable price. Besides, the microcontroller is designed to be functioned in normal house condition where standard pressure and temperature of surrounding would not affect its functionality. Next to be concluded is the uniqueness of the Smart Pet Feeding Machine itself where most of the materials used to fabricate the housing are scrap materials. This is the best approach of this project which it proves that such product like automated pet feeding device can be developed at home.

6. References

[1] William, S. (2015) US9750229. Retrieved 4 November 2017, from https://www.google.com/patents/US9750229
[2] Junior, L. A., Neto, O. T., Hernandez, M. F., Martins, P. S., Roger, L. L., & Guerra, F. A. (2013). A Low-Cost and Simple Arduino-Based Educational Robotics Kit. Cyber Journals, **3**(12), pp. 1–7
[3] Schmidt, E. (2015). US8830291. Retrieved 22 November 2017, from https://www.google.com/patents/US8830291
[4] Badamasi, Y. A. (2014). The working principle of an Arduino. In 2014 11th international conference on electronics, computer and computation (ICECCO) (pp. 1-4). IEEE.
[5] Boschetti, G. (2013). S0888327012002737. Mechanical Systems and Signal Processing, **34**(1–2), pp. 116–130
[6] Sabari Akilesh, K., Savitha, V., Vinithra, N., & Dhansekar, J. (2017). Smart Fish Feeder.
[7] Thakkar, P. K. H. (2013). Performance Evaluation of Strain Gauge Based Load Cell to Improve Weighing Accuracy. International Journal of Latest Trends in Engineering and Technology (IJLTET) Performance, **2**(1), pp. 103–107.
[8] Robertson, G. L. (2006). Food Packaging: Principles and Practice. IFLA Journal (Vol. 32).
[9] Ainsworth, J. (2015). Desktop 3D Printer. Retrieved 6 December 2017, from http://opus.ipfw.edu/cgi/viewcontent.cgi?article=1277&context=etcs_seniorproj_mcetid
[10] Olszewski, O. Z., Houlihan, R., O'Keeffe, R., O'Neill, M., Waldron, F., Mathewson, A., & Jackson, N. (2014). A MEMS silicon-based piezoelectric AC current sensor. Procedia Engineering, **87**, pp. 1457-1460.

7. Acknowledgments

This research is supported by Osaka Gas in Cultural Exchange Foundation (OGFICE) Research Grant Scheme 2017/18 under research titled "Drying Process of Lemantak (Sago Powder) using an Automated Machine for Cottage Industries in Sarawak". The authors would like to thank the Faculty of Engineering, Universiti Malaysia Sarawak (UNIMAS) for providing sponsorship and facilities for this research.

IOP Conf. Series: Materials Science and Engineering 1101 (2021) 012034 doi:10.1088/1757-899X/1101/1/012034

Wind turbine modelling and simulation using Matlab/ SIMULINK

C H Chong[1], A R H Rigit[1] and I Ali[1]

[1] Faculty of Engineering, Universiti Malaysia Sarawak (UNIMAS), Kota Samarahan 94300 Malaysia
Corresponding author: chongchihung1407@gmail.com

Abstract. Malaysia is aiming to reduce the dependency on oil and gas by increasing the utilization of renewable energy. One of the promising renewable energy available is wind energy. Malaysia is located near the equator which is situated in a low wind speed region with an annual mean wind speed of 1.2-4.1 m/s. However, Malaysia also experiences two different monsoon seasons which is the southwest monsoon and the northeast monsoon. Wind speed during the southwest monsoon season can reach 7 m/s while the wind speed during the northeast monsoon can exceed 15 m/s. Limited researches and studies had been done to investigate the potentiality of wind energy in Sarawak, Malaysia. In this paper, modelling and simulation of different wind energy conversion system (WESC) using different generators operating under the same parameters will be carried out using MATLAB/SIMULINK to investigate the efficiency of the generators. PMSG has shown to be more efficient over SCIG and DFIG for lower wind speed although all of them eventually reaches the similar efficiency. Efficiency of SCIG, DFIG and PMSG for rated wind speed are 66.25%, 69.38% and 71.88% respectively.

1. Introduction

Wind turbine is used to convert the kinetic energy of the wind into mechanical energy that rotates the rotor blades which are connected to the low speed shaft. The mechanical energy will then be transmitted into the high speed shaft via a gearbox to generate electrical energy in the generator [1]. Horizontal axis wind turbine (HAWT) rotates at an axis parallel to the direction of the wind [2]. Wind power can be harvested as long as the blades are facing the direction of the wind. Besides, HAWT usually have taller tower which allows the turbine to face stronger wind as wind speed increase with heights. With variable pitch angle turbines, maximum extractable amount of wind energy is possible by increasing or reducing the blade's pitch angle depending on the electricity generated. In this paper, modelling and simulation will be focusing on HAWT as they are more efficient.

The power, P in the wind that can be extracted by a wind turbine are given by the formula:

$$P = \frac{1}{2} . \rho . A . C_p(\beta, \lambda) . v^3 \tag{1}$$

where ρ is the air density, A is the surface area of the blade, C_p is the power coefficient of the turbine and v is the wind speed. Power coefficient of a turbine can be determined by using the formula:

$$C_p(\beta, \lambda) = c_1 \left(\frac{C_2}{Z} - c_3\beta - c_4 \right) e^{\frac{-c_5}{z}} + c_6\lambda \tag{2}$$

$$\frac{1}{z}=\frac{1}{\lambda+0.08\beta}-\frac{0.035}{1+\beta^3} \tag{3}$$

where c_1=0.5176, c_2=116, c_3=0.4, c_4=5, c_5=21, c_6=0.0068, β is the pitch angle for the blades and λ is the tip speed ratio.

Figure 1 shows the characteristic parameters of a wind turbine which is essential in choosing an appropriate turbine. Cut-in speed is the minimum wind speed required to overcome the friction of the turbine blades to rotate the blades. The rated speed is the minimum wind speed at which the turbine generate the rated power output. Rated power output is the maximum power generated by the generator without damanging the generator and batteries. Pitch angle adjustment of the blades limits the output power by varying the aerodynamic forces acting on the blades at higher wind speed [3]. Cut-out speed or furling speed is the maximum wind speed at which the generator will continue to generate electricity before shutting down to protect the turbine. Any wind speed above the cut-out speed will cause damage to the turbine and therefore the braking mechanism will be activated and shut down the turbine.

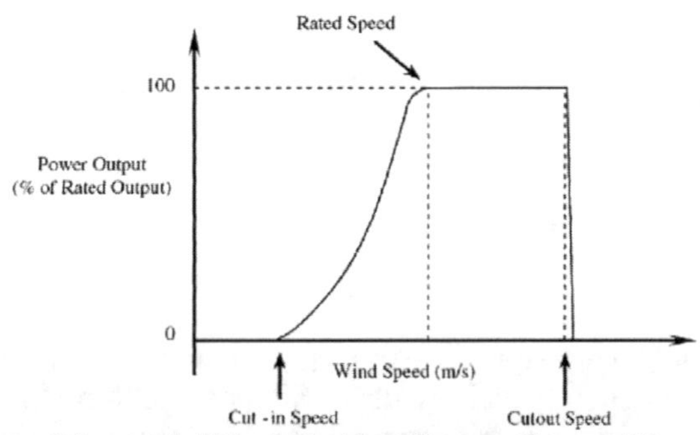

Figure 1. Wind Turbine Power Curve. [4]

1.1. Components in a WECS

Wind energy conversion system as shown in Figure 2 can be divided into the blades, generator, power converter and controller. The turbine blades are responsible of converting the kinetic energy from the wind into mechanical power that rotates the turbine shaft. The mechanical power of the shaft is converted into electrical energy via the generator. Drivetrain is used to increase the rotational speed of the turbine shaft to power the generator. The converter and inverter will convert noisy AC power generated by the variable speed generator into clean and reliable AC power [5].

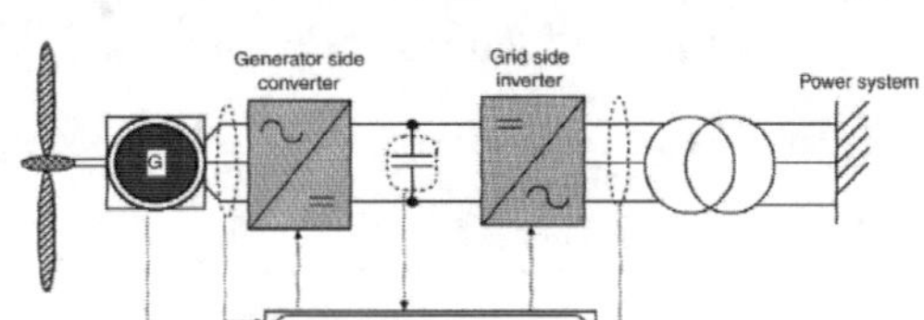

Figure 2. Wind Energy Conversion System [6]

1.2. Wind Turbine

HAWT can be further classified into fixed speed wind turbine and variable speed wind turbine [7]. One of the fixed speed wind turbine utilizes squirrel cage induction generator (SCIG). The rotor of a fixed speed turbine remains constant regardless of the wind speed. It is determined by various parameters such as gear ratio, grid frequency and generator design. Tip speed ratio, λ can be determined using $\lambda=\frac{R \bullet w_t}{v}$ where R is the turbine radius, W_t is the turbine speed and v is the wind speed. For a fixed speed wind turbine, the TSR will keep varying with the fluctuation of wind speed. This means the system will only able to achieve maximum efficiency at a certain wind speed as shown in Figure 3. SCIG is one of the most commonly used generator since this type of generator can be connected directly to the grid as the generator's frequency remains constant and same with the grid's frequency throughout the operation [8]. The advantages of this type of generators are that this system is more simple, reliable and cheap but lack in power control, uncontrollable reactive power consumption and additional mechanical stresses on the turbine due to the fixed speed generator.

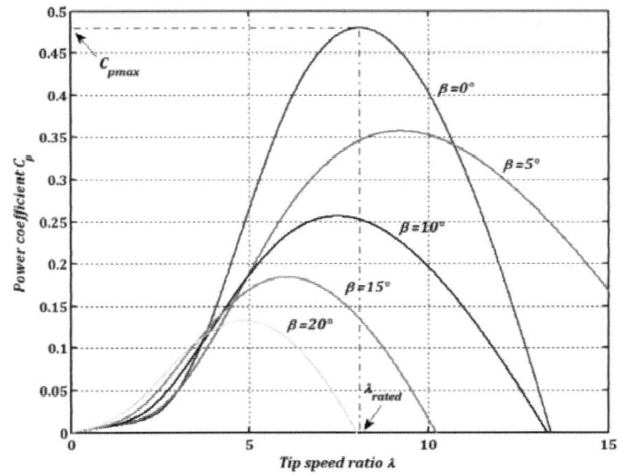

Figure 3. Power Coefficient vs TSR for Wind Turbine. [9]

Doubly-fed induction generators and permanent magnet generators are commonly used variable speed generators. For this type of generators, the rotor speed will vary proportionally with the increasing wind speed and thus the TSR and performance will remain constant throughout the operation. This will ensure that maximum efficiency of the turbine can be obtained for a wide range of wind speed. Power converters are required due to the fluctuating rotor speed of the generators and frequency of the electricity generated [10]. This system has a higher power capture efficiency, improved power quality and reduction in mechanical stress as the rotation of the turbine blades rely on the magnitude of wind speed.

1.3. Generator

Electric generators can be further divided into asynchronous and synchronous generator. The main difference between the two is that asynchronous generators require the grid supply for excitation while synchronous generators do not require grid suppy for the excitation of its winding. Gearbox is used to increase the low rotational speed of the turbine blades to a higher rotational speed of the rotor shaft to power the generator.

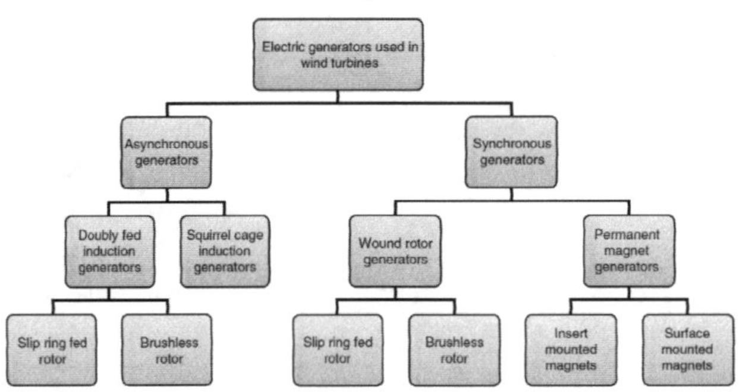

Figure 4. Types of Generator in WESC [7]

The three-phase stator windings of a squirrel cage induction generator (SCIG) can be connected directly to the utility grid by using a transformer [11]. The grid is responsible in providing the reactive

IOP Conf. Series: Materials Science and Engineering 1101 (2021) 012034 doi:10.1088/1757-899X/1101/1/012034

power required for the excitation of stator's winding. The rotor of the generator will be accelerated up to synchronous speed based on the frequency of the grid. Slip is the difference between the rotational speed of the stator's magnetic field and the rotor's speed. Slip can be calculated by using $s = \frac{n_s - n_r}{n_s}$. Negative slip means the generator is operating above the synchronous speed due to the wind power acting on the rotating blades. For SCIG which is a fixed speed turbine, the slip is very small, usually below 1%.

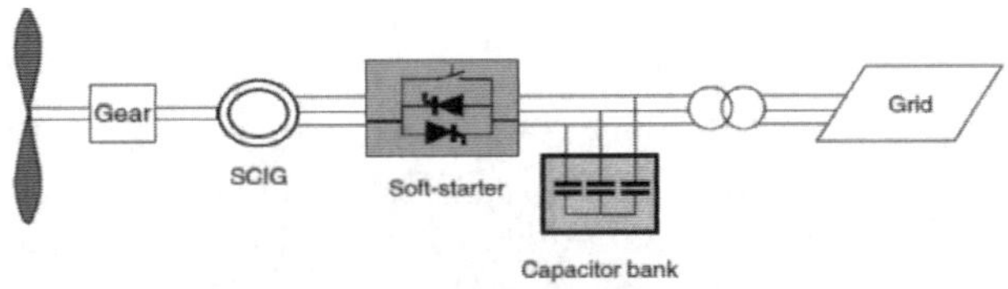

Figure 5. Squirrel Cage Induction Generator [12]

Doubly fed induction generator (DFIG) on the other hand is gaining more and more attention for applications in the range of megawatts of power [13]. This generator comprises of a three-phase stator windings and a three-phase rotor windings. Both the rotor and stator windings are connected to the grid where the rotor speed is adjusted by back to back converters. The grid will excites the rotor windings and the magnetic field of the rotor windings will rotates with the rotating blades. The rotating magnetic flux acting on the stator windings will results in the generation of alternating current. The rotating speed of the stator's magnetic field relies on the rotor speed and the frequency of the grid. In order to maintain a consistant frequency of the generated current, it is necessary to condition the frequency of the grid signal acting on the rotor windings. Back to back converter consists of a machine side converter and a grid side converter with a DC voltage line between them.

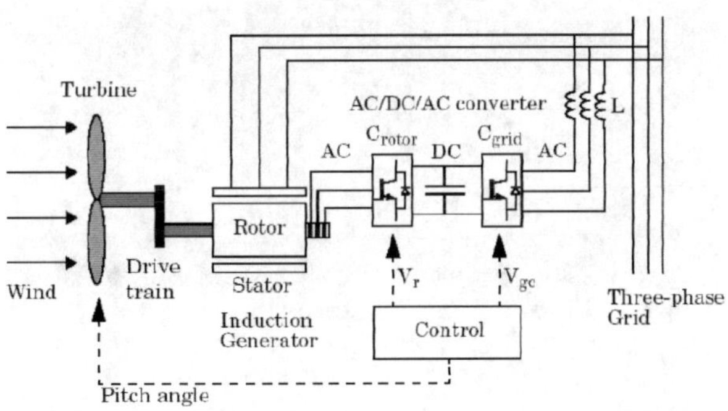

Figure 6. Doubly Fed Induction Generator [14]

Synchronous generators require external excitation either by permanent magnet or via an external DC power source. Wound rotor synchronous generator (WRSG) uses DC current, brushes and slip rings for the excitation of rotor windings [15]. Extra maintenances are required to maintain the brushes and slip rings. On the other hand, permanent magnet synchronous generator (PMSG) is a better alternative to WRSG as it eliminates the need of brushes and slip rings. For PMSG, the excitation field of the rotor windings is created by the permanent magnet or electromagnets. Three-phase power is generated in the stator windings when the rotor is rotated by the wind turbine. Figure 7 shows the PMSG wind turbine model which is popular due to their simple design. By using permanent magnets, the commutators, slip rings and brushes of a WRSG can be eliminated which will greatly improve the reliability of the system, making it a better option compared to the WRSG. PMSG requires rectifier and inverter as it is a variable speed generator. The AC-DC-AC power converter will convert the fluctuating electrical power generated into fixed DC power and then back into AC power with fixed frequency and magnitude.

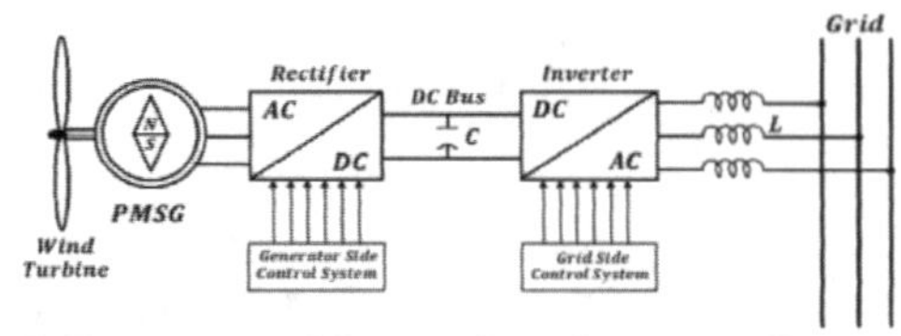

Figure 7. Permanent Magnet Synchronous Generator [16]

2. Methodology

Simulations for different WECS will be presented in this section. The wind speed for the simulation is simulated by the following blocks as shown in Figure 8. The wind speed will increase from 0m/s to 9m/s from t=1s to t=10s. It will then increase from 9m/s to 15m/s from t=15s.The wind speed will remain constant until t=30s.The purpose of this simulation is to study the behaviour of different types of generator, operating under zero wind speed, base wind speed which is 9m/s and wind speed above rated wind speed. The real power, reactive power, rotor speed, mechanical torque and pitch angle are recorded and analysed. Real power, rotor speed and pitch angle will be discussed in the following paper.

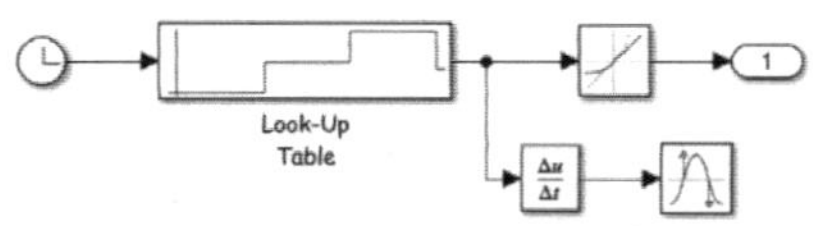

Figure 8. Wind Speed Simulation

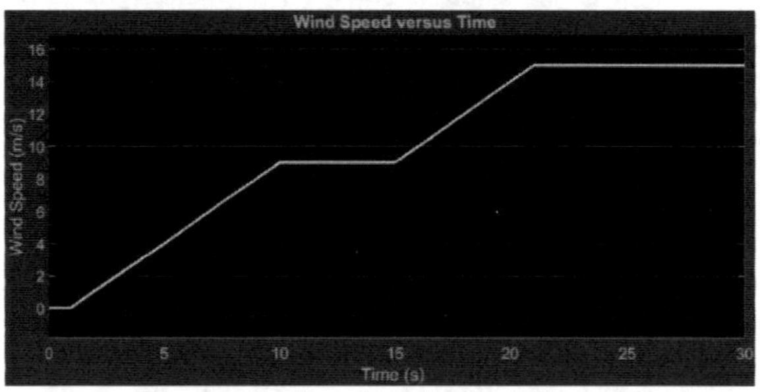

Figure 9. Simulated Wind Speed

2.1. SCIG Wind Turbine

Figure 10 shows the SCIG wind turbine modelled using SIMULINK. The 1.5MW induction generator is connected directly to the power grid via a 575V/25kV and 25kV/120kV step-up transformer and a 30km transmission line. The wind turbine model in SIMULINK will be used to provide the mechanical torque required to drive the rotor of the generator. SCIG will convert the mechanical torque into electrical power and then fed it to the grid via the stator windings. VI-measurement blocks will record the voltage, current, active and reactive power.

2.2. DFIG Wind Turbine

The DFIG model in Figure 11 is similar to the SCIG model except that the rotor windings are connected to the grid via power converters and step-up transformers. The stator windings are connected directly to the grid via transformer. Reactive power require to energize the generator stator will be provided by the grid. Simulation of DFIG wind turbine is done with the addition of rectifier and inverter using universal bridge. Rectifier will convert the fluctuating electrical power generated into fixed DC power and the inverter will convert the DC power back into AC power. The rectifier control system controls the generator speed and the inverter control system controls the power transmitted to the grid.

IOP Conf. Series: Materials Science and Engineering 1101 (2021) 012034 doi:10.1088/1757-899X/1101/1/012034

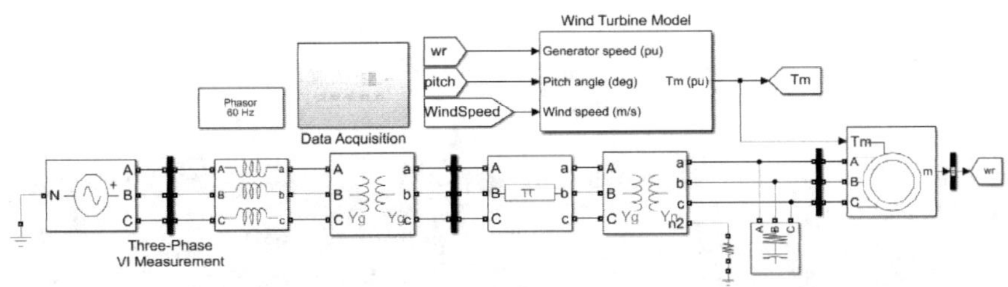

Figure 10. SIMULINK Model of SCIG Wind Turbine

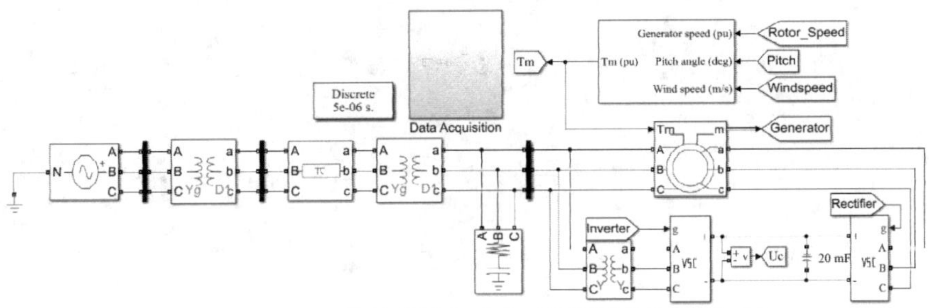

Figure 11. SIMULINK Model of DFIG Wind Turbine

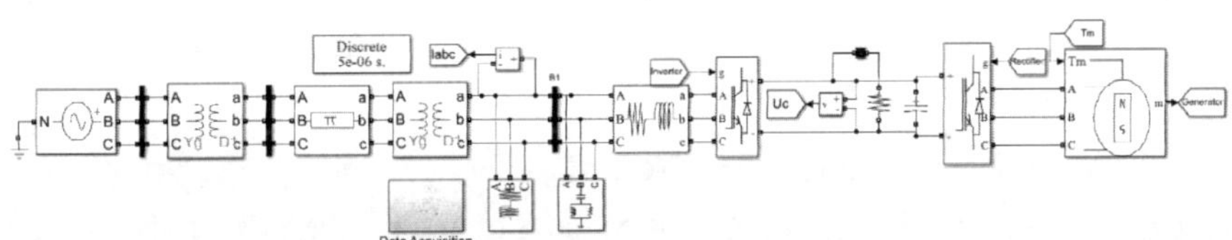

Figure 12. SIMULINK Model of PMSG Wind Turbine

2.3. PMSG

PMSG SIMULINK model in Figure 12 is also similar to the SCIG except that the generator do not require grid supply for the windings excitation. Power converters are required as PMSG is a variable speed wind turbine. Permanent magnets will excite the rotor windings while the stator windings will be connected directly to the grid via a rectifier, inverter, step up transformers and 30km transmission line.

3. Result and analysis

All of the data will be analysed in per unit(pu) for ease of comparison. Per unit is the ratio of the actual value obtained to the reference value or base value in the same unit. Analysis will be focused on power generated, rotor speed and pitch angle of the SCIG, DFIG and PMSG.

3.1 Generated Power

Figure 13 shows the data obtained from the SIMULINK models on the power generated by SCIG, DFIG and PMSG. The behaviour of each generators with varying wind speed can be observed. High starting current of the SCIG can be noticed from t=0s to t=1s as SCIG is connected directly to the power grid without any power converter. This spike in the starting current will cause severe voltage disturbances to the grid and a soft starter is required to limit the high starting current of the generator. Both stator and rotor windings for DFIG are directly connected to the grid which allow for better control of active and reactive power. Parts of the output power of DFIG are controlled by the power converter. This reduces the fluctuation in the power generated and increased in the power control. It is also worth noting that both SCIG and DFIG draw reactive power from the power grid for winding excitation. The grid will supply three-phase AC power to the stator, creating a rotating magnetic field. When the rotor driven by the wind turbine rotates at a speed higher than the synchronous speed, power will be generated.

IOP Conf. Series: Materials Science and Engineering 1101 (2021) 012034 doi:10.1088/1757-899X/1101/1/012034

Standalone induction generator wind turbines are possible with the addition of capacitors or diesel generators to provide the required current for winding excitation. PMSG do not require external excitation as the magnet or electromagnet mounted on the stator will provide the excitation required.

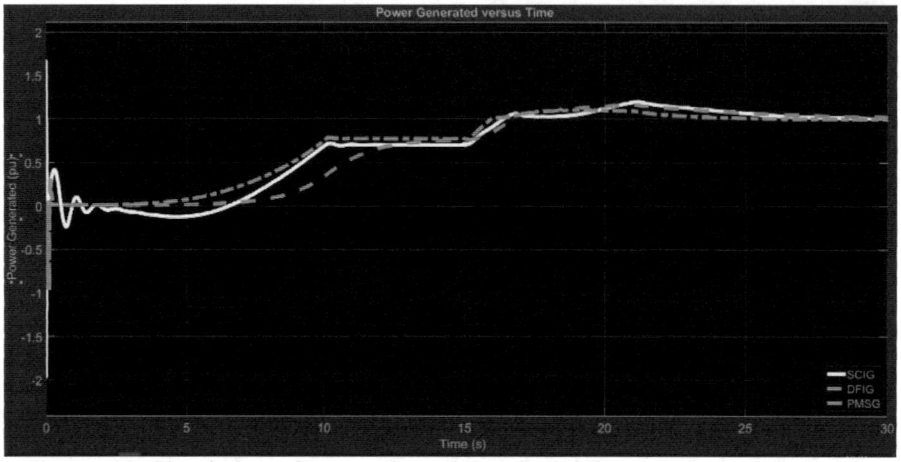

Figure 13. Simulation of Power Generation for Different Types of Generator

3.2 Generator's Efficiency

The efficiency of the generators shown in Figure 14 are obtained by using $Efficiency, \eta = \frac{Power_{out}}{Power_{in}}$. $Power_{in}$ can be determined by calculating the power of the wind, $Power_{in} = \frac{1}{2} \bullet \rho \bullet A \bullet v^3$, where ρ is the air density, *A* is the rotor swept area and *v* is the speed of wind. The efficiency of the generator are calculated until *t=15s* as the pitch controller will be activated as the wind speed rises above the rated wind speed. Efficiency for *t=0s* until *t=3s* is neglected from the study as SCIG is experiencing a current spike as it is connected to the power grid directly without any soft starter. From *t=3s* to *t=7s*, SCIG is consuming electricity from the grid for the winding excitation before the generator starts producing electricity and thus explaining the negative efficiency. PMSG can be observed to perform better under lower wind speed condition compared to SCIG and DFIG. Although all three generators eventually reach a similar efficiency, PMSG is more suitable for wind turbines application in Malaysia as it has better efficiency for lower wind speed and a lower cut-in speed.

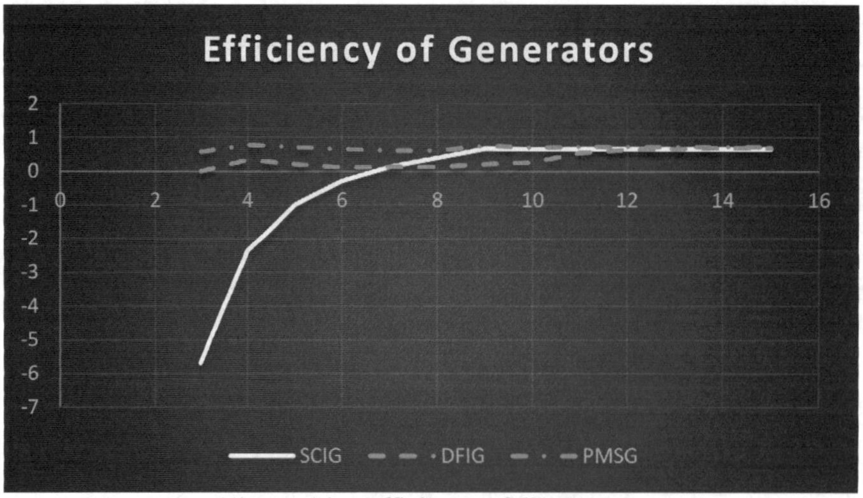

Figure 14: Efficiency of Generators

3.3 Rotor Speed

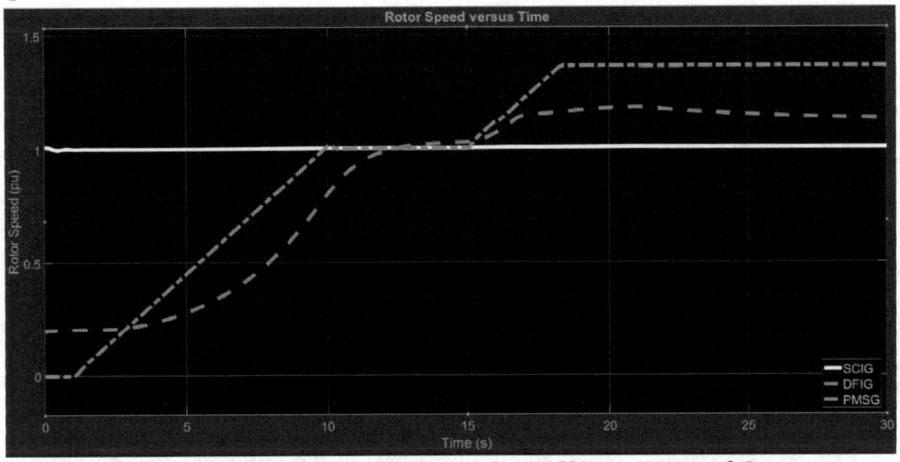

Figure 15. Simulation of Rotor Speed for Different Types of Generator

Differences between fixed and variable speed generators can be seen from Figure 15. The rotor speed for SCIG, ranges from 0.9975pu and 1.0075pu is considered to be fixed speed. On the other hand. both DFIG and PMSG required additional power converter to condition the fluctuating frequency of the generated power to a fixed frequency appropriate for the grid. This increases the complexity and cost of the system. SCIG and DFIG required additional drivetrain to increase the rotor speed of the generator while PMSG can be driven directly by the turbine. TSR for a fixed speed generator fluctuates with wind speed as the rotor speed remain constant.This results in SCIG having lower efficiency and reliability as maximum efficiency can only be achieved at a certain wind speed. Fixed rotor speed results in additional mechanical stresses, fatigue and noises of the turbine and greatly reduce the realibility of the system.

4. Conclusion

SIMULINK model for SCIG, DFIG and PMSG wind turbine had been successfully simulated in this paper. SGIG turbine design is a lot simpler compared to variable speed wind turbine using DFIG and PMSG. However, variable speed generator has the advantages over fixed speed generator as they are more efficient and reliable. From the simulation done, PMSG has shown to be more efficient over SCIG and DFIG for lower wind speed although all of them eventually reaches the similar efficiency.

Acknowledgement
This study was supported by the Sarawak Research and Development Council (SRDC) research grant ref. no. GL/F02/SRDC/11/2020.

References

[1] Saberi, Z., A. Fudholi, and K. Sopian, *Fitting of Weibull Distribution Method to Analysis Wind Energy Potential at Kuala Terengganu, Malaysia.* Journal of Advanced Research in Fluid Mechanics and Thermal Sciences, 2020. **69**: p. 13-22.

[2] Yahyaoui, I. and A.S. Cantero, *Chapter 16 - Modeling and Characterization of a Wind Turbine Emulator*, in *Advances in Renewable Energies and Power Technologies*, I. Yahyaoui, Editor. 2018, Elsevier. p. 491-508.

[3] A.R, S., et al., *Numerical study of effect of pitch angle on performance characteristics of a HAWT.* Engineering Science and Technology, an International Journal, 2016. **19**(1): p. 632-641.

[4] Katsigiannis, Y. and G. Stavrakakis, *Estimation of wind energy production in various sites in Australia for different wind turbine classes: A comparative technical and economic assessment.* Renewable Energy, 2014. **67**.

[5] Islam, M.R., Y. Guo, and J.G. Zhu, *Power converters for wind turbines: Current and future development*. 2013. p. 559–571.

[6] Funabashi, T., *Chapter 1 - Introduction*, in *Integration of Distributed Energy Resources in Power Systems*, T. Funabashi, Editor. 2016, Academic Press. p. 1-14.

[7] Sumathi, S., L. Ashok Kumar, and P. Surekha, *Wind Energy Conversion Systems*, in *Solar PV and Wind Energy Conversion Systems: An Introduction to Theory, Modeling with MATLAB/SIMULINK, and the Role of Soft Computing Techniques*, S. Sumathi, L. Ashok Kumar, and P. Surekha, Editors. 2015, Springer International Publishing: Cham. p. 247-307.

[8] Mali, S., S. James, and I. Tank, *Improving Low Voltage Ride-through Capabilities for Grid Connected Wind Turbine Generator.* Energy Procedia, 2014. **54**: p. 530-540.

[9] Laakam, M., D. Mehdi, and S. Lassaad, *Study of induction generator isolated mode.* International Journal of Research in Engineering & Advanced Technology IJARET, 2014. **Vol. 2**: p. pp. 1-14.

[10] Rawal, C.S. and A.M. Mulla, *An AC-AC converter for doubly fed induction generator driven By Wind Turbine.*

[11] Zou, Y., *Induction generator in wind power systems*, in *Induction Motors-Applications, Control and Fault Diagnostics*. 2015, IntechOpen.

[12] Shewale, A.J., A.R. Gagangras, and N.M. Lokhande, *Comparison of various Wind Turbine Generators.*

[13] Jadhav, H.T. and R. Roy, *A comprehensive review on the grid integration of doubly fed induction generator.* International Journal of Electrical Power & Energy Systems, 2013. **49**: p. 8–18.

[14] Calderaro, V., et al., *Design and implementation of a fuzzy controller for wind generators performance optimisation*. 2007. 1-10.

[15] Esterhuizen, R., *Comparative Study between Synchronous Generator and Doubly-Fed Induction Generator in Wind Energy Conversion Systems*. 2019.

IOP Conf. Series: Materials Science and Engineering 1101 (2021) 012035 doi:10.1088/1757-899X/1101/1/012035

Modelling and Control of Single Arm Exoskeleton for Upper Limb Rehabilitation Application using Hybrid Hierarchical Sliding Mode and PID Controllers

Danny Yip Wei Seng[1], Yew Ze Bing[1], Riady Siswoyo Jo[1] and Hudyjaya Siswoyo Jo[1]

[1] Faculty of Engineering, Computing and Science, Swinburne University of Technology Sarawak Campus, Kuching, Sarawak 93350, Malaysia
Corresponding author: rsiswoyo@swinburne.edu.my

Abstract. Assistive and rehabilitative robotics technologies have seen increasing adoptions in the industry and medical applications alike. This paper presents the dynamic modelling and control scheme of a single arm robotic exoskeleton that is aimed for rehabilitation and physiotherapy of upper limbs. The dynamic model was developed using Euler-Lagrange formulation and the obtained equations allow close analysis on different sources of dynamic torques, which exhibit highly non-linear behaviour. A hybrid hierarchical control scheme employing sliding mode (SMC) and PID controllers are proposed. Simulation were setup using Simulink to evaluate the ability of the proposed control scheme. The simulation results are presented and they show that the proposed control scheme was able to control and track the desired trajectory for multiple degrees-of-freedom.

1. Introduction

An exoskeleton is an electromagnetic device worn outside human limbs to augment or sense the movement of the limbs of interest. Exoskeletons have been widely used for healthcare, physiotherapy and military purposes, among a few. Based on the works of [1-2], these areas of applications can be classified into: assistive robotics [3], tele-robotic manipulation [4] and rehabilitation robotics [5].

Assistive robotics was the primary area of application explored by researchers in the early implementations of exoskeletons spearheaded by General Electric Company in the 1960s [6]. Exoskeletons assisted users (military or civilians alike) in transporting heavy loads and amplifying muscle power to perform strenuous tasks [7].

Further developments of exoskeleton expanded tele-robotic manipulation and medical applications. In tele-robotics, exoskeleton has been used to conveniently record and observe skilled and complex human movements that can be replicated and reproduced by robot manipulators, thus maximizing the functionality of industrial robots that has to otherwise be programmed through teaching pendants/devices [8].

In medical industry, traumatic circumstances lead to impairment of human motor system. The integration of bilateral movement practices to rehabilitate and repair motor function has been subject of research [9]. There has been a great potential in employing wearable exoskeletons to enable the aforementioned rehabilitative practices. For instance, [10] reported that 1130 persons out of 100000 persons suffered from upper extremity injuries per year in United States from 1 January to 31

December 2009. The same survey also reported significant injuries (dislocations, fracture, strain and sprains) at the shoulder, elbow and wrist joints.

1.1. Related Works on Upper Limb Exoskeleton Actuation Methods

Several studies were dedicated in developing exoskeletons for arm and upper limb rehabilitations, each with different actuation methods targeted at different joints. [11] uses twisted string actuation module integrated with a force sensor that is mounted at the shoulder of the patients or users. One end of twisted string is connected with the human's arm whereas the other end is attached with a DC motor inside the module. The changes of the length in twisted string are due to the rotation of the DC motor. From there, human arm's motion is controlled by the changes of length in twisted string depending on the modes of operation. The advantages of this particular actuation are its high precision and sensitivity, although in terms of installation, this actuation requires large amount of space and therefore installed at the back of users rather than attaching them directly to the targeted joints.

An arm exoskeleton based on pneumatic force-feedback system was developed by [12]. The system consists of double acting cylinders, integrated controller and a pair of on-off valves. PWM (Pulse Width Modulation) is used to control the operation of the valves. Thermodynamic and flow equations of the charging and discharging processes were employed to determine the suitable on-off time and duty cycle to obtain the required cylinder force and position. Pneumatic-driven actuations, however, still suffers from relatively low bandwidth compared to electric actuators, especially for exoskeleton applications.

Another upper limb exoskeleton was developed to demonstrate human-robot cooperation control system for augmentation purpose. The inverse dynamics produces singularities which were overcome with damped least square method to minimize output of three-axis force sensor mounted at the handle [13]. Also, [14] proposed an impedance modelling framework to design of active exoskeleton control through formulation of assistance ratio, resistance ratio and human comfort ratio in the frequency domain.

1.2. Prior Works on Control Systems of Exoskeletons

There exist also several approaches in synthesizing control systems of an exoskeleton. In general, exoskeleton control systems can be classified into model-based control systems and hierarchy-based control systems. For model-based control, the model of an exoskeleton was obtained through mathematical modelling, system identification and neural networking [1]. In hierarchy-based control systems, the controllers can be classified into three hierarchies, namely the task controller, the human-exoskeleton interaction (high-level) controller, and the low-level controller. For task controller, the highest level hierarchy, control the overall task to be performed by the exoskeleton. The high-level controller, in the context of exoskeleton control system, are impedance controller or admittance controllers to determine the level of interactions between human and exoskeleton in terms of the forces. Lastly, low-level controller is the closed-loop controller for physical parameter such as position or force of interest [15].

Many existing exoskeletons do not rely on using a single hierarchy of controller. An integrated cooperation between different hierarchies of controllers to perform better dynamic performance are used instead. In the development of exoskeleton system, most of the low-level controllers employed are PD controller, sliding mode controller [16] and H-∞ controller [17].

Sliding mode control (SMC) is a robust control system that forces a system trajectory into desirable phase trajectory and maintain system stability in the error state space. There are two transition stages of phase trajectory which are reaching phase and sliding phase. The controller consists of two control actions: the corrective controller and the equivalent controller [18]. Different reaching laws determine the chattering vulnerability of controlled system and different sliding surface dominates the stability of the system.

Recent work by [19] incorporated adaptive neural network controller and SMC to perform trajectory tracking for a 2-DOF upper limb exoskeleton. This work successfully demonstrated the robustness of SMC for trajectory tracking while using neural network controller to perform real-time estimations of unknown system parameters. A blend of PD, fuzzy logic controller and SMC was recently devised by [20] for control of wheelchair-bound 2-DOF upper limb exoskeleton. The technique presented by the authors was able to produce smooth trajectory tracking as control input chattering was successfully reduced by the fuzzy logic controller.

This work presents the dynamic modelling and control system design of a planar 3-DOF (degree-of-freedom) single arm exoskeleton using Hybrid Hierarchical PID and Sliding Mode Controllers. Section 2 discusses the dynamic modelling of the exoskeleton. Section 3 presents the proposed hierarchy-based control system design based on SMC and PID control schemes. Section 4 presents the simulation results and discussions

2. Mathematical Modelling of the Single Arm Exoskeleton Mechanical System

Considering that common upper limb injuries happen in shoulder, elbow and wrist joints, as discussed in the previous Section, this work chose the mechanical configuration of a planar 3-DOF single arm exoskeleton with shoulder, elbow and wrist joints. Figure 1 shows the free-body diagram of the proposed mechanical configuration.

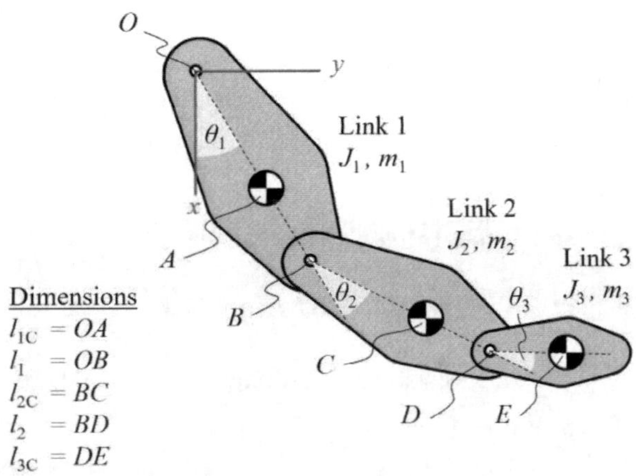

Figure 1. Free-body diagram of the 3-DOF planar single-arm exoskeleton

Point *O* refers to the shoulder joint, which is also the origin, with +*x*-axis pointing downward and +*y*-axis pointing horizontally to the right. The potential energy at point *O* is taken as zero. The elbow and wrist joints are denoted as Points *B* and *D*.

Link 1, rotating about Point *O* with a joint angular displacement θ_1, is modelled as a rigid body with a mass m_1, consisting of the mass of the upper arm and the mass of the actuator for the elbow joint and its related mechanism. The center-of-mass (COM) of Link 1 is denoted as Point *A*, with the moment of inertia J_1 taken about the link's COM.

Link 2, rotating about Point *B* with an angular displacement θ_2, has a mass of m_2, consists of the mass of the forearm and the mass of the actuator for the wrist joint and its related mechanism. Link 2 has a moment of inertia J_2 taken about its COM at Point *C*. Link 3, rotating about Point *D* with an angular displacement θ_3, has a mass of m_3 consisting of the mass of the hand. Link 3 has a moment of inertia J_3 taken about its COM at Point *E*. The dimensions associated with this model are described in figure 1.

To model the dynamic behaviour of the 3-DOF planar exoskeleton, Euler-Lagrange dynamic formulation is used. Here, the system Lagrangian *L*, is given by [21] as the difference of the total kinetic energy *K* and total potential energy *P* of the system:

$$L = K - P \tag{1}$$

The kinetic energy of each Link i is the sum of its rotational kinetic energy and translational kinetic energy:

$$K_i = \frac{1}{2} J_i \dot{\theta}_i^2 + \frac{1}{2} m_i V_{i,COM}^2 \text{ , for } i = 1, 2, 3 \tag{2}$$

The square of each translational velocity is given by:

$$V_{1,COM}^2 = \dot{x}_A^2 + \dot{y}_A^2 \text{ , } V_{2,COM}^2 = \dot{x}_C^2 + \dot{y}_C^2 \text{ , } V_{3,COM}^2 = \dot{x}_E^2 + \dot{y}_E^2 \tag{3}$$

The total potential energy from all the three links is given as below with trigonometric functions written in abbreviated forms, i.e. $S_{jkl} = \sin(\theta_j + \theta_k + \theta_l)$ and $C_{jkl} = \cos(\theta_j + \theta_k + \theta_l)$:

$$P = -m_1 g l_{1C} C_1 - m_2 g l_{2C} C_{12} - m_3 g l_{3C} C_{123} \tag{4}$$

The three joints are revolute joints and thus they are actuated by the net external torques of each joint, T_i. In Euler-Lagrange mechanics, the equations of motion for all three joints can then be written as:

$$\mathbf{T} = \begin{bmatrix} T_1 & T_2 & T_3 \end{bmatrix}^T = \begin{bmatrix} \frac{\mathrm{d}}{\mathrm{d}t}\frac{\partial L}{\partial \dot{\theta}_1} - \frac{\partial L}{\partial \theta_1} & \frac{\mathrm{d}}{\mathrm{d}t}\frac{\partial L}{\partial \dot{\theta}_2} - \frac{\partial L}{\partial \theta_2} & \frac{\mathrm{d}}{\mathrm{d}t}\frac{\partial L}{\partial \dot{\theta}_3} - \frac{\partial L}{\partial \theta_3} \end{bmatrix}^T \tag{5}$$

The solution of equation (4) takes the form of:

$$\mathbf{T} = \mathbf{H}(\mathbf{q})\ddot{\mathbf{q}} + \mathbf{C}(\mathbf{q},\dot{\mathbf{q}})\dot{\mathbf{q}} + \mathbf{G}(\mathbf{q}) \tag{6}$$

where $\mathbf{q} = \begin{bmatrix} \theta_1 & \theta_2 & \theta_3 \end{bmatrix}^T$ are the joint variables and $\mathbf{H}(\mathbf{q})$ and $\mathbf{C}(\mathbf{q},\dot{\mathbf{q}})$ are in this case 3x3 matrices. Equation (6) shows the different forms of torques in the system. The first term in the RHS of equation (6), $\mathbf{H}(\mathbf{q})\ddot{\mathbf{q}}$, signifies inertial torques, second term $\mathbf{C}(\mathbf{q},\dot{\mathbf{q}})\dot{\mathbf{q}}$ signifies torques emerging from the Coriolis and centrifugal accelerations, and third term $\mathbf{G}(\mathbf{q})$ signifies the torques due by gravitational forces. The elements of $\mathbf{H}(\mathbf{q})$, $\mathbf{C}(\mathbf{q},\dot{\mathbf{q}})$ and $\mathbf{G}(\mathbf{q})$ are derived as shown in equations (7) and (8):

$$\mathbf{H}(\mathbf{q}) = \begin{bmatrix} H_{11} & H_{12} & H_{13} \\ H_{21} & H_{22} & H_{23} \\ H_{31} & H_{32} & H_{33} \end{bmatrix} \text{ , } \mathbf{C}(\mathbf{q},\dot{\mathbf{q}}) = \begin{bmatrix} C_{11} & C_{12} & C_{13} \\ C_{21} & C_{22} & C_{23} \\ C_{31} & C_{32} & C_{33} \end{bmatrix} \text{ , } \mathbf{G}(\mathbf{q}) = \begin{bmatrix} G_1 \\ G_2 \\ G_3 \end{bmatrix} \tag{7}$$

IOP Conf. Series: Materials Science and Engineering 1101 (2021) 012035 doi:10.1088/1757-899X/1101/1/012035

$$
\begin{aligned}
&H_{11} = J_1 + m_1 l_{1C}^2 + m_2\left(l_1^2 + l_{2C}^2 + 2l_1 l_{2C} C_2\right) + m_3\left(l_1^2 + l_2^2 + l_{3C}^2 + 2l_1 l_2 C_2 + 2l_2 l_{3C} C_3 + 2l_1 l_{3C} C_{23}\right)\\
&H_{12} = H_{21} = m_2\left(l_{2C}^2 + l_1 l_{2C} C_2\right) + m_3\left(l_2^2 + l_{3C}^2 + l_1 l_2 C_2 + 2l_2 l_{3C} C_3 + l_1 l_{3C} C_{23}\right)\\
&H_{13} = H_{31} = m_3\left(l_{3C}^2 + l_2 l_{3C} C_3 + l_1 l_{3C} C_{23}\right) \quad , \quad H_{22} = J_2 + m_2 l_{2C}^2 + m_3\left(l_2^2 + l_{3C}^2 + 2l_2 l_{3C} C_3\right)\\
&H_{23} = H_{32} = m_3\left(l_{3C}^2 + l_2 l_{3C} C_3\right) \quad , \quad H_{33} = J_3 + m_3 l_{3C}^2
\end{aligned}
$$

$$
\begin{aligned}
&C_{11} = -2\dot{\theta}_2\left(m_2 l_1 l_{2C} S_2 + m_3 l_1 l_2 S_2 + m_3 l_1 l_{3C} S_{23}\right) - 2\dot{\theta}_3\left(m_3 l_2 l_{3C} S_3 + m_3 l_1 l_{3C} S_{23}\right)\\
&C_{12} = -\dot{\theta}_2\left(m_2 l_1 l_{2C} S_2 + m_3 l_1 l_2 S_2 + m_3 l_1 l_{3C} S_{23}\right) - 2\dot{\theta}_3\left(m_3 l_2 l_{3C} S_3 + m_3 l_1 l_{3C} S_{23}\right)\\
&C_{13} = -\dot{\theta}_3\left(m_3 l_2 l_{3C} S_3 + m_3 l_1 l_{3C} S_{23}\right)\\
&C_{21} = \dot{\theta}_1\left(m_2 l_1 l_{2C} S_2 + m_3 l_1 l_2 S_2 + m_3 l_1 l_{3C} S_{23}\right) - 2\dot{\theta}_3\left(m_3 l_2 l_{3C} S_3\right)\\
&C_{22} = -2\dot{\theta}_3\left(m_3 l_2 l_{3C} S_3\right) \quad , \quad C_{23} = -\dot{\theta}_3\left(m_3 l_2 l_{3C} S_3\right)\\
&C_{31} = \dot{\theta}_1\left(m_3 l_2 l_{3C} S_3 + m_3 l_1 l_{3C} S_{23}\right) + 2\dot{\theta}_2\left(m_3 l_2 l_{3C} S_3\right) \quad , \quad C_{32} = \dot{\theta}_2\left(m_3 l_2 l_{3C} S_3\right) \quad , \quad C_{33} = 0
\end{aligned}
$$

$$
\begin{aligned}
&G_1 = m_1 g l_{1C} S_1 + m_2 g\left(l_1 S_1 + l_{2C} S_{12}\right) + m_3 g\left(l_1 S_1 + l_2 S_{12} + l_{3C} S_{123}\right)\\
&G_2 = m_2 g l_{2C} S_{12} + m_3 g\left(l_2 S_{12} + l_{3C} S_{123}\right) \quad , \quad G_3 = m_3 g l_{3C} S_{123}
\end{aligned}
\tag{8}
$$

3. Hybrid Hierarchical PID and SMC Controllers

3.1. Proposed Control System

The proposed control system employs hierarchical control system to optimize controlling multiple links simultaneously to provide smooth motions according to the intention of the user (wearer) of the exoskeleton. Figure 2 illustrates the overall strategy of the control system.

In this strategy, the high-level controller that consists of trajectory generator and sliding mode controller (SMC) provide the desired torque to be produced by each link actuator on the joint. Each link actuator is then PID controlled to provide closed-loop control to track the desired torque generated by the SMC.

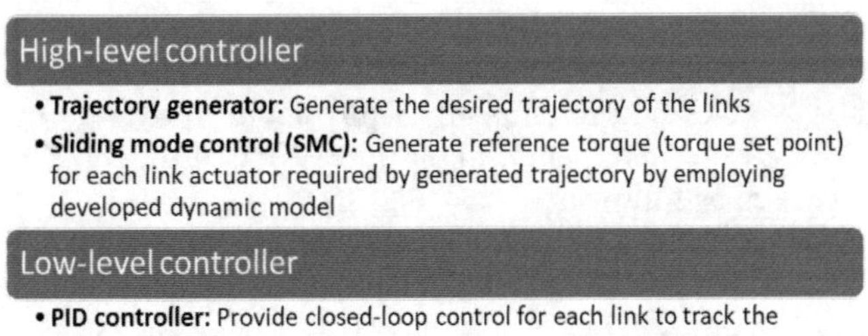

Figure 2. Proposed control strategy

3.2. High-level Controller: Trajectory Generation

The trajectory generator aims to generate optimum motion trajectory for wearer that uses the proposed exoskeleton for rehabilitation. Here, the optimization in this context covers the reduction of jerk (time derivative of acceleration) and generation of trajectory that complies with wearer's intention.

To achieve the objective above, a saturation function was applied to the domain of jerk-time function. To do so, first, the derivative of angular acceleration was evaluated to obtain the jerk of the system. Secondly, the saturation function defined was applied to the angular jerk motion. Finally, the saturated jerk was integrated accordingly to give desired angular acceleration, angular velocity and angular position. This method also has the advantage of filtering unwanted noise signal from the system and reconstructing a controlled jerk motion.

IOP Conf. Series: Materials Science and Engineering 1101 (2021) 012035 doi:10.1088/1757-899X/1101/1/012035

3.3. High-level Controller: Sliding Mode Controller

The SMC takes into account the desired angular positions $\mathbf{q}_d$ of the links from the generated trajectory and calculate the reference/desired torques for each actuator. Consider the following:

$$\mathbf{s} = \dot{\mathbf{e}} + \boldsymbol{\Lambda}\mathbf{e} \tag{9}$$

where $\mathbf{s} = \begin{bmatrix} s_1 & s_2 & s_3 \end{bmatrix}^T$, the position error $\mathbf{e} = \begin{bmatrix} e_1 & e_2 & e_3 \end{bmatrix}^T = \mathbf{q}_d - \mathbf{q}$ and a diagonal matrix $\boldsymbol{\Lambda} = \mathrm{diag}(\lambda_1, \lambda_2, \lambda_3)$. Taking the first time derivative of equation (9) gives:

$$\dot{\mathbf{s}} = \ddot{\mathbf{e}} + \boldsymbol{\Lambda}\dot{\mathbf{e}} = \ddot{\mathbf{q}}_d - \ddot{\mathbf{q}} + \boldsymbol{\Lambda}(\dot{\mathbf{q}}_d - \dot{\mathbf{q}}) \tag{10}$$

Here, we employ exponential reaching law to allow the sliding surface to reach the sliding mode function exponentially:

$$\dot{\mathbf{s}} = -\mathbf{ks} - \boldsymbol{\eta}\,\mathrm{sgn}(\mathbf{s}) \tag{11}$$

with diagonal matrices $\mathbf{k} = \mathrm{diag}(k_1, k_2, k_3)$ and $\boldsymbol{\eta} = \mathrm{diag}(\eta_1, \eta_2, \eta_3)$.

Using equations (6), (10) and (11), the reference torque for each joint can be calculated:

$$\mathbf{T}_d = \mathbf{H}(\mathbf{q})\left(\ddot{\mathbf{q}}_d + \boldsymbol{\Lambda}(\dot{\mathbf{q}}_d - \dot{\mathbf{q}}) + \mathbf{ks} + \boldsymbol{\eta}\,\mathrm{sgn}(\mathbf{s})\right) + \mathbf{C}(\mathbf{q}, \dot{\mathbf{q}})\dot{\mathbf{q}} + \mathbf{G}(\mathbf{q}) \tag{12}$$

3.4. Low-level Controller: PID Controller

The use of high-level SMC equation (12) allows for the generation of joint reference torques from the 3-DOF planar exoskeleton dynamics that is mainly non-linear. The link actuators are, however, considered as exhibiting linear behaviours and can be controlled by lower level closed-loop controllers like PID controllers.

Each proposed actuator is a DC motor that is physically actuated by varying the DC Voltage input. Here, each link i is controlled an incremental algorithm PID controller [22] in discrete mode. The Controller Output (CO) is given as follows:

$$CO_i[n] = CO_i[n-1] + K_{I,i}T_s e_{T,i}[n] + K_{P,i}\left(e_{T,i}[n] - e_{T,i}[n-1]\right) + K_{\mathrm{D},i}\frac{e_{T,i}[n] - 2e_{T,i}[n-1] + e_{T,i}[n-2]}{T_s} \tag{13}$$

where $e_{T,i} = T_{d,i} - T_i$ is the link's torque deviation from its reference/desired value, T_s is the system sampling time, n is the index of the latest sample.

4. Results and Discussions

To evaluate its effectiveness, the proposed control scheme was implemented in MATLAB Simulink environment. Here, the simulations were performed in MATLAB Simulink and simulation parameters used are based on typical weights and dimensions of adult arms. In this simulation, trajectory tracking for a step-like position were performed.

IOP Conf. Series: Materials Science and Engineering 1101 (2021) 012035 doi:10.1088/1757-899X/1101/1/012035

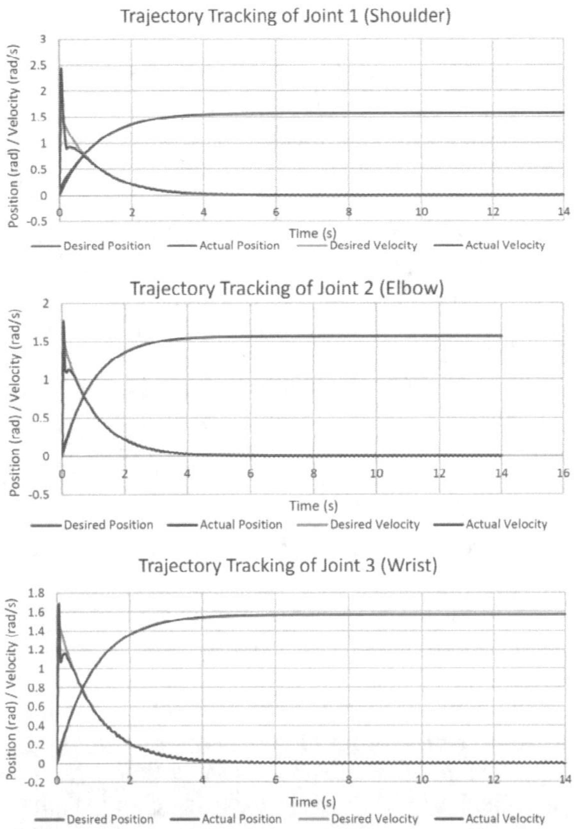

Figure 3. Simulated joint positions and velocities of step-like position tracking with the current control scheme

Figure 3 shows the simulation results for step-like position tracking for each joint. For each joint, a displacement of 90° (or $\pi/2$ radians) were applied. Due to the jerk reduction of the trajectory generation and sliding mode parameters, abrupt changes of desired angles and velocities are dampened. The position tracking of the three joints exhibit predominantly first order responses, with 10%-to-90% rise times observed to be about 2.2 seconds. It can be seen that while the position tracking performed very well, the velocity tracking took some time to track the desired velocity and exhibited small sustained oscillations even when the positions had already settled to the final values.

5. Conclusion and Future Works

In this paper, the dynamic modelling and control of 3-DOF planar upper limb exoskeleton has been discussed. The derived dynamic equations from Euler-Langrange formulation serves as the basis to analyse the behaviour of the proposed mechanical configuration. In particular, the combined non-linear effects of inertia, Coriolis, centrifugal forces and gravity to the overall joint torques allow for optimum control strategy to designed.

The proposed control scheme combined the use of smooth trajectory generation, sliding mode and PID controllers that took into consideration of human users. Simulation results show that the proposed control scheme is able to track the desired trajectories for each joint.

Future experiments to evaluate the feasibility of the proposed control scheme can be performed by developing a prototype with human users and to compare its performance with the simulated results. Based on the proposed control scheme, future prototype shall be equipped with sensory systems to measure joint kinematic states (angular positions, velocities and accelerations) as well as measurements of the joint torques.

On a practical point of view, possible challenge that may arise from developed prototype is the effects of dynamic model inaccuracies on the performance and stability of the sliding mode controllers, which is subject to further investigations.

6. References

[1] Anam K, Al-Jumaily AA 2012 Active Exoskeleton Control Systems: State of the Art. *Procedia Engineering* **41** pp 988–994.

[2] Gopura R, Bandara D, Kiguchi K, Mann GK 2016 Developments in hardware systems of active upper-limb exoskeleton robots: A review. *Robotics and Autonomous Systems* **75** pp 203–220.

[3] Bayasiri R, Gopura R, Ranaweera R 2018 A trans-humeral ortho-prosthesis: Towards power assistive prostheses. *2018 4th International Conference on Control, Automation and Robotics (Auckland, New Zealand)* pp 470–475.

[4] Lai F, Howe RD 2000 Virtual fixtures for robotic endoscopic surgery. *Harvard Univ., Cambridge, MA, Biorobotic Laboratory Rep.*

[5] Lu R, Li Z, Su C-Y, Xue A 2014 Development and learning control of a human limb with a rehabilitation exoskeleton. *IEEE Transactions on Industrial Electronics* **61** pp 3776–3785.

[6] Mosher RS 1968 Handyman to Hardiman. *SAE Transactions* pp 588–597.

[7] Cloud W 1965 Man amplifiers: Machines that let you carry a ton. *Popular Science* **187** pp 7073–204.

[8] Jeong H, Jin J, Kwon KH, Jeon JW 2013 Teaching the robot by guidance using an exoskeleton device in assembly lines. *Proceedings of the 7th International Conference on Ubiquitous Information Management and Communication (Kota Kinabalu, Malaysia)* p 85.

[9] Fikri MA, Abdullah SC, Ramli MHM 2014 Arm exoskeleton for rehabilitation following stroke by learning algorithm prediction. *Procedia Computer Science* **42** pp 357–364.

[10] Ootes D, Lambers KT, Ring DC 2012 The epidemiology of upper extremity injuries presenting to the emergency department in the United States. *Hand* **7** pp 18–22.

[11] Hosseini M, Meattini R, Palli G, Melchiorri C 2017 A Wearable Robotic Device Based on Twisted String Actuation for Rehabilitation and Assistive Applications. *Journal of Robotics* **2017** p 3036468.

[12] Ying C, Jia-fan Z, Can-jun Y, Bin N 2007 Design and hybrid control of the pneumatic force-feedback systems for Arm-Exoskeleton by using on/off valve. *Mechatronics* **17** pp 325–335.

[13] Lee H-D, Lee B-K, Kim W-S, Han J-S, Shin K-S, Han C-S 2014 Human-robot cooperation control based on a dynamic model of an upper limb exoskeleton for human power amplification. *Mechatronics* **24** pp 168–176.

[14] Nagarajan U, Aguirre-Ollinger G, Goswami A 2015 Integral admittance shaping for exoskeleton control. *2015 IEEE International Conference on Robotics and Automation (Seattle, USA)* pp 5641–5648.

[15] Minchala LI, Astudillo-Salinas F, Palacio-Baus K, Vazquez-Rodas A 2017 Mechatronic design of a lower limb exoskeleton. *Design, Control and Applications of Mechatronic Systems in Engineering* ed S Yildirim pp 110–134.

[16] Viego RAS, Ashok RS 2013 Comparative study of PD controller (vs.) integral sliding mode control on 5-link 2 DOF planar. 2013 *Proceedings of IEEE Southeastcon(Florida, USA)* pp 1–6.

[17] Gomes MA, Silveira GLM, Siqueira AA 2011 Gait pattern adaptation for an active lower-limb orthosis based on neural networks. *Advanced Robotics* **25** pp 1903–1925.

[18] Liu J, Wang X 2012 Advanced sliding mode control for mechanical systems: design, analysis and MATLAB simulation (Springer Science & Business Media)

IOP Conf. Series: Materials Science and Engineering 1101 (2021) 012035 doi:10.1088/1757-899X/1101/1/012035

[19] Jebri A, Madani T, Djouani K, Benallegue A 2020 Robust adaptive neuronal controller for exoskeletons with sliding-mode. *Neurocomputing* **399** pp 317-330.
[20] Long T, Gull MA, Bai S 2020 PD Based Fuzzy Sliding Mode Control of A Wheelchair Exoskeleton Robot. *IEEE/ASME Transactions on Mechatronics (Early Access)*
[21] Niku SB, 2001 Introduction to robotics: analysis, systems, applications (New Jersey: Prentice Hall)
[22] Lipták, BG., 2018 Instrument Engineers' Handbook, Volume Two: Process Control and Optimization. Vol. 2. (CRC press)

IOP Conf. Series: Materials Science and Engineering 1101 (2021) 012036 doi:10.1088/1757-899X/1101/1/012036

Effect of welding parameters on the properties of the Ti-6Al-4V plate resistance spot weld joint

I Fatmahardi[1], M Mustapha[1], A Ahmad[1], T L Ginta[2], I Taufiqurrahman[1] and M Danish[3]

[1] Mechanical Department, Universiti Teknologi Petronas, Persiaran UTP, 32610 Seri Iskandar, Perak, Malaysia
[2] Orca Industri Akademi, Orca Institute for Sustainable Development, Satrio Tower, Floor 26, Unit C and D, Kuningan, Jakarta 12950, Indonesia
[3] Department of Mechanical and Materials Engineering, University of Jeddah, Jeddah, 21589, Kingdom of Saudi Arabia

Email: ichwan_19001075@utp.edu.my and mazli.mustapha@utp.edu.my

Abstract. Titanium alloy Ti-6Al-4V properties advantage gives it a well-known reputation for decades as a reliably material used in a wide range and specific application of resistance spot welded joint such as automotive & aviation products. High strength joint is created depends on the welding parameters used in resistance spot welding. Particular problems occurred about the effect of welding parameters on pure titanium and titanium alloy mechanical and physical properties from previous researches. Some pores and acicular α' phase appeared in the microstructure, which caused partial interfacial failure mode in the tensile testing. This study is conducted to study the influence of welding parameters for Ti-6Al-4V weld nugget mechanical and physical properties and discover its optimum level of parameters. Tensile-shear testing is used to observed the optimum level. In the optimum level microstructure result reveals that in the weld nugget zone as a fusion zone the lamellar α+β is dominantly observed, this contrasts sharply with the base metal and the heat-affected zone where the primary α and β phase appear to be more dominant. The highest hardness value is discovered in the weld nugget area near the center proving the contribution of the lamellar α+β on this area.

1. Introduction

Resistance spot welding is one of the very effective welding method used for thin or thick metal sheets. It has a quick time of welding and also having a good quality of strength at the joint. Titanium alloy is one of the widely used materials in industrial needs, particularly in the automotive and aerospace industries [1].

In resistance spot welding, the weld joint is generated by the heat which is combined with pressure from the electrode and at a certain welding time. The name of resistance spot welding comes from the fact that the weld is not only influenced by the weld current, pressure, and time but also the material resistivity to the current flow as well, which causes a localized heating area in the material [1]. One of the best materials that had been used for so many years for industrial applications and welded using resistance spot welding (RSW) is Titanium alloy especially Ti-6Al-4V, for example in the automotive

IOP Conf. Series: Materials Science and Engineering 1101 (2021) 012036 doi:10.1088/1757-899X/1101/1/012036

industry. Titanium alloy grade 5 or Ti-6Al-4V has 130 Ksi or 895 MPa of tensile strength with 120 Ksi or 828 MPa of Yield Strength making it one of the strongest alloys among others [2-4].

Titanium has low density leading to deform easily during resistance spot welding with high force, but in another way it has a high melting point [5], so it must be investigated the optimum parameter that can be applied on its resistance spot welding. A previous study has been done with the same metal combination but with 1 mm of thickness and provides the maximum tensile-shear load of 14.3 kN [6]. Titanium alloys prove to be one of the greatest corrosion-resistant metal by creating a thin, but sufficient to protect its body as a surface oxide film. This film, which is primarily TiO2, is highly persistent, supportive, chemically stable, and can spontaneously and instantaneously recover itself if mechanically damaged [7].

Several types of research conducted a study about the effect resistance spot welding parameters on pure titanium and titanium alloy properties [6][8]. The results concluded that the changing of parameters did affect the impact on its properties. But some problems occurred, in the microstructure formed acicular α' phase [6] and appeared several pores [8], those caused partial interfacial failure mode in the weld joint tensile result. This study's objectives are to study the influence of welding parameters for titanium alloy resistance spot weld nugget mechanical and physical properties, and also to discover the optimum level of its selected parameters.

2. Experimental procedure

In this study, two-piece plates of titanium alloy (ASTM grade 5) with 3 mm of thickness were used to be welded together under resistance spot welding. The dimension of each sample was determined referred to ASTM D1002 standard and under the lap joint configuration during resistance spot welding [9]. The experiment was conducted using data acquired in our labs. The appearance of the resistance spot welding diagram and the sample can be seen in figure 1. In figure 1 (a) it can be seen that both of the material plates are pinched by the electrode tips from a different direction. The electrode tips are given force and the current flows through it to the materials and generating heat. The welding time is applied to give the heat to steadily melting down the materials and join it together forming a fusion zone called weld nugget. figure 1 (a) & (b) shown that the weld nugget shape is rounded following the shape of the electrode tips cross-sectional area. The welding experiment is performed under room temperature or normal atmospheric conditions and based on AWS D8-9M:2012 standard [10]. The chemical composition of titanium alloy (ASTM grade 5) is presented in table 1 [4].

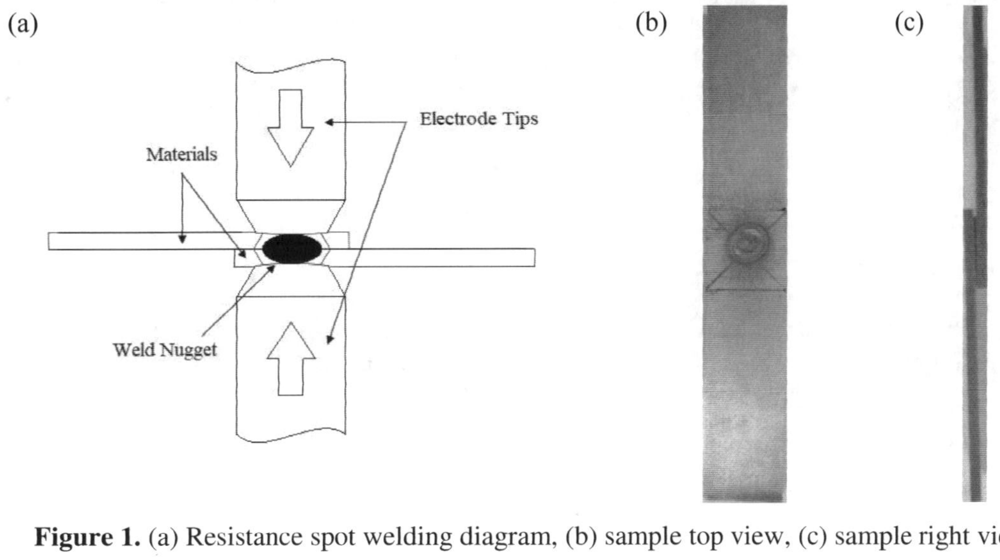

Figure 1. (a) Resistance spot welding diagram, (b) sample top view, (c) sample right view.

IOP Conf. Series: Materials Science and Engineering 1101 (2021) 012036 doi:10.1088/1757-899X/1101/1/012036

Table 1. Chemical composition of titanium alloy ti-6al-4v (ASTM grade 5) [4].

Element	Titanium alloy grade 5
Nitrogen, max	0.05
Carbon, max	0.08
Hydrogen, max	0.015
Iron, Max	0.40
Oxygen, max	0.20
Aluminum	5.5-6.5
Vanadium	3.5-4.5
Titanium	Balance

The welding parameters set up could be selected manually on the welding machine control panel for the current and welding time, while the force can be set from its pressure controller. Three current parameters are selected and varied, which are: 8, 9, and 10 kA. While the others are kept constant, such as 35 cycles of squeeze time, 30 cycles of welding time, 15 cycles of holding time, 0 cycle of off time, and 4 kN of electrode force. From the three varied current parameters, the tensile test result will show which one having the highest value. Thus, the sample with the selected parameter will be analyzed for its microstructure and hardness testing.

The maximum tensile-shear stress of the samples is defined using tensile testing referred to as ASTM E8/E8M-13a standard [11]. Microstructure analysis for the selected parameter using micro etching and optical microscope referred to the ASTM E407-07 standard [12]. Hardness distribution test using Vickers micro-hardness referred to ASTM E92-17 standard [13].

The Zwick/Roell Universal Tensil Machine which has 20 mm/min of ramp speed at room temperature and 50 kN of maximum load is utilized to conduct the tensile-shear testing. The microstructure observation is conducted in the cross-sectional area of the weld nugget with etching activities using Keller's reagents and optical observation using a LEICA optical microscope made in Germany. The hardness distribution is observed for the similar weld nugget cross-sectional area to test the hardness value using Vickers microhardness under the indentation load of 200 gf to obtain a hardness profile.

3. Result and discussion

3.1. Resistance spot welding result and tensile testing

The welding result shows a very good shape of the weld nugget. As can be seen in figure 2 the weld nuggets are perfectly circular with their fusion zone in the center and thin heat affected zone at its edge for the three samples as well. Each parameter produced different nugget diameters among each other and this can be seen in Table II. To determine the electrode diameter which will be used, it is required to calculate based on the standard that: $d > 4\sqrt{t}$, where t describes the sheet/plate thickness [14]. Hence, since this study uses 3 mm of titanium alloy Ti-6Al-4V plate, the electrode diameter should be calculated as $d > 6.9$ mm. Table 2 shows the measurement of weld diameter by calculating the average diameter of six different measurements from each sample to get the proper measurement result and also indentation depth. As can be seen in table 2 that with the constant parameter of welding time and force, the more increase of welding current the more increase weld nugget diameter. The increase of current also increasing the heat generated by the welding electrodes hit to the material. Thus, increasing the weld temperature in the material, causing a much larger weld diameter. As well as the indentation depth, because the increase of the weld temperature also increasing the material ductility during welding, so the giving electrodes penetration much deeper.

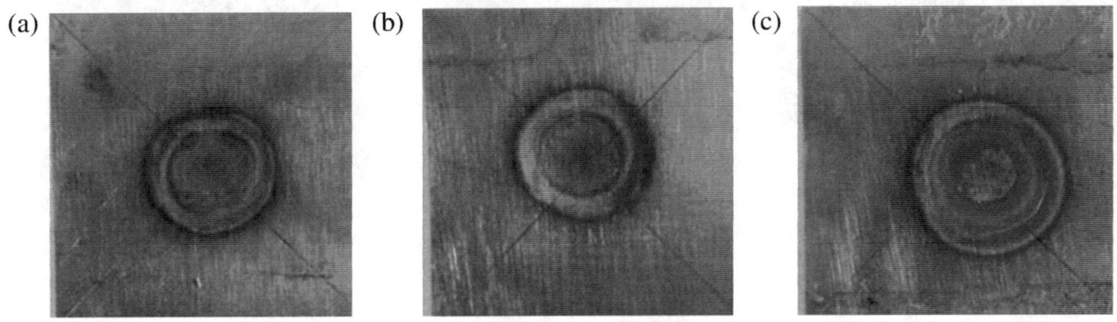

Figure 2. Weld nugget result at welding current respectively: (a) 8 kA, (b) 9 kA, (c) 10 kA.

The resistance spot welding result quality is generally implied by the weld nugget strength. Different amounts of welding parameters such as welding current will lead to a different result of weld nugget [16-31]. Tensile testing is the most commonly used to evaluate the mechanical properties in a static condition. Physical properties also could be determined by evaluating the fracture mechanism of the tensile test. According to the tensile testing result, table 2 shows that the increase of weld nugget diameter and indentation depth is not followed by the tensile stress. It is increased from 8 kA to 9 kA, but in 10 kA it is decreased, because the deeper indentation depth creating a thinner fusion zone, thus also weakening its strength.

Table 2. Weld nugget physical properties and tensile stress.

Exp.	Current (kA)	Welding time (cycle)	Welding force (kN)	Weld nugget diameter (mm)	Indentation depth (mm)	Max. tensile stress (MPa)
1	8	30	4	10.763	0.365	483.048
2	9	30	4	12.147	0.380	521.801
3	10	30	4	12.853	0.465	485.604

According to the tensile testing result, it shows that the failure modes of the welded joint are an interfacial failure for both three samples. The failure mode tends to brittle failure. It is shown in figure 3 that in welding current 8 kA, the failure occurs mostly in the weld nugget edge seems that the weld nugget is more brittle than in the HAZ and base metal. It is different in the welding current of 9 kA, the failure is occurred in outside the nugget area, mostly in the base metal. The opposite result occurs in 10 kA welding current, the failure is turned back in the nugget area mostly, meaning that the nugget zone turned weakened due to the thinner indentation depth caused by higher heat generated by higher current flow through the welding electrodes.

IOP Conf. Series: Materials Science and Engineering 1101 (2021) 012036 doi:10.1088/1757-899X/1101/1/012036

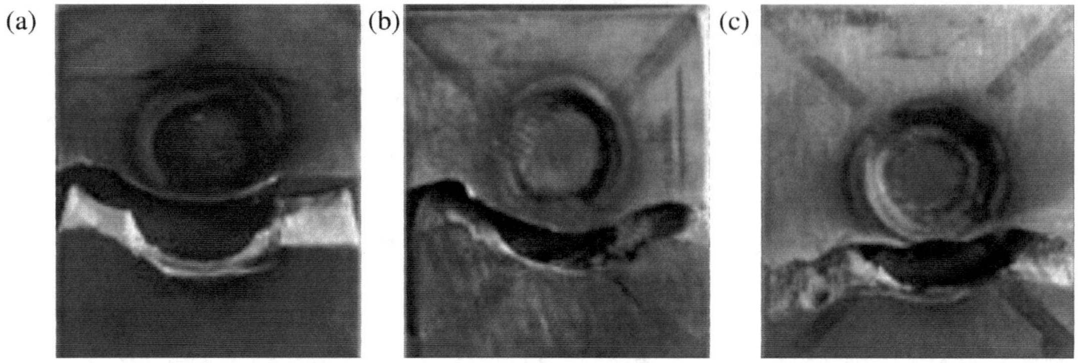

Figure 3. Failure mode at welding current: (a) 8 kA, (b) 9 kA, (c) 10 kA.

As observed in figure 3, the type of failure for the whole samples are cup and cone which are usually appeared in the failure of ductile metal. As the tensile stress in the sample approaches its yield strength, due to the sample type is a lap joint with its weld joint in both meeting endpoint, the sample is started to bend each other as can be seen in figure 4. The bending is caused as the weld joint holds the shear stress from both sides. As the load continues increasing, small cavities are formed at the bent portion of the material. The further increase of load causes the cavities to gain bigger and eventually coalesce to form an elliptical crack perpendicular to the stress direction. Eventually, the material fails, and the fractured at both materials ends to create the distinctive cup and cone shape. The failure point is the representation of the weakest area around the weld nugget zone whether it is in the weld nugget, HAZ or base metal.

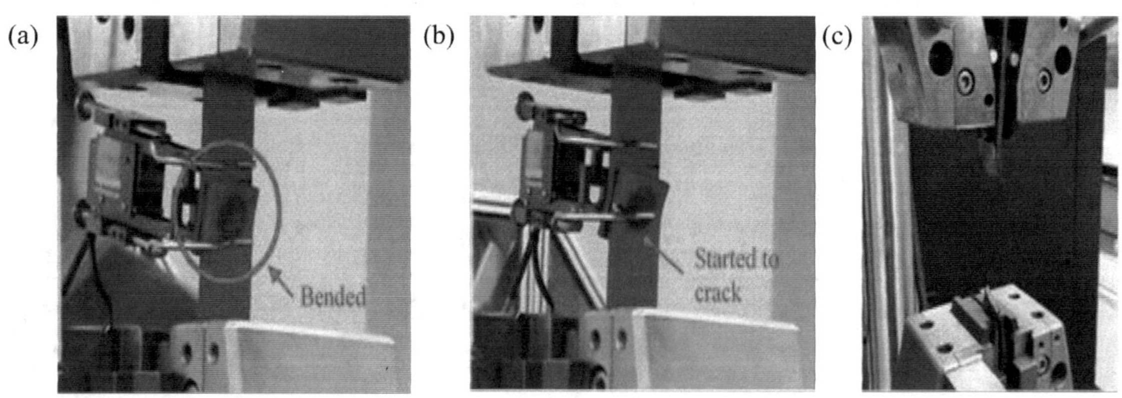

Figure 4. Failure process in tensile testing: (a) bending process, (b) cracking start, (c) sample failure form.

3.2. Microstructure analysis

In resistance spot welding, the effect of the heat input created by the welding parameters greatly affects the microstructure changes in the weld nuggets. Fusion zone (FZ), heat affected zone (HAZ), and base metal (BM) will provide different types of microstructure that will lead to mechanical performance [15].

IOP Conf. Series: Materials Science and Engineering 1101 (2021) 012036 doi:10.1088/1757-899X/1101/1/012036

The optical microscope is used to identify the microstructure and physical properties of the selected welded joint from the previous tensile test, which is welded under 9 kA of welding current, 30 cycles of welding time, and 4 kN of welding force, and the result can be seen in figure 5.

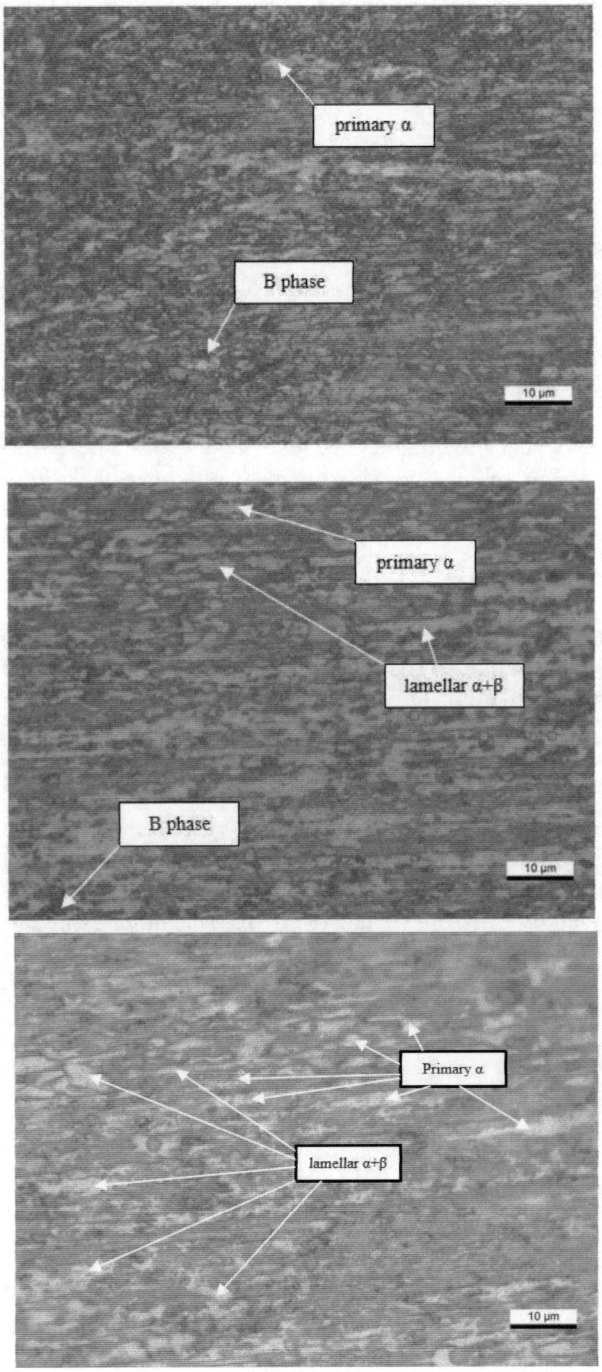

Figure 5. Microstructure analysis at 9 kA of welding current, 30 cycles of welding time, and 4 kN of welding force: (a) base metal; (b) heat-affected zone (HAZ); (c) weld nugget zone.

IOP Conf. Series: Materials Science and Engineering 1101 (2021) 012036 doi:10.1088/1757-899X/1101/1/012036

Titanium alloy Ti-6Al-4V consists of two crystallographic forms which are α which has a hexagonal close-packed (hcp) crystal structure & β phase which has a body-centered cubic (bcc) structure. Even though this particular alloy is relatively hard to form even when it is annealed, α + β alloys commonly have good formability [6]. The microstructure observation is conducted for the cross-sectional area of the welded joint perpendicular to the weld surface and mounted in a dried epoxy followed by grinding and polishing in a thorough sequence. Then, the cross-section sample is etched using the etching solution of Keller's Etch.

It is observed that the primary α phase is often represented by the bright color and the β phase is clearly shown in a dark color. Since the microstructures of the sample determined on the cross-sectional area of the weld joint as can be seen in figure 5, there are seems to be a difference in the weld nugget, HAZ, and also in the base metal. As shown in figure 5 (a) the microstructure of the base metal consists of primary α phase (colored in white) and β phase (colored in black) but in the Heat Affected Zone (HAZ) as shown in figure 5 (b), it is stated to be formed slightly of the lamellar α+β so the microstructure consists of primary α, β phase, and lamellar α+β. But in the weld nugget zone shown in figure 5 (c), as a fusion zone the lamellar α+β starting to dominantly showing themselves combine with primary α. In titanium alloy, the phase transformation is much affected by the temperature and the cooling rate. In figure 5 (c) the microstructure is called duplex because the discontinuous equiaxed α distribute less than 50% in the transformed matrix β. This microstructure is obtained in the high temperature of the α+β phase zone which can contribute to the wide range of optimum mechanical properties [32].

The cooling rate also greatly affects the changes in the alpha and beta phases. When resistance spot welding occurs at room temperature, the cooling rate is sequential from slowly in the middle of the weld nugget as the center of the heat input, to the heat-affected zone (HAZ) and base metal which are located far from the center of the heat input, so that they have a faster cooling rate. Increasing the cooling rate causes smoother on the microstructure. The rapid cooling rate will cause the growth of phase boundaries both alpha and beta to grow perpendicular to the lamellar and appear nucleated [32]. So that it does not form many new lamellar on the microstructure, especially on the base metal. The thin section in the heat-affected zone which is close enough to the fusion zone causes this area to start to accommodate the new lamellar that has formed due to the reduced cooling rate. In contrast to weld nuggets, the high fusion zone temperature as the main area of heat input and slow cooling rate provides ample opportunity for the formation of new lamellar from the alpha and beta phases. The form of lamellar shown in figure 5 increasing the weld nugget zone hardness and strength. The more lamellar structure is formed thus increasing the more hardness and strength.

3.3. Hardness testing
The application of hardness distribution testing is to indicate metal wear-resistant. The harder the metal area the more wear-resistant. Since the microstructure of the three areas is different from each other (infusion zone, HAZ, and base metal) the hardness distribution should be tested to prove that its hardness value different from each other or not. This test was using the micro-hardness Vickers test with 200 gf of load on 20 indentation points along the cross-section of the welded joint. Figure 6 is showing the hardness test positions on the cross-sectional sample, 19 points were tested across the cross-sectional area. Each point has 0.8 mm on the base metal and HAZ, and 1.5 mm on the weld nugget zone, due to the limited size of mounting.

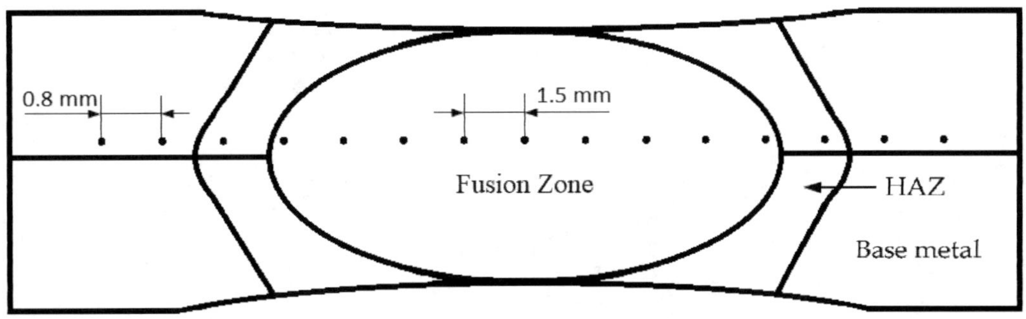

Figure 6. Hardness testing positions in the weld nugget.

The result of hardness testing can be seen in figure 7 & table 3. The result of The lowest HV is in the base metal with the value of 219.4, meanwhile, in the HAZ the HV is raising because the lamellar α+β slightly formed increasing the hardness of the material. Approaching the weld nugget zone as the fusion zone the HV is increasing until the peak of the hardness is nearly the center of the weld nugget, making the weld nugget is the hardest zone of all. The highest HV is 470.9, in the weld nugget area near the center. It is proving the result of microstructure in figure 5 which shows that in the fusion zone is observed the lamellar α+β phase dominantly appeared increasing the hardness and ductility.

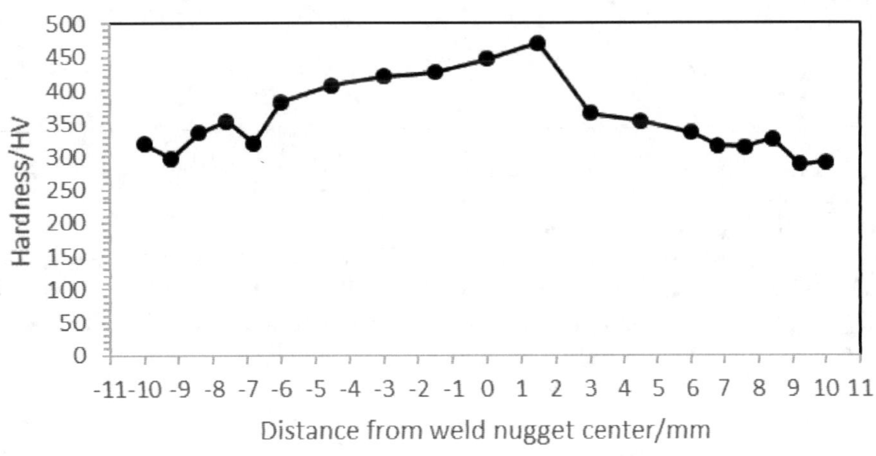

Figure 7. Hardness distribution chart.

IOP Conf. Series: Materials Science and Engineering 1101 (2021) 012036 doi:10.1088/1757-899X/1101/1/012036

Table 3. The result of Vickers microhardness measurement.

No	Position	D1	D2	HV
1	-10	35.05	33.2	318.5
2	-9.2	36.32	34.36	297
3	-8.4	33.64	32.82	335.9
4	-7.6	34.52	30.33	352.8
5	-6.8	33.74	33.74	320.3
6	-6	31.54	30.82	381.4
7	-4.5	30.03	30.37	406.6
8	-3	29.46	29.98	419.9
9	-1.5	29.49	29.52	426
10	0	28.56	29.03	447.3
11	1.5	28.56	27.57	470.9
12	3	32.16	31.68	364
13	4.5	32.45	32.38	353
14	6	32.91	33.6	335.4
15	6.8	34.39	34.18	315.5
16	7.6	34.67	34.67	313.4
17	8.4	34.04	33.42	326
18	9.2	35.9	35.91	287.7
19	10	35.89	35.46	291.4

4. Conclusion

The weld joint maximum stress from the tensile testing for welding parameters of 8 kA, 9 kA & 10 kA are respectively: 483.048 MPa, 521.801 MPa, 485.604 MPa. Then the weld joint with the welding parameters of 9 kA current, 30 cycle welding time & 4 kN force which produce 521.801 MPa maximum stress, is selected to be the strongest weld joint among others from the tensile testing result. Meanwhile, the pull-out failure with cup and cone shape is the mostly failure mode that appeared in samples and broken mostly around the HAZ and base metal area, meaning that the weld nugget is stronger than the HAZ and the base metal. The Primary α, β phase, and lamellar $\alpha+\beta$ are the microstructures that appeared in the weld nugget zone. Particularly the form of lamellar $\alpha+\beta$ making the weld nugget stronger and harder than the HAZ and the base metal. For the result of the hardness test, the highest HV is 470.9, in the weld nugget area near the center. It explains that the increase of the welding parameter particularly current will increase the weld nugget strength. But when the current is increased, the strength will be reduced because of the thinner indentation depth formed. It is caused by higher heat generated by higher current flow through the welding electrodes.

Acknowledgments

This study is fully supported by the funding given by Yayasan Universiti Teknologi PETRONAS (YUTP) no. 015LC0-052(YUTP).

5. References

[1] Miller Electric Mfg. Co 2005 *Handbook for Resistance Spot Welding* (www.MillerWelds.com)
[2] Pandey A K, Khan M I and Moeed K M 2013 Optimization of Resistance Spot Welding Parameters Using Taguchi Method *International Journal of Engineering Science and Technology (IJEST)* **5** No. 02
[3] Podržaj P, Polajnar I, Diaci J and Kariž Z 2008 Overview of resistance spot welding control *Science and Technology of Welding and Joining* **13**(3) pp 215-224.
[4] ASTM 2015 B265-15 Standard Specification for Titanium and Titanium Alloy Strip, Sheet, and Plate *ASTM Int.* doi: 10.1520/B0265-13AE01.2.
[5] RMI Titanium Company 2000 *Titanium Alloy Guide*
[6] Zhang X, Zhang J, Chen F, Yang Z and He J 2017 Characteristics of resistance spot welded Ti6Al4V titanium alloy sheets *Metals (Basel)* **7**(10), 424.
[7] ASM International Handbook Committee 2001 *ASM Handbook Vol. 2: Properties and selection--nonferrous alloys and special-purpose materials*
[8] Kahraman N 2007 The influence of welding parameters on the joint strength of resistance spot-welded titanium sheets *Mater. design* **28**(2) pp 420-7.
[9] ASTM D1002-99 Standard Test Method for Apparent Shear Strength of Single-Lap-Joint Adhesively Bonded Metal Specimens by Tension Loading (Metal-to-Metal) *ASTM International*
[10] AWS 2012 D8.9M:2012 Test methods for evaluating the resistance spot welding behavior of automotive sheet steel materials *American Welding Society*
[11] ASTM 2020 E8/E8M – 16a Standard Test Methods for Tension Testing of Metallic Materials *ASTM Stand.* doi: 10.1520/E0008_E0008M-16AE01.
[12] ASTM 2015 E407-07 Standard Practice for Microetching Metals and Alloys *ASTM Int.* doi: 10.1520/E0407-07R15E01.2.
[13] ASTM 2017 E92 Standard Test Methods for Vickers Hardness and Knoop Hardness of Metallic Materials *ASTM B. Stand.* doi: 10.1520/E0092-16.
[14] Koenigsberger F 1948 *Design for Welding in Mechanical Engineering* (London, New York: Longmans, Gree, and Co.) pp 81–91.
[15] Shirmohammadi D, Movahedi M and Pouranvari M 2017 Resistance spot welding of martensitic stainless steel: Effect of initial base metal microstructure on weld microstructure and mechanical performance *Mater. Sci. Eng. A* **703** pp 154-161.
[16] Zhou M, Hu S J and Zhang H 1999 Critical Specimen Size for Tensile-Shear Testing of Steel Sheets *Weld. J.* **78**
[17] Liu X D, Xu Y B, Misra R D K, Peng F, Wang Y and Du Y B 2019 Mechanical properties in double pulse resistance spot welding of Q&P 980 steel *J. Mater. Process. Technol.* **263** pp 186-197.
[18] Shi H, Qiu R, Zhu J, Zhang K, Yu H and Ding G 2010 Effects of welding parameters on the characteristics of magnesium alloy joint welded by resistance spot welding with cover plates *Mater.* **31** (10) pp 4853 - 57.
[19] Vural M, Akkuş A and Eryürek B 2006 Effect of welding nugget diameter on the fatigue strength of the resistance spot welded joints of different steel sheets *J. Mater. Process. Technol.* **176** pp 127-132.

[20] Mansor M S M, Yusof F, Ariga T and Miyashita Y 2018 Microstructure and mechanical properties of micro-resistance spot welding between stainless steel 316L and Ti-6Al-4V *Int. J. Adv. Manuf. Technol.* **96** pp 2567-81.

[21] Long H, Hu Y, Jin X, Shao J and Zhu H 2016 Effect of holding time on microstructure and mechanical properties of resistance spot welds between low carbon steel and advanced high strength steel *Comput. Mater. Sci.* **117** pp 556-563.

[22] Anijdan S H M, Sabzi M, Ghobeiti-Hasab M and Roshan-Ghiyas A 2018 Optimization of spot welding process parameters in dissimilar joint of dual phase steel DP600 and AISI 304 stainless steel to achieve the highest level of shear-tensile strength *Mater. Sci. Eng.* **726** pp 120-5.

[23] Nasir Z and Khan M I 2016 Resistance spot welding and optimization techniques used to optimize its process parameters *Int. Res. J. Eng. Technol.* **3** (5) pp 887 – 893.

[24] Shelly K and Sahota D 2017 A Review Paper on Resistance Spot Welding of Austenitic Stainless Steel 316 *Int. J. Eng. Trends Technol.* **47** (7) pp 424 - 9.

[25] Khan M I, Kuntz M L, Biro E and Zhou Y 2008 Microstructure and mechanical properties of resistance spot welded advanced high strength steels *Mater. Trans.* **49** (7) pp 1629-37.

[26] Jafari M V, Araee A, Senkara J, Jakubowski J and Godek J 2016 Experimental and numerical analysis of shunting effect in resistance spot welding of Al2219 sheets *Bull. Polish Acad. Sci. Tech. Sci.* 64 (2) pp 425 - 434.

[27] Khan M I 2007 *Spot Welding of Advanced High Strength Steels* A Thesis for University of Waterloo, University of Waterloo.

[28] Hou L, Qiu R, Cui L, Shen Z and Li Q 2015 Microstructure and mechanical property of resistance spot welded joint between pure titanium and stainless steel with interlayer of Nb *5th Int. Conf. on Advanced Design and Manufacturing Engineering*.

[29] Wan X, Wang Y, Zhao D and Huang Y A 2017 A comparison of two types of neural network for weld quality prediction in small scale resistance spot welding *Mech. Syst. Signal Process.* **93** pp 634 - 644.

[30] Sammaiah K, Balunaik B and Srikanth D V 2018 Evaluation of Tension - Shear Stress for zig-zag spot welds by using indirect resistance spot weld technique for fabrication industries *Materials Today: Proc.***5** (9) pp 18697 - 706.

[31] Yang T, Xu D, Chen W, Yang R and Lv S 2019 Microstructure evolution and deformation resistance of heavy-thickness Ti-6Al-4V narrow-gap welded joints *Mater. Lett.* **250** pp 116 - 118.

[32] Zhong C, Liu J, Zhao T, Schopphoven T, Fu J, Gasser A and Schleifenbaum J H 2020 Laser Metal Deposition of Ti6Al4V—A Brief Review *Appl. Sci.* **10** p 764.

IOP Conf. Series: Materials Science and Engineering 1101 (2021) 012037 doi:10.1088/1757-899X/1101/1/012037

Influence of Drop Height on the Impact Characteristics of Futsal Ball Size Four

S T Syed Shazali[1], F H P Alban Dattu[2], A M Aizuddin[1], M Andrew-Munot[1] and N H Noor Mohamed[1]

[1] Faculty of Engineering, Universiti Malaysia Sarawak, Jalan Datuk Mohammad Musa, 94300 Kota Samarahan, Sarawak, Malaysia
[2] School of Science and Technology, International College of Advanced Technology Sarawak, Jalan Stampin Timur, 93350, Kuching Sarawak, Malaysia
Corresponding author: starmizi@unimas.my

Abstract. Futsal research on the ball characteristic has received little scientific attention. Recently modern football has been developed that afford excellent ball control and delivering. Difference constructions and sizes of ball deliver different characteristics of the ball. The purpose of this study is to examine Futsal ball size four deformations and coefficient of restitution (COR) during the impact. In this study, the Futsal ball was dropped vertically under three conditions. A subject performed five repetitive drops off the ball to impact a steel plate. The ball velocity was measured by SparkVue motion sensor attached on the top of the apparatus while the deformation was recorded using a high-speed video camera, that can up to 1,000 frames per second. From the experiment, the ball deformation and COR were measured. The results obtained showed the deformation and COR value vary with different heights for the same type of Futsal ball size four. It shows that the different heights of ball drops give different ball impact characteristics. It was found that the COR depended on the amount of the deformation of the ball.

1. Introduction

In many sports, ball properties are held to strict standards to maintain a high level of consistency and performance. The ball itself is a critical component in specific sports such as football, hockey, tennis, and volleyball where those games have to meet strict requirements of game safety. To achieve better performance and quality, the sports ball design and manufacturing required a test, such as an impact test to ensure that the sports ball meets the safety requirements. The deformation of the ball itself plays an essential role in the performance of the ball. The design and development of sports balls are significant activities to gain competitive advantages within the market. Regardless of the market interest, the researchers aim to improve sports performance and to reduce injuries [1].

A head injury in football accounted as the highest number of injuries compared to neck injuries. The football-related head injuries may occur in at least four ways; head contact with the ball (heading), contact with another player, contact with the ground and contact with a stationary object. Head risks of football player injuries are increased in goalkeepers. The goalkeeper is the position that has the most risk of the casualties among the players. The goalkeeper needs to be ready with the oncoming balls, which in the position most at risk for head injury. They are also exposed to the impact of the ball at the fastball kicked by another player [2]. During the actual games, the footballer can kick the ball with speed up to 130 km/hr. Deformation plays a vital role in understanding the impact characteristics. Ball deformation results from the abrupt deceleration connected with its impact against another body. This is evident from a reduction in the ball's diameter perpendicular to the impact surface. Many studies, which are conducted shown that the deformation was increased as the velocity increase. When a ball is dropped to the ground, the potential energy from the fall is used to deform the ball, which deformation stores the energy gained from the fall. Once the velocity downward becomes zero relative to in energy contained in the deformation, the stored energy pushed the ground and launched the ball back up.

The previous studies have shown that the COR decreased as the ball speed increased, as it was affected by the object released height [3]. During the football impact, the ball will undergo a significant deformation. The coefficient of restitution depends on the amount of deformation of the foot and the ball at impact, which results in a perfect collision by the material properties of the colliding objects. The higher deformation causes increase forces and release of energy [4]. The coefficient of restitution depends on the mechanical properties of the ball, the shoe, the ankle and the foot upon contact. Therefore, the highest deformation by the foot gives a lower coefficient of restitution [5].

From the available literature, have shown that from the standard size football ball, the value of COR varies from the types of the experiment conducted. From the pendulum experiment, the COR was approximately 0.82. From the experiment, by conducting the human kicking motion, the COR is between from 0.50 to 0.65. Additionally, other studies of COR effects from shoes had been conducted. It was performed the toe kicks for the football boot. It recorded that, the COR is varied from 0.4 to 0.8 [6].

The characteristics of the Futsal ball were different from the standard ball, as it differs from other size and the bounce of the ball. The Futsal ball less bounce compares to standard ball up to 30%. The present study aimed to examine Futsal ball size four characteristic impacts between the effects of height and the rigid plate target with three different height of drop impact. High-frames images were recorded on the point of impact to capture the deformation of Futsal ball during impact. The motion sensor was employed at the top of the apparatus to determine the height of the Futsal ball bounce. The COR and deformation were then calculated using the formula. The present study aimed to examine the effect of the height of the ball drop to the characteristics of a Futsal ball size four.

2. Equipment and Methods

Adidas® FIFA standard Futsal ball size four with diameter 320 mm and weight 420g were used for the experiment [7]. Markers were placed on the specific area of the ball to measure the same face of drop test of the ball. The impact was captured using Mark IV SONY camera with 1,000 frames per second recording speed. The recorded video of the impact was then analyzed using the ImageJ software, and the normal and tangential deformation was calculated using the equation (1) and equation (2). The Impact drop apparatus was attached with the SparkVue motion sensor to capture the ball impact velocity, the velocity before and after impact.

The measurement of ball deformation is likely to influence players' perception of ball hardness softness. Deformation influences the trajectory strike of the direction of the ball [8]. To analyze the maximum ball deformation quantitatively, the maximum normal (d_n) and tangential (d_t) deformation ratios were calculated using the equations below;

$$d_n = 100|D_n - D_o|/D_o \quad (1)$$

$$D_t = 100|D_t - D_o|/D_o \quad (2)$$

Where D_o, D_n and D_t indicate the initial ball diameter, maximum ball span normal and tangential to the flat surface, respectively.

The experiment was performed using the apparatus as shown in figure 1. The drop impact apparatus was designed with the maximum height of 1,800 mm. The frames of the impact test were built from hollow steel. The base of the impact test was from a 1.5 mm thickness, steel plate. The dropped case was free dropped from three different heights, 1,500 mm, 1,000 mm, and 500 mm high onto a rigid planar surface under the earth's gravitational effect [9].

The dropped case for each height was conducted until five repetitions. The value of COR is calculated based on the height drop and bounce of the impact on the rigid plate. The COR was calculated using the equation (3). The amount of energy lost is described as a ratio between the inbound velocity and the outbound velocity, commonly known as the coefficient of restitution (COR). In the elastic of an impact, the value of COR has described the ability of a body to return to its original shape once it has been deformed.

$$e = \frac{v_2}{v_1} = \left(\frac{h_2}{h_1}\right)^{\frac{1}{2}} \quad (3)$$

The equation (3) derived related to the height by using the energy principle, from the conservation of mechanical energy. A perfectly elastic impact will result in zero energy loss, COR = 1, while a perfectly plastic impact results in a zero rebound height, COR = 0, with the COR of sports balls generally somewhere in between. The more elastic ball, the longer it bounces [10].

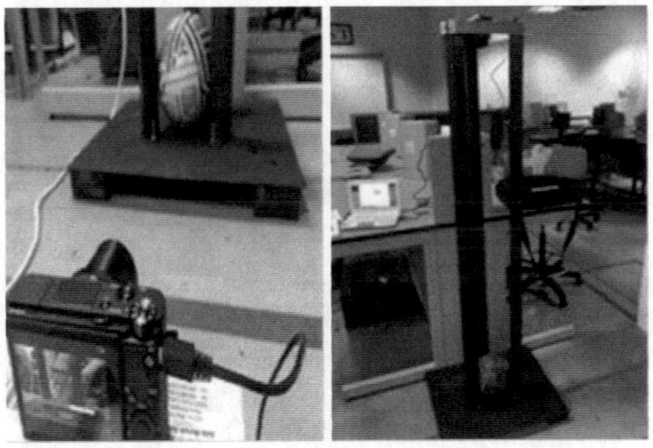

Figure 1: The experiment set-up apparatus.

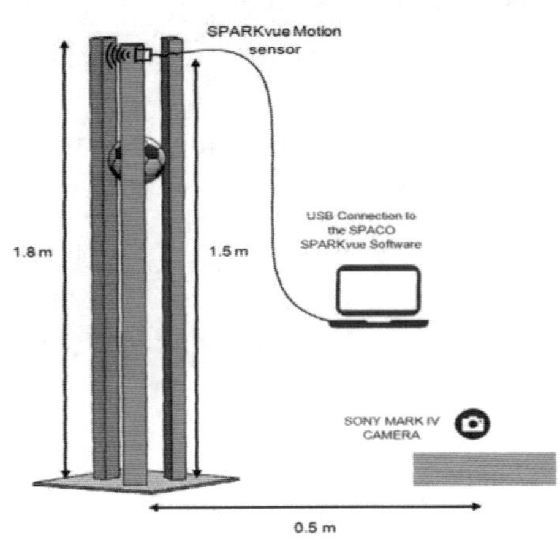

Figure 2: Schematic diagram.

3. Result and Discussions

A ball dropped vertically onto a surface experiences a vertical Impulsive Force; F is defined as the change in momentum per unit time. F is typically 100-1000 times larger than mg, in which case the gravitational force can be neglected during the impact. The assumption for this study, the air friction was neglect, as the ball drop was under the earth's gravitational effects.

The magnitude of acceleration is equal to gravitational, in the absence of air resistance. In this experiment, the Futsal ball size 4 undergoes a drop test to impact with a rigid steel target. The normal and tangential deformation for each height drop as shown in Figure 3 and Figure 4.

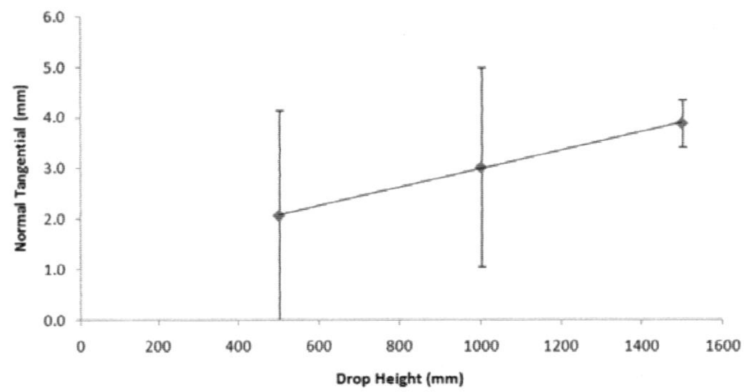

Figure 3: Calculated normal deformation with a different drop height.

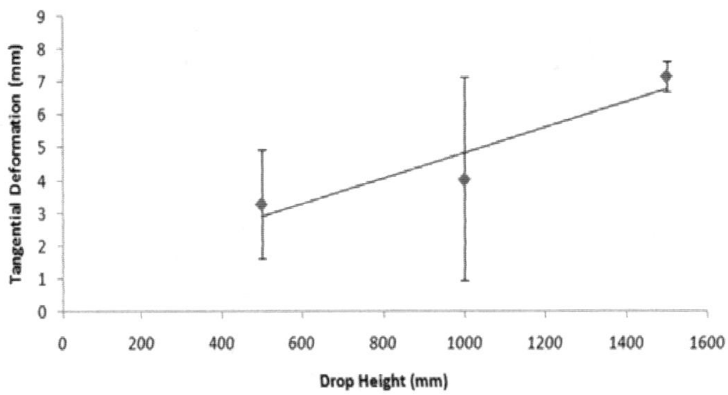

Figure 4: Calculated tangential deformation with a different drop height.

From the calculated analysis (figure 3 and figure 4), 1,500 mm recorded the highest deformation value for both the standard and tangential deformation. It had shown that the deformation of the Futsal ball increases as the drop height increases which was also ascertained by other authors [11, 12, 13].

For the normal deformation, the highest deformation shown from 1500 mm, was 3.874 mm, while 1000mm, and 500mm recorded 3.010 mm and 2.065 mm respectively. Meanwhile, for the tangential deformation, from 1500 mm was recorded at 7.142 mm, while 1000mm and 500 mm recorded 4.026 mm and 3.278 mm respectively. From this test, it is shown that the Futsal ball size 4 undergoes more tangential deformation compared to normal deformation.

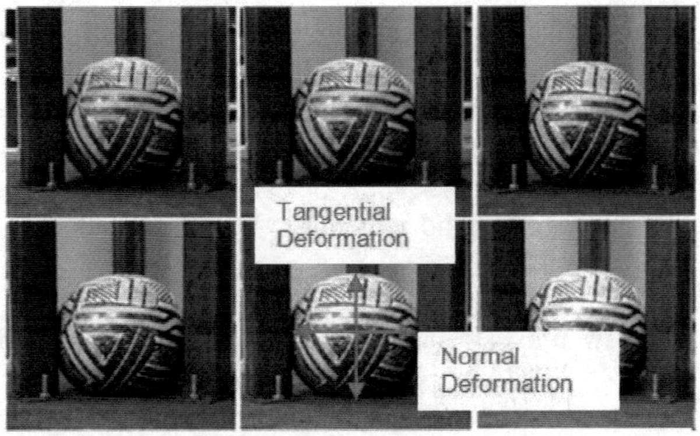

Figure 5: Video images of the Futsal ball size four during the impact at 1500 mm.

The ball drop was captured using a SONY camera which rates that up to 1,000 frames per second (Figure 5). The image of the ball drop was measured using the ImageJ software.

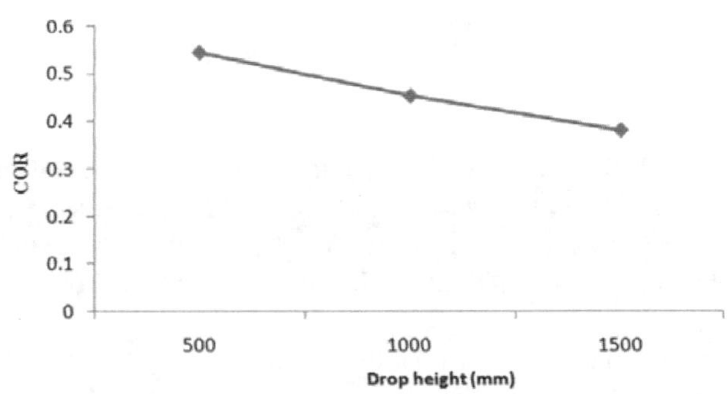

Figure 6: Calculated COR with difference drop height

From the COR graph analysis above (Figure 6), 500 mm height recorded the highest COR value 0.5448 while 1,000 mm and 1,500 mm recorded 0.4538 and 0.3812 respectively. Figure 6 shows a reduction in COR with increase drop height, which was also ascertained by other authors [14, 15, 16]. The value of the COR was between 0.3 to 0.5, which can be said as an inelastic collision, which some kinetic energy is dissipated after the impact. A summary of the ball deformation and COR are shown in table 1.

Table 1: Experiment Result summary

Characteristics	1500 mm	1000 mm	500 mm
Max ball normal deformation (mm)	3.874	3.010	2.065
Max ball tangential deformation (mm)	7.142	4.026	3.278
Coefficient of Restitution (COR)	0.3812	0.4538	0.5448

The above studies can be summaries that, the Futsal ball size 4 had difference characteristic as a standard football ball types. From the results table 1, it is shown that both normal and tangential deformation increased as the drop height increased. Compared to COR, the COR increased as the drop height decreased. It summaries that the COR depends on the deformation of the ball. The higher ball deformation that causes the increased force and also some energy dissipated.

4. Conclusion and Recommendation

From the drop test experiment, the deformation and the COR of the Futsal ball vary by the drop height of the ball. A COR depends on the amount of the deformation of the ball itself. The higher deformation of the ball, the less COR value of the ball. It can be concluded that the reduction value of COR can be related to the energy loss due to the ball deformation during impact. For the further work recommendation, an experiment with larger sample size and different models and sizes of balls could be conducted in future to obtain precisely and statistically analyze results.

5. References

[1] Niedfeldt M 2011 Head Injuries, Heading, and the use of Headgear in Soccer *Curr Sports Med Rep*. **10**(6) pp 324-9.

[2] Al-Kashmiri A and Delaney S 2006 Head and neck injuries in football (soccer) *trauma* **8** pp 189-195.

[3] Adli H and Ismail K A 2012 Coefficient of restitution of sports balls: A normal drop test *IOP Conf. Ser.: Mater. Sci. Eng.* **36** 012038.

[4] Eleftherios Kellis and Athanasios Katis 2007 Biomechanical characteristics and determinants of instep soccer kick *Journal of Sports Science and Medicine* **6** pp 154-165.

[5] Ahmad M, Ismail K A and Mat F 2016 Impact Model and Coefficient of Restitution: A Review *Journal of Engineering and Applied Science* **11** pp 6549-55.

[6] Rod Cross 1998 The bounce of a ball *American Association of Physics Teachers* pp 222-227.

[7] FIFA Law of the games (2014/2015). Retrieved on April, 2018: http://www.fifa.com/mm/Document/FootballDevelopment/Refereeing/02/36/01/11/LawsofthegamewebEN_Neutral.pdf

[8] Thompset B, Harland A and Roberts J 2016 Investigating the relationship between physical properties of football and player perceptions *Procedia Engineering* **147** pp 519-525.

[9] Arakawa K, Mada T, Komatsu H, Shimizu T, Satou M, Takehara K and Etoh G 2009 Dynamic deformation behaviour of a golf ball during normal impact *Experimental Mechanics* **49** pp 471-7.

[10] Choi J Y, Yi B J and Kwon Y 2011 Impact dynamic of a flying soccer ball with consideration of variable characteristics of the coefficient of restitution *IEEE Int. Conf. on Robotics and Automation* (New York: IEEE Xplore) pp 1689-94.

[11] Ahmad N, Taha Z, Ujihashi S and Tanaka K 2009 An experimental study of the impact of a sepak takraw ball on a flat surface *The Impact of Technology on Sport III*, (Australia: RMIT University Publisher) Chapter: Design pp 447-452.

[12] Aris M A, Taha Z and Hassan M H A 2013 The influence of football boot construction on ball velocity and deformation *IOP Conf. Ser.: Mater. Sci. Eng.* **50** 012028.

[13] Price D S, Jones R and Harland A R 2006 Computational modelling of manually stitched soccer balls *Proc. of the Institution of Mechanical Engineers, Part L: Journal of Materials: Design and Applications* **220** (4) pp 259-268

[14] Bao R H and Yu T H 2015 Collision and rebound of ping-pong ball on a Rigid Target *Materials and Design* **87** pp 278-286.

[15] Andersen T B, Sørensen H and Kristensen L B 2008 Biomechanical Differences Between Toe and Instep Kicking; Influence of Contact Area on the Coefficient of Restitution *Football Science* **5** pp 45-50.

[16] Andersen T B, Dorge, and Thomsen 1999 Collisions in Soccer Kicking *Sports Engineering* **2** pp 121-5.

Acknowledgments
The authors would like to thank Universiti Malaysia Sarawak, and the Minister of Higher Education, Malaysia for providing the funding for this research under the grant F02/SpGS/1543/2017. Also, a big thank you to the International College of Advanced Technology Sarawak (*i*-CATS) for their kind assistance in ensuring this research work a success.

IOP Conf. Series: Materials Science and Engineering 1101 (2021) 012038 doi:10.1088/1757-899X/1101/1/012038

Performance analysis of fabricated bio-filtration system for small-scaled greywater treatment: TSS and turbidity removals

N A S Abdul Samat[1*], W Renang[1], R Baini[1], J C H Lai[1] and M R Rahman[1]

[1]Department of Chemical Engineering and Energy Sustainability, Faculty of Engineering, Universiti Malaysia Sarawak (UNIMAS), Sarawak, Malaysia
[*]Corresponding author email: asnamalina@unimas.my

Abstract. River pollution, which is mainly due to excessive drainage of untreated domestic greywater, has becoming a serious concern in both poor and developing areas like Sarawak in the current years. Realizing the vital role of Sarawak rivers, the need to improve the level of pollution, demand for clean freshwater resources, as well as to save cost on household water supply, the greywater generated daily from every household needs to be pretreated for reuse, otherwise, discharged. This further demands for cost-effective pretreatment technology for domestic and residential use. This research, therefore, analyzes the efficiency of a low-cost bio-filtration system which utilizes agricultural wastes, namely rice husk and coconut coir, as bio-filter media to improve the quality of greywater effluent samples in terms of total suspended solids (TSS) and turbidity removals. This paper presents the extended results obtained from the previous research work using the same fabricated pre-treatment system, which consists of six main units i.e. wastewater storage tank, water feed tank, pre-sedimentation tank, bio-filter, post-sedimentation tank and treated water storage tank. The system is operated for six hours to complete several cycles of treatment. At every one-hour interval, both TSS and turbidity levels of the wastewater samples are recorded and the removal performances are evaluated and analyzed. Positive outcomes are attained from this research study such that the rice husk system is able to reduce the TSS and turbidity levels by 49.06% i.e. from 53.00 mg/L to 27.00 mg/L with an average rate of 4.33 mg/L.hr, as well as 57.79% i.e. from 41.70 FNU to about 17.60 FNU, at a rate of 4.02 FNU/hr respectively. Besides, the coconut coir bio-filtration system achieves TSS removal efficiency of 49.70% i.e. from 65.60 mg/L to 33.00 mg/L at mean rate of 5.43 mg/L.hr, and 63.10% turbidity removal i.e. from 48.40 FNU to 17.86 FNU, at 5.09 FNU/hr respectively.

Keywords: *bio-filtration, coconut coir, domestic greywater, rice husk*

1. Introduction

Sarawak, being one of the developing states in Malaysia, has been experiencing credibly an explosive growth in its population in the recent years and is anticipated to further expand in terms of its technology, sociology and economic status in the near future. This strongly demands for cleaner and safer environment to ensure a more sustainable and promising future. According to Kuok et al. [1], in the past, only 40% of the pollutants, excluding nutrients and bacteria, were able to be removed from the black water through the septic tank treatment whilst the greywater was directly discharged, along with the stormwater, to the Sarawak tributaries without any pretreatment. The disposal of this untreated wastewater had adversely upset the quality and ecosystem of the Sarawak rivers.

To ensure and offer better access to safe and sustainable water supplies, minimize environmental degradation, as well as, to preserve the public health and the burdening costs associated to it, strenuous efforts must be done [2]. According to Chan et al. [3], rivers contribute to 97% of Malaysia water supply, whilst the remaining 3% are from the sub-grounds aquifers. Realizing this important role of rivers as the main freshwater source and to mitigate river pollution, particularly in Sarawak, domestic greywater needs to be treated before its release to the drains and river bodies [4]. Without prompt preservation actions and proper wastewater management, various roles of Sarawak rivers will be affected and Sarawak's natural resources and water will continue to scarce [5]. Besides, the high cost of installation and operation of many water pollution treatment technologies introduced has burdened the society, thus the need to propose more cost effective and sustainable schemes.

Many previous studies had been carried out to propose potential alternative treatments and safe wastewater disposal strategies in the face of limited freshwater supply in the future. One of the most suggested and high potential approaches is to recycle and reuse the treated greywater by filtering and recycling miscellaneous drain water then using it for many secondary water applications [6]. Commonly, the treated water is used for non-potable purposes that do not require drinking water quality to be achieved. These include activities like cooling, rinsing and flushing in industries, as well as for outside use and irrigation purposes rather than wastefully discharging the greywater to Sarawak rivers [1]. Mah et al. [7] also convinced that it is a waste to dispose greywater as it is generated about the same volume daily, thus providing a constant resource of water for other secondary purposes. It can also help to lower the cost of water supply since the normal potable water is generally more expensive due to the costly treatment procedures and management.

Consequently, to recycle, pretreat and reuse the domestic greywater without incurring high cost of technology to the users, more cost effective and sustainable wastewater recovery technologies are demanded. El Nadi et al. [8] suggested the use of agricultural wastes as the natural packing media in the filtration vessel, as one of the feasible options to reduce the cost of the treatment system and control environment pollution; both river and land pollutions due to the massive accumulated amount of agricultural wastes generated and dumped. Therefore, this research aims to investigate the effectiveness of such wastes, namely rice husk and coconut coir, as the low-cost media installed in a fabricated bio-filtration system for the pretreatment of the domestic greywater effluent. While the previous work by Renang et al. [4] reported positive results of both media in improving the BOD5 and COD quality of the wastewater, this paper presents the extended results to the similar system that is, to evaluate their effectiveness in terms of total suspended solids (TSS) and turbidity removals.

2. Materials and Methods

2.1. Fabricated bio-filtration system

This research is an extension of the previous work which similarly focuses on the treatment of domestic greywater using the same continuous bio-filtration system. The previous results were reported by Renang et al. [4]. The fabricated bio-filtration system basically consists of six main unit operations, as depicted in the schematic diagram in figure 1. The effluent sample was collected from Desa Ilmu, Kota Samarahan, and the system was made to operate for six hours to complete several continuous treatment cycles. The primary treatment occurs in the baffled pre-sedimentation tank in which heavy particulates in the effluent sample settle down at the bottom of the tank.

Secondary treatment follows in the bioreactor which is designed at a capacity of 20L of wastewater and consists of a mesh net compacted with natural filters of local agricultural wastes to trap the available particulates passing through it. Coconut coir and rice husks are identified as the potential packing media for the greywater treatment for this purpose. The packing media also serves as microbial bed and support where microorganisms develop on the surface over the continuous cycles of operation and biodegrade the available organic pollutants in the wastewater. The media is designed to be fully submerged in the greywater as shown in figure 2. They were collected locally from Saratok in Betong Division, Sarawak and used in the bioreactor system without any chemical pretreatment.

The filtrate proceeds to the tertiary treatment process in the baffled post-sedimentation tank to further separate traces of other impurities in the activated sludge, which will settle at the bottom of the tank and later be disposed as waste sludge.

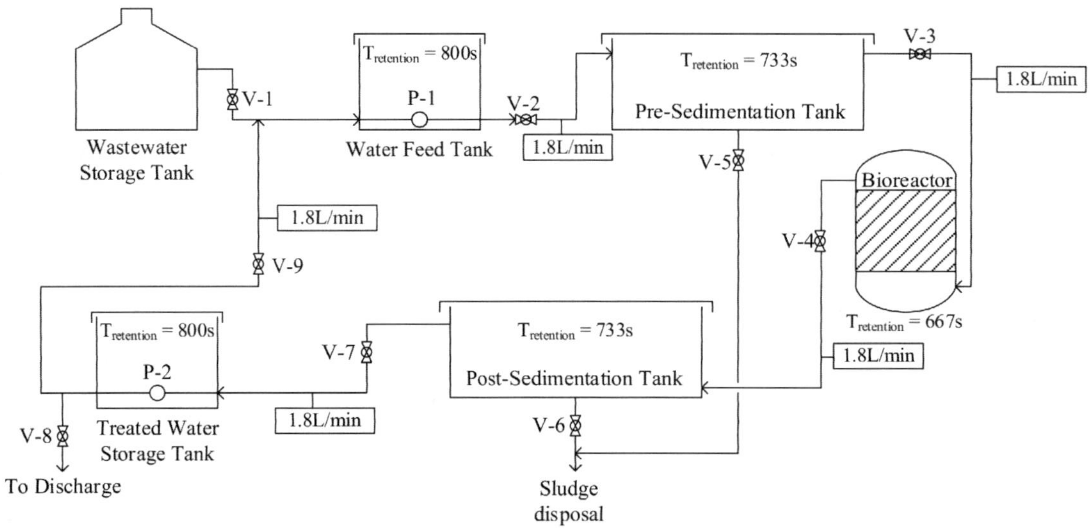

Figure 1. PFD of continuous bio-treatment system [4].

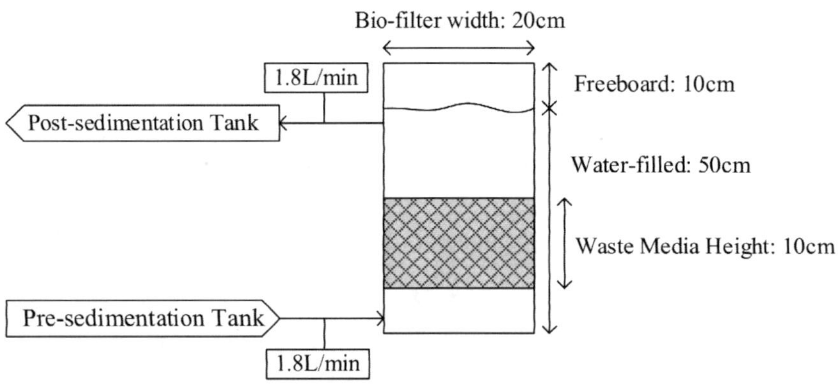

Figure 2. Design of waste-packed bio-filter unit.

2.2. Performance analysis
The performance of each type of filter media, i.e. rice husk and coconut coir, in treating the greywater effluent, are analyzed and compared based on the reduction of total suspended solids (TSS) and turbidity levels within six hours of continuous treatment cycles.

2.2.1. Total suspended solids (TSS). The amount of total suspended solids in the wastewater samples are determined based on the methods outlined by the APHA (American Public Health Association) (1992). A volume of effluent sample is stirred and pipetted on dish before dried for at least an hour in a drying oven operating at about 105°C. After cooled in a desiccator, the sample dish is weighed using an analytical balance. The steps of drying, desiccating and weighing are repeated until a constant weight is achieved. Then, the amount of total suspended solids in the sample, measured in mg/L, is determined using DR 900 multi-parameter portable colorimeter. The TSS values of the wastewater samples are analyzed at every one-hour interval for six hours and the percentage of cumulative removals are evaluated and recorded.

2.2.2. Turbidity. The technique to determine the turbidity level in the wastewater sample is also based on the APHA (American Public Health Association) (1992) standard methods. The sample is thoroughly shaken and poured slowly into the turbidimeter tubes to avoid bubbles formation. The water sample is added until the viewing disk can no longer be seen. The turbidity level is then determined directly from the instrument scale on the side of the turbidimeter tube. In this experiment, the turbidity level is measured in the unit of FNU. The turbidity levels of the wastewater samples are analyzed at every one-hour interval for six hours and the percentage of cumulative removals are evaluated and recorded.

3. Results and discussion

The performance of both rice husk and coconut coir, as natural filters of the bio-filtration system, on the TSS and turbidity levels of the sample effluents are analyzed in this section.

3.1. Total suspended solids (TSS) removal

Figure 3 illustrates the trends of the measured mean values of TSS in the sample effluents over the six-hours treatment. The difference in the TSS performances between the rice husk and coconut coir systems are comparatively insignificant whereby the prior resulted in an overall TSS removal efficiency of 49.06% with an average rate of 4.33 mg/L.hr, from 53.00 mg/L to 27.00 mg/L, whilst the latter similarly showed about the same TSS reduction rates of 49.70% and 5.43 mg/L.hr, from 65.60 mg/L to 33.00 mg/L.

The trends in figure 3 also portray that the rate of reduction of TSS are relatively higher at the early hours of treatment, before it generally decreases over time as tabulated in table 1. It is shown that the highest TSS removal rates for rice husk and coconut coir systems are obtained only within the first hour of treatment, which are at rates of 11.00 mg/L.hr with 20.75% TSS reduction from its initial value, and 14.60 mg/L.hr with 22.26% reduction, respectively. These results indicate that both rice husk and coconut coir can well behave as effective mediums for the physical filtration of suspended solids in the greywater samples, by which, the coconut coir system provides a minimally better performance than the other.

The performance of coconut coir packing media for TSS removal in this research is compared with that reported by Vinod & Mahalingegowda [9]. Significantly lower overall removal efficiency is obtained in this work, which is at 49.70%, as compared to that achieved by Vinod & Mahalingegowda [9] at 89.82%. This happens due to higher initial TSS values of the sample effluent and longer retention and operating period of six days performed by the authors which allowed higher efficiency, twice as much obtained in the current investigation. Hence, this shows that the removal efficiency will increase as the retention time in the reactor increases.

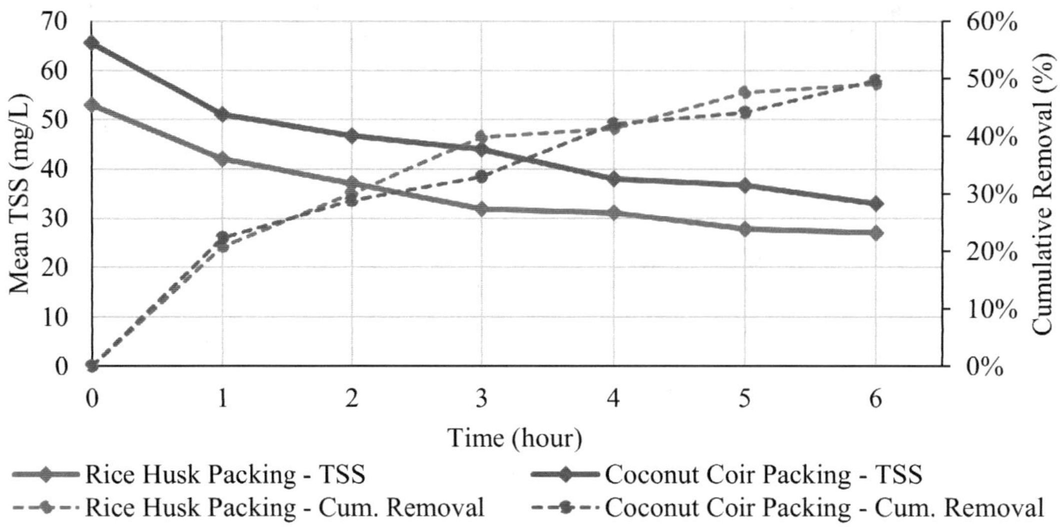

Figure 3. Performance analysis on total suspended solids (TSS).

Table 1. Rate of total suspended solids (TSS) removal.

Time (hr)	Mean TSS (mg/L)		Hourly Rate of TSS Removal (mg/L.hr)		Rate of Cumulative TSS Removal (mg/L.hr)	
	Rice Husk	Coconut Coir	Rice Husk	Coconut Coir	Rice Husk	Coconut Coir
0	53.00	65.60	-	-	-	-
1	42.00	51.00	11.00	14.60	11.00	14.60
2	37.10	46.70	4.90	4.30	7.95	9.45
3	31.90	44.00	5.20	2.70	7.03	7.20
4	31.10	38.00	0.80	6.00	5.48	6.90
5	27.80	36.70	3.30	1.30	5.04	5.78
6	27.00	33.00	0.80	3.70	4.33	5.43

Table 2. Deviation of total suspended solids (TSS) from standard A[a] and standard B[a] limits.

Time (hr)	Deviation from Standard A DOE (%)		Deviation from Standard B DOE (%)	
	Rice Husk[b]	Coconut Coir[b]	Rice Husk[b]	Coconut Coir[b]
0	6.00	31.20	-47.00	-34.40
1	-16.00	2.00	-58.00	-49.00
2	-25.80	-6.60	-62.90	-53.30
3	-36.20	-12.00	-68.10	-56.00
4	-37.80	-24.00	-68.90	-62.00
5	-44.40	-26.60	-72.20	-63.30
6	-46.00	-34.00	-73.00	-67.00

[a] TSS values for Standard A and Standard B are 50mg/L and 100mg/L respectively (based on the 8th Schedule of EQA 1974 by Department of Environment (DOE) Malaysia).
[b] Negative percentage reflects lower TSS values from the standard limits, and vice versa.

3.2. Turbidity removal

Figure 4 shows the effects of rice husk and coconut coir bio-filtration systems on the turbidity of the effluent samples over the six hours of continuous treatment. The turbidity levels are observed to decrease over time for both packing systems. For instance, the rice husk packing media system is able to reduce the initial average turbidity of sample from 41.70 FNU to about 17.60 FNU with a total removal of 24.10 FNU (57.79% efficiency) and an average rate of 4.02 FNU/hr throughout the six hours of treatment. On the other hand, over the same period of treatment, the coconut coir filtration system reduces the sample mean turbidity by 63.10%, from 48.40 FNU to 17.86 FNU with an average rate of 5.09 FNU/hr. It is hence, observed that the overall removal efficiency of effluent turbidity by the coconut coir bio-filtration system is insignificantly better than that achieved by the rice husk system.

As viewed in figure 4 and table 3, the turbidity of the domestic greywater sample decreases significantly for the first four hours of operation before attaining steady removal rates towards the end of the cyclic treatment. The highest rate is achieved at the first hour of treatment for both systems whereby the rice husk system records a rate of 7.40 FNU/hr which is 17.75% reduction from the initial average turbidity, whilst the coconut coir media records a significantly higher rate of 19.83 FNU/hr with 40.97% turbidity reduction. These rates gradually reduce over time as more substances are trapped on the packing media and the media become more spent. Similar to the performance on TSS removal, the results depict that both rice husk and coconut coir bio-filtration systems can effectively improve the turbidity of the greywater samples, in which the latter performs slightly better than the prior.

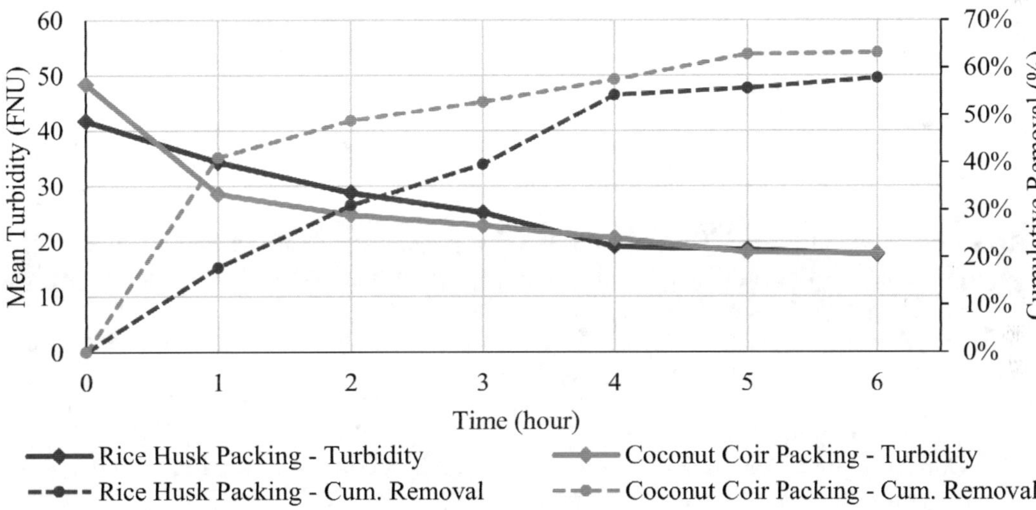

Figure 4. Performance analysis on turbidity.

Table 3. Rate of turbidity removal.

Time (hr)	Mean Turbidity (FNU)		Hourly Rate of Turbidity Removal (FNU/hr)		Rate of Cumulative Turbidity Removal (FNU/hr)	
	Rice Husk	Coconut Coir	Rice Husk	Coconut Coir	Rice Husk	Coconut Coir
0	41.70	48.40	-	-	-	-
1	34.30	28.57	7.40	19.83	7.40	19.83
2	28.80	24.80	5.50	3.77	6.45	11.80
3	25.20	22.87	3.60	1.93	5.50	8.51
4	19.10	20.60	6.10	2.27	5.65	6.95
5	18.50	18.00	0.60	2.60	4.64	6.08
6	17.60	17.86	0.90	0.14	4.02	5.09

IOP Conf. Series: Materials Science and Engineering 1101 (2021) 012038 doi:10.1088/1757-899X/1101/1/012038

4. Conclusion

The rice husk packing filters have been found to reduce the TSS and turbidity levels by 49.06% from 53.00 mg/L to 27.00 mg/L (average rate of 4.33 mg/L.hr) and 57.79% from 41.70 FNU to about 17.60 FNU (average rate of 4.02 FNU/hr) respectively, whilst, the coconut coir media packs have reduced both levels by 49.70% from 65.60 mg/L to 33.00 mg/L (mean rate of 5.43 mg/L.hr) and 63.10% from 48.40 FNU to 17.86 FNU (mean rate of 5.09 FNU/hr) respectively. These outcomes also depict that both rice husk and coconut coir systems demonstrate almost similar removal efficiencies in reducing the TSS and turbidity levels from the effluent, with the latter achieving just slightly higher removal percentage. In conclusion, both rice husk and coconut coir bio-filtration systems can perform well as effective mediums and have shown positive outcomes in improving the quality of the greywater samples, in terms of the amount of Total Suspended Solids (TSS) and turbidity level. Nevertheless, the performance of the system could be more obvious when the sample effluent has higher initial TSS values or when longer retention time is employed. The findings of this research, hence, suggest that the biofiltration pre-treatment system using natural packing filters i.e. rice husk and coconut coir can be employed for greywater pretreatment especially for use in the rural areas.

5. References

[1] Kuok K K, Chiu P C and Mah D Y S 2015 Wastewater reclamation, recycling and reuse potential: A case study for Kuching city, Sarawak *Water Utility J.* **11** 63-72

[2] Hunter P R, MacDonald A M and Carter R C 2010 Water supply and health *PLoS Medicine* **7** 11

[3] Chan C M, Kamis N S and Radin Mohamed R M S 2014 Using a peat media for laundry greywater filtration: Geochemical and water quality check *Middle East J. Sci. Res.* **21(8)** 1365-70

[4] Renang W, Abdul Samat N A S, Baini R, Rahman M R and Selaman O S 2018 The effectiveness of a fabricated bio-filtration system in treating the domestic greywater *Malays Appl Biol. J.* **47(1)** 65-71

[5] Kuok K K, Harun S and Chiu P C 2011 A Review of integrated river basin management for Sarawak river *Am. J. Environ. Sci.* **7(3)** 276-85

[6] Juan Y K, Chen Y and Lin J M 2016 Greywater reuse system design and economic analysis for residential buildings in Taiwan *Water* **8** 546.

[7] Mah D Y S, Putuhena F J, Salim S and Lai S H 2008 Modeling of greywater reuse for Kuching, Malaysia *J. Australian Water Association* 31-34

[8] El Nadi M H, Elazizy I M and Abdalla M A F 2014 Use of agricultural wastes as bio filter media in aerobic sewage treatment *Australian J. Basic Applied Sci.* **8(16)** 181-5

[9] Vinod A and Mahalingegowda R 2014 A study on suitability of selected fibrous packing media for sewage treatment *J. Int. Academic Res. Multidisciplinary* **2(2)** 487-96

Acknowledgments

The authors would like to acknowledge Universiti Malaysia Sarawak (UNIMAS) for the support and assistance throughout this research work [Research Grant: Osaka Gas in Cultural Exchange Foundation (OGFICE) Research Grant Scheme 2016/2017].

IOP Conf. Series: Materials Science and Engineering 1101 (2021) 012039 doi:10.1088/1757-899X/1101/1/012039

Development of solar power system for Sarawak peat water continuous electrocoagulation treatment process

N Abdul Rahman[1*], A Albania Linus[1], A Philip[2], E E Jihed[1], U J Gilan[1], N K M F Kumar[1] and A Yassin[3]

[1] Department of Chemical Engineering and Energy Sustainability, Faculty of Engineering, Universiti Malaysia Sarawak, 94300 Kota Samarahan, Sarawak, Malaysia
[2] Department of Electrical and Electronic Engineering, Faculty of Engineering, Universiti Malaysia Sarawak, 94300 Kota Samarahan, Sarawak, Malaysia
[3] Department of Mechanical and Manufacturing Engineering, Faculty of Engineering, Universiti Malaysia Sarawak, 94300 Kota Samarahan, Sarawak, Malaysia
*Corresponding author email: arnazeri@unimas.my

Abstract. Sarawak state government has established Sarawak Alternative Rural Water Supply (SAWAS) programme in order to serve as a purpose of providing safe and clean water to the rural communities not connected to municipal clean water supply. In the rural areas of Sarawak, particularly on the coastal region where municipal water supply is not available, the villagers are normally resorted to utilize rainwater and peat water for daily usage. Some of these rural areas are even not connected to electricity grid. Subsequently, one of the proposed methods to eradicate these problems in supplying clean water without electricity supply grid is to implement stand-alone water treatment system with solar power system. As such, the main aim of the study is to design a solar power system to support Sarawak peat water electrocoagulation treatment process. The study is divided into two stages. In the first stage, the study designs a solar power system to support the treatment process of peat water for both batch and continuous electrocoagulation systems. This includes designing and fabrication of a small-scale solar power system. The second stage of the study involved experimental studies on both batch and continuous electrocoagulation systems in order to study the effectiveness of solar power system to supply electricity for the electrocoagulation systems. Overall, the study has developed a solar power system for both batch and continuous electrocoagulation of peat water system. From the experiments conducted, the developed systems are capable to reduce 18.8% and 46.15% of peat water turbidity for batch and continuous electrocoagulation systems respectively. However, in order to meet a more stringent drinking water standards, some improvements on the designed systems are indispensable.

1. Introduction

Located in the northwest of Borneo, Sarawak has a population of about 2.6 million [1] in which electricity grid need to be extended through rugged terrain and thick jungle in order to reach isolated communities. However, extending such grid to sparse and remote communities is not always practical in economic terms [2]. Hence, there are still about 30,000 remaining rural households which are not connected to the electricity grid [3]. For villages located more than 30 km from the electricity grid, an off-grid method is preferred [1]. Sarawak is located near the equator line where the climate is hot and

IOP Conf. Series: Materials Science and Engineering 1101 (2021) 012039 doi:10.1088/1757-899X/1101/1/012039

humid throughout the year. The state ambient temperature is approximately from 22 °C to 33 °C with monthly solar radiation of 400–600 MJ/m^2 [4,5]. Correspondingly, solar energy is one of the promising renewable energy in the state due to the strategic location where solar energy can be harvested throughout the year [4].

Peat water is abundantly found in peat land due to availability of underground water and rainwater. Malaysia has approximately 2.6 Mha of peatland in which over 70% of the total peatland in Malaysia is located in Sarawak [6]. Peat water is commonly used for domestic usage in rural coastal areas households which are yet to receive clean water from the municipal water supply. However, direct consumption of peat water is not advisable, hence, a simple and versatile water treatment system such as electrocoagulation treatment is useful in this case. Electrocoagulation is a water treatment process which is an alternative to chemical coagulation process. The process is capable of reducing the need for chemical coagulants since the electrodes provide the coagulants [7]. Essentially, electrocoagulation process is an environmentally friendly process, and it is significantly less complicated unlike other conventional water treatment methods. Electrocoagulation water treatment had been known for its beneficial advantages including relatively low costs, the possibility of complete automation with high efficiency of particulate removal at a compact treatment facility [8]. This method is characterized by minimum use of chemicals in the process, easy operation and reduced sludge production [9,10]. Moreover, electrocoagulation process have the ability to treat water with low temperature and low turbidity, while it is effective in a wide range of water pH ranging from pH 4 to pH 6 due to its neutralization effect [11]. Additionally, electrocoagulation flocs are large, stable, acid resistant and contains less bound water allowing it to be separated faster by filtration; contains effluent with less total dissolved solids (TDS) that contributed to low water recovery cost; produced gas bubbles to carry pollutant upwards for easy collection and removal; and most importantly convenience for use in rural areas with no access to electricity since solar panel attached to the unit may be sufficient to run the treatment [12]. As 13%, of the Sarawak state's total area is peatland. This imposes the existence of peat water which contains some natural organic matters that causes foul smell, colour, taste and the acidity issues [13]. Consequently, this study is aimed to develop a stand-alone water treatment system which is expected to be able to mitigate this problem. The system should be cheap and could effectively treat peat water. In addition, the technology used for the fabrication process of this water treatment system should be available locally in which easy to maintain and fabricate.

2. Methodology

There are a total of three main tasks performed in this project which includes (i) design and fabrication of solar power system, (ii) experimental studies, and (iii) evaluation of solar system for electrocoagulation treatment system.

2.1. Design and fabrication

The design and fabrication process are divided into two parts, (i) design and fabrication of the solar power system and (ii) design and fabrication for the automation of both batch and continuous electrocoagulation of peat water treatment system. The design of the systems is based on these criteria: (i) Materials used are readily available locally, (ii) Simple fabrication and maintenance process, and (iii) Low fabrication and operating cost.

2.1.1. Solar power system sizing. The solar power system sizing is done by initially determining the initial value of power demand, solar panel arrangement, battery storage capacity sizing, charge controller, inverter, and load management. The sizing calculations are performed for the solar power system as depicted in figure 1.

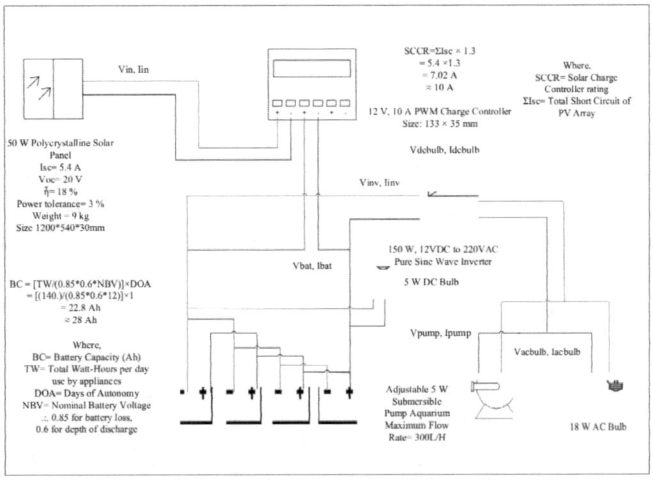

Figure 1. Schematic diagram of the solar system.

There are several equations used in solar power system components sizing. First, for Photovoltaic (PV) module sizing, it is crucial to investigate the electrical specification of the PV panel. The formula used for the calculation of PV module sizing is stated by equation (1) and the calculation of the number of PV panels needed is shown in equation (2).

$$S_p = \frac{\left[\dfrac{Avg_L}{Avg_{sun}}\right]}{\eta} \quad (1)$$

Where,

S_p	=	Sizing of Solar Panel	Avg_{sun}	=	Average daily peak sun-hours
Avg_L	=	Average daily load	η	=	Efficiency factor

$$N_{\max\_total} = \frac{V_{bat} \times I_{\max\_oc}}{P_{np\_solar} \times \eta} \quad (2)$$

Where,

$N_{\max\_total}$	=	Maximum number of PV panel	$I_{\max\_oc}$	=	Maximum output current
V_{bat}	=	Voltage of the battery	P_{np_solar}	=	Nominal power of solar

Second, charge controller is designed in order to prevent battery from overcharging in which it might affects the performance of the battery as well as reducing the battery lifespan. The formula to calculate the size for standard charge controller is stated by equation (3).

$$SCCR = \sum I_{SC} \times 1.3 \quad (3)$$

Where,

SCCR	=	Solar charge controller rating
ΣI_{SC}	=	Total short circuit current of PV array

Third, the battery is used as backup storage to overcome uninterruptible power supply (UPS). Equation (4) is used to calculate battery capacity [14] while Equation (5) is used to calculate the required number of batteries.

$$BC = \frac{TW}{0.85 \times 0.6 \times NBV} \times DOA \quad (4)$$

Where,

BC	=	Battery capacity (Ah)	NBV	=	Nominal battery voltage
TW	=	Total watt-hours per day used by appliances	DOA	=	Days of autonomy

IOP Conf. Series: Materials Science and Engineering 1101 (2021) 012039 doi:10.1088/1757-899X/1101/1/012039

$$N_{pbat} = \frac{C_{Ah\_req}}{C_{bat}} \quad (5)$$

Where,

$N_{p_{bat}}$ = Number of batteries required
C_{Ah_req} = Required charge capacity (Ah)
C_{bat} = Battery capacity (Ah)

Fourth, the inverter is used to convert DC voltage to AC voltage. The sizing of the inverter is calculated by using equation (6).

$$S_i = T_L \times [\frac{(1+Af)}{Le}] \quad (6)$$

Where,

S_i = Size of inverter
T_L = Total load
Af = Additional further load expansion
Le = Efficiency of inverter

2.1.2. Automation of the water treatment system. The automation of water treatment is done for both batch and continuous system. For batch process, water level sensor and motor stirrer are installed. The water level sensor acts as a switch in which it is able to turn on and off the pump within a range of level limit. The motor stirrer is installed to create turbulence during the electrocoagulation process. This is to provide some comparison studies of mixing and unmixing process [15]. Flow rate sensor and turbidity sensor are installed for the continuous system. Flow rate sensor is installed to measure the flow rate of the liquid entering the system as the flow rate affected the efficiency of the continuous electrocoagulation system. Turbidity sensor is installed to measure the turbidity of the treated water during the electrocoagulation process.

2.2. Experimental studies on the effect of the angle of the solar panel

The experiments are conducted on the effect of solar panel angle at 20°, 25° and 30° while being coupled with mimic loads of continuous electrocoagulation which are 5W Alternating Current (AC) pump, 18W AC bulb, 5W Direct Current (DC) bulb as shown in figure 2. The experiment is conducted for three days in a row on sunny day. The averaged three days' results obtained are then evaluated in accordance to temperature, irradiance received, power generation, voltage output and current output.

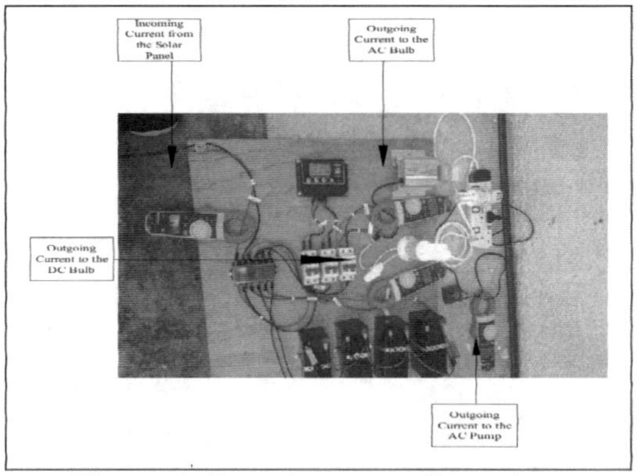

Figure 2. Experiment configuration.

2.3. Evaluation of solar system for electrocoagulation treatment system

There are two different systems for electrocoagulation treatment employed in this study which are the batch and continuous system as shown in figure 3a and figure 3b respectively. The optimum angle as investigated is used to generate electricity to power up the electrocoagulation treatment system. Table 1 shows the operating parameters for both batch and continuous system.

Figure 3a. Batch electrocoagulation system.

Figure 3b. Continuous electrocoagulation system.

Table 1. Parameters for batch and continuous system.

Parameter	Batch	Continuous
Electrode spacing	2 cm electrode spacing	0.7 cm electrode spacing
Number of electrodes	8 number of electrodes	20 number of electrodes
Treatment time	2 hours	
Amount of treated water	1 litre	

3. Results and discussion

3.1. Experimental studies on solar power system configurations

The variables affected by the solar panel angle are temperature, irradiance, power generated, voltage output and current output. The results of temperature and irradiance are depicted in figure 4 and figure 5, respectively. On clear weather, the temperature increases in corresponding to the irradiation. However, when the temperature is too high at 65 ℃ it affects the efficiency of the solar panel operation [16]. Long-time PV panels operation in high temperature can cause irreversible degradation of its electric output power [17]. This is due to the fact that the temperature decreases the value of voltage linearly while the current increases exponentially. From the result obtained, solar panel of 30° angle shows the highest amount in term of temperature (about 75°C) during the peak hours whereas the angle of 25° shows a slightly less value (about 55°C). The maximum average temperature in Sarawak is 34°C to 35°C and the minimum average temperature is 21.8 ℃ to 22.3 ℃ [18]. Irradiance, on the other hand, shows that angle of 30° to be quite consistent from 9.00 am to 1.00 pm as compared to the other angle studied. Solar irradiance is one of the factors in determining site location, where location with high solar irradiance is preferable and has high potential in generating maximum output [19].

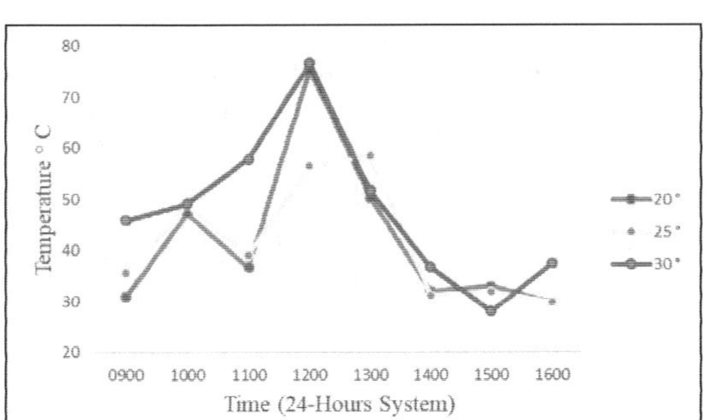

Figure 4. Temperature of solar panel.

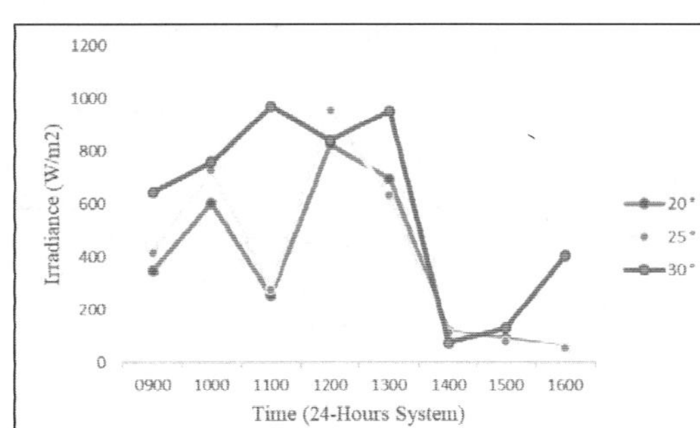

Figure 5. Irradiance received at solar panel.

Figure 6 shows the power generated by the solar panel in a specified angle of 20°, 25° and 30° while figure 7 and figure 8 show the voltage and current generated by the solar panel, respectively. The power generated for 30° angle of solar panel is the highest as compare to another angle of solar panel. For both voltage and current output, solar panel of 30° angle shows high value as compared to the other angles. However, it is noticeable that at 1200 hour, the temperature of 30° solar panel is higher than 65°C in which causing some reduction is solar panel efficiency as showed by figure 5 at 1200 hour. Despite this problem, the optimum time to harvest solar energy is from 11.00 am to 1.00 pm daily.

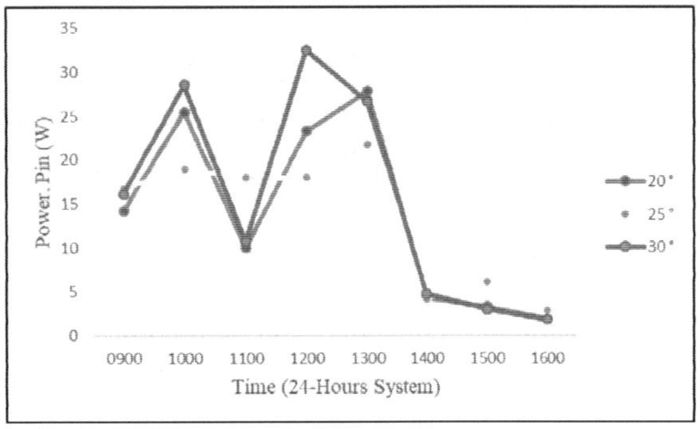

Figure 6. Power generated by solar panel.

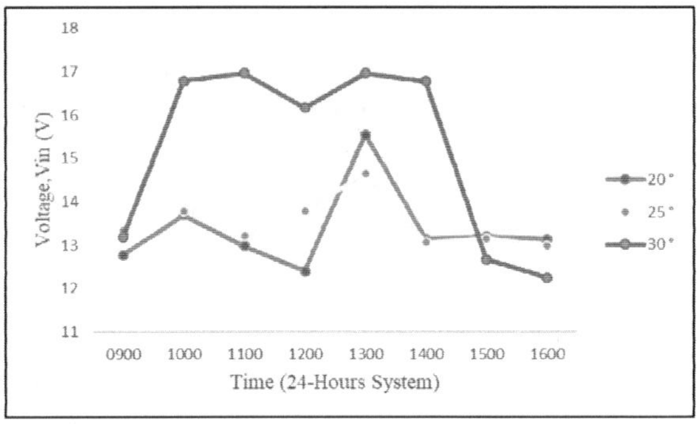

Figure 7. Voltage output from the solar panel.

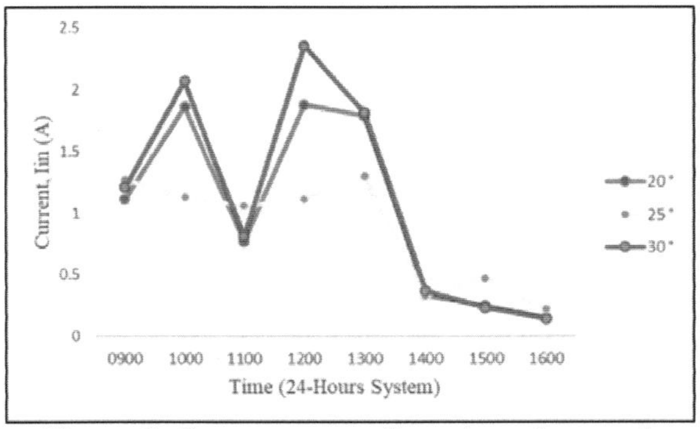

Figure 8. Current output from the solar panel.

In this solar power system, all of the power generated by solar panel is in DC form in which are used to charge the battery and from the battery, the current is converted into alternating current (AC) by the inverter before being used for the electrocoagulation system. Figure 9 shows the power output from the battery to the loads. All the loads are using their sources of power from the battery. The highest power output from the battery is at 1.00 pm of 25° instalments solar panel which is 27.657W and the lowest power output is at 4 pm for 30° instalments solar panel which is 23.907W. From the calculation, the total power needed to generate all the loads is 28W. Figure 10 shows the power input to the inverter. Figure 11, figure 12 and figure 13 show the power input to the pump, power input to the DC bulb, and power input to the AC bulb, respectively. Power output to the inverter indicated the power used by the AC loads. The AC loads compromise of the 5W AC pump and 18W AC bulb where the total AC loads is 23W. The highest power output from the inverter is on 2.00 pm for the 20° instalment of solar panel which is 25.54W and the lowest power output recorded from the inverter is at 3.00 pm for 25° instalment of solar panel which is 23.218W. There are 2.322W differences recorded for the highest and lowest power output from the inverter. Power input to the pump and the AC bulb shows a steady-state flow of bar chart where the pump recorded 4.4W power to generate pump while 16.6W is needed to generate the AC bulb. The total power generated by the AC loads is 21W according to the collection of data. Data collection for the DC bulb is still considered as steady-state as there are slight differences recorded for the power to generate the DC bulb which is still within the range of the rated power.

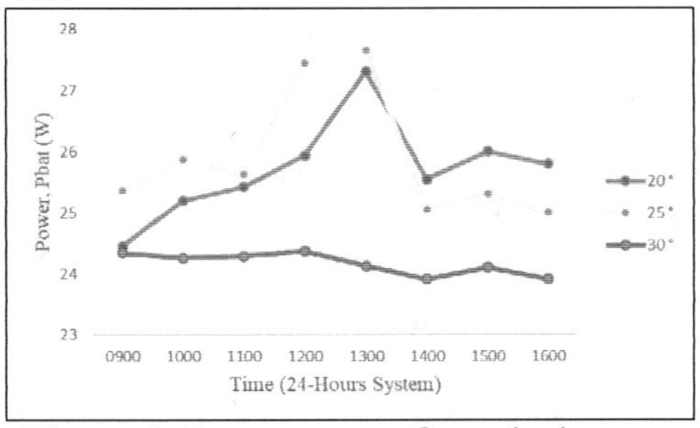

Figure 9. Power output from the battery.

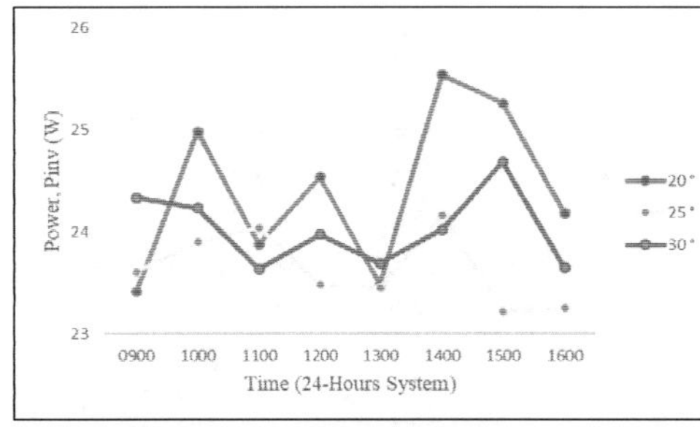

Figure 10. Power input to the inverter.

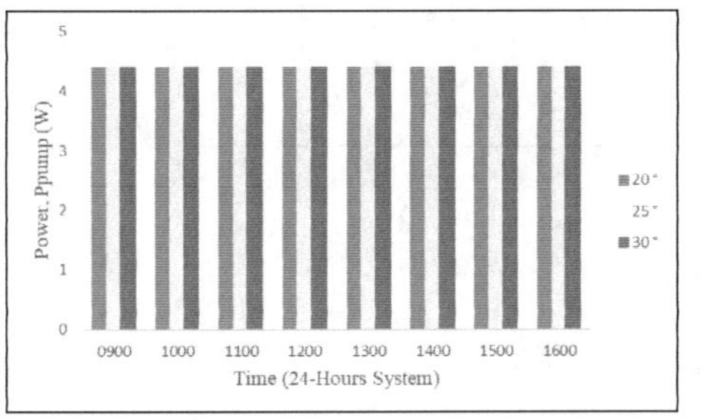

Figure 11. Power input to the pump.

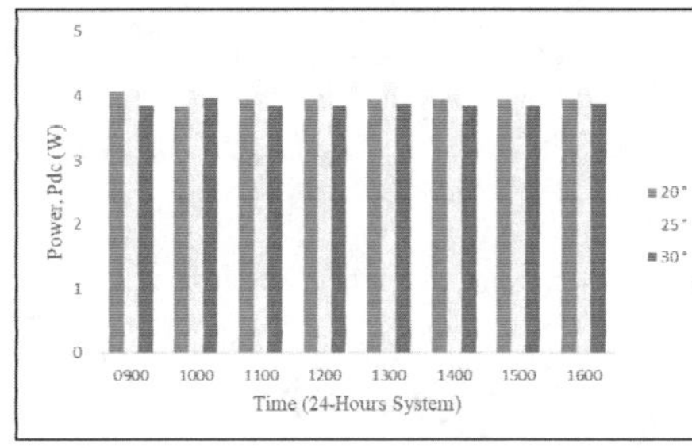

Figure 12. Power input to DC bulb.

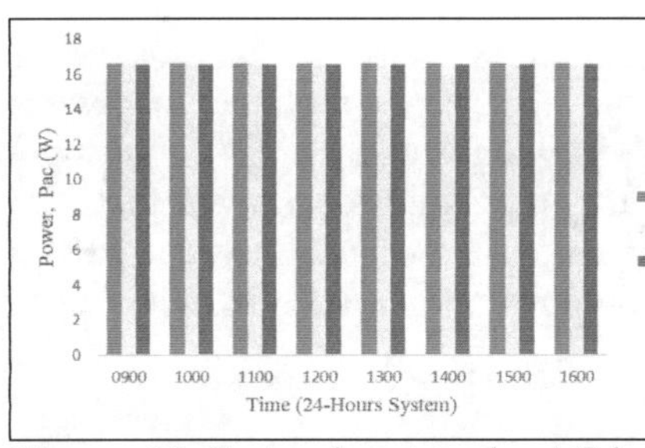

Figure 13. Power input to the AC bulb.

3.2. Suitability of solar power electrocoagulation system

Table 2 shows the comparison of the process with and without solar for both batch and continuous systems. The comparison is done based on the removal efficiency of turbidity of the peat water. Electrocoagulation treatment of peat water by using solar power system is inefficient because the removal efficiency of turbidity for both batch and continuous electrocoagulation is low which is less than 50% as compared to DC converter power grid. This is due to the low current density generated by the solar power system. Low current density leads to lower production of metal coagulant. Metal coagulant production is dependent on current density and time [19]. In addition, batch and continuous electrocoagulation water treatment used a different flow rate of submersible pump thus the power rated for the pumps are not the same. Batch electrocoagulation system uses faster flow rate which is 6.5 litres/min with a rated power of 13.2W to fill up the storage for treated peat water from raw peat water storage which is 0.44Wh energy. As for continuous electrocoagulation, it uses a slower flow rate which is 0.07 litre/min with a rated power of 4.4W with the amount of energy used of 8.8Wh. For the power usage by the electrodes for both batch and continuous are the same which is 4.949W for operating time of 2 hours. Since the rated power for both batch and continuous electrocoagulation are the same as well as the operating time, accordingly, the electricity usage for electrocoagulation process is the same which is 9.898 Wh. Both batch and continuous electrocoagulation systems are treating the same amount of treated water but the removal efficiency of turbidity are found to be different. The removal efficiency of turbidity for batch electrocoagulation is from 11 NTU to 9 NTU which is 18.8% while the removal efficiency of turbidity for continuous electrocoagulation is from 13 NTU to 7 NTU which is 46.15%. Both systems did not meet the standard set by the Ministry of Health, the Drinking Water Quality Standard where it specified the maximum allowable level of turbidity is 5 NTU. However, both systems treated water falls into Class II of National Water Quality Standard, which means it is qualified for general washing purposes. In order to improve the system, more solar panels need to be added in order to increase the power generated in which, in turn, provide sufficient electricity for the complete electrocoagulation process to occurs.

Table 2. Comparison of the process with and without solar.

Parameter	Without Solar		With solar	
	Batch System	**Continuous System**	**Batch system**	**Continuous System**
Current	5.0 A	5.0 A	0.4 A	0.4 A
Current density	25.00 A/m^2	5.99 A/m^2	2.00 A/m^2	0.50 A/m^2
Final turbidity	1 NTU	0 NTU	9 NTU	7NTU
Removal efficiency	93.35 %	100.00 %	18.80 %	46.15 %

4. Conclusion

The optimum angle for the develop solar power system for peat water electrocoagulation system is found to be 30° where it absorbed the most heat and solar irradiation from the sunlight. However, the efficiency of solar panel dropped at noon due the extreme temperature but the efficiency increases again after the temperature has reduced. The highest generated power by solar panel is at 2.00 pm while it generated the least power at 4 pm. From the experiments conducted, it is observed that the turbidity values of peat water for both batch and continuous electrocoagulation systems have been reduced. The reduction is more prominent for continuous electrocoagulation system process which is 46.15% as compare to batch system at 18.8%. However, the water treatment by mean of solar power system do not meet the Malaysian requirement for drinking water. This is due to significant low current produced by the solar power system. Nevertheless, the study has proven that it is possible to apply solar power system for electrocoagulation treatment of peat water. However, further research on the improvements of the system are needed in order to produce better quality treated water which meet the Malaysian water quality standards.

5. References

[1] van Gevelt T, Canales Holzeis C, George F and Zaman T 2017 Indigenous community preferences for electricity services: Evidence from a choice experiment in Sarawak, Malaysia *Energy Policy* **108** 102–10

[2] Anyi M, Kirke B and Ali S 2010 Remote community electrification in Sarawak, Malaysia *Renew. Energy* **35** 1609–13

[3] Sarawak Energy 2020 Power Generation

[4] Heng S Y, Asako Y, Suwa T, Tan L K, Sharifmuddin N B and Kamadinata J O 2019 Performance of a small-scale solar cogeneration system in the equatorial zone of Malaysia *Energy Convers. Manag.* **184** 127–38

[5] Mekhilef S, Safari A, Mustaffa W E S, Saidur R, Omar R and Younis M A A 2012 Solar energy in Malaysia: Current state and prospects *Renew. Sustain. Energy Rev.* **16** 386–96

[6] Melling L 2016 Peatland in Malaysia *Peatland in Malaysia* (Springer, Tokyo) pp 59–73

[7] Butler E, Hung Y, Yeh R Y, Suleiman M and Ahmad A 2011 Electrocoagulation in Wastewater Treatment *Water* **3** 495–525

[8] Cerquiera A A, Souza P S A and Marques M R C 2014 Effects of direct and alternating current on the treatment of oily water in an electroflocculation process *Brazilian J. Chem. Eng.* **31** 693–701

[9] Chen G, Chen X and Yue P L 2000 E Lectrocoagulation and E Lectroflotation *J. Environ. Eng.* **126** 858–63

[10] Vasudevan S, Lakshmi J and Sozhan G 2011 Effects of alternating and direct current in electrocoagulation process on the removal of cadmium from water *J. Hazard. Mater.* **192** 26–34

[11] Liu H, Zhao X and Qu J 2010 Electrocoagulation in Water Treatment *Electrochemistry for the Environment* (New York, NY: Springer New York) pp 245-262

[12] Mollah M, Schennach R, Parga Torres J and Cocke D 2001 Electrocoagulation (EC) – Science and Applications *J. Hazard. Mater.* **84** 29–41

[13] Ng T N 2004 *Assessment of Aluminium contaminant in public water supply in Sarawak* (Universiti Malaysia Sarawak, Malaysia)

[14] Boukenoui R, Ghanes M, Barbot J P, Bradai R, Mellit A and Salhi H 2017 Experimental assessment of maximum power point tracking methods for photovoltaic systems *Energy* **132** 324-40

[15] Pradiko H, Afiatun E and Fabian E 2018 Influence of mixing and detention time in electro coagulation process to treat raw water at Badak Singa Water Treatment Plant *Indones. J. Urban Environ. Technol.* **1** 137–50

[16] Solar Calculator 2020 The Effect of Temperature on Solar Panel Performance

[17] Amelia A R, Irwan Y M, Leow W Z, Irwanto M, Safwati I and Zhafarina M 2016 Investigation of the effect temperature on photovoltaic (PV) panel output performance *Int. J. Adv. Sci. Eng. Inf. Technol.* **6** 682–8

[18] Jakhrani A Q, Othman A, Rigit A R H and Samo S R 2012 Assessment of solar and wind energy resources at five typical locations in Sarawak *J. Energy Environ.* **4** 1–6

[19] Kuokkanen V, Kuokkanen T, Ramo J and Lassi U 2015 Electrocoagulation treatment of peat bog drainage water containing humic substances *Water Res.* **79** 79–87

Acknowledgments

The authors wish to thank and Department of Chemical Engineering and Energy Sustainability and Department of Electrical and Electronic, Faculty of Engineering, Universiti Malaysia Sarawak. Special thanks to Osaka Gas for providing Industrial Grant (Project ID: IG/F02/OSAKA/2018).

IOP Conf. Series: Materials Science and Engineering 1101 (2021) 012040 doi:10.1088/1757-899X/1101/1/012040

Dynamic Modeling of Hydrogen Production from Photo-Fermentation in Microbial Electrolysis Cell using Sago Waste

M Y Then[1], S Ramesh[1], M A Mohd Asrul[1], M F Atan[1*], H Abdul Halim Yun[1,2], J C H Lai[1], M R Rahman[1], M O Abdullah[1] and N Abdul Wahab[1,3]

[1]Department of Chemical Engineering and Energy Sustainability, Faculty of Engineering, Universiti Malaysia Sarawak, 94300 Kota Samarahan, Malaysia
[2]Centre for Applied Learning and Multimedia, Universiti Malaysia Sarawak, 94300 Kota Samarahan, Malaysia
[3]UNIMAS Community Sustainability Centre, Universiti Malaysia Sarawak, 94300 Kota Samarahan, Malaysia
*Corresponding email: amfarid@unimas.my

Abstract. Hydrogen has a great potential as an alternative energy and produces zero emissions, but most of hydrogen is produced from non-renewable fossil fuels via reforming. Thus, biomass is a promising replacement to fossil fuels where hydrogen can be sourced from. In this project, sago waste is chosen as raw material in a microbial electrolysis cell (MEC) to produce hydrogen fuel. A mathematical model with the integration of MEC with photo-fermentation has been developed and modified by using sago effluent as a substrate in a batch process. The main parameter such as concentration of microbial community has been observed in this project as it gives a huge influence on the gas product of MEC. In conclusion, the develop model was to observe the behavior of the microbial electrolysis cell where a maximum of 3.8 L/day (t = 4 days) of hydrogen production and 0.38A of MEC current were obtained.

1. Introduction

With rapid urbanization and industrialization, energy has become increasingly essential to human lives. For instance, energy is needed for various applications such as the manufacturing industry, electrical generation, and fuel for transportations. However, the major source of energy comes from non-renewable fossil fuels. As the world population continues to grow, the demand for energy is also expected to rise. This poses a challenge to the world as the number of fossil fuels available will begin to diminish and be bound to run out in the future. Furthermore, fossil fuels also bring negative impacts to the environment, releasing pollutants like nitrogen oxide (NOx), sulfur oxide (SOx), and volatile organic compounds (VOCs) which contribute to decrease air quality as well as global warming [1]. In recent years, renewable energy resources are gaining momentum in its development and have become popular because they are inexhaustible and clean. Nonetheless, there are still years to come before the world can depend on renewable energy entirely because renewable energy has not been utilized to their full potential and other limitations.

The development of renewable energy must be intensified to reduce the world's reliance on fossil fuels and eventually eliminate the usage of fossil fuels. A viable alternative to fossil fuels is hydrogen because it is widely available in nature, renewable and pollution-free. Unlike fossil fuels, hydrogen is environmentally friendly where it produces zero emissions. The only by-products from the combustion of hydrogen are steam and heat. The utilization of hydrogen as a fuel is imperative towards the goal of a sustainable future. Hydrogen has great potential as an alternative because it is energy-efficient where it has a high energy content of 122kJ/g, which is approximately 3 times greater compared to that of hydrocarbon fuels [2]. However, hydrogen is not inefficient renewable energy resource currently as most hydrogen is produced from fossil fuels, through the means of reformation. Reforming is a chemical process used to convert petroleum refinery naphtha into high-octane liquid products, producing significant amounts of by-product hydrogen gas [3]. For a sustainable future, biomass is a promising replacement to fossil fuels where hydrogen can be sourced from.

Sago exists in abundance in Sarawak as a result of the commercialization of sago through estate planning. Currently, Malaysia is the third-largest sago producer in the world, after Indonesia and Papua New Guinea [4]. Not only is sago used in the food industry, but it is also utilized in biotechnology, manufacturing industry, poultry industry, and paper industry [5]. Due to its versatility in usage and potential for export, sago's production will be expected to rise and correspondingly, the sago's waste will increase as well. If not properly treated, sago waste could cause a negative environmental impact. Therefore, a good waste management strategy for sago factories would be utilizing the sago waste to their full potential, making it as a raw material in a microbial electrolysis cell (MEC) to produce hydrogen fuel. The sago waste presents an enticing and low-cost feedstock for hydrogen production.

Until recently, MECs are emerging as a promising method for hydrogen production from the pure substrate like glucose, industrial wastewater, and solid waste. In MECs, electrochemically active bacteria are used to oxidize organic matter to produce hydrogen with the aid of externally supplied voltage [6]. Applications of MECs are vast, including an alternative to fossil fuels to run a vehicle, ammonia production, and power generation [7]. Thus, this project is conducted to establish mathematical modeling for biohydrogen production in the MECs to study the chemical reaction in-depth.

2. Overview of Microbial Electrolysis Cell (MEC)

In a MEC, electrochemically active bacteria oxidize organic matter and generate protons, electrons, and carbon dioxide. The bacteria transfer the electrons to the anode and the protons are released to the solution. The electrons then travel through a wire with the help of external voltage and combine with free protons in the electrolyte to produce hydrogen gas at cathode [8]. The current which indicates the amount of electron flow directly measures the hydrogen production rate. The membrane between the cathode and anode acts as a physical barrier to only allow free-protons (H+) to pass. Moreover, it also reduces the cross-over of bacteria and maintains the purity of the hydrogen gas evolved at the cathode. Membranes commonly used in MEC are proton-exchange membranes (PEM), anion-exchange membranes (AEM), charge-mosaic membrane (CMM), and bipolar membrane [9].

3. Integration of MEC with Photo-Fermentation

MECs alone break down carbohydrates slowly and results in low H2 production, making it not efficient as a process to treat complex wastewater. Thus, MECs are usually paired with fermentation to achieve complete and efficient conversion. Photo-fermentation is preferred over dark fermentation because dark fermentation has an intrinsic upper limit of 4 mol of H2 /mol hexose due to incomplete oxidation [10]. On the other hand, photo-fermentation produces complete conversion to H2 and CO2. Table 1 shows the comparison between fermentation, MEC, and MEC with fermentation in terms of various properties.

Table 1. The synergy between fermentation and MECs [10]

Properties	Fermentation	MEC	MEC with fermentation
H_2 yield [%]	14-27	23-71	>85
H_2 production rate [m^3 H_2/m^3 reactor/day]	0.2-10	0.1-3.4	>10
Complex biomass degradation	Yes but incomplete	No	Yes and complete
Complete organic to H_2 conversion	No	Yes	Yes
Produce H_2 at high yield	No	Yes	Yes

Produce H_2 at a high rate	Yes	No	Yes
Produce H_2 at low temperature	Difficult	Possible	Possible
Produce H_2 with high purity	No	Possible	Possible
Treat wastewater for reuse/ discharge	No	Possible	Possible
Waste biorefinery application	Long-term	Long-term	Nearer-term

4. Mathematical Model

The kinetic model proposed by Gadhamshetty et al. [11] is deemed the best kinetic model as it takes into account the inhibitory effects of high biomass concentration and excess light. Higher biomass concentration reduces the light intensity inside the photo-fermentation biohydrogen reactor (PBR), causes self-shading effects, and limits the substrate diffusion. The declining effect of excess light on biomass growth as the surplus absorbed light energy may result in damage and degradation of the reaction center involved in the photosynthetic process.

Then, MEC performance strongly depends on the activity and efficiency of the anode and cathode catalysts. The reaction occurs at the anode are as follows [12,13].

$$C_2H_4O_2 + H_2O + 4M_{ox} \rightarrow 4M_{red} + CO_2$$

$$4M_{red} \rightarrow 4M_{ox} + 8e^- + 8H^+$$

$$C_2H_4O_2 \rightarrow CH_4 + CO_2$$

$$CO_2 + 4H_2 \rightarrow CH_4 + 2H_2O$$

Among all the mathematical models, models proposed by Azwar et al. [12] and Pinto et al. [13] are adequate to describe the process in MEC. However, the model implemented by Azwar et al. [12] was derived from the model proposed by Pinto et al. [13]. Then, the model implemented by Pinto et al. [13] is more comprehensive and inclusive where it describes the MEC in three biofilm layers, and meanwhile, a model proposed by Azwar et al. [12] considered only two biofilm layers. Thus, the model proposed by Pinto et al. [13] is chosen as the best option, but since it is describing a continuous MEC operation, equations would be adjusted accordingly to fit the batch process model for this study. The substrate material balance is written as in equations (1) and (2).

$$\frac{dS}{dt} = -q_f x_f + D(S_0 - S) \quad (1)$$

$$\frac{dA}{dt} = -q_e x_e + q_m(x_{m,1} + x_{m,2}) + D(A_0 - A) + Y_{cod} q_f x_f \quad (2)$$

The outer biofilm (Layer 1) mass balance equations are formulated as in equations (3) and (4).

$$\frac{dx_f}{dt} = u_f x_f - K_{d,f} x_f - a_1 x_f \quad (3)$$

$$\frac{dx_{m,1}}{dt} = u_m x_{m,1} - K_{d,m} x_{m,1} - a_1 x_{m,1} \quad (4)$$

The inner biofilm (Layer 2) mass balance equations are written as in equations (5) and (6).

$$\frac{dx_e}{dt} = u_e x_e - K_{d,e} x_e - a_2 x_e \quad (5)$$

$$\frac{dx_{m,2}}{dt} = u_m x_{m,2} - K_{d,m} x_{m,2} - a_2 x_{m,2} \quad (6)$$

The cathodic biofilm (Layer 3) mass balance is formulated as in equation (7).

$$\frac{dx_h}{dt} = u_h x_h - K_{d,h} x_h - a_2 x_h \quad (7)$$

Meanwhile, the mass balance for the intracellular is as in equations (8) and (9).

$$M_{Total} = M_{red} + M_{ox} \quad (8)$$

$$\frac{dM_{ox}}{dt} = -Y_m q_e + \frac{\gamma}{Vx_e}\frac{I_{MEC}}{mF} \tag{9}$$

Furthermore, as referring to equations (10) – (15), the fermentative and acetoclastic methanogenic microorganisms are assumed to follow Monod kinetic equations, being limited by the organic substrate concentration while the electricigenic microorganisms are assumed to follow Monod kinetic equations, being limited by the organic substrate concentration and the oxidized form of the mediator.

$$u_f = u_{max,f}\frac{S}{K_{s,f}+S} \tag{10}$$

$$q_f = q_{max,f}\frac{S}{K_{s,f}+S} \tag{11}$$

$$u_m = u_{max,m}\frac{A}{K_{A,m}+A} \tag{12}$$

$$q_m = q_{max,m}\frac{A}{K_{A,m}+A} \tag{13}$$

$$u_e = u_{max,e}\left(\frac{A}{K_{A,e}+A}\right)\left(\frac{M_{ox}}{K_m+M_{ox}}\right) \tag{14}$$

$$q_e = q_{max,e}\left(\frac{A}{K_{A,e}+A}\right)\left(\frac{M_{ox}}{K_m+M_{ox}}\right) \tag{15}$$

The hydrogenotrophic methanogens are assumed to have a growth that is dependent on the H2 in catholyte. Since H2 has a solubility in water, zero-order growth kinetics is assumed if there is no H2 produced or no current in which can be expressed as in equation (16).

$$\mu_h \begin{cases} \mu_{max,h} & \text{if } I_{MEC} > 0 \\ 0 & \text{if } I_{MEC} = 0 \end{cases} \tag{16}$$

Apart from that, as referring to equations (17) – (19) , the biofilm retention is assumed to be limited by maximum attainable biomass concentration (X_{max}), said to be saturated when X_{max} is exceeded. The biofilm retention constants have the symbol α.

$$a_{1(x_f+x_{m,1}\geq X_{max,1}\ \&\ a_1=0)} = \frac{u_f x_f - K_{d,f}x_f + u_m x_{m,1} - K_{d,m}x_{m,1}}{x_f + x_{m,1}} \tag{17}$$

$$a_{2(x_e+x_{m,2}\geq X_{max,2}\ \&\ a_2=0)} = \frac{u_e x_e - K_{d,e}x_e + u_m x_{m,2} - K_{d,m}x_{m,2}}{x_e + x_{m,2}} \tag{18}$$

$$a_{3(x_h\geq X_{max,3}\ \&\ a_3=0)} = u_h - K_{d,h} \tag{19}$$

Then, the current is calculated by using the equations (20) – (21).

$$I_{MEC} = \frac{E_{CEF} + E_{applied} - \frac{RT}{mF}\ln\left(\frac{M_{Total}}{M_{red}}\right)}{R_{int}} - \frac{\frac{RT}{\beta mF}\sinh^{-1}\left(\frac{I_{MEC}}{A_{sur,A^{i}_{o}}}\right)}{R_{int}} \tag{20}$$

$$R_{int} = R_{MIN} + (R_{MAX} - R_{MIN})e^{-K_R X_e} \tag{21}$$

Lastly, the hydrogen production generated from this MEC is determined by using the equation (22).

$$Q_{H_2} = Y_{H_2}\left(\frac{I_{MEC}}{m_{H_2}F}\frac{RT}{P}\right) - Y_h u_h x_h V \quad (22)$$

5. Result and Discussion

As mentioned in the previous paragraph, a mathematical model developed by Pinto et al. [13] is referenced to study the behavior of the microbial electrolysis cell in biohydrogen production. The total time is set to be 20 days. The substrate used by Pinto et al. [13] was wastewater and meanwhile, for this project, the substrate is sago effluent. Due to similar nature where the substrate is highly diluted by water, the sago effluent is assumed to have the same behavior as wastewater. Apart from that, equations have been slightly modified because the model implemented by Pinto et al. [13] described a continuous process and the sago effluent is in a batch process inside the microbial electrolysis cell (MEC).

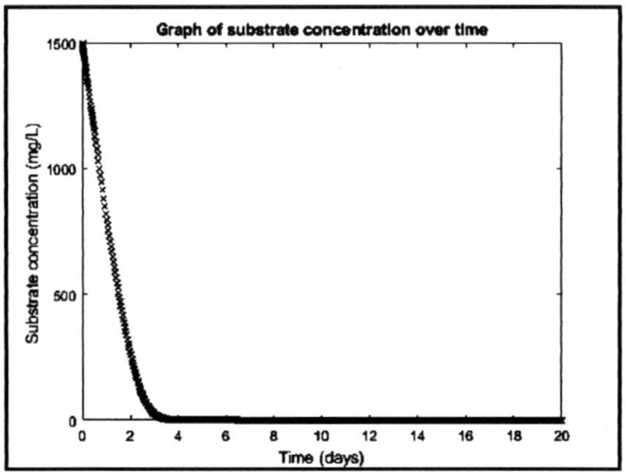

Figure 1. Substrate concentration over time

The initial substrate concentration of 1500 mg/L is chosen. The substrate concentration decreases very steeply from 1500 mg/L to 12 mg/L on the third day (t = 3 days), as seen in figure 1. Subsequently, the substrate concentration slowly decreases to 0 mg/L because the organic matter in sago is broken down to acetate by the microorganisms called fermentative microorganisms. The fermentative microbes metabolize complex carbon chains and convert them to acetate. The behavior of fermentative microorganisms is exhibited in figure 2.

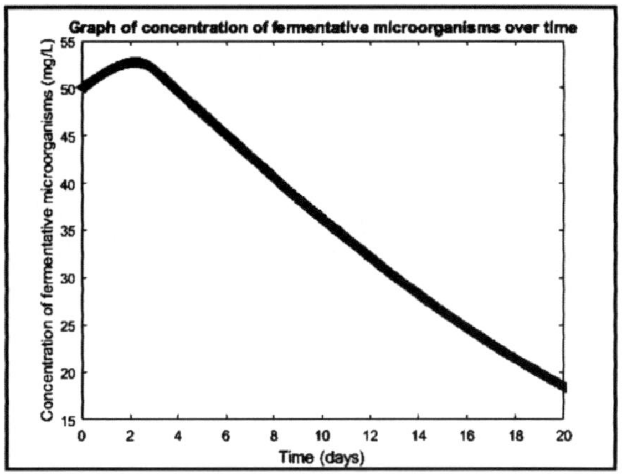

Figure 2. Concentration of fermentative microorganisms over time

IOP Conf. Series: Materials Science and Engineering 1101 (2021) 012040 doi:10.1088/1757-899X/1101/1/012040

The phenomena illustrated in figure 2 follows that the pattern of substrate concentration in figure 1. This is due to the fact that at the beginning (t = 0 day), the fermentative microorganisms are breaking down substrate rapidly, leading to the growth of fermentative microorganisms up until the third day (t = 3 days), reading its maximum concentration of 50 mg/L. On the third day (t = 3 days), the substrate has been used up by the fermentative microorganisms, thus from that point onwards, the concentration of fermentative microorganisms decreases gradually to a small number of 18 mg/L. This happens in Layer 1 which is the outer biofilm layer at the anode.

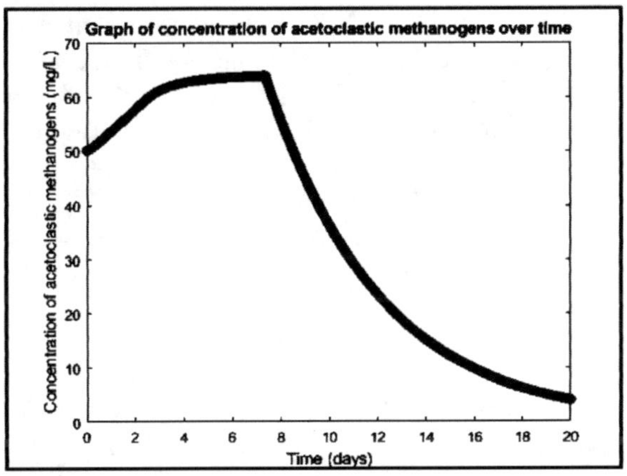

Figure 3. Acetoclastic methanogens over time

Based on figure 3, the concentration of acetoclastic methanogens, with an initial concentration of 50 mg/L, rises slowly to 63 mg/L on the seventh day (t = 7 days). This is due to the availability of acetate as a result of microbial fermentation and the lack of competition from electricigenic microorganisms in the first few days. The concentration of electricigenic microorganisms only starts to increase on around the third day (t = 3 days). Therefore, the acetoclastic methanogens are actively consuming the acetate until the seventh day (t = 7 days). From the seventh day (t = 7 days), the concentration of microbes decreases gradually to 15 mg/L at the last day (t = 20 days). This means that acetate is being converted to methane for these 20 days. However, the amount of methane production is negligible due to a small concentration of acetoclastic methanogens relative to the concentration of electrigenic microorganisms (Xmax for acetoclastic methanogens is 65 and Xmax for electricigenic microorganisms is 512.5). Based on figure 3, it can be deduced that acetoclastic methanogens have failed to compete with electrigenic microorganisms which grow at a much faster rate and experience no decline in concentration.

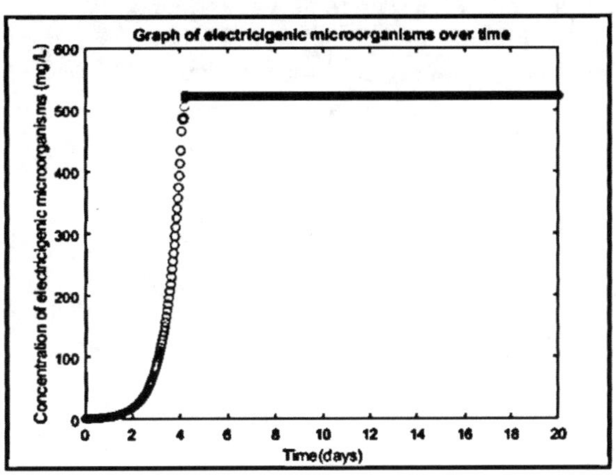

Figure 4. Electricigenic microogranism over time

In figure 4, the concentration of electrigenic microorganisms remain almost dormant in the first 2 days (t = 2 days) because of the low initial concentration. Unlike the concentration of acetoclastic methanogens which starts with 50 mg/L, the concentration of electricigenic microorganisms starts with 1 mg/L. With the initial condition of 1 mg/L, it has to take some time for growth and reproduction to increase to a substantial amount. From the second day (t = 2 days), the concentration of electricigenic microorganisms starts to increase exponentially to 512.5 mg/L, which is the maximum biofilm density after 4 days. Layer 2 represents the inner biofilm at the anode, containing electricigenic and acetoclastic methanogens. Competition exists between electrigenic and acetoclastic methanogens in consuming the same carbon source, which is the acetate. Electricigenic microorganisms convert acetate to carbon dioxide, proton, and electron while acetoclastic methanogens convert acetate to methane. From both figures 3 and 4, it can be deduced that the growth of acetoclastic methanogens is lower than that of electricigenic microorganisms. This results in a more microbial community composed of the majority of electricigenic microorganisms and little to none acetoclastic methanogens at the end (t = 20 days). This is imperative because the type of microbial community present in the MEC has a huge influence on the gas product of MEC. A MEC with a microbial population composed mainly of acetoclastic methanogens produces methane instead of hydrogen and vice versa. Thus, to make sure that the MEC produces hydrogen instead of methane, it is important to ensure that the MEC has a microbial community made up of a large proportion of electricigenic microorganisms.

Figure 5 illustrates the behavior of hydrogenotrophic methanogens in the MEC and this represents the situation in Layer 3. Layer 3 is the biofilm at the cathode, mainly populated by the hydrogenotrophic methanogens. It can be observed that the hydrogenotrophic methanogens present in a very small amount in the cathode compartment in the MEC in the beginning, with the initial value of 10 mg/L. This is due to the lack of current, which indicates the flow of electrons in the first 3 days. According to the equation (16), there will be no growth for hydrogenotrophic methanogens if no current is formed in the MEC. This is reasonable as the presence of current indicates that there is a flow of electrons and with the flow of electrons hydrogen production is feasible. From the third day (t = 3 days), there is shown to be a gradual increase of current which signifies that the electricigenic microorganisms are actively breaking down the acetate to carbon dioxide, proton, and electron, as exhibited in figure 6. With more protons flowing to the cathode and forming more hydrogen, hydrogenotrophic methanogens consume hydrogen as a source of energy will thrive in the environment. Hydrogenotrophic methanogens convert hydrogen to methane. This is an undesirable side microbial reaction arisen from the abundance of hydrogen in the cathode compartment. However, it is unavoidable in the cathode compartment in MEC which is a hydrogen-rich environment. From the third day (t = 3 days), the hydrogenotrophic methanogens slowly increase, due to the formation of current and hydrogen. At this point (t = 3 days), the concentration of hydrogenotrophic methanogens is still very low, at 10 mg/L and it takes some time for growth and reproduction to increase to a substantial amount. From the sixth day (t = 6 days), the concentration of hydrogenotrophic methanogens can be observed to grow exponentially. This shows the clear correlation between MEC current and concentration of hydrogenotrophic methanogens as well as the correlation between hydrogen production and concentration of hydrogenotrophic methanogens. Nevertheless, the biofilm density in Layer 3 will reach a maximum value of 1680 mg/L on day 16. That means the biofilm density is saturated and any growth of hydrogenotrophic microorganisms is not possible.

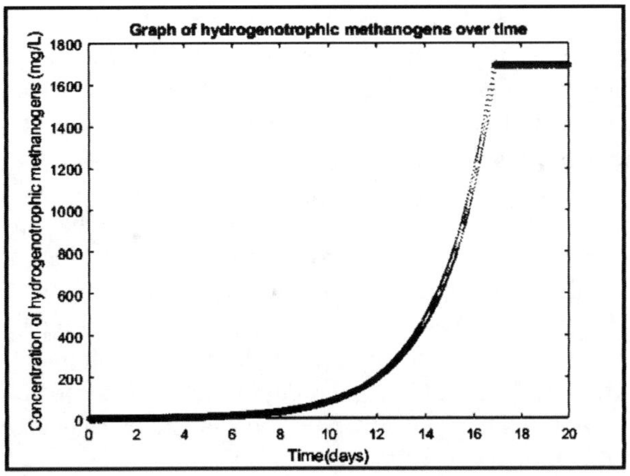

Figure 5. Hydrogenotrophic methanogens over time

The close relationship between MEC current and electricigenic microorganisms can be observed in figure 6. It can be seen that the point of ascending of MEC current (t = 3 days) is identical with that of the concentration of electricigenic microorganisms, shown in figure 4. Electricigenic microorganisms metabolize on acetate, converting it to carbon dioxide, proton, and electron. This effect of electron production has caused the flow of current in the MEC. Its growth is happening exponentially due to the abundance of acetate from the third day (t = 3 days), fermented from the sago effluent. This phenomenon is inconsistent with equation (21), which explains that the high concentration of electricigenic microorganisms, encouraging the transfer of electrons, which in turn results in the decrease in internal resistance. The decrease in internal resistance gives rise to the increase of MEC current. The current in MEC is also observed to reach a maximum value of 0.38A, due to the maximum concentration of eletricigenic microorganisms limited by its maximum biofilm density, which cannot increase further to produce more electrons. The minimum internal resistance of MEC is said to be reached. Thus, 0.38A is the maximum current attainable due to imitation of growth of electricigenic microorganisms and minimum internal resistance of MEC.

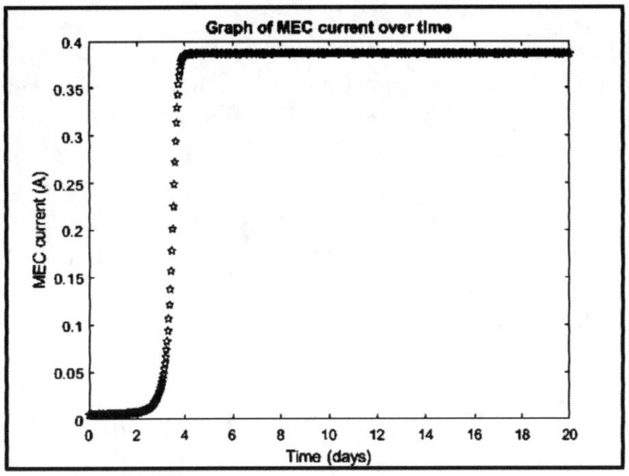

Figure 6. MEC current over time

Based on figure 7, the correlation between hydrogen production and MEC current as well as the correlation between hydrogen production and electricigenic microorganisms are observed as they have the same point of ascending (t = 3 days). The occurrence of hydrogen production is expected to follow the trend of current and electricigenic microorganisms because during the process where electricigenic microorganisms consume acetate, they do not only produce electrons, but they produce protons as well. With the combined of both electric potential that results from the microbial activity of electrigenic microorganisms and the additional voltage supplied from the outside source, the voltage is high enough to reduce the proton to hydrogen gas. In figure 7, the hydrogen production rate increases sharply from the third day (t = 3 days) until the fourth day (t = 4 days) to a value of 3.8 L/day, then it can be seen that the hydrogen production rate slowly decreases from the fourth day (t = 4 days) to sixteenth day (t = 16 days) to a value of 3 L/day. The decrease of hydrogen production can be attributed to Layer 3 biofilm, which is populated by the microbial community of hydrogenotrophic methanogens. These hydrogenotrophic methanogens, in the environment that is rich in hydrogen, aggressively consume the hydrogen in the cathode compartment and reproduces in a rapid manner, which causes the hydrogen production rate to drop further. At day 16 (t = 16 days), the biofilm at the cathode compartment has reached saturation and maximum hydrogen consumption is achieved. Therefore, from that day onwards, the hydrogen production rate is sustained at the rate of 3 L/day.

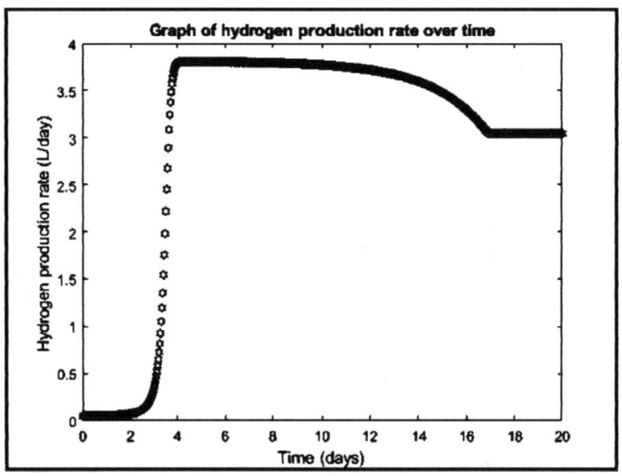

Figure 7. Hydrogen production over time

6. Conclusion

A mathematical model of the microbial and electrolysis cell in hydrogen production has been developed and modified by using sago effluent as a substrate in a batch process. Based on the results, more microbial community with a majority of electricigenic microorganisms and little to none acetoclastic methanogens which leads more hydrogen production, with a maximum of 3.8 L/day (t = 4 days) and 0.38A of MEC current compared to low concentration of hydrogenotropic methanogen with only 10 mg/L, indicating methane production. As conclusion, the model is able to predict and correlate between hydrogen production with concentration of electricigenic microorganisms and MEC current.

Acknowledgement

The authors would like to acknowledge Osaka Gas, Japan for funding this research through Osaka Gas Foundation in Cultural Exchange (OGIFICE) Research Grant Scheme. Last but not least, the authors would also to thank the Universiti Malaysia Sarawak (UNIMAS), Malaysia for providing the research facilities to execute this research.

Nomenclature

A	:	Acetate concentration in the MEC anodic compartment [mg-A/L]
A_0	:	Acetate concentration in the MEC's influent [mg-A/L]
CH_4	:	Methane [mol]
$C_2H_4O_2$	:	Methyl formate [mol]

IOP Conf. Series: Materials Science and Engineering 1101 (2021) 012040 doi:10.1088/1757-899X/1101/1/012040

CO_2	:	Carbon dioxide [mol]
D	:	Dilution rate [1/day]
e^-	:	Electron [mol]
E_{CEF}	:	Counter-electromotive force for MEC [V]
F	:	Faraday constant [Admol/e]
H^+	:	Proton [mol]
H_2O	:	Water [mol]
I_{MEC}	:	MEC current [A]
K	:	Half-saturation (Monod) constant [mg-X/L]
m	:	Number of electrons transferred mol of H_2 (mol-e/mol-H_2]
M_{ox}	:	Oxidized mediator fraction [mol]
M_{red}	:	Reduced mediator fraction [mol]
M_{Total}	:	Total mediator weight percentage in each electricigenic microorganism [mg-M/mg-X]
q_e	:	Substrate consumption rate by electricigenic microorganisms [mg-A/mgX d]
q_f	:	Substrate consumption rate by fermentative microorganism [mg-S/mgX d]
q_m	:	Substrate consumption rate by methanogenic microorganisms [mg-A/mgX d]
R	:	Ideal gas constant [J/K mol]
R_{int}	:	MEC internal resistance [ohm]
S	:	Organic substrate concentration in the MEC anodic compartment [mg-S/L]
S_0	:	Organic substrate concentration in the MEC's influent [mg-S/L]
T	:	MEC anodic compartment temperature [K]
u_e	:	Electricigenic microorganisms growth rate [1/d]
u_f	:	Fermentative microorganisms growth rate [1/d]
u_h	:	Hydrogenotrophic methanogenic microorganisms growth rate [1/d]
u_m	:	Methanogenic microorganisms growth rate [1/d]
V	:	MEC anodic compartment volume [L]
x_e	:	The concentration of electricigenic microorganisms in the MEC biofilm Layer 2 [mg-X/L]
x_f	:	The concentration of fermentative microorganisms in the MEC biofilm Later 1 [mg-X/L]
x_h	:	The concentration of hydrogenotrophic methangenic microorganisms in the MEC biofilm Layer 3 [mg-X/L]
$x_{m,1}$	:	The concentration of acetoclastic methanogenic microorganisms in MEC biofilm Layer 1 [mg-X/L]
$x_{m,2}$	:	The concentration of acetoclastic methanogenic microorganisms in MEC biofilm Layer 2 [mg-X/L]
Y_{cod}	:	Acetate yield from the organic substrate [mg-S/mg-A]
Y_m	:	Mediator yield [mg-M/mg-S]
γ	:	Mediator molar mass s[mg-M/mol-M]
α	:	Dimensionless biofilm retention constant
β	:	Dimensionless reduction or oxidation transfer coefficient

References

[1]. Meinshausen M, Meinshausen N, Hare W, Raper S C B, Frieler K, Knutti R, Frame D J and Allen M Y 2009 Greenhouse-gas emission targets for limiting global warming to 2°C. *Nature*, **458**, 1162

[2]. Edwards P P, Kuznetsov V L, David W I F and Brandon N P 2008 Hydrogen and fuel cells: Towards a sustainable energy future. *Energy Policy*, **36**, 4362

[3]. Vaidya P D and Rodrigues A E 2006 Insight into steam reforming of ethanol to produce hydrogen for fuel cells. *Chem. Eng. J.*, **117**, 49

[4]. Barton H 2012 The reversed fortunes of sago and rice, Oryza sativa, in the rainforests of Sarawak, Borneo. *Quaternary Int.*, **249**, 104

[5]. Awg-Adeni D S, Abd-Aziz S, Bujang K and Hassan M A 2009 Bioconversion of sago residue into value added products. *Afr. J. Biotechnol.*, **9**, 2021

[6]. Jafary T, Daud W R W, Ghasemi M, Kim B H, Md Jahim J, Ismail M, and Lim S S 2015 Biocathode in microbial electrolysis cell; Present status and future prospects. *Renewable and Sustainable Energy Rev.*, **47**, 33

[7]. Wrana N, Sparling R, Cicek N and Levin D B 2010 Hydrogen gas production in a microbial electrolysis cell by electrohydrogenesis. *J. Cleaner Prod.*, **18**, S111

[8]. Logan B E, Call D, Cheng S, Hamelers H V M, Sleutels T H J, Jeremiasse A W and Rozendal R A 2008 Microbial electrolysis cells for high yield hydrogen gas production from organic matter. *Environ. Sci. Technol.*, **42**, 8640

[9]. Zhou M, Wang H, Hassett D J and Gu T 2013 Recent advances in microbial fuel cells (MFCs) and microbial electrolysis cells (MECs) for wastewater treatment, bioenergy and bioproducts. *J. Chem. Technol. Biotechnol.*, **88**, 518

[10]. Lu L and Ren Z J 2016 Microbial electrolysis cells for waste biorefinery: A state of the art review. *Bioresour. Technol.*, **215**, 264

[11]. Gadhamshetty V, Sukumaran A, Nirmalakhandan N and Thein Myint M 2008 Photofermentation of malate for biohydrogen production - A modeling approach *Int. J. Hydrogen Energy*, **33**, 2146

[12]. Azwar Y, Abdul-Wahab A K and Hussain M A 2013 Optimal production of biohydrogen gas via microbial electrolysis cells (mec) in a controlled batch reactor system. *Chem. Eng. Trans.*, **32**, 732

[13]. Pinto R P, Srinivasan B, Guiot S R and Tartakovsky B 2011 The Effect of Real-Time External Resistance Optimization on Microbial Fuel Cell Performance. *Water Res.*, **45**, 1578

Characterization of Bio-adhesive derived from Sarawak Native Sago Starch

R Baini[1], M Demong[1], J L Chang Hui[1], M M Kamal[1], N A S Abdul Samat[1], M R Rahman[1], and S N L Taib[1]

[1] Faculty of Engineering, University Malaysia Sarawak (UNIMAS), Kota Samarahan 94300 Kuching Sarawak, Malaysia
Corresponding author: ruby@unimas.my

Abstract. Sago starch has multiple applications in industries such as textile, cosmetics and pharmaceutical, paper and wood. This study focuses on the utilization of sago starch to promote formaldehyde-free adhesives. Formaldehyde-based adhesive commonly used in wood industry emits formaldehyde, which is classified as carcinogen, into the air that has raised concerns over the potential hazards to human health. Sago starch-based adhesive was produced by blending the sago starch with sodium hydroxide (NaOH) and glycerine. Fourier-Transform Infrared Spectroscopy (FT-IR) was used to characterize the chemical changes in the sago starch-based adhesive. The quality of sago starch adhesive was tested according to their viscosity. The mechanical property is analyze based on lap shear (bond) strength of the sago starch adhesives according to ASTM D907 and D143 by using Model D350 Testomeric. Parameters affecting the viscosity of the sago starch-adhesive such as time, temperature, concentration of sodium hydroxide, and the amount of glycerine were studied. The viscosity of the sago starch-based adhesive decreased as the mixing time, temperature, concentration of NaOH and the amount of glycerine increased.

1. Introduction

Sago starch comes from sago palm trees such as Metroxylon Sagu as well as other species, M. longispinum, M. sylvestre, M. microcanthum and M. rumphii that can be found in South East Asia [2]. Sago palm is found abundantly in Sarawak covering an estimated area of 60,000 hectares. Sago starch export from Sarawak was estimated about 30,000 to 50, 000 tons with values around US$3.4 million to US$10.8 million [3]. Highly refined starch is used for food industry, while modified starch has a wider application in non-food industry such as textile, pharmaceutical, paper and plywood industry [15]. Even though there is some argument regarding non-food application of starch, sago starch can be considered worthy, since the sago palm has low maintenance cost, high survival rate and high productivity [8]. Plywood industry in Malaysia commonly uses formaldehyde-based adhesive as binder, and a wide application of this adhesive has raised concerns over the potential hazards to human health [14]. Emission of formaldehyde into the air is potentially ingested by human, which may affect their health even with concentration as low as 0.02 ppm. Formaldehyde is classified as a carcinogen, a contributing factor to nasal and lung cancer, and to some extent that links to brain cancer and leukemia. The risk of human exposure to formaldehyde varies depending on its duration and its concentration; long term exposure to low level of formaldehyde may cause respiration difficulty, and short-term exposure to high level formaldehyde may cause fatality [5]. Formaldehyde is a synthetic

IOP Conf. Series: Materials Science and Engineering 1101 (2021) 012041 doi:10.1088/1757-899X/1101/1/012041

chemical, commonly used, because it can be found easily in the market with proven function as adhesive. Urea formaldehyde, phenol formaldehyde, resorcinol formaldehyde and melamine formaldehyde are among the common adhesives used, and these resins are used widely due to the increasing demand of engineered wood products. The use of starch compounds such as those derived from sago palm, rice, potatoes, wheat and corn starch had been recognized as an alternative way to replace formaldehyde [17]. Starch is a polysaccharide made up from glucose unit, C6H12O6, which may consist of 20%-30% highly linear amylose and 70%-80% highly branched amylopectin [16]. In order to create adhere properties in starch, these chains are either grafted with monomer (i.e. vinyl acetate) or blended with polymer (i.e. polyvinyl alcohol) during acid hydrolysis and/or alkaline gelatinization [18]. This kind of starch bio-adhesive is non- toxic, cheap, renewable and can promote green technology, which are beneficial for both industries and consumers [22]. The characteristics of an adhesive in order to bind two surfaces together are as follows [3];

i) able to flow on the surfaces and to fill the spaces/voids of the surfaces
ii) adhere to the surfaces and stay in position
iii) solidify and stay adhered to the adherent to develop the strength holding the two surfaces
iv) durable, does not affected much by the storage period and condition, and certain environmental condition

The phenomenon of adhesion can be explained as mechanical theory [4]. Adhesion is linked to the porosity and surface roughness of the adherent, with the degree of adhesion. Material that has rough surface allows the adhesives to penetrate and to fill the voids, resulting in the anchorage between the surfaces. In this case, the adhesive has to generate adhesion anchor points, where it must have good filling power and the filling power is related to the viscosity. Therefore, the properties of adhesives can be analyzed in terms of its viscosity and its adhere strength known as bond strength. Viscosity indicates the flow-ability of the adhesives, and the right viscosity is required that allows the adhesive to spread between the surfaces well while hold the surfaces, strongly. The adhesive strength can be defined as the resistance against mechanical separation between the surfaces. Good adhesives fill the voids and develop the bond between the surfaces, which will require a higher force or energy to dismantle the interface between the two materials.

To date, many researches had been done to study the effects of various parameters on the production of starch-based adhesives including the use of acid [20], alkali [10], monomer [18] polymer, binder and adhesive additives [12] as well as other factors including the temperature [21,13], the reaction time and the type of starch origin [19,18]. However, limited studies focusing on the development and the utilization of sago starch as bio-adhesive. Therefore, the aim of this work was to analyse the effect of parameters in sago starch-adhesive process such as temperature, starch initial moisture content, starch-water ratio, Sodium Hydroxide ratio and glycerine ratio. Furthermore, the relation between each parameters and adhesive properties such as viscosity and adhere strength is suggested as empirical models. Finally, Fourier-Transform Infrared Spectroscopy (FT-IR) was used to visually compare the compounds in each of the sago starch-adhesive samples.

2. Materials and Methodology

Sago logs were collected from Kota Samarahan Sarawak, Malaysia and were kept in refrigerator at -80°C to minimise properties changes. Sago starch were prepared by extracting sago from the sago logs using the method described by [11]. The samples preparation involved chopping, blending, sieving and sedimenting. Sago logs were chopped into small pieces, blended using an electric blender and mixed with water. The slurry was sieved three times using a kitchen siever, a fabric siever and a 200 m siever. The extraction of sago starch from the blended sago logs was allowed to happen by leaving the samples at room temperature for 3 hours, 1 hour and 30 mins for the sieving using a kitchen siever, a fabric siever and a 200 m siever, respectively. The sediment collected at the bottom of the beaker after the sieving using 200 m siever was dried in an oven at 70°C for 24 hours. The adhesive was prepared by mixing 100 g of sago starch with 200 ml of deionised water in a beaker. Then, 4 grams of glycerine and 200 ml of 3M sodium hydroxide solution was added in the beaker; NaOH was used as gelatinization enhancer while glycerine was used as plasticizer. The mixture was allowed to gelatinize by heating it at room temperature, 25°C with stirring speed of 450 rpm. The viscosity of the mixture was measured using consistometer at different mixing time. The experiment was repeated for temperatures of 50°C, 75°C and 80°C. Next, the lap shear (bond) strength of the sago starch adhesives is tested according to ASTM D907 and D143 or M350 Testometric. Plywood with thickness of 3mm are cut into rectangles with dimensions 54mm × 25mm. Then, adhesives were applied to one side of each piece of sample with coated area 25mm × 25mm.

3. Results and Discussions

3.1 FT-IR profile of sago starch-based adhesive

FT-IR analysis was performed on the sago starch-adhesive sample to observe the compound changes between native starch as shown in figure 1 and that of sago starch-adhesive as shown in figure 2. The FT-IR analysis for the native starch illustrated in figure 1 shows the characteristic peaks located at 3313 cm^{-1} (-OH stretching), 2931 cm^{-1} (-CH- stretching), 1613 cm^{-1} (-OH bending), 1340 cm^{-1}, 1149 cm^{-1} and 1078 cm^{-1} (-CO stretching). Strong peak shown for –OH groups in the starch granules, correspond to high level of moisture content which is important in making good adhesive. High concentration of -OH bonding provides more availability of polymer chains to be hydrolyze resulting an increase of viscosity of the adhesive. Thus, the addition of additives will strengthen the polymer bonding and thickening ability of the adhesive.

Compared with the sago starch-adhesive, shown in figure 2, the peak at 3371 cm^{-1} becomes narrower and stronger which indicated increased number of hydroxyl group attached to the starch as the result of hydrolysis and gelatinization that had broken the inter/intra-molecular bond. The peaks at 2931 cm^{-1} and 1613 cm^{-1} in figure 1 had been distinguished and replaced by stronger and narrower peak at 1633 cm^{-1} shown in figure 2, which indicated huge replacement of glycerol ($C_3H_8O_3$) through C-O bonds. Besides, it can be seen that the intensities of peak 1020 cm^{-1}, 1082 cm^{-1} and 1153 cm^{-1} in figure 2 were decreased when compared to peak 1001 cm^{-1}, 1078 cm^{-1} and 1149 cm^{-1}, respectively, as the result of the breakage of glycosidic bond during acid hydrolysis (Yu et al., 2015).

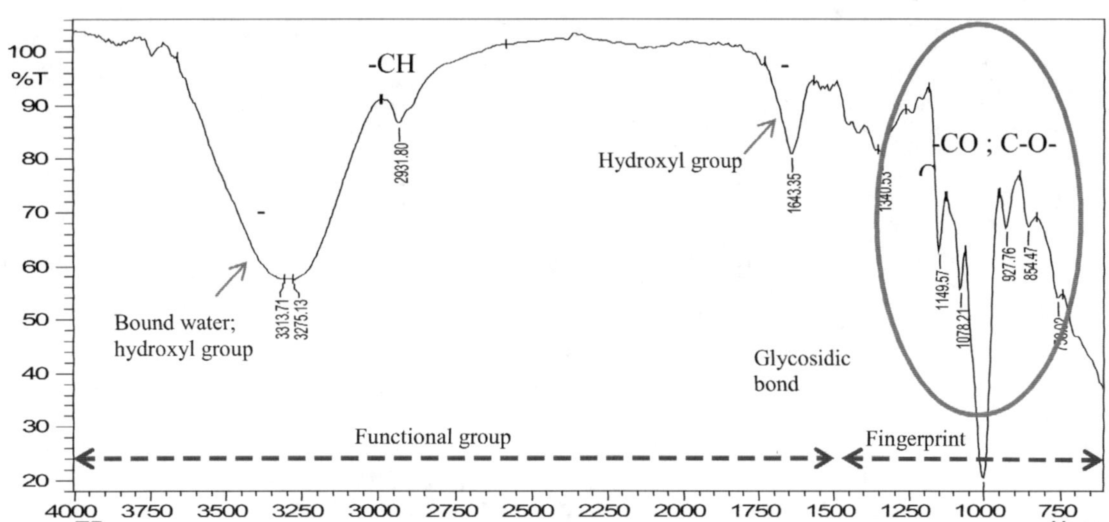

Figure 1. FT-IR result of native sago starch.

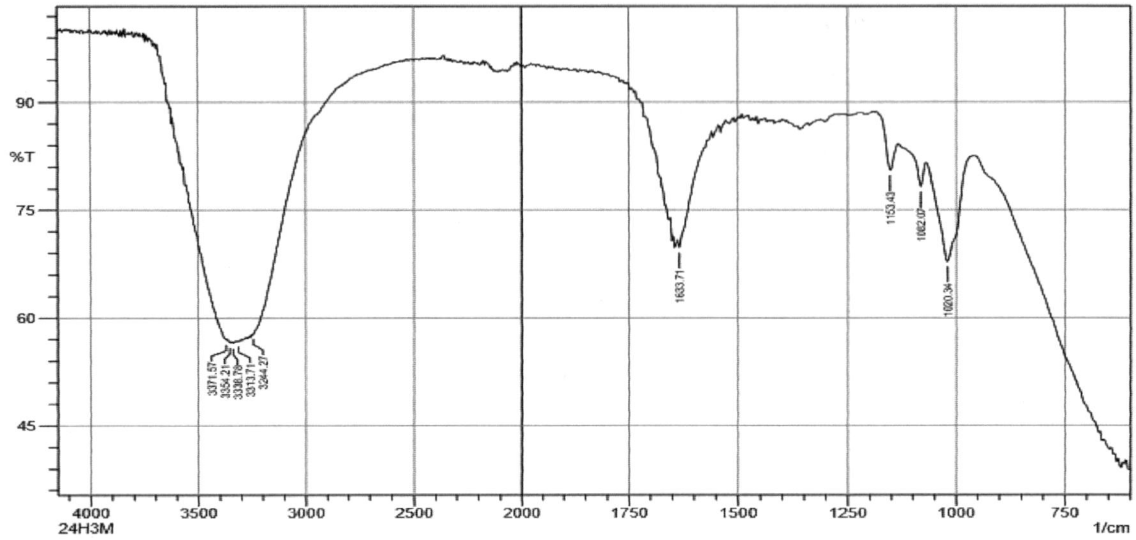

Figure 2. FTIR result of sago starch-based adhesive prepared for a mixture of 100 grams starch, 200 ml of water 3M NaOH and 4 g glycerine.

3.2 Analysis on the viscosity of the sago starch-based adhesive

Figure 3 shows the viscosity of sago starch-based adhesive as a function of time for the experiment done at 25°C.

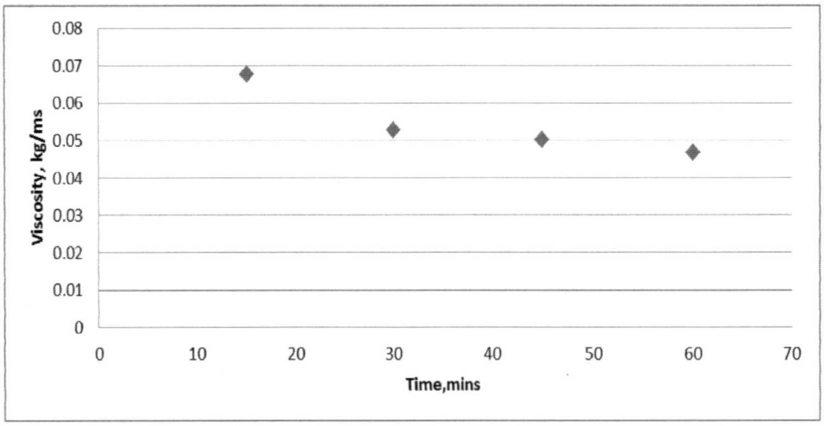

Figure 3. Viscosity of sago starch-based adhesive at 25°C.

The viscosity decreases from 0.07 at mixing time of 15 mins to 0.05 kg/ms at mixing time of 60 mins giving about 30% viscosity reduction. The addition of water and sodium hydroxide to the starch causes the starch granules to break and releasing its content that gives the gel-formation like material when the material is cooled down to room temperature. In addition, further mixing allowing NaOH to cause more starch granules to break, allowing gelatinization process to occur and more interaction of glycerine molecules with the starch decreasing the viscosity; Glycerine is used as plastizer that gives more flexible texture to the material, which may also be the reason to the decreases in the viscosity with time, shown in figure 4 below. Increasing the amount of glycerine will reduce the viscosity of the adhesive. Lower viscosity is important for the gluing process because it will easily flow into the surface pores to fill the gap between surfaces besides provide good wetting properties.

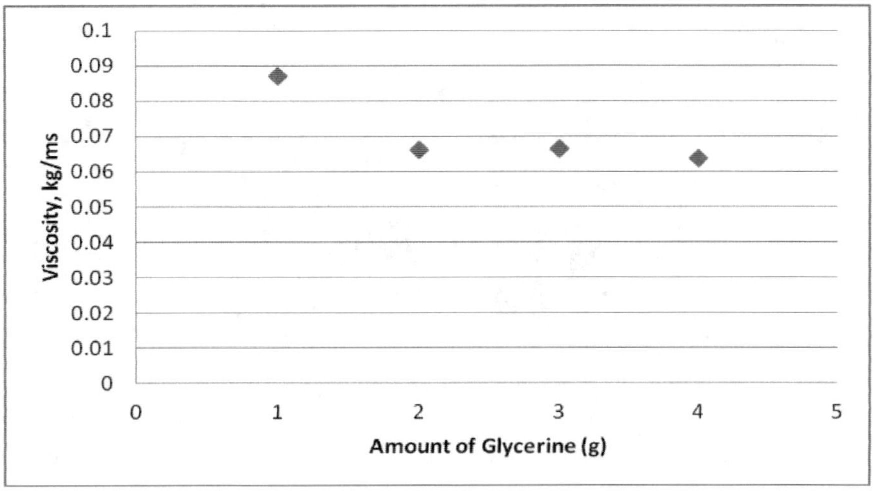

Figure 4. Viscosity of sago starch-based adhesive for different amount of glycerine.

Next, figure 5 shows the changes in the viscosity with the molarity of sodium hydroxide. The viscosity of sago starch-based adhesives showed decreasing trend as the concentration of the sodium hydroxide increases. This is due to shear thinning phenomena occurred, that was caused by sodium hydroxide. According to [6], when the concentration of NaOH increases, adhesive produced become less jelly-like and flow faster as its resistance to flow decreased. Initially, an addition of sodium hydroxide will be caused swelling of granules, but it will disperse evenly in the mixture upon stirring, which caused the adhesives to become thinner [9]. Apart, when high concentration of NaOH used, it increases the activity of NaOH during mixing then caused the starch particles begin to collapse and gradually disperse into small linear pieces results in the decreasing of starch adhesive viscosity.

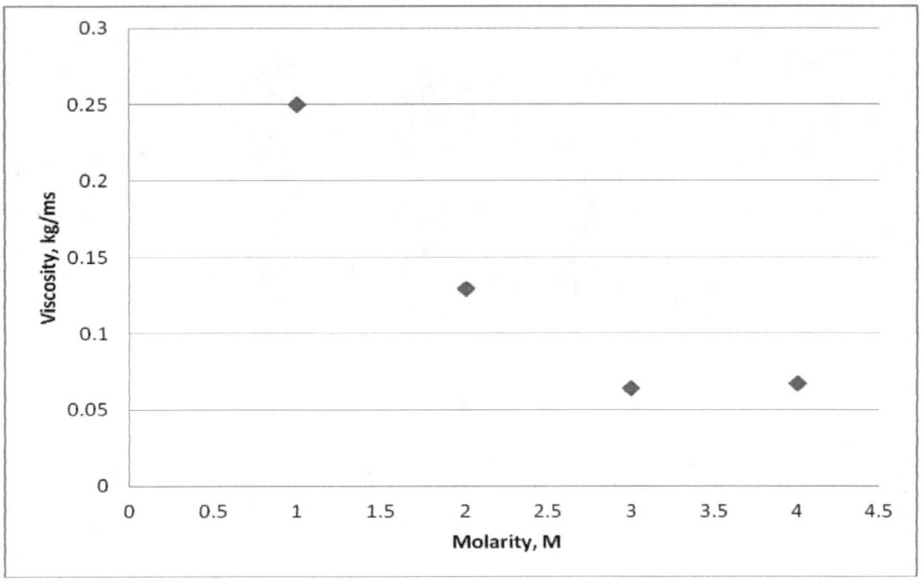

Figure 5. Viscosity versus Molarity of sodium hydroxide.

IOP Conf. Series: Materials Science and Engineering 1101 (2021) 012041 doi:10.1088/1757-899X/1101/1/012041

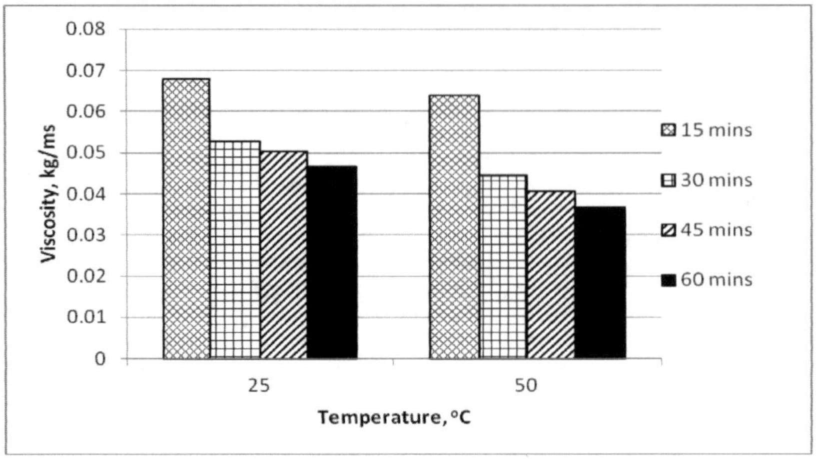

Figure 6. Viscosity of sago starch-based adhesive for temperatures of 25°C and 50°C.

Meanwhile, when the temperature was increased from 25°C to 50°C, the viscosity decreased for all the mixing time that was varied from 15 mins to 60 mins, as shown in figure 6. The starch loses its crystallinity as the temperature increases resulting to lower viscosity. With increased temperature, free radicals and propagating polymer chains become more flexible as a result of decreased viscosity and react to a greater extent and crosslinking. This increase in polymerization may lead to improved mechanical properties. Although that lower viscosity providing good wetting to the adhesive properties, however, a strong adhesive must be able to resist high temperature so that the adhesive does not melt after gluing process and maintain the bonded substrates. The effect of simultaneous mixing and heating enhances uniform heat and mass transfers within the mixture and increases the percentage change in the viscosity from 6-21%, which is shown in figure 7 below.

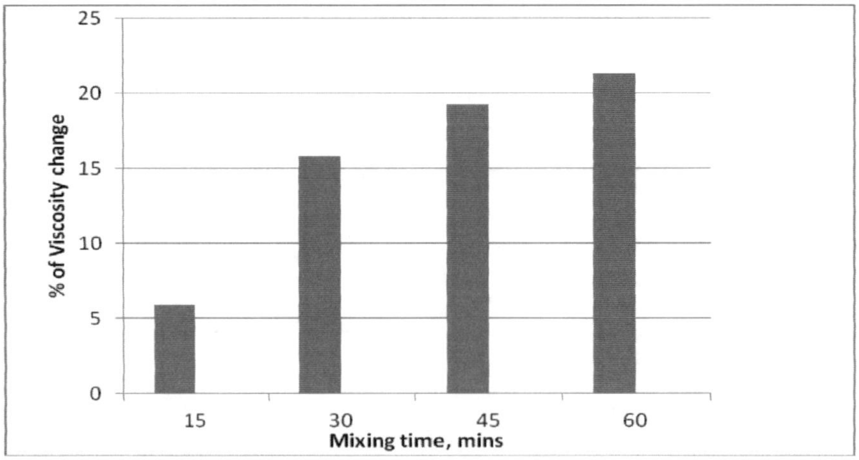

Figure 7. % change in the viscosity of sago starch-based adhesive when the temperature is increased from 25°C to 50°C.

The moisture content in sago starch adhesive also plays significant roles in making this bio-adhesive efficiently. Results shown in the following figure 8 proved that 0% of moisture content has the highest value of viscosity of 0.25 kg/ms using 1M NaOH was reduced 70% to about 0.07 kg/ms when the molarity is increased to 4M NaOH. However, at higher moisture content of 12%, the viscosity reduced to about 0.02kg/ms at 4M of NaOH.

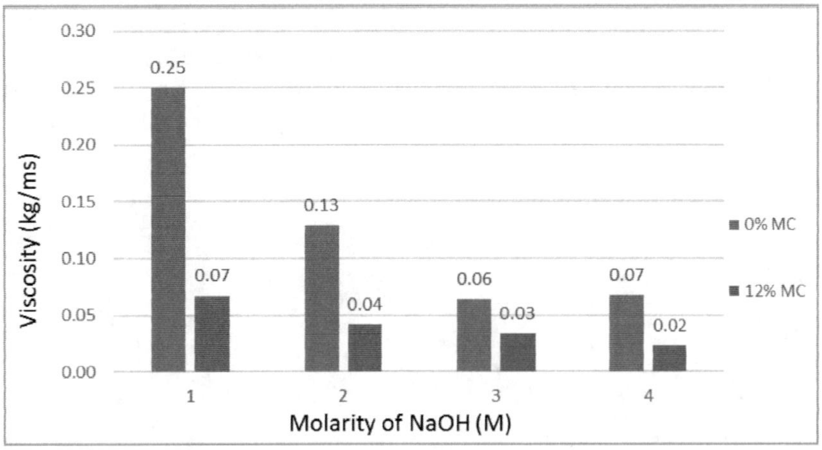

Figure 8. Viscosity of sago starch-based adhesive for moisture content of 0% and 12%.

Higher moisture content of the starch adhesive resulting to lower viscosity which makes it have high flowability particles. This prove that the level of moisture can degrade the properties of the starch adhesive and will affect the bonding strength of the adhesive. Besides, an optimum level of moisture contain in the adhesive can fills in the voids or porosity of the wood resulting anchoring points at the adhesives and adherent. Although the water molecules can lower the tensile strength and modulus, but it increases elongation at break of the adhesive which make the bonded substrates will not peel-off easily. These properties generally recover fully when the polymer is dried, unless irreversible hydrolysis has taken place. The effect of simultaneous drying and heating enhances uniform evaporation of water molecules within the starch adhesive mixture and decreased the percentage change in the viscosity from 18% - 5%, as shown in figure 9 below.

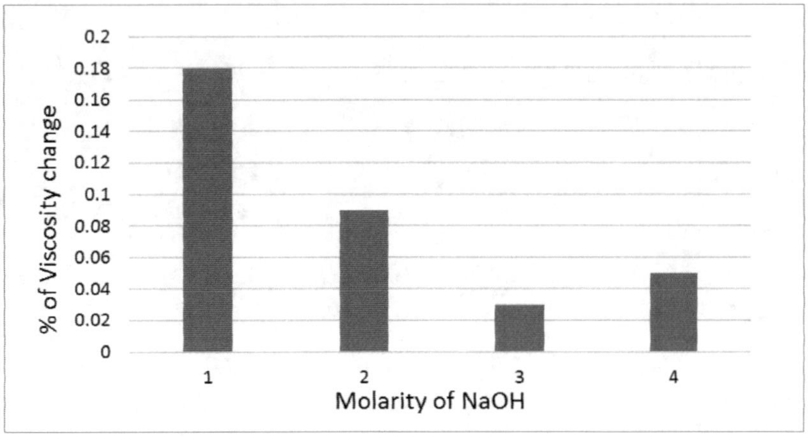

Figure 9. % change in the viscosity of sago starch-based adhesive when the moisture content is increased from 0% to 12%.

3.3 Analysis on The Strength of The Sago Starch-Based Adhesive

Figure 10 shows the strength trend of sago starch adhesive affected by the temperature when it is increasing from 25°C to 75°C. The strength of the adhesives is higher at 25°C of production due to presence of free hydroxyl groups. When more amount of available hydroxyl group presence, the cohesion force between the starch adhesive becomes stronger. However, at higher temperature, more

IOP Conf. Series: Materials Science and Engineering 1101 (2021) 012041 doi:10.1088/1757-899X/1101/1/012041

heat transfer occurred to the starch granules resulting increases in the fluidity of the adhesive, thus it weakens the bonding strength between the wood surfaces. The unreacted hydroxyl group in the adhesives enhance the interaction between the adhesives and wood surface [7]. These explain adhesive that have good wetting properties gives strong adhesion force which is between starch adhesive molecule and wood surfaces.

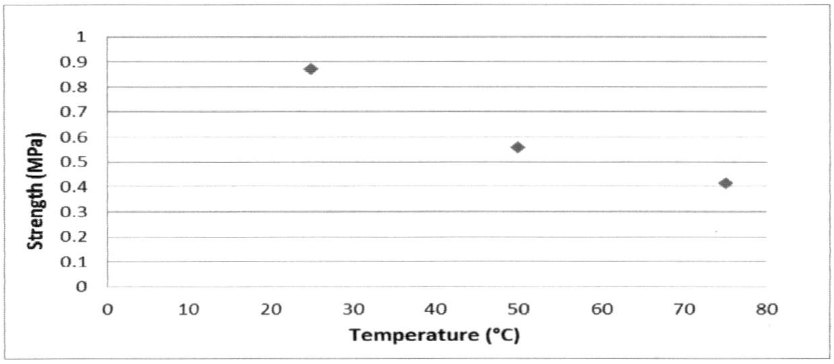

Figure 10. Strength of sago starch adhesive for temperatures of 25°C to 75°C.

Figure 11 presented the effect of molarity of NaOH to the strength of sago starch adhesive. The strength shows an increasing to 2.49 MPa when the molarity added to 3M and decreased to 2.02 MPa due to excessive ratio. The addition of sodium hydroxide accelerates the reaction between the two polymers amylose and amylopectin due to increase in pH of basic medium. Besides, NaOH improves tack properties of the starch adhesive where it opens more availability of polymer link of the starch molecule to be hydrolyze and results to strong polymer chain.

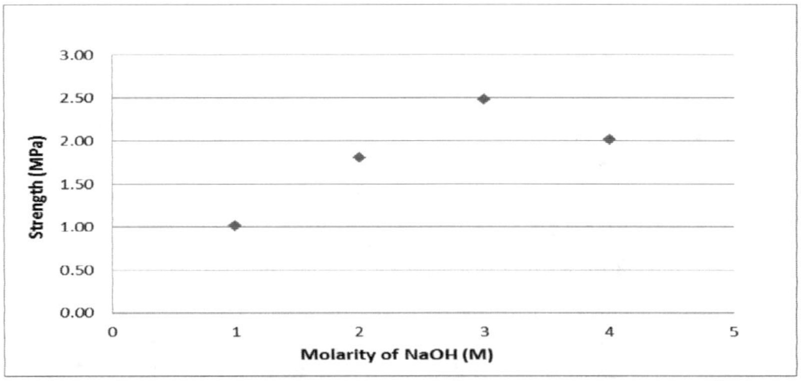

Figure 11. Strength versus molarity of NaOH.

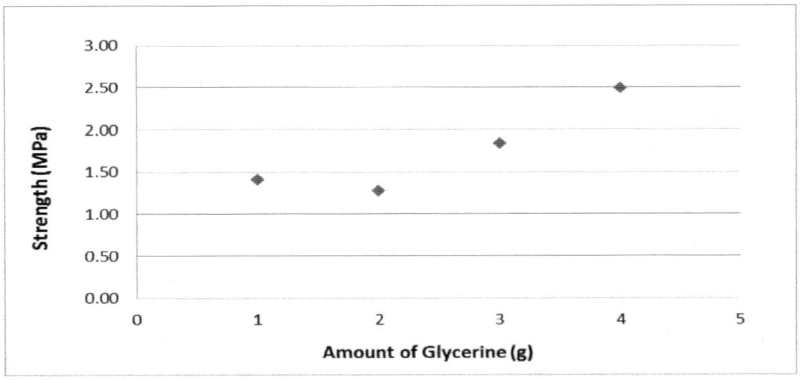

Figure 12. Strength of adhesives at different amount of glycerine.

The strength of sago starch adhesives increases as the amount of glycerine added increases as shown in figure 12. The glycerine acts as plasticizer gives more flexible bond and strong adhesion. Its strong hydrogen-bonding have been suggested as an explanation for its very good solvent activity, thus results in strong and firm starch adhesive. Glycerine also changes the composition of sago starch adhesives so as to prevent it from dry out on storage or loss of water and at the same time the elasticity of the adhesives produced increases and no adhesive power is lost. During starch adhesive formulation, addition of glycerine is due in large degree to its hygroscopic or moisture-retaining action that result in minimization of water fluctuation. Therefore, an increase of glycerine improves the flexible bond of the adhesive and provide good tackiness as well as optimum viscosity of the starch adhesive for high adhesion linking.

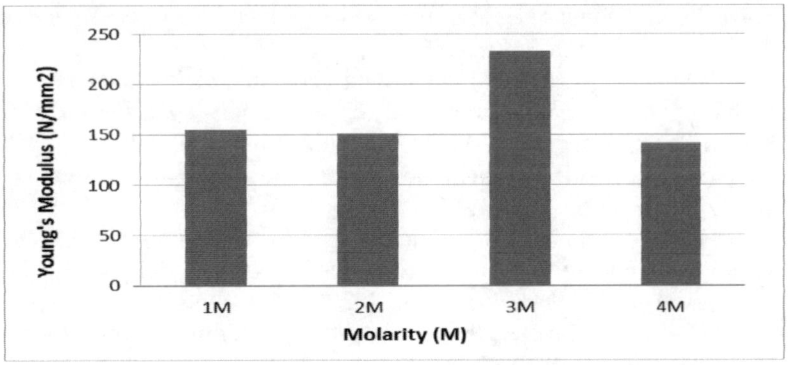

Figure 13. Modulus of Elasticity of sago starch adheive.

The increase in the molarity of the sodium hydroxide reduces the estimated modulus young as shown in figure 13. This result is similar to that shown for viscosity as described in figure 5 previously. A high Young Modulus shown in 3M corresponds to a high viscosity meaning that the adhesive is stiff and have high tensile strength. However, for 4M, the Young Modules drastically showing the adhesive has high elongation or exhibit elastic behavior due to more compound has been hydrolyzed by NaOH which resulting in low viscosity.

4. Conclusions

Findings from this work suggested that Sarawak sago starch is a potential material for bio-adhesive. The functional compounds of the sago starch-adhesive were analysed using FT-IR, and differences in the characteristics of peaks compared with that for the original sago starch was observed. The addition of sodium hydroxide and glycerine to sago starch has reduced the viscosity of the sago starch adhesive. Decrease in the viscosity of the sago starch was also observed when the mixing time and temperature increases. Overall, the addition of sodium hydroxide as gelatinization enhancer and glycerine as plasticizer to the sago starch improved the properties of the bio-adhesive produced shown by the increase in the strength of the adhesive, measured about 2.5 MPa using 3M of sodium hydroxide and 4 % by weight of glycerine.

5. References

[1] Ahmad F B, Williams P A, Doublier J L, Durand S and Buleon A 2016 Enhancing the Adhesive Strength of a Plywood Adhesive Developed from Hydrolyzed Specified Risk Materials, *Polymers* **8**(8) 285

[2] Ahmad F B, Williams P A, Doublier J L, Durand S and Buleon A 1999 Physico-chemical characterisation of sago starch, *Carbohydrate Polymers* **38**(4) 361–370

[3] Bujang K 1996 Potentials of Bioenergy From the Sago Industries in Malaysia, *Encyclopedia of Life Support System (EOLSS)* **14**

[4] Butt M A, Chughtai A, Ahmad J, Ahmad R, Majeed U and Khan I H 2008 Theory of Adhesion and its Practical Implications. A critical review, *Journal of Faculty of Engineering and Technology* **15**(1) 21–45

[5] Feinman S E 1988 Respiratory Effects from Formaldehyde, *Formaldehyde Sensitivity and Toxicity*.

[6] Gunorubon A J 2012 Production of cassava starch-based adhesive, *Research Journal in Engineering and Applied Sciences Rjeas © Emerging Academy Resources* **1**(4) 219–214

[7] He S K 2013 Research on Production and properties of crosslinked starch from potato, *Cereal and Feed Industry* **1** 39-41

[8] Karim A A, Tie A P, Manan D M A and Zaidul I S M 2008 Starch from the Sago (Metroxylon sagu) Palm Tree - Properties , Prospects , and Challenges as a New Industrial Source for Food and Other Uses, *Comprehensive Reviews in Food Science and Food Safety* **7** 215–228

[9] Kurt R and Fatih M 2008 The effect of boric acid/borac treatment on selected mechanical and combustion properties of poplar laminated veneer lumber, *Wood Research* **53**(2) 113–120

[10] Lee J S, Kumar R N, Rozman H D and Azemi B M N 2004 Flow behaviour of sago starch-g-poly(acrylic acid) in distilled water and NaOH - Effect of photografting, *Carbohydrate Polymers* **56**(3) 347–354

[11] Mustafa K M, Baini R, Mohamaddan S, Selaman O S, Ahmad Z N, Rahman M R, Othman A K 2017 Effect of temperature to the properties of sago starch, *IOP Conference Series: Materials Science and Engineering* **206** 012039.

[12] Nwokocha L M 2011 Adhesive Properties of Cyanoethyl Starch *Journal of Adhesion Science and Technology* **25**(8) 893–902

[13] Pan Z, Cathcart A and Wang D 2005 Thermal and chemical treatments to improve adhesive property of rice bran, *Industrial Crops and Products* **22**(3) 233–240

[14] Schoenberg J B and Mitchell C A 1975 Airway disease caused by phenolic (phenol-formaldehyde) resin exposure, *Archives of Environmental Health* **30**(12) 574–577

[15] Singhal R. S, Kennedy J F, Gopalakrishnan S M, Kaczmarek A, Knill C J and Faridatul P 2008 Industrial production , processing , and utilization of sago palm-derived products, *Carbohydrate Polymers* **72** 1–20

[16] Tang X., and Alavi S. 2011 Recent advances in starch, polyvinyl alcohol based polymer blends, nanocomposites and their biodegradability. *Carbohydrate Polymers* **85**(1) 7-16

[17] Gadhave V R, Mahanwar P A and Gadekar P T 2017 Starch-Based Adhesives for Wood/Wood Composite Bonding: Review, *Open Journal of Polymer Chemistry* **7**(2) 19–32

[18] Xu Q, Wen J and Wang Z 2016 Preparation and Properties of Cassava Starch-based Wood Adhesives, *BioResources* **11**(3) 6756–6767

[19] Yang L, Liu J, Du C and Qiang Y 2013 Preparation and properties of cornstarch adhesives, *Advance Journal of Food Science and Technology* **5**(8) 1068–1072
[20] Yiu P H, Loh S L, Rajan A, Wong S C and Bong C F J 2008 Physiochemical properties of sago starch modified by acid treatment in alcohol, *American Journal of Applied Sciences* **5**(4) 307–311
[21] Yu H, Cao Y, Fang Q and Liu Z 2015 Effects of treatment temperature on properties of starch-based adhesives, *BioResources* **10**(2) 3520–3530
[22] Zia F, Zia K M., Zuber M., Kamal S, and Aslam N 2015 Starch based polyurethanes: A critical review updating recent literature, *Carbohydrate Polymers* **134** 784–798

Acknowledgements
The authors would like to thank Universiti Malaysia Sarawak for providing the facilities in conducting this project. Gratitude also goes to Ministry of Higher Education for the support through the research grant FRGS/TK02(04)/1304/2015(210), as well as to CRAUN Research Sarawak for the sago logs supplied.

AUTHOR INDEX

Institute of Physics
Dirac House, Temple Back
Bristol BS1 6BE UK

ISSN: 1757-8981
ISBN 978-1-7138-3395-6

CURRAN ASSOCIATES INC.
proceedings
.com